Reliability-Based Design in Geotechnical Engineering

Also available from Taylor & Francis

Geological Hazards
Fred Bell

Hb: ISBN 0419-16970-9
Pb: ISBN 0415-31851-3

Rock Slope Engineering
Duncan Wyllie and Chris Mah

Hb: ISBN 0415-28000-1
Pb: ISBN 0415-28001-X

Geotechnical Modelling
David Muir Wood

Hb: ISBN 9780415343046
Pb: ISBN 9780419237303

Soil Liquefaction
Mike Jefferies and Ken Been

Hb: ISBN 9780419161707

Advanced Unsaturated Soil Mechanics and Engineering
Charles W.W. Ng and Bruce Menzies

Hb: ISBN 9780415436793

Advanced Soil Mechanics 3rd edition
Braja Das

Hb: ISBN 9780415420266

Pile Design and Construction Practice 5th editon
Michael Tomlinson and John Woodward

Hb: ISBN 9780415385824

Reliability-Based Design in Geotechnical Engineering

Computations and Applications

Kok-Kwang Phoon

CRC Press
Taylor & Francis Group
Boca Raton London New York

CRC Press is an imprint of the
Taylor & Francis Group, an **informa** business

A TAYLOR & FRANCIS BOOK

CRC Press
Taylor & Francis Group
6000 Broken Sound Parkway NW, Suite 300
Boca Raton, FL 33487-2742

First issued in paperback 2019

ISBN-13: 978-0-415-39630-1 (hbk)
ISBN-13: 978-0-367-86413-2 (pbk)

British Library Cataloguing in Publication Data
A catalogue record for this book is available
from the British Library

Library of Congress Cataloging in Publication Data
Phoon, Kok-Kwang.
   Reliability-based design in geotechnical engineering: computations and
applications/Kok-Kwang Phoon.
       p. cm.
   Includes bibliographical references and index.
   ISBN 978-0-415-39630-1 (hbk : alk. paper) – ISBN 978-0-203-93424-1
(e-book) 1. Rock mechanics. 2. Soil mechanics. 3. Reliability. I. Title.

   TA706.P48 2008
   624.1′51–dc22   2007034643

Visit the Taylor & Francis Web site at
http://www.taylorandfrancis.com

and the CRC Press Web site at
http://www.crcpress.com

Contents

List of contributors

Gregory B. Baecher is Glenn L Martin Institute Professor of Engineering at the University of Maryland. He holds a BSCE from UC Berkeley and PhD from MIT. His principal area of work is engineering risk management. He is co-author with J.T. Christian of *Reliability and Statistics in Geotechnical Engineering* (Wiley, 2003), and with D.N.D. Hartford of *Risk and Uncertainty in Dam Safety* (Thos. Telford, 2004). He is recipient of the ASCE Middlebrooks and State-of-the-Art Awards, and a member of the US National Academy of Engineering.

Marc Berveiller received his master's degree in mechanical engineering from the French Institute in Advanced Mechanics (Clermont-Ferrand, France, 2002) and his PhD from the Blaise Pascal University (Clermont-Ferrand, France) in 2005, where he worked on non-intrusive stochastic finite element methods in collaboration with the French major electrical company EDF. He is currently working as a research engineer at EDF on probabilistic methods in mechanics.

Denys Breysse is a Professor of Civil Engineering at Bordeaux 1 University, France. He is working on randomness in soils and building materials, teaches geotechnics, materials science and risk and safety. He is the chairman of the French Association of Civil Engineering University Members (AUGC) and of the RILEM Technical Committee TC-INR 207 on Non-Destructive Assessment of Reinforced Concrete Structures. He has also created (and chaired 2003–2007) a national research scientific network on Risk Management in Civil Engineering (MR-GenCi).

Young-Jae Choi works for Geoscience Earth & Marine Services, Inc. in Houston, Texas. He received his Bachelor of Engineering degree in 1995 from Pusan National University, his Master of Science degree from the University of Colorado at Boulder in 2002, and his PhD from The University of Texas at Austin in 2007. He also has six years of practical experience working for Daelim Industry Co., Ltd. in South Korea.

John T. Christian is a consulting geotechnical engineer in Waban, Massachusetts. He holds BSc, MSc, and PhD degrees from MIT. His principal areas of work are applied numerical methods in soil dynamics and earthquake engineering, and reliability methods. A secondary interest has been the evolving procedures and standards for undergraduate education, especially as reflected in the accreditation process. He was the 39th Karl Terzaghi Lecturer of the ASCE. He is recipient of the ASCE Middlebrooks and the BSCE Desmond Fitzgerald Medal, and a member of the US National Academy of Engineering.

George Deodatis received his Diploma in Civil Engineering from the National Technical University of Athens in Greece. He holds MS and PhD degrees in Civil Engineering from Columbia University. He started his academic career at Princeton University where he served as a Postdoctoral Fellow, Assistant Professor and eventually Associate Professor. He subsequently moved to Columbia University where he is currently the Santiago and Robertina Calatrava Family Professor at the Department of Civil Engineering and Engineering Mechanics.

Sunny Ye Fang is currently a consulting geotechnical engineer with Ardaman & Associates, Inc. She received her BS degree from Ningbo University, Ningbo, China, MSCE degree from Zhejiang University, Hangzhou, Zhejiang, China, and PhD degree in Civil Engineering from Clemson University, South Carolina. Dr Fang has extensive consulting projects experience in the field of geoenvironmental engineering.

Samuel J. Gambino has ten years of experience with URS Corporation in Oakland, California. His specialties include: foundation design, tunnel installations, and dams, reservoirs, and levees. Samuel received his bachelors degree in civil and environmental engineering from the University of Michigan at Ann Arbor in 1995, his masters degree in geotechnical engineering from the University of Texas at Austin in 1998, and has held a professional engineers license in the State of California since 2002.

Robert B. Gilbert is the Hudson Matlock Professor in Civil, Architectural and Environmental Engineering at The University of Texas at Austin. He joined the faculty in 1993. Prior to that, he earned BS (1987), MS (1988) and PhD (1993) degrees in civil engineering from the University of Illinois at Urbana-Champaign. He also practiced with Golder Associates Inc. as a geotechnical engineer from 1988 to 1993. His expertise is the assessment, evaluation and management of risk in civil engineering. Applications include building foundations, slopes, pipelines, dams and levees, landfills, and groundwater and soil remediation systems.

Yusuke Honjo is currently a professor and the head of the Civil Engineering Department at Gifu University in Japan. He holds ME from Kyoto

University and PhD from MIT. He was an associate professor and division chairman at Asian Institute of Technology (AIT) in Bangkok between 1989 and 1993, and joined Gifu University in 1995. He is currently the chairperson of ISSMGE-TC 23 'Limit state design in geotechnical engineering practice'. He has published more than 100 journal papers and international conference papers in the area of statistical analyses of geotechnical data, inverse analysis and reliability analyses of geotechnical structures.

Charng Hsein Juang is a Professor of Civil Engineering at Clemson University and an Honorary Chair Professor at National Central University, Taiwan. He received his BS and MS degrees from National Cheng Kung University and PhD degree from Purdue University. Dr Juang is a registered Professional Engineering in the State of South Carolina, and a Fellow of ASCE. He is recipient of the 1976 Outstanding Research Paper Award, Chinese Institute of Civil and Hydraulic Engineering; the 2001 TK Hsieh Award, the Institution of Civil Engineers, United Kingdom; and the 2006 Best Paper Award, the Taiwan Geotechnical Society. Dr Juang has authored more than 100 refereed papers in geotechnical related fields, and is proud to be selected by his students at Clemson University as the Chi Epsilon Outstanding Teacher in 1984.

Fred H. Kulhawy is Professor of Civil Engineering at Cornell University, Ithaca, New York. He received his BSCE and MSCE from New Jersey Institute of Technology and his PhD from University of California at Berkeley. His teaching and research focuses on foundations, soil-structure interaction, soil and rock behavior, and geotechnical computer and reliability applications, and he has authored over 330 publications. He has lectured worldwide and has received numerous awards from ASCE, ADSC, IEEE, and others, including election to Distinguished Member of ASCE and the ASCE Karl Terzaghi Award and Norman Medal. He is a licensed engineer and has extensive experience in geotechnical practice for major projects on six continents.

David Kun Li is currently a consulting geotechnical engineer with Golder Associates, Inc. He received his BSCE and MSCE degrees from Zhejiang University, Hangzhou, Zhejiang, China, and PhD degree in Civil Engineering from Clemson University, South Carolina. Dr Li has extensive consulting projects experience on retaining structures, slope stability analysis, liquefaction analysis, and reliability assessments.

Bak Kong Low obtained his BS and MS degrees from MIT, and PhD degree from UC Berkeley. He is a Fellow of the ASCE, and a registered professional engineer of Malaysia. He currently teaches at the Nanyang Technological University in Singapore. He had done research while on sabbaticals at HKUST (1996), University of Texas at Austin (1997) and Norwegian Geotechnical Institute (2006). His research interest and publications can be found at http://alum.mit.edu/www/bklow/.

Shadi S. Najjar joined the American University of Beirut (AUB) as an Assistant Professor in Civil Engineering in September 2007. Dr Najjar earned his BE and ME in Civil Engineering from AUB in 1999 and 2001, respectively. In 2005, he graduated from the University of Texas at Austin with a PhD in civil engineering. Dr Najjar's research involves analytical studies related to reliability-based design in geotechnical engineering. Between 2005 and 2007, Dr Najjar taught courses on a part-time basis in several leading universities in Lebanon. In addition, he worked with Polytechnical Inc. and Dar Al-Handasah Consultants (Shair and Partners) as a geotechnical engineering consultant.

Trevor Orr is a Senior Lecturer at Trinity College, Dublin. He obtained his PhD degree from Cambridge University. He has been involved in Eurocode 7 since the first drafting committee was established in 1981. In 2003 he was appointed Chair of the ISSMGE European Technical Committee 10 for the Evaluation of Eurocode 7 and in 2007 was appointed a member of the CEN Maintenance Group for Eurocode 7. He is the co-author of two books on Eurocode 7.

Kok-Kwang Phoon is Director of the Centre for Soft Ground Engineering at the National University of Singapore. His main research interest is related to development of risk and reliability methods in geotechnical engineering. He has authored more than 120 scientific publications, including more than 20 keynote/invited papers and edited 15 proceedings. He is the founding editor-in-chief of *Georisk* and recipient of the prestigious ASCE Normal Medal and ASTM Hogentogler Award. Webpage: http://www.eng.nus.edu.sg/civil/people/cvepkk/pkk.html.

Radu Popescu is a Consulting Engineer with URS Corporation and a Research Professor at Memorial University of Newfoundland, Canada. He earned PhD degrees from the Technical University of Bucharest and from Princeton University. He was a Visiting Research Fellow at Princeton University, Columbia University and Saitama University (Japan) and Lecturer at the Technical University of Bucharest (Romania) and Princeton University. Radu has over 25 years experience in computational and experimental soil mechanics (dynamic soil-structure interaction, soil liquefaction, centrifuge modeling, site characterization) and published over 100 articles in these areas. In his research he uses the tools of probabilistic mechanics to address various uncertainties manifested in the geologic environment.

Jean H. Prevost is presently Professor of Civil and Environmental Engineering at Princeton University. He is also an affiliated faculty at the Princeton

Materials Institute, in the department of Mechanical and Aerospace Engineering and in the Program in Applied and Computational Mathematics. He received his MSc in 1972 and his PhD in 1974 from Stanford University. He was a post-doctoral Research Fellow at the Norwegian Geotechnical Institute in Oslo, Norway (1974–1976), and a Research Fellow and Lecturer in Civil Engineering at the California Institute of Technology (1976–1978). He held visiting appointments at the Ecole Polytechnique in Paris, France (1984–1985, 2004–2005), at the Ecole Polytechnique in Lausanne, Switzerland (1984), at Stanford University (1994), and at the Institute for Mechanics and Materials at UCSD (1995). He was Chairman of the Department of Civil Engineering and Operations Research at Princeton University (1989–1994). His principal areas of interest include dynamics, nonlinear continuum mechanics, mixture theories, finite element methods, XFEM and constitutive theories. He is the author of over 185 technical papers in his areas of interest.

Bruno Sudret has a master's degree from Ecole Polytechnique (France, 1993), a master's degree in civil engineering from Ecole Nationale des Ponts et Chaussées (1995) and a PhD in civil engineering from the same institution (1999). After a post-doctoral stay at the University of California at Berkeley in 2000, he joined in 2001 the R&D Division of the French major electrical company EDF. He currently manages a research group on probabilistic engineering mechanics. He is member of the Joint Committee on Structural Safety since 2004 and member of the board of directors of the International Civil Engineering Risk and Reliability Association since 2007. He received the Jean Mandel Prize in 2005 for his work on structural reliability and stochastic finite element methods.

Thomas F. Wolff is an Associate Dean of Engineering at Michigan State University. From 1970 to 1985, he was a geotechnical engineer with the U.S. Army Corps of Engineers, and in 1986, he joined MSU. His research and consulting has focused on design and reliability analysis of dams, levees and hydraulic structures. He has authored a number of Corps of Engineers' guidance documents. In 2005, he served on the ASCE Levee Assessment Team in New Orleans, and in 2006, he served on the Internal Technical Review (ITR) team for the IPET report which analyzed the performance of the New Orleans Hurricane Protection System.

Tien H. Wu is a Professor Emeritus at Ohio State University. He received his BS from St. Johns University, Shanghai, China, and MS and PhD from University of Illinois. He has lectured and conducted research at Norwegian Geotechnical Institute, Royal Institute of Technology in Stockholm, Tonji University in Shanghai, National Polytechnical Institute in Quito, Cairo University, Institute of Forest Research in Christ Church, and others. His teaching, research and consulting activities involve geotechnical reliability, slope stability, soil properties, glaciology, and soil-bioengineering. He is the author of two books, Soil Mechanics and Soil Dynamics. He is an Honorary Member of ASCE and has received ASCE's State-of-the-Art Award, Peck Award, and Earnest Award and the US Antarctic Service Medal. He was the 2008 Peck Lecturer of ASCE.

Limin Zhang is an Associate Professor of Civil Engineering and Associate Director of Geotechnical Centrifuge Facility at the Hong Kong University of Science and Technology. His research areas include pile foundations, dams and slopes, centrifuge modelling, and geotechnical risk and reliability. He is currently secretary of Technical Committee TC23 on 'Limit State Design in Geotechnical Engineering', and member of TC18 on 'Deep Foundations' of the ISSMGE, and Vice Chair of the International Press-In Association.

Chapter 1

Numerical recipes for reliability analysis – a primer

Kok-Kwang Phoon

1.1 Introduction

Currently, the geotechnical community is mainly preoccupied with the transition from working or allowable stress design (WSD/ASD) to Load and Resistance Factor Design (LRFD). The term LRFD is used in a loose way to encompass methods that require all limit states to be checked using a specific multiple-factor format involving load and resistance factors. This term is used most widely in the United States and is equivalent to Limit State Design (LSD) in Canada. Both LRFD and LSD are philosophically akin to the partial factors approach commonly used in Europe, although a different multiple-factor format involving factored soil parameters is used. Over the past few years, Eurocode 7 has been revised to accommodate three design approaches (DAs) that allow partial factors to be introduced at the beginning of the calculations (strength partial factors) or at the end of the calculations (resistance partial factors), or some intermediate combinations thereof. The emphasis is primarily on the re-distribution of the original global factor safety in WSD into separate load and resistance factors (or partial factors).

It is well accepted that uncertainties in geotechnical engineering design are unavoidable and numerous practical advantages are realizable if uncertainties and associated risks can be quantified. This is recognized in a recent National Research Council (2006) report on *Geological and Geotechnical Engineering in the New Millennium: Opportunities for Research and Technological Innovation*. The report remarked that "paradigms for dealing with ... uncertainty are poorly understood and even more poorly practiced" and advocated a need for "improved methods for assessing the potential impacts of these uncertainties on engineering decisions ...". Within the arena of design code development, increasing regulatory pressure is compelling geotechnical LRFD to advance beyond empirical re-distribution of the original global factor of safety to a simplified reliability-based design (RBD) framework that is compatible with structural design. RBD calls for a willingness to accept the fundamental philosophy that: (a) absolute reliability is an unattainable goal in the presence of uncertainty, and (b) probability theory can provide a formal framework for developing design criteria that would ensure that the probability of "failure" (used herein to refer to exceeding any prescribed limit state) is acceptably small. Ideally, geotechnical LRFD should be derived as the logical end-product of a philosophical shift in mind-set to probabilistic design in the first instance and a simplification of rigorous RBD into a familiar "look and feel" design format in the second. The need to draw a clear distinction between accepting reliability analysis as a necessary theoretical basis for geotechnical design and downstream calibration of simplified multiple-factor design formats, with emphasis on the former, was highlighted by Phoon et al. (2003b). The former provides a consistent method for propagation of uncertainties and a unifying framework for risk assessment across disciplines (structural and geotechnical design) and national boundaries. Other competing frameworks have been suggested (e.g. λ-method by Simpson et al., 1981; worst attainable value method by Bolton, 1989; Taylor series method by Duncan, 2000), but none has the theoretical breadth and power to handle complex real-world problems that may require nonlinear 3D finite element or other numerical approaches for solution.

Simpson and Yazdchi (2003) proposed that "limit state design requires analysis of un-reality, not of reality. Its purpose is to demonstrate that limit states are, in effect, unreal, or alternatively that they are 'sufficiently unlikely,' being separated by adequate margins from expected states." It is clear that limit states are "unlikely" states and the purpose of design is to ensure that expected states are sufficiently "far" from these limit states. The pivotal point of contention is how to achieve this separation in numerical terms (Phoon et al., 1993). It is accurate to say that there is no consensus on the preferred method and this issue is still the subject of much heated debate in the geotechnical engineering community. Simpson and Yazdchi (2003) opined that strength partial factors are physically meaningful, because "it is the gradual mobilisation of strength that causes deformation." This is consistent with our prevailing practice of applying a global factor of safety to the capacity to limit deformations. Another physical justification is that variations in soil parameters can create disproportionate nonlinear variations in the response (or resistance) (Simpson, 2000). In this situation, the author felt that "it is difficult to choose values of partial factors which are applicable over the whole range of the variables." The implicit assumption here is that the resistance factor is not desirable because it is not practical to prescribe a large number of resistance factors in a design code. However, if maintaining uniform reliability is a desired goal, a single partial factor for, say, friction angle would not produce the same reliability in different problems because the relevant design equations are not equally sensitive to changes in the friction angle. For complex soil–structure interaction problems, applying a fixed partial factor can result in unrealistic failure mechanisms. In fact, given the complexity of the limit state surface, it is quite unlikely for the same partial factor to locate the most probable failure point in all design problems. These problems would not occur if resistance factors are calibrated from reliability analysis. In addition, it is more convenient to account for the bias in different calculation models using resistance factors. However, for complex problems, it is possible that a large number of resistance factors are needed and the design code becomes too unwieldy or confusing to use. Another undesirable feature is that the engineer does not develop a feel for the failure mechanism if he/she is only required to analyze the expected behavior, followed by application of some resistance factors at the end of the calculation.

Overall, the conclusion is that there are no simple methods (factored parameters or resistances) of replacing reliability analysis for sufficiently complex problems. It may be worthwhile to discuss if one should insist on simplicity despite all the known associated problems. The more recent performance-based design philosophy may provide a solution for this dilemma, because engineers can apply their own calculations methods for ensuring performance compliance, without being restricted to following rigid design codes containing a few partial factors. In the opinion of the author, the need to derive simplified RBD equations perhaps is of practical importance to maintain continuity with past practice, but it is not necessary and it is increasingly fraught with difficulties when sufficiently complex problems are posed. The limitations faced by simplified RBD have no bearing on the generality of reliability theory. This is analogous to arguing that limitations in closed-form elastic solutions are related to elasto-plastic theory. The application of finite element softwares on relatively inexpensive and powerful PCs (with gigahertz processors, a gigabyte of memory, and hundreds of gigabytes – verging on terabyte – of disk) permit real-world problems to be simulated on an unprecedented realistic setting almost routinely. It suffices to note here that RBD can be applied to complex real-world problems using powerful but practical stochastic collocation techniques (Phoon and Huang, 2007).

One common criticism of RBD is that good geotechnical sense and judgment would be displaced by the emphasis on probabilistic analysis (Boden, 1981; Semple, 1981; Bolton, 1983; Fleming, 1989). This is similar to the ongoing criticism of numerical analysis, although this criticism seems to have grown more muted in recent years with the emergence of powerful and user-friendly finite element softwares. The fact of the matter is that experience, sound judgment, and soil mechanics still are needed for all aspects of geotechnical RBD (Kulhawy and Phoon, 1996). Human intuition is not suited for reasoning with uncertainties and only this aspect has been removed from the purview of the engineer. One example in which intuition can be misleading is the common misconception that a larger partial factor should be assigned to a more uncertain soil parameter. This is not necessarily correct, because the parameter may have little influence on the overall response (e.g. capacity, deformation). Therefore, the magnitude of the partial factor should depend on the uncertainty of the parameter and the sensitivity of the response to that parameter. Clearly, judgment is not undermined; instead, it is focused on those aspects for which it is most suited.

Another criticism is that soil statistics are not readily available because of the site-specific nature of soil variability. This concern only is true for total variability analyses, but does not apply to the general approach where inherent soil variability, measurement error, and correlation uncertainty are quantified separately. Extensive statistics for each component uncertainties have been published (Phoon and Kulhawy, 1999a; Uzielli et al., 2007). For each combination of soil type, measurement technique, and correlation model, the uncertainty in the design soil property can be evaluated systematically by combining the appropriate component uncertainties using a simple second-moment probabilistic approach (Phoon and Kulhawy, 1999b).

In summary, we are now at the point where RBD really can be used as a rational and practical design mode. The main impediment is not theoretical (lack of power to deal with complex problems) or practical (speed of computations, availability of soil statistics, etc.), but the absence of simple computational approaches that can be easily implemented by practitioners. Much of the controversies reported in the literature are based on qualitative arguments. If practitioners were able to implement RBD easily on their PCs and calculate actual numbers using actual examples, they would gain a concrete appreciation of the merits and limitations of RBD. Misconceptions will be dismissed definitively, rather than propagated in the literature, generating further confusion. The author believes that the introduction of powerful but simple-to-implement approaches will bring about a greater acceptance of RBD amongst practitioners in the broader geotechnical engineering community. The Probabilistic Model Code was developed by the Joint Committee on Structural Safety (JCSS) to achieve a similar objective (Vrouwenvelder and Faber, 2007).

The main impetus for this book is to explain RBD to students and practitioners with emphasis on "how to calculate" and "how to apply." Practical computational methods are presented in Chapters 1, 3, 4 and 7. Geotechnical examples illustrating reliability analyses and design are provided in Chapters 5, 6, 8–13. The spatial variability of geomaterials is one of the distinctive aspects of geotechnical RBD. This important aspect is covered in Chapter 2. The rest of this chapter provides a primer on reliability calculations with references to appropriate chapters for follow-up reading. Simple MATLAB codes are provided in Appendix A and at http://www.eng.nus.edu.sg/civil/people/cvepkk/prob_lib.html. By focusing on demonstration of RBD through calculations and examples, this book is expected to serve as a valuable teaching and learning resource for practitioners, educators, and students.

1.2 General reliability problem

The general stochastic problem involves the propagation of input uncertainties and model uncertainties through a computation model to arrive at a random output vector (Figure 1.1). Ideally, the full finite-dimensional distribution function of the random output vector is desired, although partial solutions such as second-moment characterizations and probabilities of failure may be sufficient in some applications. In principle, Monte Carlo simulation can be used to solve this problem, regardless of the complexities underlying the computation model, input uncertainties, and/or model uncertainties. One may assume with minimal loss of generality that a complex geotechnical problem (possibly 3D, nonlinear, time and construction dependent, etc.) would only admit numerical solutions and the spatial domain can be modeled by a scalar/vector random field. Monte Carlo simulation requires a procedure to generate realizations of the input and model uncertainties and a numerical scheme for calculating a vector of outputs from each realization. The first step is not necessarily trivial and not completely solved even in the theoretical sense. Some "user-friendly" methods for simulating random variables/vectors/processes are outlined in Section 1.3, with emphasis on key calculation steps and limitations. Details are outside the scope of this chapter and given elsewhere (Phoon, 2004a, 2006a). In this chapter, "user-friendly" methods refer to those that can be implemented on a desktop PC by a non-specialist with limited programming skills; in other words, methods within reach of the general practitioner.

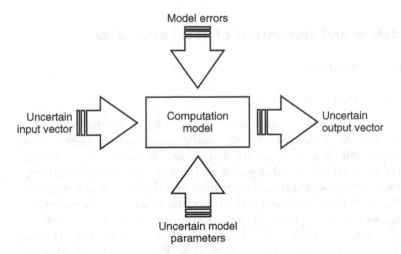

Figure 1.1 General stochastic problem.

The second step is identical to the repeated application of a deterministic solution process. The only potential complication is that a particular set of input parameters may be too extreme, say producing a near-collapse condition, and the numerical scheme may become unstable. The statistics of the random output vector are contained in the resulting ensemble of numerical output values produced by repeated deterministic runs. Fenton and Griffiths have been applying Monte Carlo simulation to solve soil-interaction problems within the context of a random field since the early 1990s (e.g. Griffiths and Fenton, 1993, 1997, 2001; Fenton and Griffiths, 1997, 2002, 2003). Popescu and co-workers developed simulation-based solutions for a variety of soil-structure interaction problems, particularly problems involving soil dynamics, in parallel. Their work is presented in Chapter 6 of this book.

For a sufficiently large and complex soil-structure interaction problem, it is computationally intensive to complete even a single run. The rule-of-thumb for Monte Carlo simulation is that $10/p_f$ runs are needed to estimate a probability of failure, p_f, within a coefficient of variation of 30%. The typical p_f for a geotechnical design is smaller than one in a thousand and it is expensive to run a numerical code more than ten thousand times, even for a modest size problem. This significant practical disadvantage is well known. At present, it is accurate to say that a computationally efficient and "user-friendly" solution to the general stochastic problem remains elusive. Nevertheless, reasonably practical solutions do exist if the general stochastic problem is restricted in some ways, for example, accept a first-order estimate of the probability of failure or accept an approximate but less costly output. Some of these probabilistic solution procedures are presented in Section 1.4.

1.3 Modeling and simulation of stochastic data

1.3.1 Random variables

Geotechnical uncertainties

Two main sources of geotechnical uncertainties can be distinguished. The first arises from the evaluation of design soil properties, such as undrained shear strength and effective stress friction angle. This source of geotechnical uncertainty is complex and depends on inherent soil variability, degree of equipment and procedural control maintained during site investigation, and precision of the correlation model used to relate field measurement with design soil property. Realistic statistical estimates of the variability of design soil properties have been established by Phoon and Kulhawy (1999a, 1999b). Based on extensive calibration studies (Phoon et al., 1995), three ranges of soil property variability (low, medium, high) were found to be sufficient to achieve reasonably uniform reliability levels for simplified RBD checks:

Geotechnical parameter	Property variability	COV (%)
Undrained shear strength	Low	10–30
	Medium	30–50
	High	50–70
Effective stress friction angle	Low	5–10
	Medium	10–15
	High	15–20
Horizontal stress coefficient	Low	30–50
	Medium	50–70
	High	70–90

In contrast, Réthàti (1988), citing the 1965 specification of the American Concrete Institute, observed that the quality of concrete can be evaluated in the following way:

Quality	COV (%)
Excellent	< 10
Good	10–15
Satisfactory	15–20
Bad	> 20

It is clear that the coefficients of variations or COVs of natural geomaterials can be much larger and do not fall within a narrow range. The ranges of quality for concrete only apply to the effective stress friction angle.

The second source arises from geotechnical calculation models. Although many geotechnical calculation models are "simple," reasonable predictions of fairly complex soil–structure interaction behavior still can be achieved through empirical calibrations. Model factors, defined as the ratio of the measured response to the calculated response, usually are used to correct for simplifications in the calculation models. Figure 1.2 illustrates model factors for capacity of drilled shafts subjected to lateral-moment loading. Note that "S.D." is the standard deviation, "n" is the sample size, and "p_{AD}" is the p-value from the Anderson–Darling goodness-of-fit test (> 0.05 implies acceptable lognormal fit). The COVs of model factors appear to fall between 30 and 50%.

It is evident that a geotechnical parameter (soil property or model factor) exhibiting a range of values, possibly occurring at unequal frequencies, is best modeled as a random variable. The existing practice of selecting one characteristic value (e.g. mean, "cautious" estimate, 5% exclusion limit, etc.) is attractive to practitioners, because design calculations can be carried out easily using only one set of input values once they are selected. However, this simplicity is deceptive. The choice of the characteristic values clearly affects

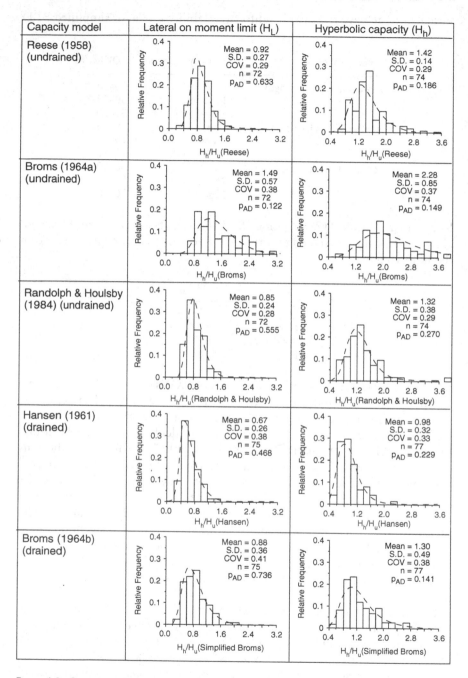

Figure 1.2 Capacity model factors for drilled shafts subjected to lateral-moment loading (modified from Phoon, 2005).

the overall safety of the design, but there are no simple means of ensuring that the selected values will achieve a consistent level of safety. In fact, these values are given by the design point of a first-order reliability analysis and they are problem-dependent. Simpson and Driscoll (1998) noted in their commentary to Eurocode 7 that the definition of the characteristic value "has been the most controversial topic in the whole process of drafting Eurocode 7." If random variables can be included in the design process with minimal inconvenience, the definition of the characteristic value is a moot issue.

Simulation

The most intuitive and possibly the most straightforward method for performing reliability analysis is the Monte Carlo simulation method. It only requires repeated execution of an existing deterministic solution process. The key calculation step is to simulate realizations of random variables. This step can be carried in a general way using:

$$Y = F^{-1}(U) \tag{1.1}$$

in which Y is a random variable following a prescribed cumulative distribution $F(\cdot)$ and U is a random variable uniformly distributed between 0 and 1 (also called a standard uniform variable). Realizations of U can be obtained from EXCEL™ under "Tools > Data Analysis > Random Number Generation > Uniform Between 0 and 1." MATLAB™ implements U using the "rand" function. For example, $U = \text{rand}(10,1)$ is a vector containing 10 realizations of U. Some inverse cumulative distribution functions are available in EXCEL (e.g. norminv for normal, loginv for lognormal, betainv for beta, gammainv for gamma) and MATLAB (e.g. norminv for normal, logninv for lognormal, betainv for beta, gaminv for gamma). More efficient methods are available for some probability distributions, but they are lacking in generality (Hastings and Peacock, 1975; Johnson *et al.*, 1994). A cursory examination of standard probability texts will reveal that the variety of classical probability distributions is large enough to cater to almost all practical needs. The main difficulty lies with the selection of an appropriate probability distribution function to fit the limited data on hand. A complete treatment of this important statistical problem is outside the scope of this chapter. However, it is worthwhile explaining the method of moments because of its simplicity and ease of implementation.

The first four moments of a random variable (Y) can be calculated quite reliably from the typical sample sizes encountered in practice. Theoretically, they are given by:

$$\mu = \int yf(y)\mathrm{d}y = E(Y) \tag{1.2a}$$

$$\sigma^2 = E[(Y-\mu)^2] \qquad (1.2b)$$

$$\gamma_1 = \frac{E[(Y-\mu)^3]}{\sigma^3} \qquad (1.2c)$$

$$\gamma_2 = \frac{E[(Y-\mu)^4]}{\sigma^4} - 3 \qquad (1.2d)$$

in which μ = mean, σ^2 = variance (or σ = standard deviation), γ_1 = skewness, γ_2 = kurtosis, $f(\cdot)$ = probability density function = dF(y)/dy, and $E[\cdot]$ = mathematical expectation. The practical feature here is that moments can be estimated directly from empirical data without knowledge of the underlying probability distribution [i.e. $f(y)$ is unknown]:

$$\bar{y} = \frac{\sum_{i=1}^{n} y_i}{n} \qquad (1.3a)$$

$$s^2 = \frac{1}{n-1} \sum_{i=1}^{n} (y_i - \bar{y})^2 \qquad (1.3b)$$

$$g_1 = \frac{n}{(n-1)(n-2)s^3} \sum_{i=1}^{n} (y_i - \bar{y})^3 \qquad (1.3c)$$

$$g_2 = \frac{n(n+1)}{(n-1)(n-2)(n-3)s^4} \sum_{i=1}^{n} (y_i - \bar{y})^4 - \frac{3(n-1)^2}{(n-2)(n-3)} \qquad (1.3d)$$

in which n = sample size, $(y_1, y_2, \ldots, y_n)$ = data points, $\bar{y}$ = sample mean, s^2 = sample variance, g_1 = sample skewness, and g_2 = sample kurtosis. Note that the MATLAB "kurtosis" function is equal to $g_2 + 3$. If the sample size is large enough, the above sample moments will converge to their respective theoretical values as defined by Equation (1.2) under some fairly general conditions. The majority of classical probability distributions can be determined uniquely by four or less moments. In fact, the Pearson system (Figure 1.3) reduces the selection of an appropriate probability distribution function to the determination of $\beta_1 = g_1^2$ and $\beta_2 = g_2 + 3$. Calculation steps with illustrations are given by Elderton and Johnson (1969). Johnson *et al.* (1994) provided useful formulas for calculating the Pearson parameters based on the first four moments. Réthári (1988) provided some β_1 and β_2 values of Szeged soils for distribution fitting using the Pearson system (Table 1.1).

Johnson system

A broader discussion of distribution systems (in which the Pearson system is one example) and distribution fitting is given by Elderton and

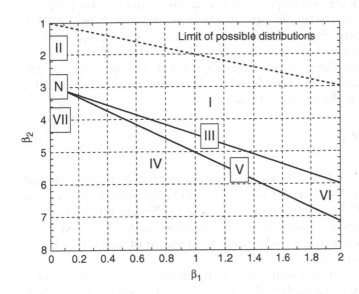

Figure 1.3 Pearson system of probability distribution functions based on $\beta_1 = g_1^2$ and $\beta_2 = g_2 + 3$; N = normal distribution.

Table 1.1 β_1 and β_2 values for some soil properties from Szeged, Hungary (modified from Rétháti, 1988).

Soil layer	Water content	Liquid limit	Plastic limit	Plasticity index	Consistency index	Void ratio	Bulk density	Unconfined compressive strength
β_1								
S1	2.76	4.12	6.81	1.93	0.13	6.50	1.28	0.02
S2	0.01	0.74	0.34	0.49	0.02	0.01	0.09	0.94
S3	0.03	0.96	0.14	0.85	0.00	1.30	2.89	5.06
S4	0.05	0.34	0.13	0.64	0.03	0.13	1.06	4.80
S5	1.10	0.02	2.92	0.00	0.98	0.36	0.10	2.72
β_2								
S1	7.39	8.10	12.30	4.92	3.34	11.19	3.98	1.86
S2	3.45	3.43	3.27	2.69	3.17	2.67	3.87	3.93
S3	7.62	5.13	4.32	4.46	3.31	5.52	11.59	10.72
S4	7.17	3.19	3.47	3.57	4.61	4.03	8.14	10.95
S5	6.70	2.31	9.15	2.17	5.13	4.14	4.94	6.74

Note
Plasticity index $(I_p) = w_L - w_P$, in which w_L = liquid limit and w_p = plastic limit; consistency index $(I_c) = (w_L - w)/I_p = 1 - I_L$, in which w = water content and I_L = liquidity index; unconfined compressive strength $(q_u) = 2s_u$, in which s_u = undrained shear strength.

Johnson (1969). It is rarely emphasized that almost all distribution functions cannot be generalized to handle correlated random variables (Phoon, 2006a). In other words, univariate distributions such as those discussed above cannot be generalized to multivariate distributions. Elderton and Johnson (1969) discussed some interesting exceptions, most of which are restricted to bivariate cases. It suffices to note here that the only convenient solution available to date is to rewrite Equation (1.1) as:

$$Y = F^{-1}[\Phi(Z)] \tag{1.4}$$

in which Z = standard normal random variable with mean = 0 and variance = 1 and $\Phi(\cdot)$ = standard normal cumulative distribution function (normcdf in MATLAB or normsdist in EXCEL). Equation (1.4) is called the *translation* model. It requires all random variables to be related to the standard normal random variable. The significance of Equation (1.4) is elaborated in Section 1.3.2.

An important practical detail here is that standard normal random variables can be simulated directly and efficiently using the Box–Muller method (Box and Muller, 1958):

$$Z_1 = \sqrt{-2\ln(U_1)}\cos(2\pi U_2) \tag{1.5}$$

$$Z_2 = \sqrt{-2\ln(U_1)}\sin(2\pi U_2)$$

in which Z_1, Z_2 = independent standard normal random variables and U_1, U_2 = independent standard uniform random variables. Equation (1.5) is computationally more efficient than Equation (1.1) because the inverse cumulative distribution function is not required. Note that the cumulative distribution function and its inverse are not available in closed-form for most random variables. While Equation (1.1) "looks" simple, there is a hidden cost associated with the numerical evaluation of $F^{-1}(\cdot)$. Marsaglia and Bray (1964) proposed one further improvement:

1. Pick V_1 and V_2 randomly within the unit square extending between -1 and 1 in both directions, i.e.:

$$V_1 = 2U_1 - 1$$
$$V_2 = 2U_2 - 1 \tag{1.6}$$

2. Calculate $R^2 = V_1^2 + V_2^2$. If $R^2 \geq 1.0$ or $R^2 = 0.0$, repeat step (1).

3. Simulate two independent standard normal random variables using:

$$Z_1 = V_1 \sqrt{\frac{-2\ln(R^2)}{R^2}}$$

$$Z_2 = V_2 \sqrt{\frac{-2\ln(R^2)}{R^2}}$$

(1.7)

Equation (1.7) is computationally more efficient than Equation (1.5) because trigonometric functions are not required!

A translation model can be constructed systematically from (β_1, β_2) using the Johnson system:

1. Assume that the random variable is lognormal, i.e.

$$\ln(Y - A) = \lambda + \xi Z \quad Y > A$$

(1.8)

   in which $\ln(\cdot)$ is the natural logarithm, $\xi^2 = \ln[1 + \sigma^2/(\mu - A)^2]$, $\lambda = \ln(\mu - A) - 0.5\xi^2$, $\mu = $ mean of Y, and $\sigma^2 = $ variance of Y. The "lognormal" distribution in the geotechnical engineering literature typically refers to the case of $A = 0$. When $A \neq 0$, it is called the "shifted lognormal" or "3-parameter lognormal" distribution.
2. Calculate $\omega = \exp(\xi^2)$.
3. Calculate $\beta_1 = (\omega - 1)(\omega + 2)^2$ and $\beta_2 = \omega^4 + 2\omega^3 + 3\omega^2 - 3$. For the lognormal distribution, β_1 and β_2 are related as shown by the solid line in Figure 1.4.
4. Calculate $\beta_1 = g_1^2$ and $\beta_2 = g_2 + 3$ from data. If the values fall close to the lognormal (LN) line, the lognormal distribution is acceptable.
5. If the values fall below the LN line, Y follows the SB distribution:

$$\ln\left(\frac{Y - A}{B - Y}\right) = \lambda + \xi Z \quad B > Y > A$$

(1.9)

   in which $\lambda, \xi, A, B = $ distribution fitting parameters.
6. If the values fall above the LN line, Y follows the SU distribution:

$$\ln\left[\left(\frac{Y - A}{B - A}\right) + \sqrt{1 + \left(\frac{Y - A}{B - A}\right)^2}\right] = \sinh^{-1}\left(\frac{Y - A}{B - A}\right) = \lambda + \xi Z$$

(1.10)

Examples of LN, SB, and SU distributions are given in Figure 1.5. Simulation of Johnson random variables using MATLAB is given in Appendix A.1. Carsel and Parrish (1988) developed joint probability distributions for

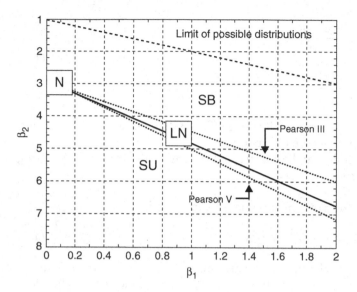

Figure 1.4 Johnson system of probability distribution functions based on $\beta_1 = g_1^2$ and $\beta_2 = g_2 + 3$; N = normal distribution; LN = lognormal distribution.

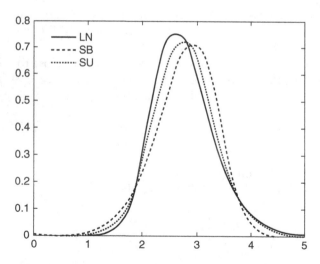

Figure 1.5 Examples of LN ($\lambda = 1.00, \xi = 0.20$), SB ($\lambda = 1.00, \xi = 0.36$, A = -3.00, B = 5.00), and SU ($\lambda = 1.00, \xi = 0.09$, A = -1.88, B = 2.08) distributions with approximately the same mean and coefficient of variation ≈ 30 %.

parameters of soil–water characteristic curves using the Johnson system. The main practical obstacle is that the SB Johnson parameters (λ, ξ, A, B) are related to the first four moments $(\bar{y}, s^2, g_1, g_2)$ in a very complicated way (Johnson, 1949). The SU Johnson parameters are comparatively easier to estimate from the first four moments (see Johnson *et al.*, 1994 for closed-form formulas). In addition, both Pearson and Johnson systems require the proper identification of the relevant region (e.g. SB or SU in Figure 1.4), which in turn determines the distribution function [Equation (1.9) or (1.10)], before one can attempt to calculate the distribution parameters.

Hermite polynomials

One-dimensional Hermite polynomials are given by:

$$
\begin{aligned}
H_0(Z) &= 1 \\
H_1(Z) &= Z \\
H_2(Z) &= Z^2 - 1 \\
H_3(Z) &= Z^3 - 3Z \\
H_{k+1}(Z) &= ZH_k(Z) - kH_{k-1}(Z)
\end{aligned}
\tag{1.11}
$$

in which Z is a standard normal random variable (mean = 0 and variance = 1). Hermite polynomials can be evaluated efficiently using the recurrence relation given in the last row of Equation (1.11). It can be proven rigorously (Phoon, 2003) that any random variable Y (with finite variance) can be expanded as a series:

$$
Y = \sum_{k=0}^{\infty} a_k H_k(Z)
\tag{1.12}
$$

The numerical values of the coefficients, a_k, depend on the distribution of Y. The key practical advantage of Equation (1.12) is that the randomness of Y is completely accounted for by the randomness of Z, which is a *known* random variable. It is useful to observe in passing that Equation (1.12) may not be a monotonic function of Z when it is truncated to a finite number of terms. The extrema are located at points with zero first derivatives but non-zero second derivatives. Fortunately, derivatives of Hermite polynomials can be evaluated efficiently as well:

$$
\frac{dH_k(Z)}{dZ} = kH_{k-1}(Z)
\tag{1.13}
$$

The Hermite polynomial expansion can be modified to fit the first four moments as well:

$$Y = \bar{y} + s \, k[Z + h_3(Z^2 - 1) + h_4(Z^3 - 3Z)] \tag{1.14}$$

in which $\bar{y}$ and s = sample mean and sample standard deviation of Y, respectively and k = normalizing constant = $1/(1 + 2h_3^2 + 6h_4^2)^{0.5}$. It is important to note that the coefficients h_3 and h_4 can be calculated from the sample skewness (g_1) and sample kurtosis (g_2) in a relatively straightforward way (Winterstein *et al.*, 1994):

$$h_3 = \frac{g_1}{6}\left(\frac{1 - 0.015|g_1| + 0.3g_1^2}{1 + 0.2g_2}\right) \tag{1.15a}$$

$$h_4 = \frac{(1 + 1.25g_2)^{1/3} - 1}{10}\left(1 - \frac{1.43g_1^2}{g_2}\right)^{1 - 0.1(g_2+3)^{0.8}} \tag{1.15b}$$

The theoretical skewness (γ_1) and kurtosis (γ_2) produced by Equation (1.14) are (Phoon, 2004a):

$$\gamma_1 = k^3(6h_3 + 36h_3h_4 + 8h_3^3 + 108h_3h_4^2) \tag{1.16a}$$

$$\gamma_2 = k^4(3 + 48h_3^4 + 3348h_4^4 + 24h_4 + 1296h_3^3 + 60h_3^2 + 252h_4^2$$
$$+ 2232h_3^2h_4^2 + 576h_3^2h_4) - 3 \tag{1.16b}$$

Equation (1.15) is determined empirically by minimizing the error $[(\gamma_1 - g_1)^2 + (\gamma_2 - g_2)^2]$ subjected to the constraint that Equation (1.14) is a monotonic function of Z. It is intended for cases with $0 < g_2 < 12$ and $0 \leq g_1^2 < 2g_2/3$ (Winterstein *et al.*, 1994). It is possible to minimize the error in skewness and kurtosis numerically using the SOLVER function in EXCEL, rather than applying Equation (1.15).

In general, the entire cumulative distribution function of Y, $F(y)$, can be fully described by the Hermite expansion using the following simple stochastic collocation method:

1. Let $(y_1, y_2, \ldots, y_n)$ be n realizations of Y. The standard normal data is calculated from:

$$z_i = \Phi^{-1}F(y_i) \tag{1.17}$$

2. Substitute y_i and z_i into Equation (1.12). For a third-order expansion, we obtain:

$$y_i = a_0 + a_1 z_i + a_2(z_i^2 - 1) + a_3(z_i^3 - 3z_i) = a_0 + a_1 h_{i1}$$
$$+ a_2 h_{i2} + a_3 h_{i3} \tag{1.18}$$

in which $(1, h_{i1}, h_{i2}, h_{i3})$ are Hermite polynomials evaluated at z_i.

3. The four unknown coefficients (a_0, a_1, a_2, a_3) can be determined using four realizations of Y, (y_1, y_2, y_3, y_4). In matrix notation, we write:

$$\mathbf{H}\mathbf{a} = \mathbf{y} \tag{1.19}$$

in which $\mathbf{H}$ is a 4×4 matrix with ith row given by $(1, h_{i1}, h_{i2}, h_{i3})$, $\mathbf{y}$ is a 4×1 vector with ith component given by y_i and $\mathbf{a}$ is a 4×1 vector containing the unknown numbers $(a_0, a_1, a_2, a_3)'$. This is known as the stochastic collocation method. Equation (1.19) is a linear system and efficient solutions are widely available.

4. It is preferable to solve for the unknown coefficients using regression by using more than four realizations of Y:

$$(\mathbf{H}'\mathbf{H})\mathbf{a} = \mathbf{H}'\mathbf{y} \tag{1.20}$$

in which $\mathbf{H}$ is an $n \times 4$ matrix and n is the number of realizations. Note that Equation (1.20) is a linear system amenable to fast solution as well.

Calculation of Hermite coefficients using MATLAB is given in Appendix A.2. Hermite coefficients can be calculated with ease using EXCEL as well (Chapter 9). Figure 1.6 demonstrates that Equation (1.12) is quite efficient – it is possible to match very small probabilities (say 10^{-4}) using a third-order Hermite expansion (four terms).

Phoon (2004a) pointed out that Equations (1.4) and (1.12) are theoretically identical. Equation (1.12) appears to present a rather circuitous route of achieving the same result as Equation (1.4). The key computational difference is that Equation (1.4) requires the costly evaluation of $F^{-1}(\cdot)$ thousands of times in a low-probability simulation exercise (typical of civil engineering problems), while Equation (1.17) requires less than 100 costly evaluations of $\Phi^{-1}F(\cdot)$ followed by less-costly evaluation of Equation (1.12) thousands of times. Two pivotal factors govern the computational efficiency of Equation (1.12): (a) cheap generation of standard normal random variables using the Box–Muller method [Equation (1.5) or (1.7)] and (b) relatively small number of Hermite terms.

The one-dimensional Hermite expansion can be easily extended to random vectors and processes. The former is briefly discussed in the next

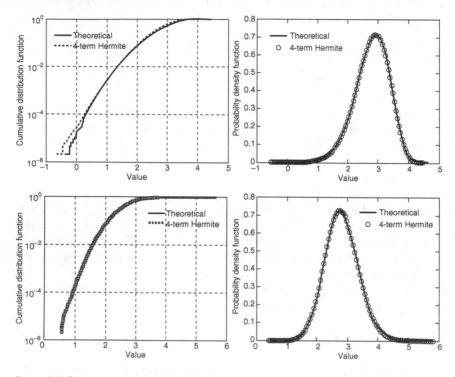

Figure 1.6 Four-term Hermite expansions for: Johnson SB distribution with $\lambda = 1.00$, $\xi = 0.36$, A = −3.00, B = 5.00 (top row) and Johnson SU distribution with $\lambda = 1.00$, $\xi = 0.09$, A = −1.88, B = 2.08 (bottom row).

section while the latter is given elsewhere (Puig *et al.*, 2002; Sakamoto and Ghanem, 2002; Puig and Akian, 2004). Chapters 5, 7, 9 and elsewhere (Sudret and Der Kiureghian, 2000; Sudret, 2007) present applications of Hermite polynomials in more comprehensive detail.

1.3.2 Random vectors

The multivariate normal probability density function is available analytically and can be defined uniquely by a mean vector and a covariance matrix:

$$f(\mathbf{X}) = |\mathbf{C}|^{-\frac{1}{2}} (2\pi)^{-\frac{n}{2}} \exp\left[-0.5(\mathbf{X} - \mathbf{\mu})'\mathbf{C}^{-1}(\mathbf{X} - \mathbf{\mu})\right] \qquad (1.21)$$

in which $\mathbf{X} = (X_1, X_2, \ldots, X_n)'$ is a normal random vector with n components, μ is the mean vector, and $\mathbf{C}$ is the covariance matrix. For the bivariate (simplest) case, the mean vector and covariance matrix are given by:

$$\mu = \left\{ \begin{array}{c} \mu_1 \\ \mu_2 \end{array} \right\}$$

$$\mathbf{C} = \left[\begin{array}{cc} \sigma_1^2 & \rho\sigma_1\sigma_2 \\ \rho\sigma_1\sigma_2 & \sigma_2^2 \end{array} \right]$$

(1.22)

in which μ_i and $\sigma_i =$ mean and standard deviation of X_i, respectively, and $\rho =$ product-moment (Pearson) correlation between X_1 and X_2.

The practical usefulness of Equation (1.21) is not well appreciated. First, the full multivariate dependency structure of a normal random vector only depends on a covariance matrix ($\mathbf{C}$) containing bivariate information (correlations) between all possible pairs of components. The practical advantage of capturing multivariate dependencies in any dimension (i.e. any number of random variables) using only bivariate dependency information is obvious. Note that coupling two random variables is the most basic form of dependency and also the simplest to evaluate from empirical data. In fact, there are usually insufficient data to calculate reliable dependency information beyond correlations in actual engineering practice. Second, fast simulation of correlated normal random variables is possible because of the elliptical form $(\mathbf{X} - \boldsymbol{\mu})'\mathbf{C}^{-1}(\mathbf{X} - \boldsymbol{\mu})$ appearing in the exponent of Equation (1.21). When the random dimension is small, the following Cholesky method is the most efficient and robust:

$$\mathbf{X} = L\mathbf{Z} + \boldsymbol{\mu}$$

(1.23)

in which $\mathbf{Z} = (Z_1, Z_2, ..., Z_n)'$ contains uncorrelated normal random components with zero means and unit variances. These components can be simulated efficiently using the Box–Muller method [Equations (1.5) or (1.7)]. The lower triangular matrix $\mathbf{L}$ is the Cholesky factor of $\mathbf{C}$, i.e.:

$$\mathbf{C} = \mathbf{LL}'$$

(1.24)

Cholesky factorization can be roughly appreciated as taking the "square root" of a matrix. The Cholesky factor can be calculated in EXCEL using the array formula MAT_CHOLESKY, which is provided by a free add-in at http://digilander.libero.it/foxes/index.htm. MATLAB produces $\mathbf{L}'$ using chol ($\mathbf{C}$) [note: $\mathbf{L}'$ (transpose of $\mathbf{L}$) is an upper triangular matrix]. Cholesky factorization fails if $\mathbf{C}$ is not "positive definite." The important practical ramifications here are: (a) the correlation coefficients in the covariance matrix $\mathbf{C}$ cannot be selected independently and (b) an erroneous $\mathbf{C}$ is automatically flagged when Cholesky factorization fails. When the random dimension is high, it is preferable to use the fast fourier transform (FFT) as described in Section 1.3.3.

As mentioned in Section 1.3.1, the multivariate normal distribution plays a central role in the modeling and simulation of correlated non-normal random variables. The translation model [Equation (1.4)] for one random variable (Y) can be generalized in a straightforward to a random vector $(Y_1, Y_2, ..., Y_n)$:

$$Y_i = F_i^{-1}[\Phi(X_i)] \tag{1.25}$$

in which $(X_1, X_2, ..., X_n)$ follows a multivariate normal probability density function [Equation (1.21)] with:

$$\mu = \begin{Bmatrix} 0 \\ 0 \\ \vdots \\ 0 \end{Bmatrix} \qquad C = \begin{bmatrix} 1 & \rho_{12} & \cdots & \rho_{1n} \\ \rho_{21} & 1 & \cdots & \rho_{2n} \\ \vdots & \vdots & \ddots & \vdots \\ \rho_{n1} & \rho_{n2} & \cdots & 1 \end{bmatrix}$$

Note that $\rho_{ij} = \rho_{ji}$, i.e. C is a symmetric matrix containing only $n(n-1)/2$ distinct entries. Each non-normal component, Y_i, can follow any arbitrary cumulative distribution function $F_i(\cdot)$. The cumulative distribution function prescribed to each component can be different, i.e. $F_i(\cdot) \neq F_j(\cdot)$. The simulation of correlated non-normal random variables using MATLAB is given in Appendix A.3. It is evident that "translation.m" can be modified to simulate any number of random components as long as a compatible size covariance matrix is specified. For example, if there are three random components $(n = 3)$, C should be a 3×3 matrix such as [1 0.8 0.4; 0.8 1 0.2; 0.4 0.2 1]. This computational simplicity explains the popularity of the translation model. As mentioned previously, not all 3×3 matrices are valid covariance matrices. An example of an invalid covariance matrix is $C = [10.8 - 0.8; 0.81 - 0.2; -0.8 - 0.21]$. An attempt to execute "translation.m" produces the following error messages:

??? Error using ==> chol
Matrix must be positive definite.

Hence, simulation from an invalid covariance matrix will not take place. The left panel of Figure 1.7 shows the scatter plot of two correlated normal random variables with $\rho = 0.8$. An increasing linear trend is apparent. A pair of uncorrelated normal random variables will not exhibit any trend in the scatter plot. The right panel of Figure 1.7 shows the scatter plot of two correlated non-normal random variables. There is an increasing nonlinear trend. The nonlinearity is fully controlled by the non-normal distribution functions (SB and SU in this example). In practice, there is no reason for the scatter plot (produced by two columns of numbers) to be related to the

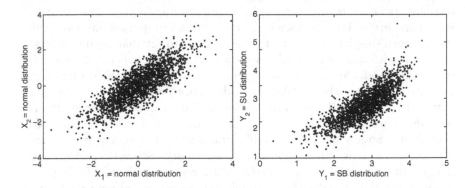

Figure 1.7 Scatter plots for: correlated normal random variables with zero means and unit variances (left) and correlated non-normal random variables with 1st component = Johnson SB distribution ($\lambda = 1.00, \xi = 0.36, A = -3.00, B = 5.00$) and 2nd component = Johnson SU distribution ($\lambda = 1.00, \xi = 0.09, A = -1.88, B = 2.08$) (right).

distribution functions (produced by treating each column of numbers separately). Hence, it is possible for the scatter plot produced by the translation model to be unrealistic.

The simulation procedure described in "translation.m" cannot be applied directly to practice, because it requires the covariance matrix of **X** as an input. The empirical data can only produce an estimate of the covariance matrix of **Y**. It can be proven theoretically that these covariance matrices are not equal, although they can be approximately equal in some cases. The simplest solution available so far is to express Equation (1.25) as Hermite polynomials:

$$Y_1 = a_{10}H_0(X_1) + a_{11}H_1(X_1) + a_{12}H_2(X_1) + a_{13}H_3(X_1) + \cdots$$
$$Y_2 = a_{20}H_0(X_2) + a_{21}H_1(X_2) + a_{22}H_2(X_2) + a_{23}H_3(X_2) + \cdots$$

(1.26)

The relationship between the observed correlation coefficient ($\rho_{Y_1 Y_2}$) and the underlying normal correlation coefficient (ρ) is:

$$\rho_{Y_1 Y_2} = \frac{\sum_{k=1}^{\infty} k! a_{1k} a_{2k} \rho^k}{\sqrt{\left(\sum_{k=1}^{\infty} k! a_{1k}^2\right)\left(\sum_{k=1}^{\infty} k! a_{2k}^2\right)}}$$

(1.27)

This complication is discussed in Chapter 9 and elsewhere (Phoon, 2004a, 2006a).

From a practical viewpoint [only bivariate information needed, correlations are easy to estimate, voluminous n-dimensional data compiled into mere $n(n-1)/2$ coefficients, etc.] and a computational viewpoint (fast, robust, simple to implement), it is accurate to say that the multivariate normal model and multivariate non-normal models generated using the translation model are already very good and sufficiently general for many practical scenarios. The translation model is not perfect – there are important and fundamental limitations (Phoon, 2004a, 2006a). For stochastic data that cannot be modeled using the translation model, the hunt for probability models with comparable practicality, theoretical power, and simulation speed is still on-going. Copula theory (Schweizer, 1991) produces a more general class of multivariate non-normal models, but it is debatable at this point if these models can be estimated empirically and simulated numerically with equal ease for high random dimensions. The only exception is a closely related but non-translation approach based on the multivariate normal copula (Phoon, 2004b). This approach is outlined in Chapter 9.

1.3.3 Random processes

A natural probabilistic model for correlated spatial data is the random field. A one-dimensional random field is typically called a random process. A random process can be loosely defined as a random vector with an infinite number of components that are indexed by a real number (e.g. depth coordinate, z). We restrict our discussion to a normal random process, $X(z)$. Non-normal random processes can be simulated from a normal random process using the same translation approach described in Section 1.3.2.

The only computational aspect that requires some elaboration is that simulation of a process is usually more efficient in the frequency domain. Realizations belonging to a zero-mean stationary normal process $X(z)$ can be generated using the following spectral approach:

$$X(z) = \sum_{k=1}^{\infty} \sigma_k (Z_{1k} \sin 2\pi f_k z + Z_{2k} \cos 2\pi f_k z) \tag{1.28}$$

in which $\sigma_k = \sqrt{2S(f_k)\Delta f}$, Δf is the interval over which the spectral density function $S(f)$ is discretized, $f_k = (2k-1)\Delta f/2$, and Z_{1k} and Z_{2k} are uncorrelated normal random variables with zero means and unit variances. The single exponential autocorrelation function is commonly used in geostatistics: $R(\tau) = exp(-2|\tau|/\delta)$, in which τ is the distance between data points and δ the scale of fluctuation. $R(\tau)$ can be estimated from a series of numbers, say cone tip resistances sampled at a vertical spacing of Δz, using the method of moments:

$$R(\tau = j\Delta z) \approx \frac{1}{(n-j-1)s^2} \sum_{i=1}^{n-j} (x_i - \bar{x})(x_{i+j} - \bar{x}) \tag{1.29}$$

in which $x_i = x(z_i), z_i = i(\Delta z), \bar{x}$ = sample mean [Equation (1.3a)], s^2 = sample variance [Equation (1.3b)], and n = number of data points. The scale of fluctuation is estimated by fitting an exponential function to Equation (1.29). The spectral density function corresponding to the single exponential correlation function is:

$$S(f) = \frac{4\delta}{(2\pi f \delta)^2 + 4} \tag{1.30}$$

Other common autocorrelation functions and their corresponding spectral density functions are given in Table 1.2. It is of practical interest to note that $S(f)$ can be calculated numerically from a given target autocorrelation function or estimated directly from $x(z_i)$ using the FFT. Analytical solutions such as those shown in Table 1.2 are convenient but unnecessary. The simulation of a standard normal random process with zero mean and unit variance using MATLAB is given in Appendix A.4. Note that the main input to "ranprocess.m" is $R(\tau)$, which can be estimated empirically from Equation (1.29). No knowledge of $S(f)$ is needed. Some realizations based on the five common autocorrelation functions shown in Table 1.2 with $\delta = 1$ are given in Figure 1.8.

Table 1.2 Common autocorrelation and two-sided power spectral density functions.

Model	Autocorrelation, $R(\tau)$	Two-sided power spectral density, $S(f)$*	Scale of fluctuation, δ
Single exponential	$\exp(-a\lvert\tau\rvert)$	$\dfrac{2a}{\omega^2 + a^2}$	$\dfrac{2}{a}$
Binary noise	$\begin{array}{ll} 1 - a\lvert\tau\rvert & \lvert\tau\rvert \leq 1/a \\ 0 & \text{otherwise} \end{array}$	$\dfrac{\sin^2(\omega/2a)}{a(\omega/2a)^2}$	$\dfrac{1}{a}$
Cosine exponential	$\exp(-a\lvert\tau\rvert)\cos(a\tau)$	$a\left(\dfrac{1}{a^2 + (\omega+a)^2} + \dfrac{1}{a^2 + (\omega-a)^2}\right)$	$\dfrac{1}{a}$
Second-order Markov	$(1 + a\lvert\tau\rvert)\exp(-a\lvert\tau\rvert)$	$\dfrac{4a^3}{(\omega^2 + a^2)^2}$	$\dfrac{4}{a}$
Squared exponential	$\exp[-(a\tau)^2]$	$\dfrac{\sqrt{\pi}}{a}\exp\left(-\dfrac{\omega^2}{4a^2}\right)$	$\dfrac{\sqrt{\pi}}{a}$

* $\omega = 2\pi f$.

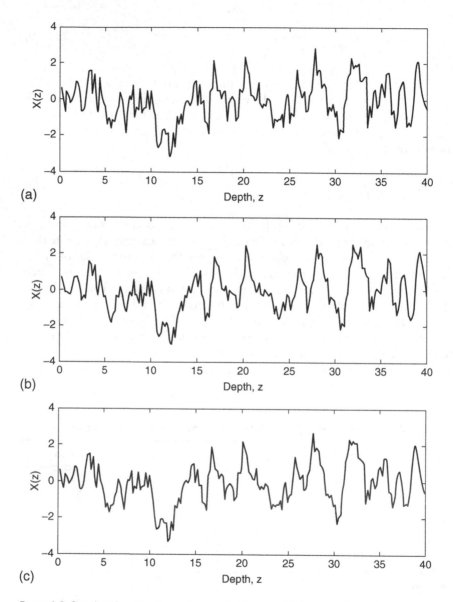

Figure 1.8 Simulated realizations of normal process with mean = 0, variance = 1, scale of fluctuation = 1 based on autocorrelation function = (a) single exponential, (b) binary noise, (c) cosine exponential, (d) second-order Markov, and (e) square exponential.

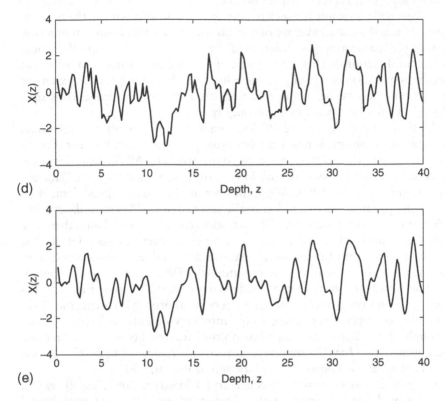

(d)

(e)

Figure 1.8 Cont'd.

In geotechnical engineering, spatial variability is frequently modeled using random processes/fields. This practical application was first studied in geology and mining under the broad area of geostatistics. The physical premise that makes estimation of spatial patterns possible is that points close in space tend to assume similar values. The autocorrelation function (or variogram) is a fundamental tool describing similarities in a statistical sense as a function of distance [Equation (1.29)]. The works of G. Matheron, D.G. Krige, and F.P. Agterberg are notable. Parallel developments also took place in meteorology (L.S. Gandin) and forestry (B. Matérn). Geostatistics is mathematically founded on the theory of random processes/fields developed by A.Y. Khinchin, A.N. Kolmogorov, P. Lévy, N. Wiener, and A.M. Yaglom, among others. The interested reader can refer to books by Cressie (1993) and Chilès and Delfiner (1999) for details. VanMarcke (1983) remarked that all measurements involve some degree of local averaging and that random field models do not need to consider variations below a finite scale because they are smeared by averaging. VanMarcke's work is incomplete in one crucial aspect. This crucial aspect is that actual failure surfaces in 2D or 3D problems will automatically seek to connect "weakest" points in the random domain. It is the spatial average of this failure surface that counts in practice; not simple spatial averages along *pre-defined* surfaces (or pre-specified measurement directions). This is an exceptionally difficult problem to solve – at present simulation is the only viable solution. Chapter 6 presents extensive results on emergent behaviors resulting from spatially heterogeneous soils that illustrate this aspect quite thoroughly.

Although the random process/field provides a concise mathematical model for spatial variability, it poses considerable practical difficulties for statistical inference in view of its complicated data structure. All classical statistical tests are invariably based on the important assumption that the data are independent (Cressie, 1993). When they are applied to correlated data, large bias will appear in the evaluation of the test statistics (Phoon *et al.*, 2003a). The application of standard statistical tests to correlated soil data is therefore potentially misleading. Independence is a very convenient assumption that makes a large part of mathematical statistics tractable. Statisticians can go to great lengths to remove this dependency (Fisher, 1935) or be content with less-powerful tests that are robust to departures from the independence assumption. In recent years, an alternate approach involving the direct modeling of dependency relationships into very complicated test statistics through Monte Carlo simulation has become feasible because desktop computing machines have become very powerful (Phoon, *et al.*, 2003a; Phoon and Fenton, 2004; Phoon 2006b; Uzielli and Phoon, 2006).

Chapter 2 and elsewhere (Baecher and Christian, 2003; Uzielli *et al.*, 2007) provide extensive reviews of geostatistical applications in geotechnical engineering.

1.4 Probabilistic solution procedures

The practical end point of characterizing uncertainties in the design input parameters (geotechnical, geo-hydrological, geometrical, and possibly thermal) is to evaluate their impact on the performance of a design. Reliability analysis focuses on the most important aspect of performance, namely the probability of failure ("failure" is a generic term for non-performance). This probability of failure clearly depends on both parametric and model uncertainties. The probability of failure is a more consistent and complete measure of safety because it is invariant to all mechanically equivalent definitions of safety and it incorporates additional uncertainty information. There is a prevalent misconception that reliability-based design is "new." All experienced engineers would conduct parametric studies when confidence in the choice of deterministic input values is lacking. Reliability analysis merely allows the engineer to carry out a much broader range of parametric studies without actually performing thousands of design checks with different inputs one at a time. This sounds suspiciously like a "free lunch," but exceedingly clever probabilistic techniques do exist to calculate the probability of failure efficiently. The chief drawback is that these techniques are difficult to understand for the non-specialist, but they are not necessarily difficult to implement computationally. There is no consensus within the geotechnical community whether a more consistent and complete measure of safety is worth the additional efforts, or if the significantly simpler but inconsistent global factor of safety should be dropped. Regulatory pressure appears to be pushing towards RBD for non-technical reasons. The literature on reliability analysis and RBD is voluminous. Some of the main developments in geotechnical engineering are presented in this book.

1.4.1 Closed-form solutions

There is a general agreement in principle that limit states (undesirable states in which the system fails to perform satisfactorily) should be evaluated explicitly and separately, but there is no consensus on how to verify that exceedance of a limit state is "sufficiently unlikely" in numerical terms. Different opinions and design recommendations have been made, but there is a lack of discussion on basic issues relating to this central idea of "exceeding a limit state." A simple framework for discussing such basic issues is to imagine the limit state as a boundary surface dividing sets of design parameters (soil, load, and/or geometrical parameters) into those that result in satisfactory and unsatisfactory designs. It is immediately clear that this surface can be very complex for complex soil-structure interaction problems. It is also clear without any knowledge of probability theory that likely failure sets of design parameters (producing likely failure mechanisms) cannot be discussed without characterizing the uncertainties in the design parameters explicitly or implicitly. Assumptions that "values are physically bounded," "all values are likely in the absence of information," etc., are probabilistic assumptions, regardless of whether or not this probabilistic nature is acknowledged explicitly. If the engineer is 100% sure of the design parameters, then only one design check using these fully deterministic parameters is necessary to ensure that the relevant limit state is not exceeded. Otherwise, the situation is very complex and the only rigorous method available to date is reliability analysis. The only consistent method to control exceedance of a limit state is to control the reliability index. It would be very useful to discuss at this stage if this framework is logical and if there are alternatives to it. In fact, it has not been acknowledged explicitly if problems associated with strength partial factors or resistance factors are merely problems related to simplification of reliability analysis. If there is no common underlying purpose (e.g. to achieve uniform reliability) for applying partial factors and resistance factors, then the current lack of consensus on which method is better cannot be resolved in any meaningful way. Chapter 8 provides some useful insights on the reliability levels implied by the empirical soil partial factors in Eurocode 7.

Notwithstanding the above on-going debate, it is valid to question if reliability analysis is too difficult for practitioners. The simplest example is to consider a foundation design problem involving a random capacity (Q) and a random load (F). The ultimate limit state is defined as that in which the capacity is equal to the applied load. Clearly, the foundation will fail if the capacity is less than this applied load. Conversely, the foundation should perform satisfactorily if the applied load is less than the capacity. These three situations can be described concisely by a single performance function P, as follows:

$$P = Q - F \tag{1.31}$$

Mathematically, the above three situations simply correspond to the conditions of $P = 0, P < 0$, and $P > 0$, respectively.

The basic objective of RBD is to ensure that the probability of failure does not exceed an acceptable threshold level. This objective can be stated using the performance function as follows:

$$p_f = \text{Prob}(P < 0) \leq p_T \tag{1.32}$$

in which $\text{Prob}(\cdot) =$ probability of an event, $p_f =$ probability of failure, and $p_T =$ acceptable target probability of failure. A more convenient alternative to the probability of failure is the reliability index (β), which is defined as:

$$\beta = -\Phi^{-1}(p_f) \tag{1.33}$$

in which $\Phi^{-1}(\cdot) =$ inverse standard normal cumulative function. The function $\Phi^{-1}(\cdot)$ can be obtained easily from EXCEL using normsinv (p_f) or MATLAB using norminv (p_f). For sufficiently large β, simple approximate closed-form solutions for $\Phi(\cdot)$ and $\Phi^{-1}(\cdot)$ are available (Appendix B).

The basic reliability problem is to evaluate p_f from some pertinent statistics of F and Q, which typically include the mean (μ_F or μ_Q) and the standard deviation (σ_F or σ_Q), and possibly the probability density function. A simple closed-form solution for p_f is available if Q and F follow a *bivariate* normal distribution. For this condition, the solution to Equation (1.32) is:

$$p_f = \Phi \left(-\frac{\mu_Q - \mu_F}{\sqrt{\sigma_Q^2 + \sigma_F^2 - 2\rho_{QF}\sigma_Q\sigma_F}} \right) = \Phi(-\beta) \tag{1.34}$$

in which $\rho_{QF} =$ product-moment correlation coefficient between Q and F. Numerical values for $\Phi(\cdot)$ can be obtained easily using the EXCEL function normsdist($-\beta$) or the MATLAB function normcdf($-\beta$). The reliability indices for most geotechnical components and systems lie between 1 and 5, corresponding to probabilities of failure ranging from about 0.16 to 3×10^{-7}, as shown in Figure 1.9.

Equation (1.34) can be generalized to a linear performance function containing any number of normal components $(X_1, X_2, \ldots, X_n)$ as long as they follow a *multivariate* normal distribution function [Equation (1.21)]:

$$P = a_0 + \sum_{i=1}^{n} a_i X_i \tag{1.35}$$

$$p_f = \Phi \left(-\frac{a_0 + \sum_{i=1}^{n} a_i \mu_i}{\sqrt{\sum_{i=1}^{n} \sum_{j=1}^{n} a_i a_j \rho_{ij} \sigma_i \sigma_j}} \right) \tag{1.36}$$

in which a_i = deterministic constant, μ_i = mean of X_i, σ_i = standard deviation of X_i, ρ_{ij} = correlation between X_i and X_j (note: $\rho_{ii} = 1$). Chapters 8, 11, and 12 present some applications of these closed-form solutions.

Equation (1.34) can be modified for the case of translation lognormals, i.e. $\ln(Q)$ and $\ln(F)$ follow a bivariate normal distribution with mean of $\ln(Q) = \lambda_Q$, mean of $\ln(F) = \lambda_F$, standard deviation of $\ln(Q) = \xi_Q$, standard deviation

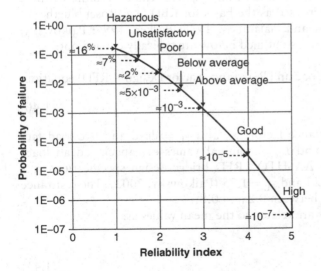

Figure 1.9 Relationship between reliability index and probability of failure (classifications proposed by US Army Corps of Engineers, 1997).

of $\ln(F) = \xi_F$, and correlation between $\ln(Q)$ and $\ln(F) = \rho'_{QF}$:

$$p_f = \Phi\left(-\frac{\lambda_Q - \lambda_F}{\sqrt{\xi_Q^2 + \xi_F^2 - 2\rho'_{QF}\xi_Q\xi_F}}\right) \qquad (1.37)$$

The relationships between the mean (μ) and standard deviation (σ) of a lognormal and the mean (λ) and standard deviation (ξ) of the equivalent normal are given in Equation (1.8). The correlation between Q and F (ρ_{QF}) is related to the correlation between $\ln(Q)$ and $\ln(F)$ (ρ'_{QF}) as follows:

$$\rho_{QF} = \frac{\exp(\xi_Q\xi_F\rho'_{QF}) - 1}{\sqrt{[\exp(\xi_Q^2) - 1][\exp(\xi_F^2) - 1]}} \qquad (1.38)$$

If $\rho'_{QF} = \rho_{QF} = 0$ (i.e. Q and F are independent lognormals), Equation (1.37) reduces to the following well-known expression:

$$\beta = \frac{\ln\left(\frac{\mu_Q}{\mu_F}\sqrt{\frac{1+COV_F^2}{1+COV_Q^2}}\right)}{\sqrt{\ln\left[\left(1 + COV_Q^2\right)\left(1 + COV_F^2\right)\right]}} \qquad (1.39)$$

in which $COV_Q = \sigma_Q/\mu_Q$ and $COV_F = \sigma_F/\mu_F$. If there are physical grounds to disallow negative values, the translation lognormal model is more sensible. Equation (1.39) has been used as the basis for RBD (e.g. Rosenblueth and Esteva, 1972; Ravindra and Galambos, 1978; Becker, 1996; Paikowsky, 2002). The calculation steps outlined below are typically carried out:

1. Consider a typical Load and Resistance Factor Design (LRFD) equation:

    $$\phi Q_n = \gamma_D D_n + \gamma_L L_n \qquad (1.40)$$

    in which ϕ = resistance factor, γ_D and γ_L = dead and live load factors, and Q_n, D_n and L_n = nominal values of capacity, dead load, and live load. The AASHTO LRFD bridge design specifications recommended $\gamma_D = 1.25$ and $\gamma_L = 1.75$ (Paikowsky, 2002). The resistance factor typically lies between 0.2 and 0.8.

2. The nominal values are related to the mean values as:

    $$\mu_Q = b_Q Q_n$$
    $$\mu_D = b_D D_n \qquad (1.41)$$
    $$\mu_L = b_L L_n$$

in which b_Q, b_D, and b_L are bias factors and μ_Q, μ_D, and μ_L are mean values. The bias factors for dead and live load are typically 1.05 and 1.15, respectively. Figure 1.2 shows that b_Q (mean value of the histogram) can be smaller or larger than one.

3. Assume typical values for the mean capacity (μ_Q) and D_n/L_n. A reasonable range for D_n/L_n is 1 to 4. Calculate the mean live load and mean dead load as follows:

$$\mu_L = \frac{(\phi b_Q \mu_Q) b_L}{\left(\gamma_D \frac{D_n}{L_n} + \gamma_L\right)} \tag{1.42}$$

$$\mu_D = \left(\frac{D_n}{L_n}\right) \frac{\mu_L b_D}{b_L} \tag{1.43}$$

4. Assume typical coefficients of variation for the capacity, dead load, and live load, say $COV_Q = 0.3$, $COV_D = 0.1$, and $COV_L = 0.2$.

5. Calculate the reliability index using Equation (1.39) with $\mu_F = \mu_D + \mu_L$ and:

$$COV_F = \frac{\sqrt{(\mu_D COV_D)^2 + (\mu_L COV_L)^2}}{\mu_D + \mu_L} \tag{1.44}$$

Details of the above procedure are given in Appendix A.4. For $\phi = 0.5$, $\gamma_D = 1.25$, $\gamma_L = 1.75$, $b_Q = 1$, $b_D = 1.05$, $b_L = 1.15$, $COV_Q = 0.3$, $COV_D = 0.1$, $COV_L = 0.2$, and $D_n/L_n = 2$, the reliability index from Equation (1.39) is 2.99. Monte Carlo simulation in "LRFD.m" validates the approximation given in Equation (1.44). It is easy to show that an alternate approximation $COV_F^2 = COV_D^2 + COV_L^2$ is erroneous. Equation (1.39) is popular because the resistance factor in LRFD can be back-calculated from a target reliability index (β_T) easily:

$$\phi = \frac{b_Q(\gamma_D D_n + \gamma_L L_n)\sqrt{\frac{1+COV_F^2}{1+COV_Q^2}}}{(b_D D_n + b_L L_n)\exp\left\{\beta_T \sqrt{\ln\left[\left(1+COV_Q^2\right)\left(1+COV_F^2\right)\right]}\right\}} \tag{1.45}$$

In practice, the statistics of Q are determined by comparing load test results with calculated values. Phoon and Kulhawy (2005) compared the statistics of Q calculated from laboratory load tests with those calculated from full-scale load tests. They concluded that these statistics are primarily influenced by model errors, rather than uncertainties in the soil parameters. Hence, it is likely that the above lumped capacity approach can only accommodate the "low" COV range of parametric uncertainties mentioned in Section 1.3.1. To accommodate larger uncertainties in the soil parameters ("medium" and "high" COV ranges), it is necessary to expand Q as a function of the governing soil parameters. By doing so, the above closed-form solutions are no longer applicable.

1.4.2 First-order reliability method (FORM)

Structural reliability theory has a significant impact on the development of modern design codes. Much of its success could be attributed to the advent of the first-order reliability method (FORM), which provides a practical scheme of computing small probabilities of failure at high dimensional space spanned by the random variables in the problem. The basic theoretical result was given by Hasofer and Lind (1974). With reference to time-invariant reliability calculation, Rackwitz (2001) observed that: "For 90% of all applications this simple first-order theory fulfills all practical needs. Its numerical accuracy is usually more than sufficient." Ang and Tang (1984) presented numerous practical applications of FORM in their well known book, *Probability Concepts in Engineering Planning and Design.*

The general reliability problem consists of a performance function $P(y_1, y_2, \ldots, y_n)$ and a multivariate probability density function $f(y_1, y_2, \ldots, y_n)$. The former is defined to be zero at the limit state, less than zero when the limit state is exceeded ("fail"), and larger than zero otherwise ("safe"). The performance function is nonlinear for most practical problems. The latter specifies the likelihood of realizing any one particular set of input parameters $(y_1, y_2, \ldots, y_n)$, which could include material, load, and geometrical parameters. The objective of reliability analysis is to calculate the probability of failure, which can be expressed formally as follows:

$$p_f = \int_{P<0} f(y_1, y_2, \ldots, y_n) \, dy_1 dy_2 \ldots dy_n \qquad (1.46)$$

The domain of integration is illustrated by a shaded region in the left panel of Figure 1.10a. Exact solutions are not available even if the multivariate probability density function is normal, unless the performance function is linear or quadratic. Solutions for the former case are given in Section 1.4.1. Other exact solutions are provided in Appendix C. Exact solutions are very useful for validation of new reliability codes or calculation methods. The only general solution to Equation (1.46) is Monte Carlo simulation. A simple example is provided in Appendix A.5. The calculation steps outlined below are typically carried out:

1.   Determine $(y_1, y_2, \ldots, y_n)$ using Monte Carlo simulation. Section 1.3.2 has presented fairly general and "user friendly" methods of simulating correlated non-normal random vectors.

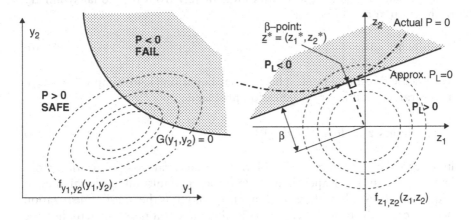

Figure 1.10 (a) General reliability problem, and (b) solution using FORM.

2. Substitute $(y_1, y_2, \ldots, y_n)$ into the performance function and count the number of cases where $P < 0$ ("failure").
3. Estimate the probability of failure using:

$$\hat{p}_f = \frac{n_f}{n} \tag{1.47}$$

in which $n_f =$ number of failure cases and $n =$ number of simulations.
4. Estimate the coefficient of variation of $\hat{p}_f$ using:

$$\mathrm{COV}_{p_f} = \sqrt{\frac{1 - p_f}{p_f n}} \tag{1.48}$$

For civil engineering problems, $p_f \approx 10^{-3}$ and hence, $(1 - p_f) \approx 1$. The sample size (n) required to ensure COV_{p_f} is reasonably small, say 0.3, is:

$$n = \frac{1 - p_f}{p_f \mathrm{COV}_{p_f}^2} \approx \frac{1}{p_f(0.3)^2} \approx \frac{10}{p_f} \tag{1.49}$$

It is clear from Equation (1.49) that Monte Carlo simulation is not practical for small probabilities of failure. It is more often used to *validate* approximate but more efficient solution methods such as FORM.

The approximate solution obtained from FORM is easier to visualize in a standard space spanned by uncorrelated Gaussian random variables with zero mean and unit standard deviation (Figure 1.10b). If one replaces the actual limit state function ($P = 0$) by an approximate *linear* limit state function ($P_L = 0$) that passes through the most likely failure point (also called design point or β-point), it follows immediately from rotational symmetry of the circular contours that:

$$p_f \approx \Phi(-\beta) \tag{1.50}$$

The practical result of interest here is that Equation (1.46) simply reduces to a constrained nonlinear optimization problem:

$$\beta = \min\sqrt{z'z} \quad \text{for}\{z : G(z) \le 0\} \tag{1.51}$$

in which $z = (z_1, z_2, \ldots, z_n)'$. The solution of a constrained optimization problem is significantly cheaper than the solution of a multi-dimensional integral [Equation (1.46)]. It is often cited that the β-point is the "best" linearization point because the probability density is highest at that point. In actuality, the choice of the β-point requires asymptotic analysis (Breitung, 1984). In short, FORM works well only for sufficiently large β – the usual rule-of-thumb is $\beta > 1$ (Rackwitz, 2001).

Low and co-workers (e.g. Low and Tang, 2004) demonstrated that the SOLVER function in EXCEL can be easily implemented to calculate the first-order reliability index for a range of practical problems. Their studies are summarized in Chapter 3. The key advantages to applying SOLVER for the solution of Equation (1.51) are: (a) EXCEL is available on almost all PCs, (b) most practitioners are familiar with the EXCEL user interface, and (c) no programming skills are needed if the performance function can be calculated using EXCEL built-in mathematical functions.

FORM can be implemented easily within MATLAB as well. Appendix A.6 demonstrates the solution process for an infinite slope problem (Figure 1.11). The performance function for this problem is:

$$P = \frac{[\gamma(H-h) + h(\gamma_{sat} - \gamma_w)]\cos\theta\tan\phi}{[\gamma(H-h) + h\gamma_{sat}]\sin\theta} - 1 \tag{1.52}$$

in which H = depth of soil above bedrock, h = height of groundwater table above bedrock, γ and γ_{sat} = moist unit weight and saturated unit weight of the surficial soil, respectively, γ_w = unit weight of water (9.81 kN/m^3), ϕ = effective stress friction angle, and θ = slope inclination. Note that the height of the groundwater table (h) cannot exceed the depth of surficial soil (H) and cannot be negative. Hence, it is modeled by $h = H \times U$, in which U = standard uniform variable. The moist and saturated soil unit weights are not independent, because they are related to the specific gravity of the soil solids (G_s) and the void ratio (e). The uncertainties in γ and γ_{sat} are characterized by modeling G_s and e as two independent uniform random variables. There are six independent random variables in this problem (H, U, ϕ, β, e, and G_s)

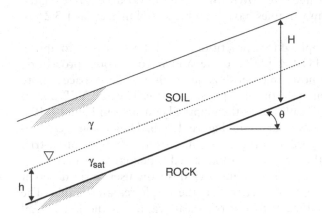

Figure 1.11 Infinite slope problem.

Table 1.3 Summary of probability distributions for input random variables.

Variable	Description	Distribution	Statistics
H	Depth of soil above bedrock	Uniform	[2,8] m
$h = H \times U$	Height of water table	U is uniform	[0, 1]
ϕ	Effective stress friction angle	Lognormal	mean = 35° cov = 8%
θ	Slope inclination	Lognormal	mean = 20° cov = 5%
γ	Moist unit weight of soil	*	*
γ_{sat}	Saturated unit weight of soil	**	**
γ_w	Unit weight of water	Deterministic	9.81 kN/m³

* $\gamma = \gamma_w (G_s + 0.2e)/(1 + e)$ (assume degree of saturation = 20% for "moist").
** $\gamma_{sat} = \gamma_w (G_s + e)/(1 + e)$ (degree of saturation = 100%).
Assume specific gravity of solids = G_s = uniformly distributed [2.5, 2.7] and void ratio = e = uniformly distributed [0.3, 0.6].

and their probability distributions are summarized in Table 1.3. The first-order reliability index is 1.43. The reliability index calculated from Monte Carlo simulation is 1.57.

Guidelines for modifying the code to solve other problems can be summarized as follows:

1. Specify the number of random variables in the problem using the parameter "m" in "FORM.m".
2. The objective function, "objfun.m," is independent of the problem.
3. The performance function, "Pfun.m," can be modified in a straight-forward way. The only slight complication is that the physical variables (e.g. H, U, ϕ, β, e, and G_s) must be expressed in terms of the standard normal variables (e.g. $z_1, z_2 ..., z_6$). Practical methods for

converting uncorrelated standard normal random variables to correlated non-normal random variables have been presented in Section 1.3.2.

Low and Tang (2004) opined that practitioners will find it easier to appreciate and implement FORM without the above conversion procedure. Nevertheless, it is well-known that optimization in the standard space is more stable than optimization in the original physical space. The SOLVER option "Use Automatic Scaling" only scales the elliptical contours in Figure 1.10a, but cannot make them circular as in Figure 1.10b. Under some circumstances, SOLVER will produce different results from different initial trial values. Unfortunately, there are no automatic and dependable means of flagging this instability. Hence, it is extremely vital for the user to try different initial trial values and partially verify that the result remains stable. The assumption that it is sufficient to use the mean values as the initial trial values is untrue.

In any case, it is crucial to understand that a multivariate probability model is necessary for any probabilistic analysis involving more than one random variable, FORM or otherwise. It is useful to recall that most multivariate non-normal probability models are related to the multivariate normal probability model in a fundamental way as discussed in Section 1.3.2. Optimization in the original physical space does not eliminate the need for the underlying multivariate normal probability model if the non-normal physical random vector is produced by the translation method. It is clear from Section 1.3.2 that non-translation methods exist and non-normal probability models cannot be constructed uniquely from correlation information alone. Chapters 9 and 13 report applications of FORM for RBD.

1.4.3 System reliability based on FORM

The first-order reliability method (FORM) is capable of handling any nonlinear performance function and any combination of correlated non-normal random variables. Its accuracy depends on two main factors: (a) the curvature of the performance function at the design point and (b) the number of design points. If the curvature is significant, the second-order reliability method (SORM) (Breitung, 1984) or importance sampling (Rackwitz, 2001) can be applied to improve the FORM solution. Both methods are relatively easy to implement, although they are more costly than FORM. The calculation steps for SORM are given in Appendix D. Importance sampling is discussed in Chapter 4. If there are numerous design points, FORM can underestimate the probability of failure significantly. At present, no solution method exists that is of comparable simplicity to FORM. Note that problems containing multiple failure modes are likely to produce more than one design point.

Figure 1.12 illustrates a simple system reliability problem involving two linear performance functions, P_1 and P_2. A common geotechnical engineering example is a shallow foundation subjected to inclined loading. It is governed by bearing capacity (P_1) and sliding (P_2) modes of failure. The system reliability is formally defined by:

$$p_f = \text{Prob}[(P_1 < 0) \cup (P_2 < 0)] \tag{1.53}$$

There are no closed-form solutions, even if P_1 and P_2 are linear and if the underlying random variables are normal and uncorrelated. A simple estimate based on FORM and second-order probability bounds is available and is of practical interest. The key calculation steps can be summarized as follows:

1. Calculate the correlation between failure modes using:

$$\rho_{P_2,P_1} = \boldsymbol{\alpha}_1 \cdot \boldsymbol{\alpha}_2 = \alpha_{11}\alpha_{21} + \alpha_{12}\alpha_{22} = \cos\theta \tag{1.54}$$

in which $\boldsymbol{\alpha}_i = (\alpha_{i1}, \alpha_{i2}) = $ unit normal at design point for ith performance function. Referring to Figure 1.10b, it can be seen that this unit normal can be readily obtained from FORM as: $\alpha_{i1} = z_{i1}^*/\beta_i$ and $\alpha_{i2} = z_{i2}^*/\beta_i$, with $(z_{i1}^*, z_{i2}^*) = $ design point for ith performance function.

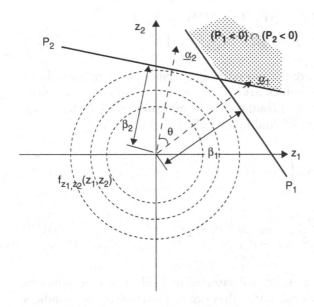

Figure 1.12 Simple system reliability problem.

2. Estimate $p_{21} = \text{Prob}[(P_2 < 0) \cap (P_1 < 0)]$ using first-order probability bounds:

$$p_{21}^- \leq p_{21} \leq p_{21}^+$$
$$\max[P(B_1), P(B_2)] \leq p_{21} \leq P(B_1) + P(B_2) \tag{1.55}$$

in which $P(B_1) = \Phi(-\beta_1)\Phi[(\beta_1 \cos\theta - \beta_2)/\sin\theta]$ and $P(B_2) = \Phi(-\beta_2)\Phi[(\beta_2 \cos\theta - \beta_1)/\sin\theta]$. Equation (1.55) only applies for $\rho_{P_2, P_1} > 0$. Failure modes are typically positively correlated because they depend on a common set of loadings.

3. Estimate p_f using second-order probability bounds:

$$p_1 + \max[(p_2 - p_{21}^+), 0] \leq p_f \leq \min[p_1 + (p_2 - p_{21}^-), 1] \tag{1.56}$$

in which $p_i = \Phi(-\beta_i)$.

The advantages of the above approach are that it does not require information beyond what is already available in FORM, and generalization to n failure modes is quite straightforward:

$$p_f \geq p_1 + \max[(p_2 - p_{21}^+), 0] + \max[(p_3 - p_{31}^+ - p_{32}^+, 0] + \cdots$$
$$+ \max[(p_n - p_{n1}^+ - p_{n2}^+ - \cdots - p_{n,n-1}^+), 0] \tag{1.57a}$$

$$p_f \leq p_1 + \min\{p_1 + [p_2 - p_{21}^-] + [p_3 - \max(p_{31}^-, p_{32}^-)] + \cdots$$
$$+ [p_n - \max(p_{n1}^-, p_{n2}^-, \cdots, p_{n,n-1}^-)], 1\} \tag{1.57b}$$

The clear disadvantages are that no point probability estimate is available and calculation becomes somewhat tedious when the number of failure modes is large. The former disadvantage can be mitigated using the following point estimate of p_{21} (Mendell and Elston, 1974):

$$a_1 = \frac{1}{\Phi(-\beta_1)\sqrt{2\pi}} \exp\left(-\frac{\beta_1^2}{2}\right) \tag{1.58}$$

$$p_{21} \approx \Phi\left[\frac{\rho_{P_1 P_2} a_1 - \beta_2}{\sqrt{1 - \rho_{P_1 P_2}{}^2 a_1 (a_1 - \beta_1)}}\right] \Phi(-\beta_1) \tag{1.59}$$

The accuracy of Equation (1.59) is illustrated in Table 1.4. Bold numbers shaded in gray are grossly inaccurate – they occur when the reliability indices are significantly different and the correlation is high.

Table 1.4 Estimation of p_{21} using probability bounds, point estimate, and simulation.

Performance functions	Correlation	Probability bounds	Point estimate	Simulation*
$\beta_1 = 1$	0.1	$(2.9, 5.8) \times 10^{-2}$	3.1×10^{-2}	3.1×10^{-2}
$\beta_2 = 1$	0.5	$(4.5, 8.9) \times 10^{-2}$	6.3×10^{-2}	6.3×10^{-2}
	0.9	$(0.6, 1.3) \times 10^{-1}$	1.2×10^{-1}	1.2×10^{-1}
$\beta_1 = 1$	0.1	$(4.8, 9.3) \times 10^{-3}$	5.0×10^{-3}	5.0×10^{-3}
$\beta_2 = 2$	0.5	$(1.1, 1.8) \times 10^{-2}$	1.3×10^{-2}	1.3×10^{-2}
	0.9	$(2.2, 2.3) \times 10^{-2}$	2.3×10^{-2}	2.3×10^{-2}
$\beta_1 = 1$	0.1	$(3.3, 6.1) \times 10^{-4}$	3.4×10^{-4}	3.5×10^{-4}
$\beta_2 = 3$	0.5	$(1.0, 1.3) \times 10^{-3}$	1.0×10^{-3}	1.0×10^{-3}
	0.9	$(1.3, 1.3) \times 10^{-3}$	0.5×10^{-3}	1.3×10^{-3}
$\beta_1 = 1$	0.1	$(8.6, 15.7) \times 10^{-6}$	8.9×10^{-6}	8.5×10^{-6}
$\beta_2 = 4$	0.5	$(2.8, 3.2) \times 10^{-5}$	2.3×10^{-5}	2.8×10^{-5}
	0.9	$(3.2, 3.2) \times 10^{-5}$	0.07×10^{-5}	$3.2 \times 10s^{-5}$
$\beta_1 = 2$	0.1	$(8.0, 16.0) \times 10^{-4}$	8.7×10^{-4}	8.8×10^{-4}
$\beta_2 = 2$	0.5	$(2.8, 5.6) \times 10^{-3}$	4.1×10^{-3}	4.1×10^{-3}
	0.9	$(0.7, 1.5) \times 10^{-2}$	1.4×10^{-2}	1.3×10^{-2}
$\beta_1 = 2$	0.1	$(5.9, 11.5) \times 10^{-5}$	6.3×10^{-5}	6.0×10^{-5}
$\beta_2 = 3$	0.5	$(3.8, 6.2) \times 10^{-4}$	4.5×10^{-4}	4.5×10^{-4}
	0.9	$(1.3, 1.3) \times 10^{-3}$	1.2×10^{-3}	1.3×10^{-3}
$\beta_1 = 2$	0.1	$(1.7, 3.2) \times 10^{-6}$	1.8×10^{-6}	2.5×10^{-6}
$\beta_2 = 4$	0.5	$(1.6, 2.2) \times 10^{-5}$	1.6×10^{-5}	1.5×10^{-5}
	0.9	$(3.2, 3.2) \times 10^{-5}$	0.5×10^{-5}	3.2×10^{-5}
$\beta_1 = 3$	0.1	$(4.5, 9.0) \times 10^{-6}$	4.9×10^{-6}	4.5×10^{-6}
$\beta_2 = 3$	0.5	$(5.6, 11.2) \times 10^{-5}$	8.2×10^{-5}	8.0×10^{-5}
	0.9	$(3.3, 6.6) \times 10^{-4}$	6.3×10^{-4}	6.0×10^{-4}

*Sample size = 5,000,000.

1.4.4 Collocation-based stochastic response surface method

The system reliability solution outlined in Section 1.4.3 is reasonably practical for problems containing a few failure modes that can be individually analyzed by FORM. A more general approach that is gaining wider attention is the spectral stochastic finite element method originally proposed by Ghanem and Spanos (1991). The key element of this approach is the expansion of the unknown random output vector using multi-dimensional Hermite polynomials as basis functions (also called a polynomial chaos expansion). The unknown deterministic coefficients in the expansion can be solved using the Galerkin or collocation method. The former method requires significant modification of the existing deterministic numerical code and is impossible to apply for most engineers with no access to the source code of their commercial softwares. The collocation method can be implemented using a small number of fairly simple computational steps and does not require the modification of existing deterministic numerical code. Chapter 7 and elsewhere (Sudret and Der Kiureghian, 2000) discussed this important class of methods in detail. Verhoosel and Gutiérrez (2007) highlighted challenging difficulties in applying these methods to nonlinear finite element problems involving discontinuous fields. It is interesting to observe in passing that the output vector can be expanded using the more well-known Taylor series expansion as well. The coefficients of the expansion (partial derivatives) can be calculated using the perturbation method. This method can be applied relatively easily to finite element outputs (Phoon *et al.*, 1990; Quek *et al.*, 1991, 1992), but is not covered in this chapter.

This section briefly explains the key computational steps for the more practical collocation approach. We recall in Section 1.3.2 that a vector of correlated non-normal random variables $\mathbf{Y} = (Y_1, Y_2, \dots, Y_n)'$ can be related to a vector of correlated standard normal random variables $\mathbf{X} = (X_1, X_2, \dots, X_n)'$ using one-dimensional Hermite polynomial expansions [Equation (1.26)]. The correlation coefficients for the normal random vector are evaluated from the correlation coefficients of the non-normal random vector using Equation (1.27). This method can be used to construct any non-normal random vector as long as the correlation coefficients are available. This is indeed the case if $\mathbf{Y}$ represents the input random vector. However, if $\mathbf{Y}$ represents the output random vector from a numerical code, the correlation coefficients are unknown and Equation (1.26) is not applicable. Fortunately, this practical problem can be solved by using multi-dimensional Hermite polynomials, which are supported by a *known* normal random vector with zero mean, unit variance, and uncorrelated or independent components.

The multi-dimensional Hermite polynomials are significantly more complex than the one-dimensional version. For example, the second-order and third-order forms can be expressed, respectively, as follows (Isukapalli, 1999):

$$Y \approx a_o + \sum_{i=1}^{n} a_i Z_i + \sum_{i=1}^{n} a_{ii}\left(Z_i^2 - 1\right) + \sum_{i=1}^{n-1}\sum_{j>i}^{n} a_{ij} Z_i Z_j \tag{1.60a}$$

$$Y \approx a_o + \sum_{i=1}^{n} a_i Z_i + \sum_{i=1}^{n} a_{ii}\left(Z_i^2 - 1\right) + \sum_{i=1}^{n} a_{iii}\left(Z_i^3 - 3Z_i\right) + \sum_{i=1}^{n-1}\sum_{j>i}^{n} a_{ij} Z_i Z_j$$

$$+ \sum_{i=1}^{n}\sum_{\substack{j=1 \\ j\neq i}}^{n} a_{ijj}\left(Z_i Z_j^2 - Z_i\right) + \sum_{i=1}^{n-2}\sum_{j>i}^{n-1}\sum_{k>j}^{n} a_{ijk} Z_i Z_j Z_k \tag{1.60b}$$

For $n = 3$, Equation (1.60) produces the following expansions (indices for coefficients are re-labeled consecutively for clarity):

$$Y \approx a_o + a_1 Z_1 + a_2 Z_2 + a_3 Z_3 + a_4(Z_1^2 - 1) + a_5(Z_2^2 - 1) + a_6(Z_3^2 - 1)$$
$$+ a_7 Z_1 Z_2 + a_8 Z_1 Z_3 + a_9 Z_2 Z_3 \tag{1.61a}$$

$$Y \approx a_o + a_1 Z_1 + a_2 Z_2 + a_3 Z_3 + a_4(Z_1^2 - 1) + a_5(Z_2^2 - 1) + a_6(Z_3^2 - 1)$$
$$+ a_7 Z_1 Z_2 + a_8 Z_1 Z_3 + a_9 Z_2 Z_3 + a_{10}(Z_1^3 - 3Z_1) + a_{11}(Z_2^3 - 3Z_2)$$
$$+ a_{12}(Z_3^3 - 3Z_3) + a_{13}(Z_1 Z_2^2 - Z_1) + a_{14}(Z_1 Z_3^2 - Z_1)$$
$$+ a_{15}(Z_2 Z_1^2 - Z_2) + a_{16}(Z_2 Z_3^2 - Z_2) + a_{17}(Z_3 Z_1^2 - Z_3)$$
$$+ a_{18}(Z_3 Z_2^2 - Z_3) + a_{19} Z_1 Z_2 Z_3 \tag{1.61b}$$

In general, N_2 and N_3 terms are respectively required for the second-order and third-order expansions (Isukapalli, 1999):

$$N_2 = 1 + 2n + \frac{n(n-1)}{2} \tag{1.62a}$$

$$N_3 = 1 + 3n + \frac{3n(n-1)}{2} + \frac{n(n-1)(n-2)}{6} \tag{1.62b}$$

For a fairly modest random dimension of $n = 5$, N_2 and N_3 terms are respectively equal to 21 and 56. Hence, fairly tedious algebraic expressions are incurred even at a third-order truncation. One-dimensional Hermite polynomials can be generated easily and efficiently using a three-term recurrence relation [Equation (1.11)]. No such simple relation is available for multi-dimensional Hermite polynomials. They are usually generated using symbolic algebra, which is possibly out of reach of the general practitioner. This major practical obstacle is currently being addressed by Liang *et al.* (2007). They have developed a user-friendly EXCEL add-in to generate tedious multi-dimensional Hermite expansions automatically. Once the multi-dimensional Hermite expansions are established, their coefficients can be calculated following the steps described in Equations (1.18)–(1.20). Two practical aspects are noteworthy:

1. The random dimension of the problem should be minimized to reduce the number of terms in the polynomial chaos expansion. The spectral decomposition method can be used:

$$\mathbf{X} = \mathbf{P} \mathbf{D}^{1/2} \mathbf{Z} + \mathbf{\mu} \tag{1.63}$$

in which $\mathbf{D}$ = diagonal matrix containing eigenvalues in the leading diagonal and $\mathbf{P}$ = matrix whose columns are the corresponding eigenvectors.

If C is the covariance matrix of $\mathbf{X}$, the matrices $\mathbf{D}$ and $\mathbf{P}$ can be calculated easily using $[\mathbf{P}, \mathbf{D}] = \text{eig}(\mathbf{C})$ in MATLAB. The key advantage of replacing Equation (1.23) by Equation (1.63) is that an n-dimensional correlated normal vector $\mathbf{X}$ can be simulated using an uncorrelated standard normal vector $\mathbf{Z}$ with a dimension less than n. This is achieved by discarding eigenvectors in P that correspond to small eigenvalues. A simple example is provided in Appendix A.7. The objective is to simulate a 3D correlated normal vector following a prescribed covariance matrix:

$$C = \begin{bmatrix} 1 & 0.9 & 0.2 \\ 0.9 & 1 & 0.5 \\ 0.2 & 0.5 & 1 \end{bmatrix}$$

Spectral decomposition of the covariance matrix produces:

$$\mathbf{D} = \begin{bmatrix} 0.045 & 0 & 0 \\ 0 & 0.832 & 0 \\ 0 & 0 & 2.123 \end{bmatrix} \quad \mathbf{P} = \begin{bmatrix} 0.636 & 0.467 & 0.614 \\ -0.730 & 0.108 & 0.675 \\ 0.249 & -0.878 & 0.410 \end{bmatrix}$$

Realizations of $\mathbf{X}$ can be obtained using Equation (1.63). Results are shown as open circles in Figure 1.13. The random dimension can be reduced from three to two by ignoring the first eigenvector corresponding to a small eigenvalue of 0.045. Realizations of $\mathbf{X}$ are now simulated using:

$$\begin{Bmatrix} X_1 \\ X_2 \\ X_3 \end{Bmatrix} = \begin{bmatrix} 0.467 & 0.614 \\ 0.108 & 0.675 \\ -0.878 & 0.410 \end{bmatrix} \begin{bmatrix} \sqrt{0.832} & 0 \\ 0 & \sqrt{2.123} \end{bmatrix} \begin{Bmatrix} Z_1 \\ Z_2 \end{Bmatrix}$$

Results are shown as crosses in Figure 1.13. Note that three correlated random variables can be represented reasonably well using only two uncorrelated random variables by neglecting the smallest eigenvalue and the corresponding eigenvector. The exact error can be calculated theoretically by observing that the covariance of $\mathbf{X}$ produced by Equation (1.63) is:

$$\mathbf{C_X} = (\mathbf{PD}^{1/2})(\mathbf{D}^{1/2}\mathbf{P'}) = \mathbf{PDP'} \tag{1.64}$$

If the matrices $\mathbf{P}$ and $\mathbf{D}$ are exact, it can be proven that $\mathbf{PDP'} = \mathbf{C}$, i.e. the target covariance is reproduced. For the truncated $\mathbf{P}$ and $\mathbf{D}$ matrices

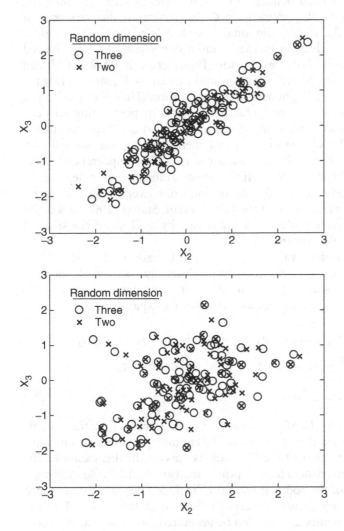

Figure 1.13 Scatter plot between X_1 and X_2 (top) and X_2 and X_3 (bottom).

discussed above, the covariance of **X** is:

$$\mathbf{C_X} = \begin{bmatrix} 0.467 & 0.614 \\ 0.108 & 0.675 \\ -0.878 & 0.410 \end{bmatrix} \begin{bmatrix} 0.832 & 0 \\ 0 & 2.123 \end{bmatrix}$$

$$\times \begin{bmatrix} 0.467 & 0.108 & -0.878 \\ 0.614 & 0.675 & 0.410 \end{bmatrix} = \begin{bmatrix} 0.982 & 0.921 & 0.193 \\ 0.921 & 0.976 & 0.508 \\ 0.193 & 0.508 & 0.997 \end{bmatrix}$$

The exact error in each element of $\mathbf{C_X}$ can be clearly seen by comparing with the corresponding element in $\mathbf{C}$. In particular, the variances of $\mathbf{X}$ (elements in the leading diagonal) are slightly reduced. For random processes/fields, reduction in the random dimension can be achieved using the Karhunen–Loeve expansion (Phoon et al., 2002, 2004). It can be shown that the spectral representation given by Equation (1.28) is a special case of the Karhunen–Loeve expansion (Huang et al., 2001). The random dimension can be further reduced by performing sensitivity analysis and discarding input parameters that are "unimportant." For example, the square of the components of the unit normal vector, α, shown in Figure 1.12, are known as FORM importance factors (Ditlevsen and Madsen, 1996). If the input parameters are independent, these factors indicate the degree of influence exerted by the corresponding input parameters on the failure event. Sudret (2007) discussed the application of Sobol' indices for sensitivity analysis of the spectral stochastic finite element method.

2. The number of output values $(y_1, y_2, \ldots)$ in Equation (1.20) should be minimized, because they are usually produced by costly finite element calculations. Consider a problem containing two random dimensions and approximating the output as a third-order expansion:

$$y_i = a_o + a_1 z_{i1} + a_2 z_{i2} + a_3(z_{i1}^2 - 1) + a_4(z_{i2}^2 - 1) + a_5 z_{i1} z_{i2}$$
$$+ a_6(z_{i1}^3 - 3z_{i1}) + a_7(z_{i2}^3 - 3z_{i2}) + a_8(z_{i1} z_{i2}^2 - z_{i1})$$
$$+ a_9(z_{i2} z_{i1}^2 - z_{i2}) \tag{1.65}$$

Phoon and Huang (2007) demonstrated that the collocation points (z_{i1}, z_{i2}) are best sited at the roots of the Hermite polynomial that is one order higher than that of the Hermite expansion. In this example, the roots of the fourth-order Hermite polynomial are $\pm\sqrt{(3\pm\sqrt{6})}$. Although zero is not one of the roots, it should be included because the standard normal probability density function is highest at the origin. Twenty-five collocation points (z_{i1}, z_{i2}) can be generated by combining the roots and zero in two dimensions, as illustrated in Figure 1.14. The roots of Hermite polynomials (up to order 15) can be calculated numerically as shown in Appendix A.8.

1.4.5 Subset simulation method

The collocation-based stochastic response surface method is very efficient for problems containing a small number of random dimensions and performance functions that can be approximated quite accurately using low-order Hermite expansions. However, the method suffers from a rapid proliferation

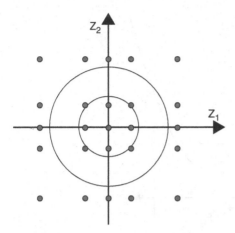

Figure 1.14 Collocation points for a third-order expansion with two random dimensions.

of Hermite expansion terms when the random dimension and/or order of expansion increase. A critical review of reliability estimation procedures for high dimensions is given by Schuëller *et al.* (2004). One potentially practical method that is worthy of further study is the subset simulation method (Au and Beck, 2001). It appears to be more robust than the importance sampling method. Chapter 4 and elsewhere (Au, 2001) present a more extensive study of this method.

This section briefly explains the key computational steps involved in the implementation. Consider the failure domain F_m defined by the condition $P(y_1, y_2, \ldots, y_n) < 0$, in which P is the performance function and $(y_1, y_2, \ldots, y_n)$ are realizations of the uncertain input parameters. The "failure" domain F_i defined by the condition $P(y_1, y_2, \ldots, y_n) < c_i$, in which c_i is a positive number, is larger by definition of the performance function. We assume that it is possible to construct a nested sequence of failure domains of increasing size by using an increasing sequence of positive numbers, i.e. there exists $c_1 > c_2 > \ldots > c_m = 0$ such that $F_1 \supset F_2 \supset \ldots F_m$. As shown in Figure 1.15 for the one-dimensional case, it is clear that this is always possible as long as one value of y produces only one value of $P(y)$. The performance function will satisfy this requirement. If one value of y produces two values of $P(y)$, say a positive value and a negative value, then the physical system is simultaneously "safe" and "unsafe," which is absurd.

The probability of failure (p_f) can be calculated based on the above nested sequence of failure domains (or subsets) as follows:

$$p_f = \mathrm{Prob}(F_m) = \mathrm{Prob}(F_m|F_{m-1})\mathrm{Prob}(F_{m-1}|F_{m-2})$$

$$\times \ldots \mathrm{Prob}(F_2|F_1)\mathrm{Prob}(F_1) \tag{1.66}$$

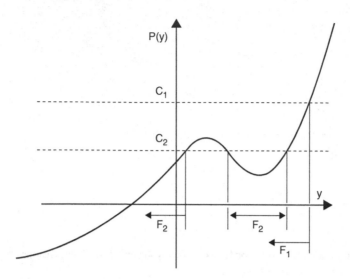

Figure 1.15 Nested failure domains.

At first glance, Equation (1.66) appears to be an indirect and more tedious method of calculating p_f. In actuality, it can be more efficient because the probability of each subset conditional on the previous (larger) subset can be selected to be sufficiently large, say 0.1, such that a significantly smaller number of realizations is needed to arrive at an acceptably accurate result. We recall from Equation (1.49) that the rule of thumb is $10/p_f$, i.e. only 100 realizations are needed to estimate a probability of 0.1. If the actual probability of failure is 0.001 and the probability of each subset is 0.1, it is apparent from Equation (1.66) that only three subsets are needed, implying a total sample size $= 3 \times 100 = 300$. In contrast, direct simulation will require $10/0.001 = 10,000$ realizations!

The typical calculation steps are illustrated below using a problem containing two random dimensions:

1. Select a subset sample size (n) and prescribe $p = \text{Prob}(F_i|F_{i-1})$. We assume $p = 0.1$ and $n = 500$ from hereon.
2. Simulate $n = 500$ realizations of the uncorrelated standard normal vector $(Z_1, Z_2)'$. The physical random vector $(Y_1, Y_2)'$ can be determined from these realizations using the methods described in Section 1.3.2.
3. Calculate the value of the performance function $g_i = P(y_{i1}, y_{i2})$ associated with each realization of the physical random vector $(y_{i1}, y_{i2})', i = 1, 2, \ldots, 500$.
4. Rank the values of $(g_1, g_2, \ldots, g_{500})$ in ascending order. The value located at the $(np + 1) = 51$st position is c_1.

5. Define the criterion for the first subset (F_1) as $P < c_1$. By construction at step (4), $P(F_1) = np/n = p$. The realizations contained in F_1 are denoted by $z_j = (z_{j1}, z_{j2})', j = 1, 2, \ldots, 50$.

6. Simulate $1/p = 10$ new realizations from z_j using the following Metropolis-Hastings algorithm:

    a. Simulate 1 realization using a uniform proposal distribution with mean located at $\mu = z_j$ and range $= 1$, i.e. bounded by $\mu \pm 0.5$. Let this realization be denoted by $u = (u_1, u_2)'$.

    b. Calculate the acceptance probability:

    $$\alpha = \min\left(1.0, \frac{I(u)\phi(u)}{\phi(\mu)}\right) \qquad (1.67)$$

    in which $I(u) = 1$ if $P(u_1, u_2) < c_1$, $I(u) = 0$ if $P(u_1, u_2) \geq c_1$, $\phi(u) = \exp(-0.5u'u)$, and $\phi(\mu) = \exp(-0.5\mu'\mu)$.

    c. Simulate 1 realization from a standard uniform distribution bounded between 0 and 1, denoted by v.

    d. The first new realization is given by:

    $$\begin{aligned} w &= u \qquad \text{if } v < \alpha \\ w &= \mu \qquad \text{if } v \geq \alpha \end{aligned} \qquad (1.68)$$

    Update the mean of the uniform proposal distribution in step (a) as $\mu = w$ (i.e. centred about the new realization) and repeat the algorithm to obtain a "chain" containing 10 new realizations. Fifty chains are obtained in the same way with initial seeds at $z_j, j = 1, \ldots, 50$. It can be proven that these new realizations would follow $\text{Prob}(\cdot | F_1)$ (Au, 2001).

7. Convert these realizations to their physical values and repeat Step (3) until c_i becomes negative (note: the smallest subset should correspond to $c_m = 0$).

It is quite clear that the above procedure can be extended to any random dimension in a trivial way. In general, the accuracy of this subset simulation method depends on "tuning" factors such as the choice of n, p, and the proposal distribution in step 6(a) (assumed to be uniform with range $= 1$). A study of some of these factors is given in Chapter 4. The optimal choice of these factors to achieve minimum runtime appears to be problem-dependent. Appendix A.9 provides the MATLAB code for subset simulation. The performance function is specified in "Pfun.m" and the number of random variables is specified in parameter "m." Figure 1.16 illustrates the behavior of the subset simulation method for the following performance function:

$$P = 2 + 3\sqrt{2} - Y_1 - Y_2 = 2 + 3\sqrt{2} + \ln[\Phi(-Z_1)] + \ln[\Phi(-Z_2)] \quad (1.69)$$

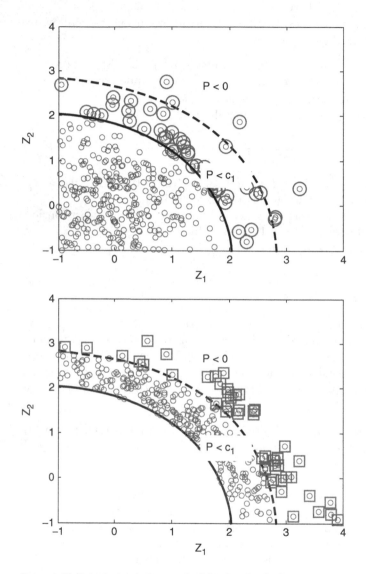

Figure 1.16 Subset simulation method for $P = 2 + 3\sqrt{2} - Y_1 - Y_2$, in which Y_1 and Y_2 are exponential random variables with mean = 1: first subset (top) and second subset (bottom).

in which Y_1, Y_2 = exponential random variables with mean = 1 and Z_1, Z_2 = uncorrelated standard normal random variables. The exact solution for the probability of failure is 0.0141 (Appendix C). The solution based on subset simulation method solution is achieved as follows:

1. Figure 1.16a: Simulate 500 realizations (small open circles) and identify the first 50 realizations in order of increasing P value (big open circles). The 51st smallest P value is $c_1 = 2.192$.
2. These realizations satisfy the criterion, $P < c_1$ (solid line), by construction. The domain above the solid line is F_1.
3. Figure 1.16b: Simulate 10 new realizations from each big open circle in Fig 1.16a. The total number is $10 \times 50 = 500$ realizations (small open circles).
4. There are 66 realizations (big open squares) satisfying $P < 0$ (dashed line). The domain above the dashed line is F_2. F_2 is the actual failure domain. Hence, $\text{Prob}(F_2|F_1) = 66/500 = 0.132$.
5. The estimated probability of failure is: $p_f = \text{Prob}(F_2|F_1)P(F_1) = 0.132 \times 0.1 = 0.0132$.

Note that only 7 realizations in Figure 1.16a lie in F_2. In contrast, 66 realizations in Figure 1.16b lie in F_2, providing a more accurate p_f estimate.

1.5 Conclusions

This chapter presents general and user-friendly computational methods for reliability analysis. "User-friendly" methods refer to those that can be implemented on a desktop PC by a non-specialist with limited programming skills; in other words, methods within reach of the general practitioner. This chapter is organized under two main headings describing: (a) how to simulate uncertain inputs numerically, and (b) how to propagate uncertain inputs through a physical model to arrive at the uncertain outputs. Although some of the contents are elaborated in greater detail in subsequent chapters, the emphasis on numerical implementations in this chapter should shorten the learning curve for the novice. The overview will also provide a useful roadmap to the rest of this book.

For the simulation of uncertain inputs following arbitrary non-normal probability distribution functions and correlation structure, the translation model involving memoryless transform of the multivariate normal probability distribution function can cater for most practical scenarios. Implementation of the translation model using one-dimensional Hermite polynomials is relatively simple and efficient. For stochastic data that cannot be modeled using the translation model, the hunt for probability models with comparable practicality, theoretical power, and simulation speed is still on-going. Copula theory produces a more general class of multivariate non-normal models, but it is debatable at this point if these models can be estimated empirically and simulated numerically with equal ease for high random dimensions. The only exception is a closely related but non-translation approach based on the multivariate normal copula.

The literature is replete with methodologies on the determination of uncertain outputs from a given physical model and uncertain inputs. The most general method is Monte Carlo simulation, but it is notoriously tedious. One may assume with minimal loss of generality that a complex geotechnical problem (possibly 3D, nonlinear, time-and construction-dependent, etc.) would only admit numerical solutions and the spatial domain can be modeled by a scalar/vector random field. For a sufficiently large and complex problem, it is computationally intensive to complete even a single run. At present, it is accurate to say that a computationally efficient and "user-friendly" solution to the general stochastic problem remains elusive. Nevertheless, reasonably practical solutions do exist if the general stochastic problem is restricted in some ways, for example, accept a first-order estimate of the probability of failure or accept an approximate but less costly output. The FORM is by far the most efficient general method for estimating the probability of failure of problems involving one design point (one dominant mode of failure). There are no clear winners for problems beyond the reach of FORM. The collocation-based stochastic response surface method is very efficient for problems containing a small number of random dimensions and performance functions that can be approximated quite accurately using low-order Hermite expansions. However, the method suffers from a rapid proliferation of Hermite expansion terms when the random dimension and/or order of expansion increase. The subset simulation method shares many of the advantages of the Monte Carlo simulation method, such as generality and tractability at high dimensions, without requiring too many runs. At low random dimensions, it does require more runs than the collocation-based stochastic response surface method but the number of runs is probably acceptable up to medium-scale problems. The subset simulation method has numerous "tuning" factors that are not fully studied for geotechnical problems.

Simple MATLAB codes are provided in the Appendix A to encourage practitioners to experiment with reliability analysis numerically so as to gain a concrete appreciation of the merits and limitations. Theory is discussed where necessary to furnish sufficient explanations so that users can modify codes correctly to suit their purpose and to highlight important practical limitations.

Appendix A – MATLAB codes

MATLAB codes are stored in text files with extension ".m" – they are called M-files. M-files can be executed easily by typing the filename (e.g. hermite) within the command window in MATLAB. The location of the M-file should be specified using "File > Set Path ...." M-files can be opened and edited using "File > Open ...." Programming details are given under "Help > Contents > MATLAB > Programming." The M-files provided below are available at http://www.eng.nus.edu.sg/civil/people/cvepkk/prob_lib.html.

A.1 Simulation of Johnson distributions

```
% Simulation of Johnson distributions
% Filename: Johnson.m
%
% Simulation sample size, n
n = 100000;
%
% Lognormal with lambda, xi
lambda = 1;
xi = 0.2;
Z = normrnd(0, 1, n, 1);
X = lambda + Z*xi;
LNY = exp(X);
%
% SB with lambda, xi, A, B
lambda = 1;
xi = 0.36;
A = -3;
B = 5;
X = lambda + Z*xi;
SBY = (exp(X)*B+A)./(exp(X)+1);
%
% SU with lambda, xi, A, B
lambda = 1;
xi = 0.09;
A = -1.88;
B = 2.08;
X = lambda + Z*xi;
SUY = sinh(X)*(B-A)+A;
%
% Plot probability density functions
[f, x] = ksdensity(LNY);
plot(x,f);
hold;
[f, x] = ksdensity(SBY);
plot(x,f,'red');
[f, x] = ksdensity(SUY);
plot(x,f,'green');
```

A.2 Calculation of Hermite coefficients using stochastic collocation method

```
% Calculation of Hermite coefficients using stochastic
collocation method
% Filename: herm_coeffs.m
%
% Number of realizations, n
n = 20;
%
% Example: Lognormal with lambda, xi
lambda = 0;
xi = 1;
Z = normrnd(0, 1, n, 1);
X = lambda + Z*xi;
LNY = exp(X);
%
% Order of Hermite expansion, m
m = 6;
%
% Construction of Hermite matrix, H
H = zeros(n,m+1);
H(:,1) = ones(n,1);
H(:,2) = Z;
for k = 3:m+1;
H(:,k) = Z.*H(:,k-1) - (k-2)*H(:,k-2);
end;
%
% Hermite coefficients stored in vector a
K = H'*H;
f = H'*LNY;
a = inv(K)*f;
```

A.3 Simulation of correlated non-normal random variables using translation method

```
% Simulation of correlated non-normal random variables
using translation method
% Filename: translation.m
%
% Number of random dimension, n
n = 2;
%
% Normal covariance matrix
```

```
C = [1 0.8; 0.8 1];
%
% Number of realizations, m
m = 100000;
%
% Simulation of 2 uncorrelated normal random variables
with
% mean = 0 and variance = 1
Z = normrnd(0, 1, m, n);
%
% Cholesky factorization
CF = chol(C);
%
% Simulation of 2 correlated normal random variables
with
% with mean = 0 and covariance = C
X = Z*CF;
%
% Example: simulation of correlated non-normal with
% Component 1 = Johnson SB
lambda = 1;
xi = 0.36;
A = -3;
B = 5;
W(:,1) = lambda + X(:,1)*xi;
Y(:,1) = (exp(W(:,1))*B+A)./(exp(W(:,1))+1);
% Component 2 = Johnson SU
lambda = 1;
xi = 0.09;
A = -1.88;
B = 2.08;
W(:,2) = lambda + X(:,2)*xi;
Y(:,2) = sinh(W(:,2))*(B-A)+A;
```

A.4 Simulation of normal random process using the two-sided power spectral density function, S(f)

```
% Simulation of normal random process using the
two-sided power spectral
% density function, S(f)
% Filename: ranprocess.m
%
% Number of data points based on power of 2, N
N = 512;
```

```
%
% Depth sampling interval, delz
delz = 0.2;
%
% Frequency interval, delf
delf = 1/N/delz;
%
% Discretization of autocorrelation function, R(tau)
% Example: single exponential function
%
tau = zeros(1,N);
tau = -(N/2-1)*delz:delz:(N/2)*delz;
d = 2; % scale of fluctuation
a = 2/d;
R = zeros(1,N);
R = exp(-a*abs(tau));
%
% Numerical calculation of S(f) using FFT
H = zeros(1,N);
H = fft(R);
% Notes:
% 1. f=0, delf, 2*delf, ... (N/2-1)*delf corresponds
to H(1), H(2), ... H(N/2)
% 2. Maximum frequency is Nyquist frequency
= N/2*delf= 1/2/delz
% 3. Multiply H (discrete transform) by delz to get
continuous transform
% 4. Shift R back by tau0, i.e. multiply H by
exp(2*pi*i*f*tau0)
f = zeros(1,N);
S = zeros(1,N);
%
% Shuffle f to correspond with frequencies ordering
implied in H
f(1) = 0;
for k = 2:N/2+1;
f(k)= f(k-1)+delf;
end;
f(N/2+2) = -f(N/2+1)+delf;
for k = N/2+3:N;
f(k) = f(k-1)+delf;
end;
%
tau0 = (N/2-1)*delz;
```

```
%
for k = 1:N;
S(k) = delz*H(k)*exp(2*pi*i*f(k)*tau0); % i is
imaginary number
end;
S = real(S); % remove possible imaginary parts due to
roundoff errors
%
% Shuffle S to correspond with f in increasing order
f = -(N/2-1)*delf:delf:(N/2)*delf;
temp = zeros(1,N);
for k = 1:N/2-1;
temp(k) = S(k+N/2+1);
end;
for k = N/2:N;
temp(k)=S(k-N/2+1);
end;
S = temp;
clear temp;
%
% Simulation of normal process using spectral
representation
% mean of process = 0 and variance of process = 1
%
% Maximum possible non-periodic process length,
Lmax = 1/2/fmin = 1/delf
% Minimum frequency, fmin = delf/2
Lmax = 1/delf;
%
% Simulation length, L = b*Lmax with b < 1
L = 0.5*Lmax;
%
% Number of simulated data points, nz
% Depth sampling interval, dz
% Depth coordinates, z(1), z(2) ... z(nz)
nz = round(L/delz);
dz = L/nz;
z = dz:dz:L;
%
% Number of realisations, m
m = 10000;
%
% Number of positive frequencies in the spectral
expansion, nf = N/2
```

```
nf = N/2;
%
% Simulate uncorrelated standard normal random
variables
randn('state', 1);
Z = randn(m,2*nf);
sigma = zeros(1,nf);
wa = zeros(1,nf);
%
% Calculate energy at each frequency using trapezoidal
rule, 2*S(f) for k = 1:nf;
sigma(k) = 2*0.5*(S(k+nf-1)+S(k+nf))*delf;
wa(k) = 0.5*2*pi*(f(k+nf-1)+f(k+nf));
end;
sigma = sigma.^0.5;
%
% Calculate realizations with mean = 0 and
variance = 1
X = zeros(m,nz);
X = Z(:,1:nf)*diag(sigma)*cos(wa'*z)+Z(:,nf+1:2*nf)
*diag(sigma)*sin(wa'*z);
```

A.5 Reliability analysis of Load and Resistance Factor Design (LRFD)

```
% Reliability analysis of Load and Resistance Factor
Design (LRFD)
% Filename: LRFD.m
%
% Resistance factor, phi
phi = 0.5;
% Dead load factor, gD; Live load factor, gL
gD = 1.25;
gL = 1.75;
% Bias factors for Q, D, L
bR = 1;
bD = 1.05;
bL = 1.15;
%
% Ratio of nominal dead to live load, loadratio
loadratio = 2;
%
% Assume mR = bR*Rn; mD = bD*Dn; mL = bL*Ln
% Design equation: phi(Rn) = gD(Dn) + gL(Ln)
```

```
mR = 1000;
mL = phi*bR*mR*bL/(gD*loadratio+gL);
mD = loadratio*mL*bD/bL;
%
% Coefficients of variation for R, D, L
cR = 0.3;
cD = 0.1;
cL = 0.2;
%
% Lognormal X with mean = mX and coefficent of
variation = cX
xR = sqrt(log(1+cR^2));
lamR = log(mR)-0.5*xR^2;
xD = sqrt(log(1+cD^2));
lamD = log(mD)-0.5*xD^2;
xL = sqrt(log(1+cL^2));
lamL = log(mL)-0.5*xL^2;
%
% Simulation sample size, n
n = 500000;
%
Z = normrnd(0, 1, n, 3);
LR = lamR + Z(:,1)*xR;
R = exp(LR);
LD = lamD + Z(:,2)*xD;
D = exp(LD);
LL = lamL + Z(:,3)*xL;
L = exp(LL);
%
% Total load = Dead load + Live Load
F = D + L;
%
% Mean of F
mF = mD + mL;
%
% Coefficient of variation of F based on second-moment
approximation
cF = sqrt((cD*mD)^2+(cL*mL)^2)/mF;
%
% Failure occurs when R < F
failure = 0;
for i = 1:n;
if (R(i) < F(i)) failure = failure+1; end;
end;
```

```
%
% Probability of failure = no. of failures/n
pfs = failure/n;
%
% Reliability index
betas = -norminv(pfs);
%
% Closed-form lognormal solution
a1 = sqrt(log((1+cR^2)*(1+cF^2)));
a2 = sqrt((1+cF^2)/(1+cR^2));
beta = log(mR/mF*a2)/a1;
pf = normcdf(-beta);
```

A.6 First-order reliability method

```
% First-order reliability method
% Filename: FORM.m
%
% Number of random variables, m
m = 6;
%
% Starting guess is z0 = 0
z0 = zeros(1, m);
%
% Minimize objective function
% exitflag = 1 for normal termination
options = optimset('LargeScale','off');
[z,fval,exitflag,output] =
fmincon(@objfun,z0,[],[],[],[],[],[],@Pfun,options);
%
% First-order reliability index
beta1 = fval;
pf1 = normcdf(-beta1);

% Objective function for FORM
% Filename: objfun.m
%
function f = objfun(z)
%
% Objective function = distance from the origin
% z = vector of uncorrelated standard normal random
variables f = norm(z);
```

```
% Definition of performance function, P
% Filename: Pfun.m
%
function [c, ceq] = Pfun(z)
%
% Convert standard normal random variables to physical
variables
%
% Depth of rock, H
y(1) = 2+6*normcdf(z(1));
%
% Height of water table
y(2) = y(1)*normcdf(z(2));
%
% Effective stress friction angle (radians)
xphi = sqrt(log(1+0.08^2));
lamphi = log(35)-0.5*xphi^2;
y(3) = exp(lamphi+xphi*z(3))*pi/180;
%
% Slope inclination (radians)
xbeta = sqrt(log(1+0.05^2));
lambeta = log(20)-0.5*xbeta^2;
y(4) = exp(lambeta+xbeta*z(4))*pi/180;
%
% Specific gravity of solids
Gs = 2.5+0.2*normcdf(z(5));
%
% Void ratio
e = 0.3+0.3*normcdf(z(6));
%
% Moist unit weight
y(5) = 9.81*(Gs+0.2*e)/(1+e);
%
% Saturated unit weight
y(6) = 9.81*(Gs+e)/(1+e);
%
% Nonlinear inequality constraints, c < 0
c = (y(5)*(y(1)-y(2))+y(2)*(y(6)-9.81))*
cos(y(4))*tan(y(3))/((y(5)*(y(1)-y(2))+y(2)*y(6))*
sin(y(4)))-1;
%
% Nonlinear equality constraints
ceq = [];
```

A.7 Random dimension reduction using spectral decomposition

```
% Random dimension reduction using spectral
decomposition
% Filename: eigendecomp.m
%
% Example: Target covariance
C = [1 0.9 0.2; 0.9 1 0.5; 0.2 0.5 1];
%
% Spectral decomposition
[P,D] = eig(C);
%
% Simulation sample size, n
n = 10000;
%
% Simulate standard normal random variables
Z = normrnd(0,1,n,3);
%
% Simulate correlated normal variables following C
X = P*sqrt(D)*Z';
X = X';
%
% Simulate correlated normal variables without 1st
eigen-component
XT = P(1:3,2:3)*sqrt(D(2:3,2:3))*Z(:,2:3)';
XT=XT';
%
% Covariance produced by ignoring 1st eigen-component
CX = P(1:3,2:3)*D(2:3,2:3)*P(1:3,2:3)';
```

A.8 Calculation of Hermite roots using polynomial fit

```
% Calculation of Hermite roots using polynomial fit
% Filename: herm_roots.m
%
% Order of Hermite polynomial, m < 15
m = 5;
%
% Specification of z values for fitting
z = (-4:1/2/m:4)';
%
% Number of fitted points, n
n = length(z);
%
```

```
% Construction of Hermite matrix, H
H = zeros(n,m+1);
H(:,1) = ones(n,1);
H(:,2) = z;
for k = 3:m+1;
H(:,k) =z.*H(:,k-1)-(k-2)*H(:,k-2);
end;
%
% Calculation of Hermite expansion
y = H(:,m+1);
%
% Polynomial fit
p = polyfit(z,y,m);
%
% Roots of hermite polynomial
r = roots(p);
%
% Validation of roots
nn = length(r);
H = zeros(nn,m+1);
H(:,1) = ones(nn,1);
H(:,2) = r;
for k = 3:m+1;
H(:,k) =r.*H(:,k-1)-(k-2)*H(:,k-2);
end;
%
norm(H(:,m+1))
```

A.9 Subset Markov Chain Monte Carlo method

```
% Subset Markov Chain Monte Carlo method
% Filename: subset.m
% © (2007) Kok-Kwang Phoon
%
% Number of random variables, m
m = 2;
%
% Sample size, n
n = 500;
%
% Probability of each subset
psub = 0.1;
%
% Number of new samples from each seed sample
ns = 1/psub;
```

```
%
% Simulate standard normal variable, z
z = normrnd(0,1,n,m);
%
stopflag = 0;
pfss = 1;
while (stopflag == 0)
  %
  % Values of performance function
  for i=1:n; g(i) = Pfun(z(i,:)); end;
  %
  % Sort vector g to locate n*psub smallest values
  [gsort,index]=sort(g);
  %
  % Subset threshold, gt = n*psub+1 smallest value of g
  % if gt < 0, exit program
  gt = gsort(n*psub+1);
  if (gt < 0)
      i=1;
      while gsort(i)<0
          i=i+1;
      end;
      pfss=pfss*(i-1)/n;
      stopflag = 1;
      break;
  else
      pfss = pfss*psub;
  end;
  %
  % n*psub seeds satisfying Pfun < gt stored in
  z(index(1:n*psub),:)
  % Simulate 1/psub samples from each seed to get n
  samples for next subset
  w = zeros(n,m);
  for i=1:n*psub
     seed = z(index(i),:);
     for j=1:ns
       % Proposal density: standard uniform with mean =
       seed
       u = rand(1,m);
       u = (seed-0.5)+ u;
       % Calculate acceptance probability
       pdf2 = exp(-0.5*sum(seed.^2));
       pdf1 = exp(-0.5*sum(u.^2));
```

```
      I = 0;
      if Pfun(u)<gt; I=1; end;
      alpha = min(1,I*pdf1/pdf2);
      % Accept u with probability = alpha
      V = rand;
      if V < alpha;
         w(ns*(i-1)+j,:)= u;
      else
         w(ns*(i-1)+j,:)= seed;
      end;
   seed = w(ns*(i-1)+j,:);
   end;
 end;
 %
z=w;
end;
% Reliability index
betass = -norminv(pfss);
```

Appendix B – Approximate closed-form solutions for $\Phi(\cdot)$ and $\Phi^{-1}(\cdot)$

The standard normal cumulative distribution function, $\Phi(\cdot)$ can be calculated using normcdf in MATLAB. For sufficiently large β, it is also possible to estimate $\Phi(-\beta)$ quite accurately using the following asymptotic approximation (Breitung, 1984):

$$\Phi(-\beta) \sim (2\pi)^{-1/2}\exp(-\beta^2/2)\beta^{-1} \tag{B.1}$$

An extensive review of estimation methods for the standard normal cumulative distribution function is given by Johnson *et al.* (1994). Comparison between Equation (B.1) and normcdf($\cdot$) is given in Figure B.1. It appears that $\beta > 2$ is sufficiently "large." Press *et al.* (1992) reported an extremely accurate empirical fit with an error less than 1.2×10^{-7} everywhere:

$$t = \frac{1}{1+0.25\sqrt{2}\beta}0.5t \times \exp(-0.5\beta^2 - 1.26551223 + t$$

$$\times (1.00002368 + t \times (0.37409196 + \Phi(-\beta) \approx t \times (0.09678418 + t$$

$$\times (-0.18628806 + t \times (0.27886807 + t \times (-1.13520398 + t$$

$$\times (1.48851587 + t \times (-0.82215223 + t \times 0.17087277)))))))) \tag{B.2}$$

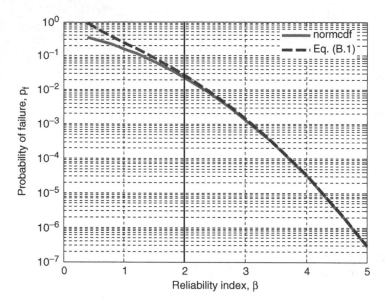

Figure B.1 Comparison between normcdf(·) and asymptotic approximation of the standard normal cumulative distribution function.

However, it seems difficult to obtain $\beta = -\Phi^{-1}(p_f)$, which is of significant practical interest as well. Equation (B.1) is potentially more useful in this regard. To obtain the inverse standard normal cumulative distribution function, we rewrite Equation (B.1) as follows:

$$\beta^2 + 2\ln(\sqrt{2\pi}\,p_f) + 2\ln(\beta) = 0 \tag{B.3}$$

To solve Equation (B.3), it is natural to linearize $\ln(\beta)$ using Taylor series expansion: $\ln(\beta) = -\ln[1 + (1/\beta - 1)] \approx 1 - 1/\beta$ with $\beta > 0.5$. Using this linearization, Equation (B.3) simplifies to a cubic equation:

$$\beta^3 + 2[\ln(\sqrt{2\pi}\,p_f) + 1]\beta - 2 = 0 \tag{B.4}$$

The solution is given by:

$$Q = \frac{2[\ln(\sqrt{2\pi}\,p_f) + 1]}{3}$$

$$R = 1$$

$$\beta = \left(R + \sqrt{Q^3 + R^2}\right)^{1/3} + \left(R - \sqrt{Q^3 + R^2}\right)^{1/3}, \quad p_f < \exp(-1)/\sqrt{2\pi} \tag{B.5}$$

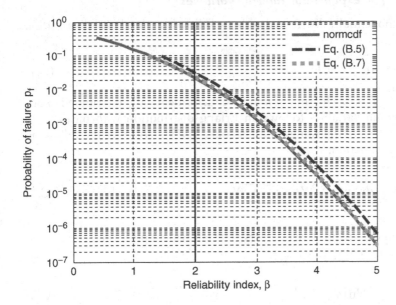

Figure B.2 Approximate closed-form solutions for inverse standard normal cumulative distribution function.

Equation (B.5) is reasonably accurate, as shown in Figure B.2. An extremely accurate and simpler linearization can be obtained by fitting $\ln(\beta) = b_0 + b_1\beta$ over the range of interest using linear regression. Using this linearization, Equation (B.3) simplifies to a quadratic equation:

$$\beta^2 + 2b_1\beta + 2[\ln(\sqrt{2\pi}p_f) + b_0] = 0 \qquad (B.6)$$

The solution is given by:

$$\beta = -b_1 + \sqrt{b_1^2 - 2[\ln(\sqrt{2\pi}p_f) + b_0]}, \quad p_f < \exp(b_1^2/2 - b_0)/\sqrt{2\pi} \quad (B.7)$$

The parameters $b_0 = 0.274$ and $b_1 = 0.266$ determined from linear regression over the range $2 \le \beta \le 6$ were found to produce extremely good results, as shown in Figure B.2. To the author's knowledge, Equations (B.5) and (B.7) appear to be original.

Appendix C – Exact reliability solutions

It is well known that exact reliability solutions are available for problems involving the sum of normal random variables or the product of lognormal random variables. This appendix provides other exact solutions that are useful for validation of new reliability codes/calculation methods.

C.1 Sum of n exponential random variables

Let Y be exponentially distributed with the following cumulative distribution function:

$$F(y) = 1 - \exp(-y/b) \tag{C.1}$$

The mean of Y is b. The sum of n independent identically distributed exponential random variables follows an Erlang distribution with mean $= nb$ and variance $= nb^2$ (Hastings and Peacock, 1975). Consider the following performance function:

$$P = nb + \alpha b\sqrt{n} - \sum_{i=1}^{n} Y_i = c - \sum_{i=1}^{n} Y_i \tag{C.2}$$

in which c and α are positive numbers. The equivalent performance function in standard normal space is:

$$P = c + \sum_{i=1}^{n} b\ln\left[\Phi(-Z_i)\right] \tag{C.3}$$

The probability of failure can be calculated exactly based on the Erlang distribution (Hastings and Peacock, 1975):

$$p_f = \text{Prob}(P < 0)$$

$$= \text{Prob}\left(\sum_{i=1}^{n} Y > c\right) \tag{C.4}$$

$$= \exp\left(-\frac{c}{b}\right) \sum_{i=0}^{n-1} \frac{(c/b)^i}{i!}$$

Equation (C.2) with $b = 1$ is widely used in the structural reliability community to validate FORM/SORM (Breitung, 1984; Rackwitz, 2001).

C.2 Probability content of an ellipse in n-dimensional standard normal space

Consider the "safe" elliptical domain in 2D standard normal space shown in Figure C.1. The performance function is:

$$P = \alpha^2 - \left(\frac{Z_1}{b_1}\right)^2 - \left(\frac{Z_2}{b_2}\right)^2 \tag{C.5}$$

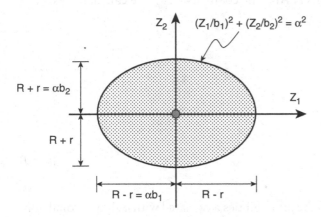

Figure C.1 Elliptical safe domain in 2D standard normal space.

in which $b_2 > b_1 > 0$. Let $R = \alpha(b_1 + b_2)/2$ and $r = \alpha(b_2 - b_1)/2$. Johnson and Kotz (1970), citing Ruben (1960), provided the following the exact solution for the safe domain:

$$\text{Prob}(P > 0) = \text{Prob}(\chi_2'^2(r^2) \leq R^2) - \text{Prob}(\chi_2'^2(R^2) \leq r^2) \qquad (C.6)$$

in which $\chi_2'^2(r^2) = $ non-central chi-square random variable with two degrees of freedom and non-centrality parameter r^2. The cumulative distribution function, $\text{Prob}(\chi_2'^2(r^2) \leq R^2)$, can be calculated using ncx2cdf($R^2, 2, r^2$) in MATLAB.

There is no simple generalization of Equation (C.6) to higher dimensions. Consider the following performance function involving n uncorrelated standard normal random variables $(Z_1, Z_2, ..., Z_n)$:

$$P = \alpha^2 - \left(\frac{Z_1}{b_1}\right)^2 - \left(\frac{Z_2}{b_2}\right)^2 - \cdots \left(\frac{Z_n}{b_n}\right)^2 \qquad (C.7)$$

in which $b_n > b_{n-1} > \cdots > b_1 > 0$. Johnson and Kotz (1970), citing Ruben (1963), provided a series solution based on the sum of central chi-square cumulative distribution functions:

$$\text{Prob}(P > 0) = \sum_{r=0}^{\infty} e_r \, \text{Prob}\left[\chi_{n+2r}^2 \leq (b_n \alpha)^2\right] \qquad (C.8)$$

in which $\chi_v^2(\cdot) = $ central chi-square random variable with v degrees of freedom. The cumulative distribution function, $\text{Prob}(\chi_v^2 \leq y)$, can be calculated using chi2cdf(y, v) in MATLAB. The coefficients, e_r, are calculated as:

$$e_0 = \prod_{j=1}^{n} \left(\frac{b_j}{b_n} \right)$$

$$e_r = (2r)^{-1} \sum_{j=0}^{r-1} e_j h_{r-j} \quad r \geq 1$$

(C.9)

$$h_r = \sum_{j=1}^{n} \left(1 - \frac{b_j^2}{b_n^2} \right)^r$$

(C.10)

It is possible to calculate the probabilities associated with n-dimensional non-central ellipsoidal domains as well. The performance function is given by:

$$P = \alpha^2 - \left(\frac{Z_1 - a_1}{b_1} \right)^2 - \left(\frac{Z_2 - a_2}{b_2} \right)^2 - \cdots \left(\frac{Z_n - a_n}{b_n} \right)^2$$

(C.11)

in which $b_n > b_{n-1} > \ldots > b_1 > 0$. Johnson and Kotz (1970), citing Ruben (1962), provided a series solution based on the sum of non-central chi-square cumulative distribution functions:

$$\text{Prob}(P > 0) = \sum_{r=0}^{\infty} e_r \text{Prob} \left[\chi_{n+2r}^{'2} \left(\sum_{j=1}^{n} a_j^2 \right) \leq (b_n \alpha)^2 \right]$$

(C.12)

The coefficients, e_r, are calculated as:

$$e_0 = \prod_{j=1}^{n} \left(\frac{b_j}{b_n} \right)$$

$$e_r = (2r)^{-1} \sum_{j=0}^{r-1} e_j h_{r-j} \quad r \geq 1$$

(C.13)

$$h_1 = \sum_{j=1}^{n} (1 - a_j^2) \left(1 - \frac{b_j^2}{b_n^2} \right)$$

$$h_r = \sum_{j=1}^{n} \left[\left(1 - \frac{b_j^2}{b_n^2} \right)^r + \frac{r}{b_n^2} a_j^2 b_j^2 \left(1 - \frac{b_j^2}{b_n^2} \right)^{r-1} \right] \quad r \geq 2$$

(C.14)

Remark: It is possible to create a very complex performance function by scattering a large number of ellipsoids in nD space. The parameters in

Equation (C.11) can be visualized geometrically even in nD space. Hence, they can be chosen in a relatively simple way to ensure that the ellipsoids are disjoint. One simple approach is based on nesting: (a) encase the first ellipsoid in a hyper-box, (b) select a second ellipsoid that is outside the box, (c) encase the first and second ellipsoids in a larger box, and (d) select a third ellipsoid that is outside the larger box. Repeated application will yield disjoint ellipsoids. The exact solution for such a performance function is merely the sum of Equation (C.12), but numerical solution can be very challenging. An example constructed by this "cookie-cutter" approach may be contrived, but it has the advantage of being as complex as needed to test the limits of numerical methods while retaining a relatively simple exact solution.

Appendix D – Second-order reliability method (SORM)

The second-order reliability method was developed by Breitung and others in a series of papers dealing with asymptotic analysis (Breitung, 1984; Breitung and Hohenbichler, 1989; Breitung, 1994). The calculation steps are illustrated below using a problem containing three random dimensions:

1. Let the performance function be defined in the standard normal space, i.e. $P(z_1, z_2, z_3)$ and let the first-order reliability index calculated from Equation (1.50) be $\beta = (\mathbf{z}^{*\prime}\mathbf{z}^*)^{1/2}$, in which $\mathbf{z}^* = (z_1^*, z_2^*, z_3^*)'$ is the design point.

2. The gradient vector at $\mathbf{z}^*$ is calculated as:

$$\nabla P(\mathbf{z}^*) = \left\{ \begin{array}{l} \partial P(\mathbf{z}^*)/\partial z_1 \\ \partial P(\mathbf{z}^*)/\partial z_2 \\ \partial P(\mathbf{z}^*)/\partial z_3 \end{array} \right\} \tag{D.1}$$

   The magnitude of the gradient vector is:

$$\|\nabla P(\mathbf{z}^*)\| = [\nabla P(\mathbf{z}^*)' \nabla P(\mathbf{z}^*)]^{1/2} \tag{D.2}$$

3. The probability of failure is estimated as:

$$p_f = \Phi(-\beta)|\mathbf{J}|^{-1/2} \tag{D.3}$$

   in which $\mathbf{J}$ is the following 2×2 matrix:

$$\mathbf{J} = \begin{bmatrix} 1 & 0 \\ 0 & 1 \end{bmatrix} + \frac{\beta}{\|\nabla P(\mathbf{z}^*)\|} \begin{bmatrix} \partial^2 P(\mathbf{z}^*)/\partial z_1^2 & \partial^2 P(\mathbf{z}^*)/\partial z_1 \partial z_2 \\ \partial^2 P(\mathbf{z}^*)/\partial z_2 \partial z_1 & \partial^2 P(\mathbf{z}^*)/\partial z_2^2 \end{bmatrix} \tag{D.4}$$

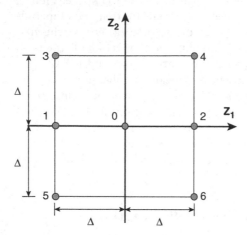

Figure D.1 Central finite difference scheme.

Equations (D.1) and (D.4) can be estimated using the central finite difference scheme shown in Figure D.1. Let P_i be the value of the performance function evaluated at node i. Then, the first derivative is estimated as:

$$\frac{\partial P}{\partial z_1} \approx \frac{P_2 - P_1}{2\Delta} \tag{D.5}$$

The second derivative is estimated as:

$$\frac{\partial^2 P}{\partial z_1^2} \approx \frac{P_2 - 2P_0 + P_1}{\Delta^2} \tag{D.6}$$

The mixed derivative is estimated as:

$$\frac{\partial^2 P}{\partial z_1 \partial z_2} \approx \frac{P_4 - P_3 - P_6 + P_5}{4\Delta^2} \tag{D.7}$$

References

Ang, A. H-S. and Tang, W. H. (1984). *Probability Concepts in Engineering Planning and Design*, Vol. 2 (Decision, Risk, and Reliability). John Wiley and Sons, New York.

Au, S. K. (2001). On the solution of first excursion problems by simulation with applications to probabilistic seismic performance assessment. PhD thesis, California Institute of Technology, Pasadena.

Au, S. K. and Beck, J. (2001). Estimation of small failure probabilities in high dimensions by subset simulation. *Probabilistic Engineering Mechanics*, 16(4), 263–77.

Baecher, G. B. and Christian, J. T. (2003). *Reliability and Statistics in Geotechnical Engineering*. John Wiley & Sons, New York.

Becker, D. E. (1996). Limit states design for foundations, Part II – Development for National Building Code of Canada. *Canadian Geotechnical Journal*, 33(6), 984–1007.

Boden, B. (1981). Limit state principles in geotechnics. *Ground Engineering*, 14(6), 2–7.

Bolton, M. D. (1983). Eurocodes and the geotechnical engineer. *Ground Engineering*, 16(3), 17–31.

Bolton, M. D. (1989). Development of codes of practice for design. In Proceedings of the 12th International Conference on Soil Mechanics and Foundation Engineering, Rio de Janeiro, Vol. 3. Balkema. A. A. Rotterdam, pp. 2073–6.

Box, G. E. P. and Muller, M. E. (1958). A note on the generation of random normal deviates. *Annals of Mathematical Statistics*, 29, 610–1.

Breitung, K. (1984). Asymptotic approximations for multinormal integrals. *Journal of Engineering Mechanics*, ASCE, 110(3), 357–66.

Breitung, K. (1994). *Asymptotic Approximations for Probability Integrals*. Lecture Notes in Mathematics, 1592. Springer, Berlin.

Breitung, K. and Hohenbichler, M. (1989). Asymptotic approximations to multivariate integrals with an application to multinormal probabilities. *Journal of Multivariate Analysis*, 30(1), 80–97.

Broms, B. B. (1964a). Lateral resistance of piles in cohesive soils. *Journal of Soil Mechanics and Foundations Division*, ASCE, 90(SM2), 27–63.

Broms, B. B. (1964b). Lateral resistance of piles in cohesionless soils. *Journal of Soil Mechanics and Foundations Division*, ASCE, 90(SM3), 123–56.

Carsel, R. F. and Parrish, R. S. (1988). Developing joint probability distributions of soil water retention characteristics. *Water Resources Research*, 24(5), 755–69.

Chilès, J-P. and Delfiner, P. (1999). *Geostatistics – Modeling Spatial Uncertainty*. John Wiley & Sons, New York.

Cressie, N. A. C. (1993). *Statistics for Spatial Data*. John Wiley & Sons, New York.

Ditlevsen, O. and Madsen, H. (1996). *Structural Reliability Methods*. John Wiley & Sons, Chichester.

Duncan, J. M. (2000). Factors of safety and reliability in geotechnical engineering. *Journal of Geotechnical and Geoenvironmental Engineering*, ASCE, 126(4), 307–16.

Elderton, W. P. and Johnson, N. L. (1969). *Systems of Frequency Curves*. Cambridge University Press, London.

Fenton, G. A. and Griffiths, D. V. (1997). Extreme hydraulic gradient statistics in a stochastic earth dam. *Journal of Geotechnical and Geoenvironmental Engineering*, ASCE, 123(11), 995–1000.

Fenton, G. A. and Griffiths, D. V. (2002). Probabilistic foundation settlement on spatially random soil. *Journal of Geotechnical and Geoenvironmental Engineering*, ASCE, 128(5), 381–90.

Fenton, G. A. and Griffiths, D. V. (2003). Bearing capacity prediction of spatially random c-ϕ soils. *Canadian Geotechnical Journal*, 40(1), 54–65.

Fisher, R. A. (1935). *The Design of Experiments*. Oliver and Boyd, Edinburgh.

Fleming, W. G. K. (1989). Limit state in soil mechanics and use of partial factors. *Ground Engineering*, 22(7), 34–5.

Ghanem, R. and Spanos, P. D. (1991). *Stochastic Finite Element: A Spectral Approach*. Springer-Verlag, New York.

Griffiths, D. V. and Fenton, G. A. (1993). Seepage beneath water retaining structures founded on spatially random soil. *Geotechnique*, 43(6), 577–87.

Griffiths, D. V. and Fenton, G. A. (1997). Three dimensional seepage through spatially random soil. *Journal of Geotechnical and Geoenvironmental Engineering, ASCE*, 123(2), 153–60.

Griffiths, D. V. and Fenton, G. A. (2001). Bearing capacity of spatially random soil: the undrained clay Prandtl problem revisited. *Geotechnique*, 54(4), 351–9.

Hansen, J. B. (1961). Ultimate resistance of rigid piles against transversal forces. *Bulletin 12*, Danish Geotechnical Institute, Copenhagen, 1961, 5–9.

Hasofer, A. M. and Lind, N. C. (1974). An exact and invariant first order reliability format. *Journal of Engineering Mechanics Division, ASCE*, 100(EM1), 111–21.

Hastings, N. A. J. and Peacock, J. B. (1975). *Statistical Distributions. A Handbook for Students and Practitioners*. Butterworths, London.

Huang, S. P., Quek, S. T. and Phoon K. K. (2001). Convergence study of the truncated Karhunen–Loeve expansion for simulation of stochastic processes. *International Journal of Numerical Methods in Engineering*, 52(9), 1029–43.

Isukapalli, S. S. (1999). Uncertainty analysis of transport – transformation models. PhD thesis, The State University of New Jersey, New Brunswick.

Johnson, N. L. (1949). Systems of frequency curves generated by methods of translation. *Biometrika*, 73, 387–96.

Johnson, N. L., Kotz, S. and Balakrishnan, N. (1970). *Continuous Univariate Distributions*, Vol. 2. Houghton Mifflin Company, Boston.

Johnson, N. L., Kotz, S. and Balakrishnan, N. (1994). *Continuous Univariate Distributions*, Vols. 1 and 2. John Wiley and Sons, New York.

Kulhawy, F. H. and Phoon, K. K. (1996). Engineering judgment in the evolution from deterministic to reliability-based foundation design. In *Uncertainty in the Geologic Environment – From Theory to Practice (GSP 58)*, Eds. C. D. Shackelford, P. P. Nelson and M. J. S. Roth. ASCE, New York, pp. 29–48.

Liang, B., Huang, S. P. and Phoon, K. K. (2007). An EXCEL add-in implementation for collocation-based stochastic response surface method. In *Proceedings of the 1st International Symposium on Geotechnical Safety and Risk*. Tongji University, Shanghai, China, pp. 387–98.

Low, B. K. and Tang, W. H. (2004). Reliability analysis using object-oriented constrained optimization. *Structural Safety*, 26(1), 69–89.

Marsaglia, G. and Bray, T. A. (1964). A convenient method for generating normal variables. *SIAM Review*, 6, 260–4.

Mendell, N. R. and Elston, R. C. (1974). Multifactorial qualitative traits: genetic analysis and prediction of recurrence risks. *Biometrics*, 30, 41–57.

National Research Council (2006). *Geological and Geotechnical Engineering in the New Millennium: Opportunities for Research and Technological Innovation*. National Academies Press, Washington, D. C.

Paikowsky, S. G. (2002). Load and resistance factor design (LRFD) for deep foundations. In *Proceedings International Workshop on Foundation Design Codes and Soil Investigation in view of International Harmonization and Performance Based Design*, Hayama. Balkema. A. A. Lisse, pp. 59–94.

Phoon, K. K. (2003). Representation of random variables using orthogonal polynomials. In *Proceedings of the 9th International Conference on Applications of Statistics and Probability in Civil Engineering*, Vol. 1. Millpress, Rotterdam, Netherlands, pp. 97–104.

Phoon, K. K. (2004a). General non-Gaussian probability models for first-order reliability method (FORM): a state-of-the-art report. *ICG Report 2004-2-4* (NGI Report 20031091-4), International Centre for Geohazards, Oslo.

Phoon, K. K. (2004b). Application of fractile correlations and copulas to non-Gaussian random vectors. In *Proceedings of the 2nd International ASRANet (Network for Integrating Structural Safety, Risk, and Reliability) Colloquium*, Barcelona (CDROM).

Phoon, K. K. (2005). Reliability-based design incorporating model uncertainties. In *Proceedings of the 3rd International Conference on Geotechnical Engineering*. Diponegoro University, Semarang, Indonesia, pp. 191–203.

Phoon, K. K. (2006a). Modeling and simulation of stochastic data. In *Geo-Congress 2006: Geotechnical Engineering in the Information Technology Age*, Eds. D. J. DeGroot, J. T. DeJong, J. D. Frost and L. G. Braise. ASCE, Reston (CDROM).

Phoon K. K. (2006b). Bootstrap estimation of sample autocorrelation functions. In *GeoCongress 2006: Geotechnical Engineering in the Information Technology Age*, Eds. D. J. DeGroot, J. T. DeJong, J. D. Frost and L. G. Braise. ASCE, Reston (CDROM).

Phoon, K. K. and Fenton, G. A. (2004). Estimating sample autocorrelation functions using bootstrap. In *Proceedings of the 9th ASCE Specialty Conference on Probabilistic Mechanics and Structural Reliability*, Albuquerque (CDROM).

Phoon, K. K. and Huang, S. P. (2007). Uncertainty quantification using multidimensional Hermite polynomials. In *Probabilistic Applications in Geotechnical Engineering (GSP 170)*, Eds. K. K. Phoon, G. A. Fenton, E. F. Glynn, C. H. Juang, D. V. Griffiths, T. F. Wolff and L. M. Zhang. ASCE, Reston (CDROM).

Phoon, K. K. and Kulhawy, F. H. (1999a). Characterization of geotechnical variability. *Canadian Geotechnical Journal*, 36(4), 612–24.

Phoon, K. K. and Kulhawy, F. H. (1999b). Evaluation of geotechnical property variability. *Canadian Geotechnical Journal*, 36(4), 625–39.

Phoon K. K. and Kulhawy, F. H. (2005). Characterization of model uncertainties for laterally loaded rigid drilled shafts. *Geotechnique*, 55(1), 45–54.

Phoon, K. K., Becker, D. E., Kulhawy, F. H., Honjo, Y., Ovesen, N. K. and Lo, S. R. (2003b). Why consider reliability analysis in geotechnical limit state design? In *Proceedings of the International Workshop on Limit State design in Geotechnical Engineering Practice (LSD2003)*, Cambridge (CDROM).

Phoon, K. K., Kulhawy, F. H., and Grigoriu, M. D. (1993). Observations on reliability-based design of foundations for electrical transmission line structures. In *Proceedings International Symposium on Limit State Design in Geotechnical Engineering*, Vol. 2. Danish Geotechnical Institute, Copenhagen, pp. 351–62.

Phoon, K. K., Kulhawy, F. H. and Grigoriu, M. D. (1995). Reliability-based design of foundations for transmission line structures. *Report TR-105000*, Electric Power Research Institute, Palo Alto.

Phoon, K. K., Huang S. P. and Quek, S. T. (2002). Implementation of Karhunen–Loeve expansion for simulation using a wavelet-Galerkin scheme. *Probabilistic Engineering Mechanics*, 17(3), 293–303.

Phoon, K. K., Huang H. W. and Quek, S. T. (2004). Comparison between Karhunen–Loeve and wavelet expansions for simulation of Gaussian processes. *Computers and Structures*, 82(13–14), 985–91.

Phoon, K. K., Quek, S. T. and An, P. (2003a). Identification of statistically homogeneous soil layers using modified Bartlett statistics. *Journal of Geotechnical and Geoenvironmental Engineering, ASCE*, 129(7), 649–59.

Phoon, K. K., Quek, S. T., Chow, Y. K. and Lee, S. L. (1990). Reliability analysis of pile settlement. *Journal of Geotechnical Engineering, ASCE*, 116(11), 1717–35.

Press, W. H., Teukolsky, S. A., Vetterling, W. T. and Flannery, B. P. (1992). *Numerical Recipes in C: The Art of Scientific Computing*. Cambridge University Press, New York.

Puig, B. and Akian, J-L. (2004). Non-Gaussian simulation using Hermite polynomials expansion and maximum entropy principle. *Probabilistic Engineering Mechanics*, 19(4), 293–305.

Puig, B., Poirion, F. and Soize, C. (2002). Non-Gaussian simulation using Hermite polynomial expansion: convergences and algorithms. *Probabilistic Engineering Mechanics*, 17(3), 253–64.

Quek, S. T., Chow, Y. K., and Phoon, K. K. (1992). Further contributions to reliability-based pile settlement analysis. *Journal of Geotechnical Engineering, ASCE*, 118(5), 726–42.

Quek, S. T., Phoon, K. K. and Chow, Y. K. (1991). Pile group settlement: a probabilistic approach. *International Journal of Numerical and Analytical Methods in Geomechanics*, 15(11), 817–32.

Rackwitz, R. (2001). Reliability analysis – a review and some perspectives. *Structural Safety*, 23(4), 365–95.

Randolph, M. F. and Houlsby, G. T. (1984). Limiting pressure on a circular pile loaded laterally in cohesive soil. *Geotechnique*, 34(4), 613–23.

Ravindra, M. K. and Galambos, T. V. (1978). Load and resistance factor design for steel. *Journal of Structural Division, ASCE*, 104(ST9), 1337–53.

Reese, L. C. (1958). Discussion of "Soil modulus for laterally loaded piles." *Transactions, ASCE*, 123, 1071–74.

Réthàti, L. (1988). *Probabilistic Solutions in Geotechnics*. Elsevier, New York.

Rosenblueth, E. and Esteva, L. (1972). *Reliability Basis for some Mexican Codes*. Publication SP-31, American Concrete Institute, Detroit.

Ruben, H. (1960). Probability constant of regions under spherical normal distribution, I. *Annals of Mathematical Statistics*, 31, 598–619.

Ruben, H. (1962). Probability constant of regions under spherical normal distribution, IV. *Annals of Mathematical Statistics*, 33, 542–70.

Ruben, H. (1963). A new result on the distribution of quadratic forms. *Annals of Mathematical Statistics*, 34, 1582–4.

Sakamoto, S. and Ghanem, R. (2002). Polynomial chaos decomposition for the simulation of non-Gaussian non-stationary stochastic processes. *Journal of Engineering Mechanics, ASCE*, 128(2), 190–200.

Schuëller, G. I., Pradlwarter, H. J. and Koutsourelakis, P. S. (2004). A critical appraisal of reliability estimation procedures for high dimensions. *Probabilistic Engineering Mechanics*, 19(4), 463–74.

Schweizer, B. (1991). Thirty years of copulas. In *Advances in Probability Distributions with Given Marginals*. Kluwer, Dordrecht, pp. 13–50.

Semple, R. M. (1981). Partial coefficients design in geotechnics. *Ground Engineering*, 14(6), 47–8.

Simpson, B. (2000). Partial factors: where to apply them? In *Proceedings of the International Workshop on Limit State Design in Geotechnical Engineering (LSD2000)*, Melbourne (CDROM).

Simpson, B. and Driscoll, R. (1998). *Eurocode 7: A Commentary*. Construction Research Communications Ltd, Watford, Herts.

Simpson, B. and Yazdchi, M. (2003). Use of finite element methods in geotechnical limit state design. In *Proceedings of the International Workshop on Limit State design in Geotechnical Engineering Practice (LSD2003)*, Cambridge (CDROM).

Simpson, B., Pappin, J. W. and Croft, D. D. (1981). An approach to limit state calculations in geotechnics. *Ground Engineering*, 14(6), 21–8.

Sudret, B. (2007). Uncertainty propagation and sensitivity analysis in mechanical models: Contributions to structural reliability and stochastic spectral methods. 1'Habilitation à Diriger des Recherches, Université Blaise Pascal.

Sudret, B. and Der Kiureghian, A. 2000. Stochastic finite elements and reliability: a state-of-the-art report. *Report UCB/SEMM-2000/08*, University of California, Berkeley.

US Army Corps of Engineers (1997). Engineering and design introduction to probability and reliability methods for use in geotechnical engineering. *Technical Letter No. 1110-2-547*, Department of the Army, Washington, D. C.

Uzielli, M. and Phoon K. K. (2006). Some observations on assessment of Gaussianity for correlated profiles. In *GeoCongress 2006: Geotechnical Engineering in the Information Technology Age*, Eds. D. J. DeGroot, J. T. DeJong, J. D. Frost and L. G. Braise. ASCE, Reston (CDROM).

Uzielli, M., Lacasse, S., Nadim, F. and Phoon, K. K. (2007). Soil variability analysis for geotechnical practice. In *Proceedings of the 2nd International Workshop on Characterisation and Engineering Properties of Natural Soils*, Vol. 3. Taylor and Francis, Singapore, pp. 1653–752.

VanMarcke, E. H. (1983). *Random Field: Analysis and Synthesis*. MIT Press, Cambridge, MA.

Verhoosel, C. V. and Gutiérrez, M. A. (2007). Application of the spectral stochastic finite element method to continuum damage modelling with softening behaviour. In *Proceedings of the 10th International Conference on Applications of Statistics and Probability in Civil Engineering*, Tokyo (CDROM).

Vrouwenvelder, T. and Faber, M. H. (2007). Practical methods of structural reliability. In Proceedings of the 10th International Conference on Applications of Statistics and Probability in Civil Engineering, Tokyo (CDROM).

Winterstein, S. R., Ude, T. C. and Kleiven, G. (1994). Springing and slow drift responses: predicted extremes and fatigue vs. simulation. Proceedings, Behaviour of Offshore Structures, Vol. 3. Elsevier, Cambridge, pp. 1–15.

Chapter 2

Spatial variability and geotechnical reliability

Gregory B. Baecher and John T. Christian

Quantitative measurement of soil properties differentiated the new discipline of soil mechanics in the early 1900s from the engineering of earth works practiced since antiquity. These measurements, however, uncovered a great deal of variability in soil properties, not only from site to site and stratum to stratum, but even within what seemed to be homogeneous deposits. We continue to grapple with this variability in current practice, although new tools of both measurement and analysis are available for doing so. This chapter summarizes some of the things we know about the variability of natural soils and how that variability can be described and incorporated in reliability analysis.

2.1 Variability of soil properties

Table 2.1 illustrates the extent to which soil property data vary, according to Phoon and Kulhawy (1996), who have compiled coefficients of variation for a variety of soil properties. The coefficient of variation is the standard deviation divided by the mean. Similar data have been reported by Lumb (1966, 1974), Lee *et al.* (1983), and Lacasse and Nadim (1996), among others. The ranges of these reported values are wide and are only suggestive of conditions at a specific site.

It is convenient to think about the impact of variability on safety by formulating the reliability index:

$$\beta = \frac{E[MS]}{SD[MS]} \ or \ \frac{E[FS] - 1}{SD[FS]} \tag{2.1}$$

in which β = reliability index, MS = margin of safety (resistance minus load), FS = factor of safety (resistance divided by load), $E[\cdot]$ = expectation, and $SD[\cdot]$ = standard deviation. It should be noted that the two definitions of β are not identical unless $MS = 0$ or $FS = 1$. Equation 2.1 expresses the number of standard deviations separating expected performance from a failure state.

Table 2.1 Coefficient of variation for some common field measurements (Phoon and Kulhawy, 1996).

Test type	Property	Soil type	Mean	Units	Cov(%)
	q_T	Clay	0.5–2.5	MN/m^2	< 20
CPT	q_c	Clay	0.5–2	MN/m^2	20–40
	q_c	Sand	0.5–30	MN/m^2	20–60
VST	s_u	Clay	5–400	kN/m^2	10–40
SPT	N	Clay and sand	10–70	blows/ft	25–50
	A reading	Clay	100–450	kN/m^2	10–35
	A reading	Sand	60–1300	kN/m^2	20–50
	B reading	Clay	500–880	kN/m^2	10–35
DMT	B Reading	Sand	350–2400	kN/m^2	20–50
	I_D	Sand	1–8		20–60
	K_D	Sand	2–30		20–60
	E_D	Sand	10–50	MN/m^2	15–65
	p_L	Clay	400–2800	kN/m^2	10–35
PMT	p_L	Sand	1600–3500	kN/m^2	20–50
	E_{PMT}	Sand	5–15	MN/m^2	15–65
	w_n	Clay and silt	13–100	%	8–30
	w_L	Clay and silt	30–90	%	6–30
	w_P	Clay and silt	15–15	%	6–30
Lab Index	PI	Clay and silt	10–40	%	_^{a}
	LI	Clay and silt	10	%	_^{a}
	γ, γ_d	Clay and silt	13–20	KN/m^3	< 10
	D_r	Sand	30–70	%	10–40; 50–70^b

Notes
^{a}COV = (3–12%)/mean.
^{b}The first range of variables gives the total variability for the direct method of determination, and the second range of values gives the total variability for the indirect determination using SPT values.

The important thing to note in Table 2.1 is how large are the reported coefficients of variations of soil property measurements. Most are tens of percent, implying reliability indices between one and two even for conservative designs. Probabilities of failure corresponding to reliability indices within this range – shown in Figure 2.1 for a variety of common distributional assumptions – are not reflected in observed rates of failure of earth structures and foundations. We seldom observe failure rates this high.

The inconsistency between the high variability of soil property data and the relatively low rate of failure of prototype structures is usually attributed to two things: spatial averaging and measurement noise. Spatial averaging means that, if one is concerned about average properties within some volume of soil (e.g. average shear strength or total compression), then high spots balance low spots so that the variance of the average goes down as that volume of mobilized soil becomes larger. Averaging reduces uncertainty.[1]

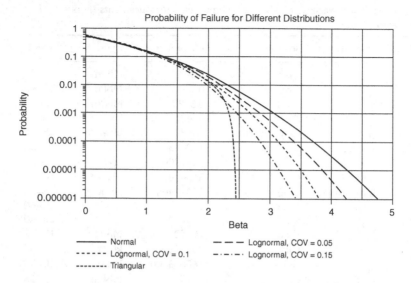

Figure 2.1 Probability of failure as a function of reliability index for a variety of common probability distribution forms.

Measurement noise means that the variability in soil property data reflects two things: real variability and random errors introduced by the process of measurement. Random errors reduce the precision with which estimates of average soil properties can be made, but they do not affect the in-field variation of actual properties, so the variability apparent in measurements is larger – possibly substantially so – than actual *in situ* variability.[2]

2.1.1 Spatial variation

Spatial variation in a soil deposit can be characterized in detail, but only with a great number of observations, which normally are not available. Thus, it is common to model spatial variation by a smooth deterministic trend combined with residuals about that trend, which are described probabilistically. This model is:

$$z(x) = t(x) + u(x) \tag{2.2}$$

in which $z(x)$ is the actual soil property at location x (in one or more dimensions), $t(x)$ is a smooth trend at x, and $u(x)$ is residual deviation from the trend. The residuals are characterized as a random variable of zero-mean and some variance:

$$Var(u) = E[\{z(x) - t(x)\}^2] \tag{2.3}$$

in which $Var(x)$ is the variance. The residuals are characterized as random because there are too few data to do otherwise. This does not presume that soil properties actually are random. The variance of the residuals reflects uncertainty about the difference between the fitted trend and the actual value of soil properties at particular locations. Spatial variation is modeled stochastically not because soil properties are random but because information is limited.

2.1.2 Trend analysis

Trends are estimated by fitting lines, curves, or surfaces to spatially referenced data. The easiest way to do this is by regression analysis. For example, Figure 2.2 shows maximum past pressure measurements as a function of depth in a deposit of Gulf of Mexico clay. The deposit appears homogeneous

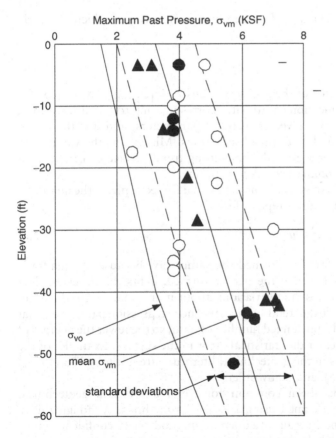

Figure 2.2 Maximum past pressure measurements as a function of depth in Gulf of Mexico clays, Mobile, Alabama.

and mostly normally consolidated. The increase of maximum past pressure with depth might be expected to be linear. Data from an over-consolidated desiccated crust are not shown. The trend for the maximum past pressure data, σ'_{vm}, with depth x is:

$$\sigma'_{vm} = t(x) + u(x) = \alpha_0 + \alpha_1 x = u \tag{2.4}$$

in which $t(x)$ is the trend of maximum past pressure with depth, x; α_0 and α_1 are regression coefficients; and u is residual variation about the trend taken to be constant with depth (i.e. it is not a function of x). Applying standard least squares analysis, the regression coefficients minimizing $Var[u]$, are $\alpha_0 = 3$ ksf (0.14 KPa) and $\alpha_1 = 0.06$ ksf/ft (1.4×10^{-3} KPa/m), yielding $Var(u) = 1.0$ ksf (0.05 KPa), for which the corresponding trend line is shown. The trend $t(x) = 3 + 0.06x$ is the best estimate or mean of the maximum past pressure as a function of depth. (NB: ksf = kip per square foot.)

The analysis can be made of data in higher dimensions, which in matrix notation becomes:

$$\mathbf{z} = \mathbf{X}\alpha + \mathbf{u} \tag{2.5}$$

in which $\mathbf{z}$ is the vector of the n observations $\mathbf{z} = \{z_1, \ldots, z_n\}$, $\mathbf{X} = \{\mathbf{x}_1, \mathbf{x}_2\}$ is the $2 \times n$ matrix of location coordinates corresponding to the observations, $\alpha = \alpha\{\alpha_1, \ldots, \alpha_n\}$ is the vector of trend parameters, and $\mathbf{u}$ is the vector of residuals corresponding to the observations. Minimizing the variance of the residuals $\mathbf{u}(\mathbf{x})$ over α gives the best-fitting trend surface in a frequentist sense, which is the common regression surface.

The trend surface can be made more flexible; for example, in the quadratic case, the linear expression is replaced by:

$$z = \alpha_0 + \alpha_1 x + \alpha_2 x^2 + u \tag{2.6}$$

and the calculation for α performed the same way. Because the quadratic surface is more flexible than the planar surface, it fits the observed data more closely, and the residual variations about it are smaller. On the other hand, the more flexible the trend surface, the more regression parameters that need to be estimated from a fixed number of data, so the fewer the degrees of freedom, and the greater the statistical error in the surface. Examples of the use of trend surfaces in the geotechnical literature are given by Wu (1974), Ang and Tang (1975), and many others.

Historically, it has been common for trend analysis in geotechnical engineering to be performed using frequentist methods. Although this is theoretically improper, because frequentist methods yield confidence intervals rather than probability distributions on parameters, the numerical error is negligible. The Bayesian approach begins with the same model.

However, rather than defining an estimator such as the least squares coefficients, the Bayesian approach specifies an a priori probability distribution on the coefficients of the model, and uses Bayes's Theorem to update that distribution in light of the observed data.

The following summarizes Bayesian results from Zellner (1971) for the one-dimensional linear case of Equation (2.4). Let $Var(u) = \sigma^2$, so that $Var(\mathbf{u}) = \mathbf{I}\sigma^2$, in which $\mathbf{I}$ is the identity matrix. The prior probability density function (pdf) of the parameters $\{\alpha, \sigma\}$ is represented as $f(\alpha, \sigma)$. Given a set of observations $\mathbf{z} = \{z_1, ..., z_n\}$, the updated or posterior pdf of $\{\alpha, \sigma\}$ is found from Bayes's Theorem as $f(\alpha, \sigma | \mathbf{z}) \propto f(\alpha, \sigma) L(\alpha, \sigma | \mathbf{z})$, in which $L(\alpha, \sigma | \mathbf{z})$ is the Likelihood of the data (i.e. the conditional probability of the observed data for various values of the parameters). If variations about the trend line or surface are jointly Normal, the likelihood function is:

$$L(\alpha, \sigma | \mathbf{z}) = MN(\mathbf{z} | \alpha, \sigma) \propto, \exp\{-(\mathbf{z} - \mathbf{X}\alpha)' \mathbf{\Sigma}^{-1}(\mathbf{z} - \mathbf{X}\alpha)\} \qquad (2.7)$$

in which $MN(.)$ is the Multivariate-Normal distribution having mean $\mathbf{X}\alpha$ and covariance matrix $\mathbf{\Sigma} = \mathbf{I}\sigma$.

Using a non-informative prior, $f(\alpha, \sigma) \propto \sigma^{-1}$, and measurements y made at depths x, the posterior pdf of the regression parameters is:

$$f(\alpha_0, \alpha_1, \sigma | \mathbf{x}, \mathbf{y}) \propto \frac{1}{\sigma^{n+1}} \exp\left[-\frac{1}{2\sigma^2} \sum_{i=1}^{n} (y_1 - (\alpha_0 + \alpha_1 x_1))^2\right] \qquad (2.8)$$

The marginal distributions are:

$$f(\alpha_0, \alpha_1 | \mathbf{x}, \mathbf{y}) \propto [vs^2 + n(\alpha_0 - \overline{\alpha}_0) + 2(\alpha_0 - \overline{\alpha}_0)(\alpha_1 - \overline{\alpha}_1)\Sigma x_i$$
$$+ (\alpha_1 - \overline{\alpha}_1)^2 \Sigma x_1^2]^{-n/2}$$

$$f(\alpha_0 | \mathbf{x}, \mathbf{y}) \propto [v + \frac{\Sigma(x_i - \overline{x})^2}{s^2 \Sigma x_i^2 / n}(\alpha_0 - \overline{\alpha}_0)^2]^{-(v-1)/2} \qquad (2.9)$$

$$f(\alpha_1 | \mathbf{x}, \mathbf{y}) \propto [v + \frac{\Sigma(x_i - \overline{x})^2}{s^2}(\alpha_1 - \overline{\alpha}_1)^2]^{-(v-1)/2}$$

$$f(\sigma | \mathbf{x}, \mathbf{y}) \propto \frac{1}{\sigma^{v-1}} \exp\left(-\frac{vs^2}{2\sigma^2}\right)$$

in which,

$$v = n - 2$$

$$\overline{\alpha}_0 = \overline{y} - \alpha_1 \overline{x}, \qquad \overline{\alpha}_1 = \left[\sum(x_i - \overline{x})(y_i - \overline{y})\right] / \left[\sum(x_i - \overline{x})\right]$$

$$s^2 = v^{-1} \sum(y_i - \overline{\alpha}_0 - \overline{\alpha}_1 x_i)^2$$

$$\bar{y} = n^{-1} \sum y_i$$

$$\bar{x} = n^{-1} \sum x_i$$

The joint and marginal pdf's of the regression coefficients are Student-t distributed.

2.1.3 Autocorrelation

In fitting trends to data, as noted above, the decision is made to divide the total variability of the data into two parts: one part explained by the trend and the other as variation about the trend. Residual variations not accounted for by the trend are characterized by a residual variance. For example, the overall variance of the blow count data of Figure 2.3 is 45 bpf² (475 bpm²). Removing a linear trend reduces this total to a residual variance of about 11 bpf²(116 bpm²). The trend explains 33 bpf² (349 bpm²), or about 75% of the spatial variation, and 25% is unexplained by the trend.

The spatial structure remaining after a trend is removed usually displays correlations among the residuals. That is, the residuals off the trend are not statistically independent of one another. Positive residuals tend to clump together, as do negative residuals. Thus, the probability of encountering a

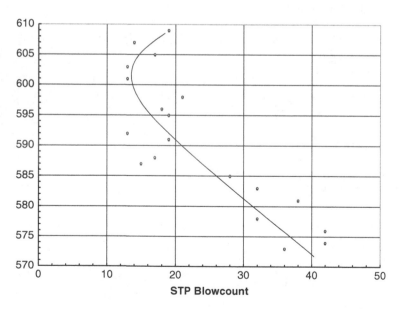

Figure 2.3 Spatial variation of SPT blow count data in a silty sand (data from Hilldale, 1971).

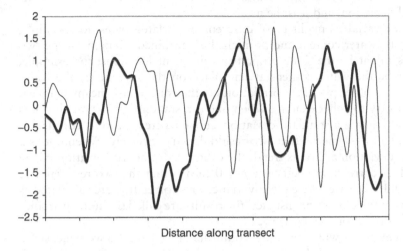

Figure 2.4 Residual variations of SPT blow counts.

continuous zone of weakness or high compressibility is greater than would be predicted if the residuals were independent.

Figure 2.4 shows residual variations of SPT blow counts measured at the same elevation every 20 m beneath a horizontal transect at a site. The data are normalized to zero mean and unit standard deviation. The dark line is a smooth curve drawn through the observed data. The light line is a smooth curve drawn through artificially simulated data having the same mean and same standard deviation, but probabilistically independent. Inspection shows the natural data to be smoothly varying, whereas the artificial data are much more erratic.

The remaining spatial structure of variation not accounted for by the trend can be described by its spatial correlation, called *autocorrelation*. Formally, autocorrelation is the property that residuals off the mean trend are not probabilistically independent but display a degree of association among themselves that is a function of their separation in space. This degree of association can be measured by a correlation coefficient, taken as a function of separation distance.

Correlation is the property that, on average, two variables are linearly associated with one another. Knowing the value of one provides information on the probable value of the other. The strength of this association is measured by a correlation coefficient ρ that ranges between -1 and $+1$. For two scalar variables z_1 and z_2, the correlation coefficient is defined as:

$$\rho = \frac{Cov(z_1, z_2)}{\sqrt{Var(z_1)Var(z_2)}} = \frac{1}{\sigma_{z_1}\sigma_{z_2}} \mathrm{E}[(z_1 - \mu_{z_1})(z_2 - \mu_{z_2})] \qquad (2.10)$$

in which $Cov(z_1, z_2)$ is the covariance, $Var(z_i)$ is the variance, σ is the standard deviation, and μ is the mean.

The two variables might be of different but related types; for example, z_1 might be water content and z_2 might be undrained strength, or the two variables might be the same property at different locations; for example, z_1 might be the water content at one place on the site and z_2 the water content at another place. A correlation coefficient $\rho = +1$ means that two residuals vary together exactly. When one is a standard deviation above its trend, the other is a standard deviation above its trend, too. A correlation coefficient $\rho = -1$ means that two residuals vary inversely. When one is a standard deviation above its trend, the other is a standard deviation below its trend. A correlation coefficient $\rho = 0$ means that the two residuals are unrelated. In the case where the covariance and correlation are calculated as functions of the separation distance, the results are called the autocovariance and autocorrelation, respectively.

The locations at which the blow count data of Figure 2.3 were measured are shown in Figure 2.5. In Figure 2.6 these data are used to estimate autocovariance functions for blow count. The data pairs at close separation exhibit a high degree of correlation; for example, those separated by 20 m have a correlation coefficient of 0.67. As separation distance increases, correlation drops, although at large separations, where the numbers of data pairs are

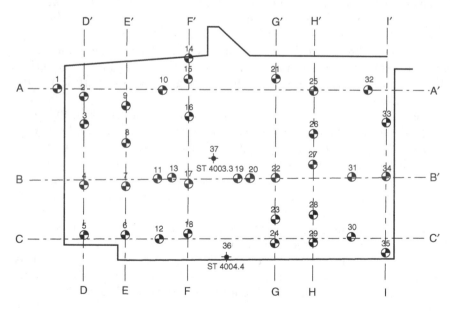

Figure 2.5 Boring locations of blow count data used to describe the site (T.W. Lambe and Associates, 1982. Earthquake Risk at Patio 4 and Site 400, Longboat Key, FL, reproduced by permission of T.W. Lambe).

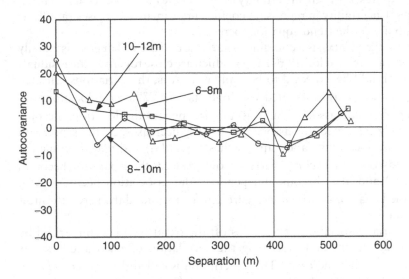

Figure 2.6 Autocorrelation functions for SPT data at at Site 400 by depth interval (T.W. Lambe and Associates, 1982. Earthquake Risk at Patio 4 and Site 400, Longboat Key, FL, reproduced by permission of T.W. Lambe).

smaller, there is much statistical fluctuation. For zero separation distance, the correlation coefficient must equal 1.0. For large separation distances, the correlation coefficient approaches zero. In between, the autocorrelation usually falls monotonically from 1.0 to zero.

An important point to note is that the division of spatial variation into a trend and residuals about the trend is an assumption of the analysis; it is not a property of reality. By changing the trend model – for example, by replacing a linear trend with a polynomial trend – both the variance of the residuals and their autocorrelation function are changed. As the flexibility of the trend increases, the variance of the residuals goes down, and in general the extent of correlation is reduced. From a practical point of view, the selection of a trend line or curve is in effect a decision on how much of the data scatter to model as a deterministic function of space and how much to treat probabilistically.

As a rule of thumb, trend surfaces should be kept as simple as possible without doing injustice to a set of data or ignoring the geologic setting. The problem with using trend surfaces that are very flexible (e.g. high-order polynomials) is that the number of data from which the parameters of those equations are estimated is limited. The sampling variance of the trend coefficients is inversely proportional to the degrees of freedom involved, $v = (n - k - 1)$, in which n is the number of observations and k is the number of parameters in the trend. The more parameter estimates that a trend surface requires, the more uncertainty there is in the numerical values of those estimates. Uncertainty in regression coefficient estimates increases rapidly as the flexibility of the trend equation increases.

If $z(x_i) = t(x_i) + u(x_i)$ is a continuous variable and the soil deposit is zonally homogeneous, then at locations i and j, which are close together, the residuals u_i and u_j should be expected to be similar. That is, the variations reflected in $u(x_i)$ and $u(x_j)$ are associated with one another. When the locations are close together, the association is usually strong. As the locations become more widely separated, the association usually decreases. As the separation between two locations i and j approaches zero, $u(x_i)$ and $u(x_j)$ become the same, the association becomes perfect. Conversely, as the separation becomes large, $u(x_i)$ and $u(x_j)$ become independent, the association becomes zero. This is the behavior observed in Figure 2.6 for the Standard Peretrationtest (SPT) data.

This spatial association of residuals off the trend $t(x_i)$ is summarized by a mathematical function describing the correlation of $u(x_i)$ and $u(x_j)$ as separation distance increases. This description is called the *autocorrelation function*. Mathematically, the autocorrelation function is:

$$R_z(\delta) = \frac{1}{Var\{u(x)\}} E[u(x_i)u(x_{i+\delta})] \tag{2.11}$$

in which $R_z(\delta)$ is the autocorrelation function, $Var[u(x)]$ is the variance of the residuals across the site, and $E[u(x_i)u(x_{i+\delta})] = Cov[u(x_i)u(x_{i+\delta})]$ is the covariance of the residuals spaced at separation distance, δ. By definition, the autocorrelation at zero separation is $R_z(0) = 1.0$; and empirically, for most geotechnical data, autocorrelation decreases to zero as δ increases.

If $R_z(\delta)$ is multiplied by the variance of the residuals, $Var[u(x)]$, the autocovariance function, $C_z(\delta)$, is obtained:

$$C_z(\delta) = E[u(x_i)u(x_{i+\delta})] \tag{2.12}$$

The relationship between the autocorrelation function and the autocovariance function is the same as that between the correlation coefficient and the covariance, except that autocorrelation and autocovariance are functions of separation distance, δ.

2.1.4 Example: TONEN refinery, Kawasaki, Japan

The SPT data shown earlier come from a site overlying hydraulic bay fill in Kawasaki (Japan). The SPT data were taken in a silty fine sand between elevations +3 and −7 m, and show little if any trend horizontally, so a constant horizontal trend at the mean of the data was assumed. Figure 2.7 shows the means and variability of the SPT data with depth. Figure 2.6 shows

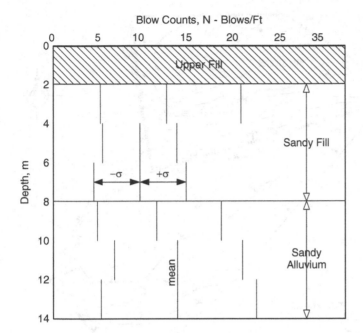

Figure 2.7 Soil model and the scatter of blow count data (T.W. Lambe and Associates, 1982. Earthquake Risk at Patio 4 and Site 400, Longboat Key, FL, reproduced by permission of T.W. Lambe).

autocovariance functions in the horizontal direction estimated for three intervals of elevation. At short separation distances the data show distinct association, i.e. correlation. At large separation distances the data exhibit essentially no correlation.

In natural deposits, correlations in the vertical direction tend to have much shorter distances than in the horizontal direction. A ratio of about 1 to 10 for these correlation distances is common. Horizontally, autocorrelation may be isotropic (i.e. $R_z(\delta)$ in the northing direction is the same as $R_z(\delta)$ in the easting direction) or anisotropic, depending on geologic history. However, in practice, isotropy is often assumed. Also, autocorrelation is typically assumed to be the same everywhere within a deposit. This assumption, called stationarity, to which we will return, is equivalent to assuming that the deposit is statistically homogeneous.

It is important to emphasize, again, that the autocorrelation function is an artifact of the way soil variability is separated between trend and residuals. Since there is nothing innate about the chosen trend, and since changing the trend changes $R_z(\delta)$, the autocorrelation function reflects a modeling decision. The influence of changing trends on $R_z(\delta)$ is illustrated in data

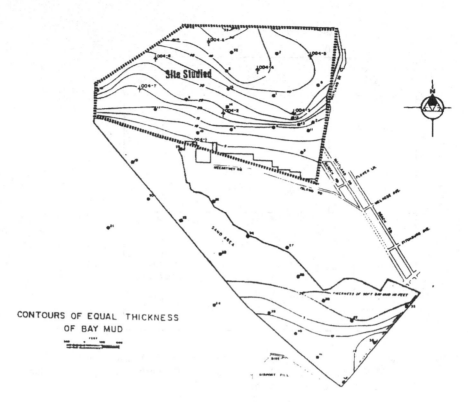

CONTOURS OF EQUAL THICKNESS
OF BAY MUD

Figure 2.8 Study area for San Francisco Bay Mud consolidation measurements (Javete, 1983) (reproduced with the author's permission).

analyzed by Javete (1983) (Figure 2.8). Figure 2.9 shows autocorrelations of water content in San Francisco Bay Mud within an interval of 3 ft (1 m). Figure 2.10 shows the autocorrelation function when the entire site is considered. The difference comes from the fact that in the first figure the mean trend is taken locally within the 3 ft (1 m) interval, and in the latter the mean trend is taken globally across the site.

Autocorrelation can be found in almost all spatial data that are analyzed using a model of the form of Equation (2.5). For example, Figure 2.11 shows the autocorrelation of rock fracture density in a copper porphyry deposit, Figure 2.12 shows autocorrelation of cone penetration resistance in North Sea Clay, and Figure 2.13 shows autocorrelation of water content in the compacted clay core of a rock-fill dam. An interesting aspect of the last data is that the autocorrelations they reflect are more a function of the construction process through which the core of the dam was placed than simply of space, per se. The time stream of borrow materials, weather, and working conditions at the time the core was

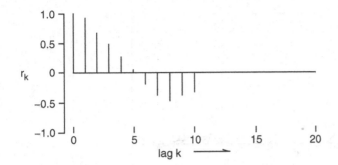

Figure 2.9 Autocorrelations of water content in San Francisco Bay Mud within an interval of 3 ft (1 m) (Javete, 1983) (reproduced with the author's permission).

Figure 2.10 Autocorrelations of water content in San Francisco Bay Mud within entire site expressed in lag intervals of 25 ft (Javete, 1983) (reproduced with the author's permission).

placed led to trends in the resulting physical properties of the compacted material.

For purposes of modeling and analysis, it is usually convenient to approximate the autocorrelation structure of the residuals by a smooth function. For example, a commonly used function is the exponential:

$$R_z(\delta) = \exp(-\delta/\delta_0) \qquad (2.13)$$

in which δ_0 is a constant having units of length. Other functions commonly used to represent autocorrelation are shown in Table 2.2. The distance at which $R_z(\delta)$ decays to 1/e (here δ_0) is sometimes called the *autocorrelation* (or *autocovariance*) *distance*.

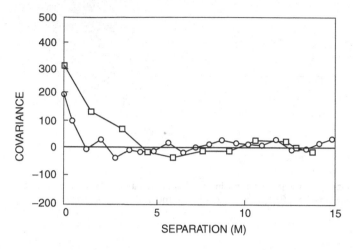

Figure 2.11 Autocorrelation of rock fracture density in a copper porphyry deposit (Baecher, 1980).

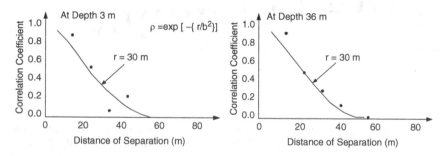

Figure 2.12 Autocorrelation of cone penetration resistance in North Sea Clay (Tang, 1979).

2.1.5 Measurement noise

Random measurement error is that part of data scatter attributable to instrument- or operator-induced variations from one test to another. This variability may sometimes increase or decrease a measurement, but its effect on any one specific measurement is unknown. As a first approximation, instrument and operator effects on measured properties of soils can be represented by a frequency diagram. In repeated testing – presuming that repeated testing is possible on the same specimen – measured values differ. Sometimes the measurement is higher than the real value of the property, sometimes it is lower, and on average it may systematically

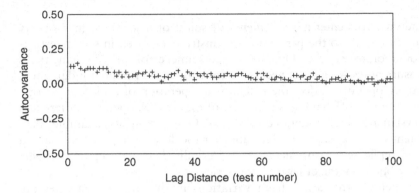

Figure 2.13 Autocorrelation of water content in the compacted clay core of a rock-fill dam (Beacher, 1987).

Table 2.2 One-dimensional autocorrelation models.

Model	Equation	Limits of validity (dimension of relevant space)
White noise	$R_x(\delta) = \begin{cases} 1 & \text{if } \delta = 0 \\ 0 & \text{otherwise} \end{cases}$	R^n
Linear	$R_x(\delta) = \begin{cases} 1 - \lvert\delta\rvert/\delta_0 & \text{if } \delta \le \delta_0 \\ 0 & \text{otherwise} \end{cases}$	R^1
Exponential	$R_x(\delta) = \exp(-\delta/\delta_0)$	R^1
Squared exponential (Gaussian)	$R_x(\delta) = \exp^2(-\delta/\delta_0)$	R^d
Power	$C_z(\delta) = \sigma^2 \{1(\lvert\delta\rvert^2/\delta_0^2)^{-\beta}$	$R^d, \beta > 0$

differ from the real value. This is usually represented by a simple model of the form:

$$z = bx + e \tag{2.14}$$

in which z is a measured value, b is a bias term, x is the actual property, and e is a zero-mean independent and identically distributed (IID) error. The systematic difference between the real value and the average of the measurements is said to be measurement bias, while the variability of the measurements about their mean is said to be random measurement error. Thus, the error terms are b and e. The bias is often assumed to be uncertain, with mean μ_b and standard deviation σ_b. The IID random perturbation is usually assumed to be Normally distributed with zero mean and standard deviation σ_e.

Random errors enter measurements of soil properties through a variety of sources related to the personnel and instruments used in soil investigations or laboratory testing. Operator or personnel errors arise in many types of measurements where it is necessary to read scales, personal judgment is needed, or operators affect the mechanical operation of a piece of testing equipment (*e.g.* SPT hammers). In each of these cases, operator differences have systematic and random components. One person, for example, may consistently read a gage too high, another too low. If required to make a series of replicate measurements, a single individual may report numbers that vary one from the other over the series.

Instrumental error arises from variations in the way tests are set up, loads are delivered, or soil response is sensed. The separation of measurement errors between operator and instrumental causes is not only indistinct, but also unimportant for most purposes. In triaxial tests, soil samples may be positioned differently with respect to loading platens in succeeding tests. Handling and trimming may cause differing amounts of disturbance from one specimen to the next. Piston friction may vary slightly from one movement to another, or temperature changes may affect fluids and solids. The aggregate result of all these variables is a number of differences between measurements that are unrelated to the soil properties of interest.

Assignable causes of minor variation are always present because a very large number of variables affect any measurement. One attempts to control those that have important effects, but this leaves uncontrolled a large number that individually have only small effects on a measurement. If not identified, these assignable causes of variation may influence the precision and possibly the accuracy of measurements by biasing the results. For example, hammer efficiency in the SPT test strongly affects measured blow counts. Efficiency with the same hammer can vary by 50% or more from one blow to the next. Hammer efficiency can be controlled, but only at some cost. If uncontrolled, it becomes a source of random measurement error and increases the scatter in SPT data.

Bias error in measurement arises from a number of reasonably well-understood mechanisms. Sample disturbance is among the more important of these mechanisms, usually causing a systematic degradation of average soil properties along with a broadening of dispersion. The second major contributor to measurement bias is the phenomenological model used to interpret the measurements made in testing, and especially the simplifying assumptions made in that model. For example, the physical response of the tested soil element might be assumed linear when in fact this is only an approximation, the reversal of principal stress direction might be ignored, intermediate principal stresses might be assumed other than they really are, and so forth.

The list of possible discrepancies between model assumptions and the real test conditions is long.

Model bias is usually estimated empirically by comparing predictions made from measured values of soil engineering parameters against observed performance. Obviously, such calibrations encompass a good deal more than just the measurement technique; they incorporate the models used to make predictions of field performance, inaccuracies in site characterization, and a host of other things.

Bjerrum's (1972, 1973) calibration of field vein test results for the undrained strength, s_u, of clay is a good example of how measurement bias can be estimated in practice. This calibration compares values of s_u measured with a field vane against back-calculated values of s_u from large-scale failures in the field. In principle, this calibration is a regression analysis of back-calculated s_u against field vane s_u, which yields a mean trend plus residual variance about the trend. The mean trend provides an estimate of μ_b while the residual variance provides an estimate of σ_b. The residual variance is usually taken to be the same regardless of the value of x, a common assumption in regression analysis.

Random measurement error can be estimated in a variety of ways, some direct and some indirect. As a general rule, the direct techniques are difficult to apply to the soil measurements of interest to geotechnical engineers, because soil tests are destructive. Indirect methods for estimating V_e usually involve correlations of the property in question, either with other properties such as index values, or with itself through the autocorrelation function.

The easiest and most powerful methods involve the autocorrelation function. The autocovariance of z after the trend has been removed becomes:

$$C_z(\delta) = C_x(\delta) + C_e(\delta) \tag{2.15}$$

in which $C_x(\delta)$ is from Equation (2.12) and $C_x(\delta)$ is the autocovariance function of e. However, since e_i and e_j are independent except when $i = j$, the autocovariance function of e is a spike at $\delta = 0$ and zero elsewhere. Thus, $C_x(\delta)$ is composed of two functions. By extrapolating the observed autocovariance function to the origin, an estimate is obtained of the fraction of data scatter that comes from random error. In the "geostatistics" literature this is called the nugget effect.

2.1.6 Example: Settlement of shallow footings on sand, Indiana (USA)

The importance of random measurement errors is illustrated by a case involving a large number of shallow footings placed on approximately 10 m of uniform sand (Hilldale, 1971). The site was characterized by Standard

Penetration blow count measurements, predictions were made of settlement, and settlements were subsequently measured.

Inspection of the SPT data and subsequent settlements reveals an interesting discrepancy. Since footing settlements on sand tend to be proportional to the inverse of average blow count beneath the footing, it would be expected that the coefficient of variation of the settlements equaled approximately that of the vertically averaged blow counts. Mathematically, settlement is predicted by a formula of the form, $\rho \propto \Delta q / \overline{N}_c$, in which $\rho =$ settlement, $\Delta q =$ net applied stress at the base of the footing, and $\overline{N}_c =$ average corrected blow count (Lambe and Whitman, 1979). Being multiplicative, the coefficient of variation of ρ should be the same as that of $\overline{N}_c$.

In fact, the coefficient of variation of the vertically averaged blow counts is about $\Omega_{\overline{N}_c} = 0.45$, while the observed values of total settlements for 268 footings have mean 0.35 inches and standard deviation 0.12 inches; so, $\Omega_\rho = (0.12/0.35) = 0.34$. Why the difference? The explanation may be found in estimates of the measurement noise in the blow count data. Figure 2.14 shows the horizontal autocorrelation function for the blow count data. Extrapolating this function to the origin indicates that the noise (or small scale) content of the variability is about 50% of the data scatter variance. Thus, the actual variability of the vertically averaged blow counts is about $\sqrt{\frac{1}{2}\Omega_N^2} = \sqrt{\frac{1}{2}(0.45)^2} = 0.32$, which is close to the observed variability

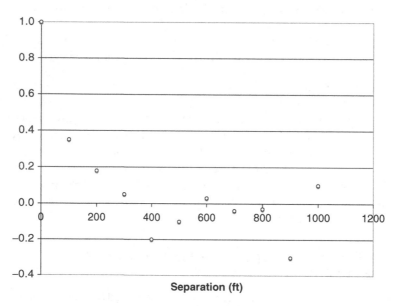

Figure 2.14 Autocorrelation function for SPT blow count in sand (Adapted from Hilldale, 1971).

of the footing settlements. Measurement noise of 50% or even more of the observed scatter of in situ test data, particularly the SPT, has been noted on several projects.

While random measurement error exhibits itself in the autocorrelation or autocovariance function as a spike at $\delta = 0$, real variability of the soil at a scale smaller than the minimum boring spacing cannot be distinguished from measurement error when using the extrapolation technique. For this reason, the "noise" component estimated in the horizontal direction may not be the same as that estimated in the vertical direction.

For many, but not all, applications the distinction between measurement error and small-scale variability is unimportant. For any engineering application in which average properties within some volume of soil are important, the small-scale variability averages quickly and therefore has little effect on predicted performance. Thus, for practical purposes it can be treated as if it were a measurement error. On the other hand, if performance depends on extreme properties – no matter their geometric scale – the distinction between measurement error and small scale is important. Some engineers think that piping (internal erosion) in dams is such a phenomenon. However, few physical mechanisms of performance easily come to mind that are strongly affected by small-scale spatial variability, unless those anomalous features are continuous over a large extent in at least one dimension.

2.2 Second-moment soil profiles

Natural variability is one source of uncertainty in soil properties, the other important source is limited knowledge. Increasingly, these are referred to as *aleatory* and *epistemic* uncertainty, respectively (Hartford, 1995).[3] Limited knowledge usually causes systematic errors. For example, limited numbers of tests lead to statistical errors in estimating a mean trend, and if there is an error in average soil strength it does not average out. In geotechnical reliability, the most common sources of knowledge uncertainty are model and parameter selection (Figure 2.15). Aleatory and epistemic uncertainties can be combined and represented in a second-moment soil profile. The second-moment profile shows means and standard deviations of soil properties with depth in a formation. The standard deviation at depth has two components, natural variation and systematic error.

2.2.1 Example: SHANSEP analysis of soft clays, Alabama (USA)

In the early 1980s, Ideal Basic Industries, Inc. (IDEAL) constructed a cement manufacturing facility 11 miles south of Mobile, Alabama, abutting a ship channel running into Mobile Bay (Baecher *et al.*, 1997). A gantry crane at the

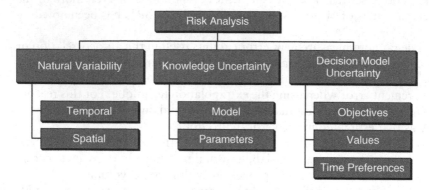

Figure 2.15 Aleatory, epistemic, and decision model uncertainty in geotechnical reliability analysis.

facility unloaded limestone ore from barges moored at a relieving platform and place the ore in a reserve storage area adjacent to the channel. As the site was underlain by thick deposits of medium to soft plastic deltaic clay, concrete pile foundations were used to support all facilities of the plant except for the reserve limestone storage area. This 220 ft (68 m) wide by 750 ft (230 m) long area provides limestone capacity over periods of interrupted delivery. Although the clay underlying the site was too weak to support the planned 50 ft (15 m) high stockpile, the cost of a pile supported mat foundation for the storage area was prohibitive.

To solve the problem, a foundation stabilization scheme was conceived in which limestone ore would be placed in stages, leading to consolidation and strengthening of the clay, and this consolidation would be hastened by vertical drains. However, given large scatter in engineering property data for the clay, combined with low factors of safety against embankment stability, field monitoring was essential.

The uncertainty in soil property estimates was divided between that caused by data scatter and that caused by systematic errors (Figure 2.16). These were separated into four components:

- spatial variability of the soil deposit,
- random measurement noise,
- statistical estimation error, and
- measurement or model bias.

The contributions were mathematically combined by noting that the variances of these nearly independent contributions are approximately additive:

$$V[x] \approx \{V_{\text{spatial}}[x] + V_{\text{noise}}[x]\} + \{V_{\text{statistical}}[x] + V_{\text{bias}}[x]\} \qquad (2.16)$$

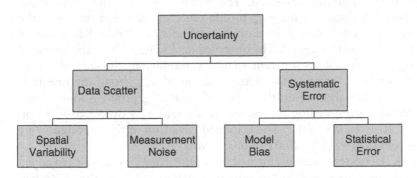

Figure 2.16 Sources of uncertainty in geotechnical reliability analysis.

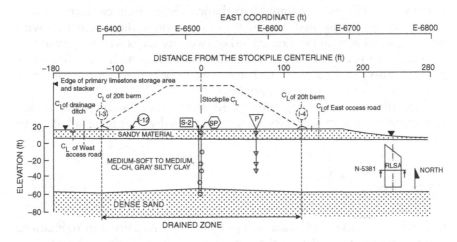

Figure 2.17 West-east cross-section prior to loading.

in which $V[x]$ = variance of total uncertainty in the property x, $V_{\text{spatial}}[x]$ = variance of the spatial variation of x, $V_{\text{noise}}[x]$ = variance of the measurement noise in x, $V_{\text{statistical}}[x]$ = variance of the statistical error in the expected value of x, and $V_{\text{bias}}[x]$ = variance of the measurement or model bias in x. It is easiest to think of spatial variation as scatter around the mean trend of the soil property and systematic error as uncertainty in the mean trend itself. The first reflects soil variability after random measurement error has been removed; the second reflects statistical error plus measurement bias associated with the mean value.

Initial vertical effective stresses, $\overline{\sigma}_{vo}$, were computed using total unit weights and initial pore pressures. Figure 2.2 shows a simplified profile prior to loading (Figure 2.17). The expected value $\overline{\sigma}_{vm}$ profile versus elevation was obtained by linear regression. The straight, short-dashed lines show the standard deviation of the $\overline{\sigma}_{vm}$ profile reflecting data scatter about the expected value. The curved, long-dashed lines show the standard deviation of the expected value trend itself. The observed data scatter about the expected value $\overline{\sigma}_{vm}$ profile reflects inherent spatial variability of the clay plus random measurement error in the determination of $\overline{\sigma}_{vm}$ from any one test. The standard deviation about the expected value $\overline{\sigma}_{vm}$ profile is about 1 ksf (0.05 MPa), corresponding to a standard deviation in over-consolidation ratio (OCR) from 0.8 to 0.2. The standard deviation of the expected value ranges from 0.2 to 0.5 ksf (0.01 to 0.024 MPa).

Ten CK_oUDSS tests were performed on undisturbed clay samples to determine undrained stress–strain–strength parameters to be used in the stress history and normalized soil engineering properties (SHANSEP) procedure of Ladd and Foott (1974). Reconsolidation beyond the *in situ* $\overline{\sigma}_{vm}$ was used to minimize the influence of sample disturbance. Eight specimens were sheared in a normally consolidated state to assess variation in the parameter s with horizontal and vertical locations. The last two specimens were subjected to a second shear to evaluate the effect of OCR. The direct simple shear (DSS) test program also provided undrained stress–strain parameters for use in finite element undrained deformation analyses.

Since there was no apparent trend with elevation, expected value and standard deviation values were computed by averaging all data to yield:

$$s_u = \overline{\sigma}_{vo} s \left(\frac{\overline{\sigma}_{vm}}{\overline{\sigma}_{vo}} \right)^m \tag{2.17}$$

in which $s = (0.213 \pm 0.028)$ and $m = (0.85 \pm 0.05)$. As a first approximation, it was assumed that 50% of the variation in s was spatial and 50% was noise. The uncertainty in m estimated from data on other clays is primarily due to variability from one clay type to another and hence was assumed purely systematic. It was assumed that the uncertainty in m estimated from only two tests on the storage area clay resulted from random measurement error.

The SHANSEP s_u profile was computed using Equation (2.17). If $\overline{\sigma}_{vo}, \overline{\sigma}_{vm}$, s and m are independent, and $\overline{\sigma}_{vo}$ is deterministic (i.e. there is no uncertainty in $\overline{\sigma}_{vo}$), first-order, second-moment error analysis leads to the expressions:

$$E[s_u] = \overline{\sigma}_{vo} E[s] \left(\frac{E[\overline{\sigma}_{vo}]}{\overline{\sigma}_{vo}} \right)^{E[m]} \tag{2.18}$$

$$\Omega^2[s_u] = \Omega^2[s] + E^2[m]\Omega^2[\overline{\sigma}_{vm}] + 1n^2 \left(\frac{E[\overline{\sigma}_{vm}]}{\overline{\sigma}_{vo}} \right) V[m] \tag{2.19}$$

in which $E[X] =$ expected value of X, $V[X] =$ variance of X, and $\Omega[X] = \sqrt{V[X]}/E[X] =$ coefficient of variation of X. The total coefficient of variation of s_u is divided between spatial and systematic uncertainty such that:

$$\Omega^2[s_u] = \Omega_{sp}^2[s_u] + \Omega_{sy}^2[s_u] \tag{2.20}$$

Figure 2.18 shows the expected value s_u profile and the standard deviation of s_u divided into spatial and systematic components.

Stability during initial undrained loading was evaluated using two-dimensional (2D) circular arc analyses with SHANSEP DSS undrained shear strength profiles. Since these analyses were restricted to the east and west slopes of the stockpile, 2D analyses assuming plane strain conditions appeared justified. Azzouz et al. (1983) have shown that this simplified approach yields factors of safety that are conservative by 10–15% for similar loading geometries.

Because of differences in shear strain at failure for different modes of failure along a failure arc, "peak" shear strengths are not mobilized simultaneously all along the entire failure surface. Ladd (1975) has proposed a procedure accounting for strain compatibility that determines an average shear strength to be used in undrained stability analyses. Fuleihan and Ladd (1976) showed that, in the case of the normally consolidated Atchafalaya Clay, the CK_0UDSS SHANSEP strength was in agreement with the average shear strength computed using the above procedure. All the 2D analyses used the Modified Bishop method.

To assess the importance of variability in s_u to undrained stability, it is essential to consider the volume of soil of importance to the performance prediction. At one extreme, if the volume of soil involved in a failure were infinite, spatial uncertainty would completely average out, and the systematic component uncertainty would become the total uncertainty. At the other extreme, if the volume of soil involved were infinitesimal, spatial and systematic uncertainties would both contribute fully to total uncertainty. The uncertainty for intermediate volumes of soil depends on the character of spatial variability in the deposit, specifically, on the rapidity with which soil properties fluctuate from one point to another across the site. A convenient index expressing this scale of variation is the autocorrelation distance, δ_0, which measures the distance to which fluctuations of soil properties about their expected value are strongly associated.

Too few data were available to estimate autocorrelation distance for the storage area, thus bounding calculations were made for two extreme cases in the 2D analyses, $L/\delta_0 \to 0$ (i.e. "small" failure surface) and $L/\delta_0 \to \infty$ (i.e. "large" failure surface), in which L is the length of the potential failure surface. Undrained shear strength values corresponding to significant averaging were used to evaluate uncertainty in the factor of safety for large failure surfaces and values corresponding to little averaging for small

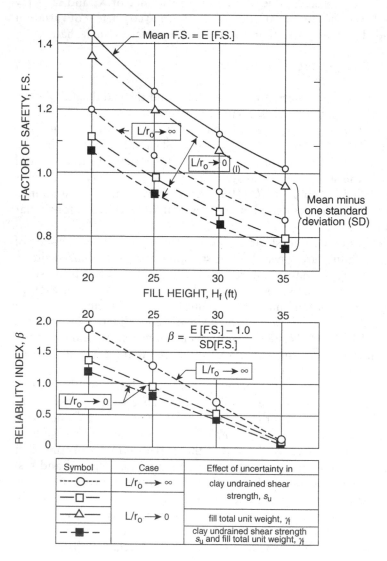

Figure 2.18 Expected value s_u profile and the standard deviation of s_u divided into spatial and systematic components.

failure surface. The results of the 2D stability analysis were plotted as a function of embankment height.

Uncertainty in the FS was estimated by performing stability analyses using the procedure of Christian *et al.* (1994) with expected value and expected value minus standard deviation values of soil properties. For a given expected value of FS, the larger the standard deviation of FS, the higher the chance that the realized FS is less than unity and thus the lower the actual safety of the facility. The second-moment reliability index [Equation (2.1)] was used to combine $E[FS]$ and $SD[FS]$ in a single measure of safety and related to a "nominal" probability of failure by assuming FS Normally distributed.

2.3 Estimating autocovariance

Estimating autocovariance from sample data is the same as making any other statistical estimate. Sample data differ from one set of observations to another, and thus the estimates of autocovariance differ. The important questions are, how much do these estimates differ, and how much might one be in error in drawing inferences? There are two broad approaches: Frequentist and Bayesian. The Frequentist approach is more common in geotechnical practice. For discussion of Bayesian approaches to estimating autocorrelation see Zellner (1971), Cressie (1991), or Berger *et al.* (2001).

In either case, a mathematical function of the sample observations is used as an estimate of the true population parameters, θ. One wishes to determine $\hat{\theta} = g(z_1, \ldots, z_n)$, in which $\{z_1, \ldots, z_n\}$ is the set of sample observations and $\hat{\theta}$, which can be a scalar, vector, or matrix. For example, the sample mean might be used as an estimator of the true population mean. The realized value of $\hat{\theta}$ for a particular sample $\{z_1 \ldots, z_n\}$ is an *estimate*. As the probabilistic properties of the $\{z_1, \ldots, z_n\}$ are assumed, the corresponding probabilistic properties of $\hat{\theta}$ can be calculated as functions of the true population parameters. This is called the *sampling distribution* of $\hat{\theta}$. The standard deviation of the sampling distribution is called the *standard error*.

The quality of the estimate obtained in this way depends on how variable the estimator $\hat{\theta}$ is about the true value θ. The sampling distribution, and hence the goodness of an estimate, has to do with how the estimate might have come out if another sample and therefore another set of observations had been made. Inferences made in this way do not admit of a probability distribution directly on the true population parameter. Put another way, the Frequentist approach presumes the state of nature θ to be a constant, and yields a probability that one would observe those data that actually were observed. The probability distribution is on the data, not on θ. Of course, the engineer or analyst wants the reverse: the probability of θ, given the data. For further discussion, see Hartford and Baecher (2004).

Bayesian estimation works in a different way. Bayesian theory allows probabilities to be assigned directly to states of nature such as θ. Thus, Bayesian methods start with an a priori probability distribution, $f(\theta)$, which is updated by the likelihood of observing the sample, using Bayes's Theorem:

$$f(\theta|z_1, \ldots, z_n) \propto f(\theta)L(\theta|z_1, \ldots, z_n) \qquad (2.21)$$

in which $f(\theta|z_1,...,z_n)$ is the a posteriori pdf of θ conditioned on the observations, and $L(\theta|z_1, ... , zn)$ is the likelihood of θ, which is the conditional probability of $\{z_1, ..., z_n\}$ as a function of θ. Note, the Fisherian concept of a maximum likelihood estimator is mathematically related to Bayesian estimation in that both adopt the *likelihood principle* that all information in the sample relevant to making an estimate is contained in the Likelihood function; however, the maximum likelihood approach still ends up with a probability statement on the variability of the estimator and not on the state of nature, which is an important distinction.

2.3.1 Moment estimation

The most common (Frequentist) method of estimating autocovariance functions for soil and rock properties is the method of moments. This uses the statistical moments of the observations (e.g. sample means, variances, and covariances) as estimators of the corresponding moments of the population being sampled.

Given the measurements $\{z_1,...,z_n\}$ made at equally spaced locations $\{x_1,...,x_n\}$ along a line, as for example in a boring, the sample autocovariance of the measurements for separation is:

$$\hat{C}_z(\delta) = \frac{1}{(n-\delta)} \sum_{i=1}^{n-\delta} [\{z(x_i) - t(x_i)\}\{z(x_{i+\delta}) - t(x_{i+\delta})\}] \tag{2.22}$$

in which $\hat{C}_z(\delta)$ is the estimator of the autocovariance function at δ, $(n-\delta)$ is the number of data pairs having separation distance δ, and $t(x_i)$ is the trend removed from the data at location x_i.

Often, $t(x_i)$ is simply replaced by the spatial mean, estimated by the mean of the sample. The corresponding moment estimator of the autocorrelation, $\hat{R}(\delta)$, is obtained by dividing both sides by the sample variance:

$$\hat{R}_z(\delta) = \frac{1}{s_z^2(n-\delta)} \sum_{i=1}^{n-\delta} [\{z(x_i) - t(x_i)\}\{z(x_{i+\delta}) - t(x_{i+\delta})\}] \tag{2.23}$$

in which s_z is the sample standard deviation. Computationally, this simply reduces to taking all data pairs of common separation distance d, calculating the correlation coefficient of that set, then plotting the result against separation distance.

In the general case, measurements are seldom uniformly spaced, at least in the horizontal plane and seldom lie on a line. For such situations the sample autocovariance can still be used as an estimator, but with some modification. The most common way to accommodate non-uniformly placed measurements is by dividing separation distances into bands, and then taking the averages within those bands.

The moment estimator of the autocovariance function requires no assumptions about the shape of the autocovariance function, except that second moments exist. The moment estimator is consistent, in that as the sample size becomes large, $E[(\hat{\theta} - \theta)^2] \to 0$. On the other hand, the moment estimator is only asymptotically unbiased. *Unbiasedness* means that the expected value of the estimator over all ways the sample might have been taken equals the actual value of the function being estimated. For finite sample sizes, the expected values of the sample autocovariance can differ significantly from the actual values, yielding negative values beyond the autocovariance distance (Weinstock, 1963).

It is well known that the sampling properties of the moment estimator of autocorrelation are complicated, and that large sampling variances (and thus poor confidence) are associated with estimates at large separation distances. Phoon and Fenton (2004) and Phoon (2006a) have experimented with bootstrapping approaches to estimate autocorrelation functions with promising success. These and similar approaches from statistical signal processing should be exploited more thoroughly in the future.

2.3.2 Example: James Bay

The results of Figure 2.19 were obtained from the James Bay data of Christian *et al.* (1994) using this moment estimator. The data are from an investigation into the stability of dykes on a soft marine clay at the James Bay Project, Québec (Ladd *et al.*, 1983). The marine clay at the site is

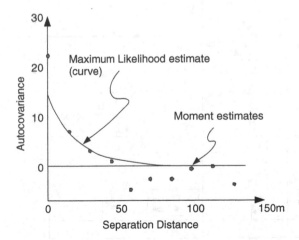

Figure 2.19 Autocovariance of field vane clay strength data, James Bay Project (Christian *et al.*, 1994, reproduced with the permission of the American Society of Civil Engineers).

approximately 8 m thick and overlies a lacustrine clay. The depth-averaged results of field vane tests conducted in 35 borings were used for the correlation analysis. Nine of the borings were concentrated in one location (Figures 2.20 and 2.21).

First, a constant mean was removed from the data. Then, the product of each pair of residuals was calculated and plotted against separation distance. A moving average of these products was used to obtain the estimated points. Note the drop in covariance in the neighborhood of the origin, and also the negative sample moments in the vicinity of 50–100 m separation. Note, also, the large scatter in the sample moments at large separation distance. From these estimates a simple exponential curve was fitted by inspection, intersecting the ordinate at about 60% of the sample variance. This yields an autocovariance function of the form:

$$C_z(\delta) = \begin{cases} 22 \text{ kPa}^2, \text{ for } \delta = 0 \\ 13\exp\{-\delta/23 \text{ m }\}, \text{ for } \delta > 0 \end{cases} \quad (2.24)$$

in which variance is in kPa² and distance in m. Figure 2.22 shows variance components for the factor of safety for various size failures.

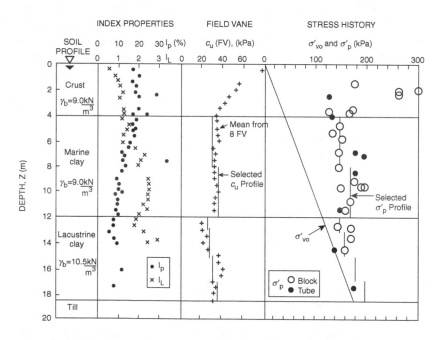

Figure 2.20 Soil property data summary, James Bay (Christian *et al.*, 1994, reproduced with the permission of the American Society of Civil Engineers).

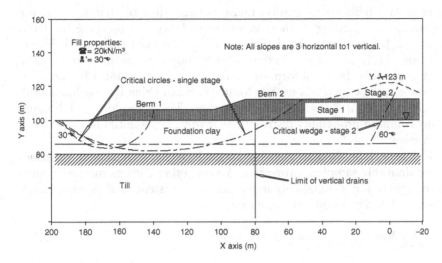

Figure 2.21 Assumed failure geometries for embankments of three heights (Christian et al., 1994, reproduced with the permission of the American Society of Civil Engineers).

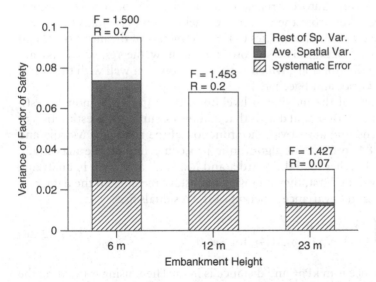

Figure 2.22 Variance components of the factor of safety for three embankment heights (Christian et al., 1994, reproduced with the permission of the American Society of Civil Engineers).

2.3.3 Maximum likelihood estimation

Maximum likelihood estimation takes as the estimator that value of the parameter(s) θ leading to the greatest probability of observing the data, $\{z_1, \ldots, z_n\}$, actually observed. This is found by maximizing the likelihood function, $L(\theta|z_1, \ldots, z_n)$. Maximum likelihood estimation is parametric because the distributional form of the pdf $f(z_1, \ldots, z_n|\theta)$ must be specified. In practice, the estimate is usually found by maximizing the log-likelihood, which, because it deals with a sum rather than a product and because many common probability distributions involve exponential terms, is more convenient.

The appeal of the maximum likelihood estimator is that it possesses many desirable sampling properties. Among others, it has minimum variance (although not necessarily unbiased), is consistent, and asymptotically Normal. The asymptotic variance of $\hat{\theta}_{ML}$ is:

$$\lim_{n \to \infty} \text{Var}[\hat{\theta}_{ML}] = I_z(\theta) = nE[-\delta^2 LL/\partial\theta^2] \tag{2.25}$$

in which $I_z(\theta)$ is Fisher's Information (Barnett, 1982) and LL is the log-likelihood.

Figure 2.23 shows the results of simulated sampling experiments in which spatial fields were generated from a multivariate Gaussian pdf with specified mean trend and autocovariance function. Samples of sizes $n = 36, 64$, and 100 were taken from these simulated fields, and maximum likelihood estimators used to obtain estimates of the parameters of the mean trend and autocovariance function. The smooth curves show the respective asymptotic sampling distributions, which in this case conform well with the actual estimates (DeGroot and Baecher, 1993).

An advantage of the maximum likelihood estimator over moment estimates in dealing with spatial data is that it allows simultaneous estimation of the spatial trend and autocovariance function of the residuals. Mardia and Marshall (1984) provide an algorithmic procedure finding the maximum. DeGroot and Baecher used the Mardia and Marshall approach in analyzing the James Bay data. First, they removed a constant mean from the data, and estimated the autocovariance function of the residuals as:

$$C_z(\delta) = \begin{cases} 23 \text{ for } \delta = 0 \\ 13.3 \exp\{-\delta/21.4\}, \text{ for } \delta > 0 \end{cases} \tag{2.26}$$

in which variance is in kPa^2 and distance is in m. Then, using estimating the trend implicitly:

$$\hat{\beta}_0 = 40.7 \text{ kPa}$$

$$\hat{\beta}_1 = -2.0 \times 10^{-3} \text{ kPa/m}$$

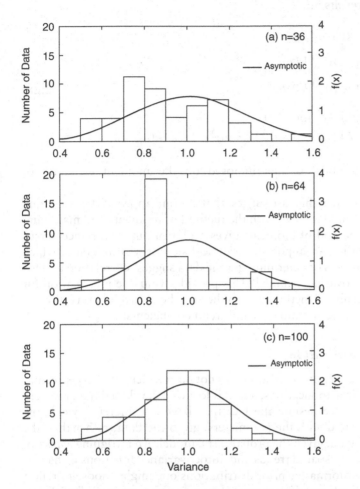

Figure 2.23 Simulated sampling experiments in which spatial fields were generated from a multivariate Gaussian pdf with specified mean trend and autocovariance function (DeGroot and Baecher, 1993, reproduced with the permission of the American Society of Civil Engineers).

$$\hat{\beta}_2 = -5.9 \times 10^{-3} \text{ kPa/m} \tag{2.27}$$

$$C_z(\delta) = \begin{cases} 23 \text{ kPa}^2 \text{ for } \delta = 0 \\ 13.3 \text{ kPa}^2 \exp\{-(\delta/21.4 \text{ m})\}, \text{ for } \delta > 0 \end{cases}$$

The small values of $\hat{\beta}_1$ and $\hat{\beta}_2$ suggest that the assumption of constant mean is reasonable. Substituting a squared-exponential model for the autocovariance results in:

$$\hat{\beta}_0 = 40.8 \text{ kPa}$$

$$\hat{\beta}_1 = -2.1 \times 10^{-3} \text{ kPa/m}$$

$$\hat{\beta}_2 = -6.1 \times 10^{-3} \text{ kPa/m} \tag{2.28}$$

$$C_z(\delta) = \begin{cases} 22.9 \text{ kPa}^2 \text{ for } \delta = 0 \\ 12.7 \text{ kPa}^2 \exp\{-(\delta/37.3 \text{ m})^2\}, \text{ for } \delta > 0 \end{cases}$$

The exponential model is superimposed on the moment estimates of Figure 2.19.

The data presented in this case suggest that a sound approach to estimating autocovariance should involve both the method of moments and maximum likelihood. The method of moments gives a plot of autocovariance versus separation, providing an important graphical summary of the data, which can be used as a means of determining if the data suggest correlation and for selecting an autocovariance model. This provides valuable information for the maximum likelihood method, which then can be used to obtain estimates of both autocovariance parameters and trend coefficients.

2.3.4 Bayesian estimation

Bayesian inference for autocorrelation has not been widely used in geotechnical and geostatistical applications, and it is less well developed than moment estimates. This is true despite the fact that Bayesian inference yields the probability associated with the parameters, given the data, rather than the confidence in the data, given the probabilistic model. An intriguing aspect of Bayesian inference of spatial trends and autocovariance functions is that for many of the non-informative prior distributions one might choose to reflect little or no prior information about process parameters (e.g. the Jeffreys prior, the Laplace prior, truncated parameter spaces), the posterior pdf's calculated through Bayesian theorem are themselves improper, usually in the sense that they do not converge toward zero at infinity, and thus the total probability or area under the posterior pdf is infinite.

Following Berger (1993), Boger *et al.* (2001) and Kitanidis (1985, 1997), the spatial model is typically written as a Multinomial random process:

$$z(\mathbf{x}) = \sum_{i=1}^{k} f_i(\mathbf{x})\beta + \varepsilon(\mathbf{x}) \tag{2.29}$$

in which $f_i(\mathbf{x})$ are unknown deterministic functions of the spatial locations x, and $\varepsilon(\mathbf{x})$ is a zero-mean spatial random function. The random term is spatially correlated, with an isotropic autocovariance function. The autocovariance function is assumed to be non-negative and to decrease monotonically with distance to zero at infinite separation. These assumptions fit most common autocovariance functions in geotechnical applications. The Likelihood of a set of observations, $z = \{z_1, ..., z_n\}$, is then:

$$L(\beta, \sigma \,|\, z) = (2\pi\sigma^2)^{-n/2} |R_\theta|^{-1/2} \exp\left\{-\frac{1}{2\sigma^2}(z - X\beta)^t R_\theta^{-1}(z - X\beta)\right\}$$

(2.30)

in which X is the $(n \times k)$ matrix defined by $X_{ij}=f_j(X_i)$, R_θ is the matrix of correlations among the observations dependent on the parameters, and $|R_\theta|$ is the determinant of the correlation matrix of the observations.

In the usual fashion, a prior non-informative distribution on the parameters (β, σ, θ) might be represented as $f(\beta, \sigma, \theta) \propto (\sigma^2)^{-a} f(\theta)$ for various choices of the parameter a and of the marginal pdf $f(\theta)$. The obvious choices might be $\{a = 1, f(\theta) = 1\}$, $\{a = 1, f(\theta) = 1/\theta\}$, or $\{a = 1, f(\theta) = 1\}$; but each of these leads to an improper posterior pdf, as does the well-known Jeffreys prior. A proper, informative prior does not share this difficulty, but it is correspondingly hard to assess from usually subjective opinion. Given this problem, Berger *et al.* (2001) suggest the reference non-informative prior:

$$f(\beta, \sigma, \theta) \propto \frac{1}{\sigma^2}\left(|W_\theta^2| - \frac{|W_\theta^2|}{(n - k)}\right)^{1/2}$$

(2.31)

in which,

$$W_\theta^2 = \frac{\partial R_\theta}{\partial \theta} R_\theta^{-1}\{I - X(X'R_\theta^{-1}X)^{-1}X'R_\theta^{-1}\}$$

(2.32)

This does lead to a proper posterior. The posterior pdf is usually evaluated numerically, although, depending on the choice of autocovariance function model and the extent to which certain of the parameters of that model are known, closed-form solutions can be obtained. Berger *et al.* (2001) present a numerically calculated example using terrain data from Davis (1986).

2.3.5 Variograms

In mining, the importance of autocorrelation for estimating ore reserves has been recognized for many years. In mining *geostatistics*, a function related to the autocovariance, called the *variogram* (Matheron, 1971), is commonly used to express the spatial structure of data. The variogram requires a less-restrictive statistical assumption on stationarity than does the autocovariance function and it is therefore sometimes preferred for inference problems. On the other hand, the variogram is more difficult to use in spatial interpolation and engineering analysis, and thus for geotechnical purposes the autocovariance is used more commonly. In practice, the two ways of characterizing spatial structure are closely related.

Whereas the autocovariance is the expected value of the product of two observations, the variogram 2γ is the expected value of the squared difference:

$$2\gamma = E[\{z(x_i) - z(x_j)\}^2] = Var[z(x_i) - z(x_j)] \tag{2.33}$$

which is a function of only the increments of the spatial properties, not their absolute values. Cressie (1991) points out that, in fact, the common definition of the variogram as the mean squared difference – rather than as the variance of the difference – limits applicability to a more restrictive class of processes than necessary, and thus the latter definition is to be preferred. None the less, one finds the former definition more commonly referred to in the literature. The term γ is referred to as the *semivariogram*, although caution must be exercised because different authors interchange the terms. The concept of average mean-square difference has been used in many applications, including turbulence (Kolmogorov, 1941) and time series analysis (Jowett, 1952), and is alluded to in the work of Matérn (1960).

The principal advantage of the variogram over the autocovariance is that it makes less restrictive assumptions on the stationarity of the spatial properties being sampled; specifically, only that their increment and not their mean is stationary. Furthermore, the use of geostatistical techniques has expanded broadly, so that a great deal of experience has been accumulated with variogram analysis, not only in mining applications, but also in environmental monitoring, hydrology, and even geotechnical engineering (Chiasson *et al.*, 1995; Soulie and Favre, 1983; Soulie *et al.*, 1990).

For spatial variables with stationary means and autocovariances (i.e. second-order stationary processes), the variogram and autocovariance function are directly related by:

$$\gamma(\delta) = C_z(0) - C_z(\delta) \tag{2.34}$$

Common analytical forms for one-dimensional variograms are given in Table 2.3.

For a stationary process, as $|\delta| \rightarrow \infty, C_z(\delta) \rightarrow 0$; thus, $\gamma(\delta) \rightarrow C_z(0) = Var(z(x))$. This value at which the variogram levels off, $2C_z(\delta)$, is called the *sill* value. The distance at which the variogram approaches the sill is called the *range*. The sampling properties of the variogram are summarized by Cressie (1991).

Table 2.3 One-dimensional variogram models.

Model	Equation	Limits of validity
Nugget	$g(\delta) = \begin{cases} 0 & \text{if } \delta = 0 \\ 1 & \text{otherwise} \end{cases}$	R^n
Linear	$g(\delta) = \begin{cases} 0 & \text{if } \delta = 0 \\ c_0 + b\|\delta\| & \text{otherwise} \end{cases}$	R^1
Spherical	$g(\delta) = \begin{cases} (1.5)(\delta/a) - (1/2)(\delta/a)^3 & \text{if } \delta = 0 \\ 1 & \text{otherwise} \end{cases}$	R^n
Exponential	$g(\delta) = 1 - \exp(-3\delta/a)$	R^1
Gaussian	$g(\delta) = 1 - \exp(-3\delta^2/a^2)$	R^n
Power	$g(\delta) = h^\omega$	$R^n, 0 < \delta < 2$

2.4 Random fields

The application of random field theory to spatial variation is based on the assumption that the property of concern, $z(\mathbf{x})$, is the realization of a random process. When this process is defined over the space $\mathbf{x} \in S$, the variable $z(\mathbf{x})$ is said to be a *stochastic process*. In this chapter, when S has dimension greater than one, $z(\mathbf{x})$ is said to be a *random field*. This usage is more or less consistent across civil engineering, although the geostatistics literature uses a vocabulary all of its own, to which the geotechnical literature occasionally refers.

A random field is defined as the joint probability distribution:

$$F_{x_1,\cdots,x_n}(z_1,\ldots,z_n) = P\{z(x_1) \le z_1, \ldots, z(x_n) \le z_n\} \tag{2.35}$$

This joint probability distribution describes the simultaneous variation of the variables $\mathbf{z}$ within a space S_x. Let, $E[z(\mathbf{x})] = \mu(\mathbf{x})$ be the mean or trend of $z(\mathbf{x})$, and let $Var[z(\mathbf{x})] = \sigma^2(\mathbf{x})$ be the variance. The covariances of $z(\mathbf{x}_1), \ldots, z(\mathbf{x}_n)$ are defined as:

$$Cov[z(\mathbf{x}_i), z(\mathbf{x}_j)] = E[(z(\mathbf{x}_i) - \mu(\mathbf{x}_i)) \cdot (z(\mathbf{x}_j) - \mu(\mathbf{x}_j))] \tag{2.36}$$

A random field is said to be *second-order stantionary* (weak or wide-sense stationary) if $E[z(\mathbf{x})] = \mu$ for all $\mathbf{x}$, and $Cov[z(\mathbf{x}_i), z(\mathbf{x}_j)]$ depends only on vector separation of $\mathbf{x}_i$, and $\mathbf{x}_j$, and not on location, $Cov[z(\mathbf{x}_i), z(\mathbf{x}_j)] = C_z(\mathbf{x}_i - \mathbf{x}_j)$, in which $C_z(\mathbf{x}_i - \mathbf{x}_j)$ is the autocovariance function. The random field is said to be *stationary* (strong or strict stationarity) if the complete probability distribution, $F_{x_1,\ldots,x_n}(z_1,\ldots,z_n)$, is independent of absolute location, depending only on vector separations among the $x_i...x_n$. Strong stationarity implies second-order stationarity. In the geotechnical literature, stationarity is sometimes referred to as *statistical homogeneity*. If the autocovariance function depends only on the absolute separation distance and not direction, the random field is said to be *isotropic*.

Ergodicity is a concept originating in the study of time series, in which one observes individual time series of data, and wishes to infer properties of an ensemble of all possible time series. The meaning and practical importance of ergodicity to the inference of unique realizations of spatial random fields is less well-defined and is debated in the literature. Simply, ergodicity means that the probabilistic properties of a random process (field) can be completely estimated from observing one realization of that process. For example, the stochastic time series $z(t) = v + \varepsilon(t)$, in which v is discrete random variable and $\varepsilon(t)$ is an autocorrelated random process of time, is non-ergotic. In one realization of the process there is but one value of v, and thus the probability distribution of v cannot be estimated. One would need to observe many realizations of zt, in order to have sufficiently many observations of v, to estimate $F_v(v)$. Another non-ergotic process of more relevance to the spatial processes in geotechnical engineering is $z(x) = v + \varepsilon(x)$, in which the mean of $z(x)$ varies linearly with location x. In this case, $z(x)$ is non-stationary; $Var[z(x)]$ increases without limit as the window within which $z(x)$ is observed increases, and the mean m_z of the sample of $z(x)$ is a function of the location of the window.

The meaning of ergodicity for spatial fields of the sort encountered in geotechnical engineering is less clear, and has not been widely discussed in the literature. An assumption weaker than full erogodicity, which none the less should apply for spatial fields, is that the observed sample mean m_z and sample autocovariance function $\hat{C}_z(\delta)$ converge in mean-squared error to the respective random field mean and autocovariance function as the volume of space within which they are sampled increases. This means that, as the volume of space increases, $E[(m_z - \mu)^2] \to 0$, and $E[\{(\hat{C}_z(\delta) - \hat{C}_z(\delta)\}^2] \to 0$. Soong and Grigoriu (1993) provide conditions for checking ergodicity in mean and autocorrelation.

When the joint probability distribution $F_{x_1,...,x_n}(z_1,...,z_n)$ is multivariate Normal (Gaussian), the process $z(x)$ is said to be a Gaussian random field. A sufficient condition for ergodicity of a Gaussian random field is that $\lim_{|\delta| \to \infty} C_z(\delta) = 0$. This can be checked empirically by inspecting the sample moments of the autocovariance function to ensure they converge to 0. Cressie (1991) notes limitations of this procedure. Essentially all the analytical autocovariance functions common in the geotechnical literature obey this condition, and few practitioners appear concerned about verifying ergodicity. Christakos (1992) suggests that, in practical situations, it is difficult or impossible to verify ergodicity for spatial fields.

A random field that does not meet the conditions of stationarity is said to be *non-stationary*. Loosely speaking, a non-stationary field is statistically heterogeneous. It can be heterogeneous in a number of ways. In the simplest case, the mean may be a function of location, for example, if there is a spatial trend that has not been removed. In a more complex case, the variance or autocovariance function may vary in space. Depending on the way in which the random field is non-stationary, sometimes a transformation of variables can convert a non-stationary field to a stationary or nearly stationary field. For example, if the mean varies with location, perhaps a trend can be removed.

In the field of geostatistics, a weaker assumption is made on stationarity than that described above. Geostatisticians usually assume only that increments of a spatial process are stationary (i.e. differences $|z_1 - z_2|$) and then operate on the probabilistic properties of those increments. This leads to the use of the variogram rather than the autocovariance function. Stationarity of the autocovariance function implies stationarity of the variogram, but the reverse is not true.

Like most things in the natural sciences, stationarity is an assumption of the model and may only be approximately true in the world. Also, stationarity usually depends on scale. Within a small region soil properties may behave as if drawn from a stationary process, whereas the same properties over a larger region may not be so well behaved.

2.4.1 Permissible autocovariance functions

By definition, the autocovariance function is symmetric, meaning:

$$C_z(\delta) = C_z(-\delta) \tag{2.37}$$

and bounded, meaning:

$$C_z(\delta) \le C_z(0) = \sigma_z^2 \tag{2.38}$$

In the limit, as distance becomes large:

$$\lim_{|\delta| \to \infty} \frac{C_z(\delta)}{|\delta|^{-(n-1)/2}} = 0 \tag{2.39}$$

In general, in order for $C_z(\delta)$ to be a permissible autocovariance function, it is necessary and sufficient that a continuous mathematical expression of the form:

$$\sum_{i=1}^{m} \sum_{j=1}^{m} k_i k_j C_z(\delta) \ge 0 \tag{2.40}$$

be non-negative-definite for all integers m, scalar coefficients $k_1,...k_m$, and δ. This condition follows from the requirement that variances of linear combinations of the $z(x_i)$, of the form:

$$var\left[\sum_{i=1}^{m}k_i z(x_i)\right] = \sum_{i=1}^{m}\sum_{j=1}^{m}k_i k_j C_z(\delta) \geq 0 \tag{2.41}$$

be non-negative, i.e. the matrix is positive definite (Cressie, 1991). Christakos (1992) discusses the mathematical implications of this condition on selecting permissible forms for the autocovariance. Suffice it to say that analytical models of autocovariance common in the geotechnical literature usually satisfy the condition.

Autocovariance functions valid in a space of dimension d are valid in spaces of lower dimension, but the reverse is not necessarily true. That is, a valid autocovariance function in 1D is not necessarily valid in 2D or 3D. Christakos (1992) gives the example of the linearly declining autcovariance:

$$C_z(\delta) = \begin{cases} \sigma^2(1 - \delta/\delta_0), \ for \ 0 \leq \delta \leq \delta_0 \\ 0, \ for \ \delta > \delta_0 \end{cases} \tag{2.42}$$

which is valid in 1D, but not in higher dimensions.

Linear sums of valid autocovariance functions are also valid. This means that if $C_{z1}(\delta)$ and $C_{z2}(\delta)$ are valid, then the sum $C_{z1}(\delta) + C_{z2}(\delta)$ is also a valid autocovariance function. Similarly, if $C_z(\delta)$ is valid, then the product with a scalar, $\alpha C_z(\delta)$, is also valid.

An autocovariance function in d-dimensional space is *separable* if

$$C_z(\delta) = \prod_{i=1}^{d} C_{zi}(\delta_i) \tag{2.43}$$

in which δ is the d-dimensioned vector of orthogonal separation distances $\{\delta_1,...,\delta_d\}$, and $C_i(\delta_i)$ is the one-dimensional autocovariance function in direction i. For example, the autocovariance function:

$$C_z(\delta) = \sigma^2 \exp\{-a^2|\delta|^2\}$$

$$= \sigma^2 \exp\{-a^2(\delta_1^2 + \cdots + \delta_d^2)\} \tag{2.44}$$

$$= \sigma^2 \prod_{i=1}^{d} \exp\{-a^2 \delta_i^2\}$$

is separable into its one-dimensional components.

The function is *partially separable* if

$$C_z(\delta) = C_z(\delta_i) C_z(\delta_{j \neq i}) \tag{2.45}$$

in which $C_z(\delta_{j \neq i})$ is a $(d-1)$ dimension autocovariance function, implying that the function can be expressed as a product of autocovariance functions of lower dimension fields. The importance of partial separability to geotechnical applications, as noted by VanMarcke (1983), is the 3D case of separating autocorrelation in the horizontal plane from that with depth:

$$C_z(\delta_1, \delta_2, \delta_3) = C_z(\delta_1, \delta_2) C_z(\delta_3) \tag{2.46}$$

in which δ_1, δ_2, are horizontal distances, and δ_3 is depth.

2.4.2 Gaussian random fields

The Gaussian random field is an important special case because it is widely applicable due to the Central Limit Theorem, has mathematically convenient properties, and is widely used in practice. The probability density distribution of the Gaussian or Normal variable is:

$$f_z(z) = -\frac{1}{\sqrt{2\pi}\sigma} \exp\left\{ -\frac{1}{2}\left(\frac{x-\mu}{\sigma}\right)^2 \right\} \tag{2.47}$$

for $-\infty \leq z \leq \infty$. The mean is $E[z] = \mu$, and variance $Var[z] = \sigma^2$. For the multivariate case of vector z, of dimension n, the correponding pdf is:

$$f_z(\mathbf{z}) = (2\pi)^{-n/2} |\mathbf{\Sigma}|^{-1/2} \exp\left\{ -\frac{1}{2}(\mathbf{z} - \boldsymbol{\mu})' \mathbf{\Sigma}^{-1} (\mathbf{z} - \boldsymbol{\mu}) \right\} \tag{2.48}$$

in which $\boldsymbol{\mu}$ is the mean vector, and $\mathbf{\Sigma}$ the covariance matrix:

$$\mathbf{\Sigma}_{ij} = \left\{ Cov[z_i(\mathbf{x}), z_j(\mathbf{x})] \right\} \tag{2.49}$$

Gaussian random fields have the following convenient properties (Adler, 1981): (1) they are completely characterized by the first- and second-order moments: the mean and autocovarinace function for the univariate case, and mean vector and autocovariance matrix (function) for the multivariate case; (2) any subset of variables of the vector is also jointly Gaussian; (3) the conditional probability distributions of any two variables or vectors are also Gaussian distributed; (4) if two variables, z_1 and z_2, are bivariate Gaussian, and if their covariance $Cov[z_1, z_2]$ is zero, then the variables are independent.

2.4.3 Interpolating random fields

A problem common in site characterization is interpolating among spatial observations to estimate soil or rock properties at specific locations where they have not been observed. The sample observations themselves may have been taken under any number of sampling plans: random, systematic, cluster, and so forth. What differentiates this spatial estimation question from the sampling theory estimates in preceding sections of this chapter is that the observations display spatial correlation. Thus, the assumption of IID observations underlying the estimator results is violated in an important way. This question of spatial interpolation is also a problem common to the natural resources industries such as forestry (Matérn, 1986) and mining (Matheron, 1971), but also geohydrology (Kitanidis, 1997), and environmental monitoring (Switzer, 1995).

Consider the case for which the observations are sampled from a spatial population with constant mean, μ, and autocovariance function $C_z(\delta) = E[z(x_i)z(x_{i+\delta})]$. The set of observations $\mathbf{z}=\{z_i,\ldots,z_n\}$ therefore has mean vector $\mathbf{m}$ in which all the terms are equal, and covariance matrix:

$$\Sigma = \begin{bmatrix} Var(z_1) & \cdots & Cov(z_1, z_n) \\ \vdots & \ddots & \vdots \\ Cov(z_n, z_1) & \cdots & Var(z_n) \end{bmatrix} \tag{2.50}$$

in which the terms $z(x_i)$ are replaced by z_i for convenience. These terms are found from the autocovariance function as $Cov(z(x_i)z(x_j)) = C_z(\delta_{ij})$, in which δ_{ij} is the (vector) separation between locations x_i and x_j.

In principle, we would like to estimate the full distribution of $z(x_0)$ at an unobserved location x_0, but in general this is computationally intensive if a large grid of points is to be interpolated. Instead, the most common approach is to construct a simple linear unbiased estimator based on the observations:

$$\hat{z}(x_0) = \sum_{i=1}^{n} w_i z(x_i) \tag{2.51}$$

in which the weights $\mathbf{w}=\{w_1,\ldots, w_n\}$ are scalar values chosen to make the estimate in some way optimal. Usually, the criteria of optimality are unbiasedness and minimum variance, and the result is sometimes called the *best linear unbiased estimator* (BLUE).

The BLUE estimator weights are found by expressing the variance of the estimate $\hat{z}(x_0)$ using a first-order second-moment formulation, and minimizing the variance over $\mathbf{w}$ using a Lagrange multiplier approach subject to the condition that the sum of the weights equals one. The solution in matrix form is:

$$\mathbf{w} = \mathbf{G}^{-1}\mathbf{h} \tag{2.52}$$

in which $\mathbf{w}$ is the vector of optimal weights, and the matrices $\mathbf{G}$ and $\mathbf{h}$ relate the covariance matrix of the observations and the vector of covariances of the observations to the value of the spatial variable at the interpolated location, x_0, respectively:

$$\mathbf{G} = \begin{bmatrix} Var(z_1) & \cdots & Cov(z_1, z_n) & 1 \\ \vdots & \ddots & \vdots & 1 \\ Cov(z_n, z_1) & \cdots & Var(z_n) & 1 \\ 1 & 1 & 1 & 0 \end{bmatrix} \tag{2.53}$$

$$\mathbf{h} = \begin{bmatrix} Cov(z_1, z_0) \\ \vdots \\ Cov(z_n, z_0) \\ 1 \end{bmatrix}$$

The resulting estimator variance is:

$$Var(\hat{z}_0) = E[(z_0 - \hat{z}_0)^2]$$

$$= Var(z_0) - \sum_{i=1}^{n} w_i Cov(z_0, z_i) - \lambda \tag{2.54}$$

in which λ is the Lagrange multiplier resulting from the optimization. This is a surprisingly simple and convenient result, and forms the basis of the increasingly vast literature on the subject of so-called *kriging* in the field of geostatistics. For regular grids of observations, such as a grid of borings, an algorithm can be established for the points within an individual grid cell, and then replicated for all cells to form an interpolated map of the larger site or region (Journel and Huijbregts, 1978). In the mining industry, and increasingly in other applications, it has become common to replace the auto-covariance function as a measure of spatial association with the variogram.

2.4.4 Functions of random fields

Thus far, we have considered the properties of random fields themselves. In this section, we consider the extension to properties of functions of random fields. Spatial averaging of random fields is among the most important considerations for geotechnical engineering. Limiting equilibrium stability of slopes depends on the average strength across the failure surface.

Settlements beneath foundations depend on the average compressibility of the subsurface soils. Indeed, many modes of geotechnical performance of interest to the engineer involve spatial averages – or differences among spatial averages – of soil and rock properties. Spatial averages also play a significant role in mining geostatistics, where average ore grades within blocks of rock have important implications for planning. As a result, there is a rich literature on the subject of averages of random fields, only a small part of which can be reviewed here.

Consider the one-dimensional case of a continuous, scalar stochastic process (1D random field), $z(x)$, in which x is location, and $z(x)$ is a stochastic variable with mean μ_z, assumed to be constant, and autocovariance function $C_z(r)$, in which r is separation distance, $r = (x_1 - x_2)$. The spatial average or mean of the process within the interval $[0, X]$ is:

$$M_X\{z(x)\} = \frac{1}{X} \int_0^X z(x)\mathrm{d}x \tag{2.55}$$

The integral is defined in the common way, as a limiting sum of $z(x)$ values within infinitesimal intervals of x, as the number of intervals increases. We assume that $z(x)$ converges in a mean square sense, implying the existence of the first two moments of $z(x)$. The weaker assumption of convergence in probability, which does not imply existence of the moments, could be made, if necessary (see Parzen 1964, 1992) for more detailed discussion).

If we think of $M_X\{z(\mathbf{x})\}$ as a sample observation within one interval of the process $z(x)$, then, over the set of possible intervals that we might observe, $M_X\{z(\mathbf{x})\}$ becomes a random variable with mean, variance, and possibly other moments. Consider first the integral of $z(x)$ within intervals of length X. Parzen (1964) shows that the first two moments of $\int_0^X z(x)\mathrm{d}x$ are:

$$E\left[\int_0^X z(x)\mathrm{d}x\right] = \int_0^X \mu(x)\mathrm{d}x = \mu X \tag{2.56}$$

$$Var\left[\int_0^X z(x)\mathrm{d}x\right] = \int_0^X \int_0^X C_z(x_i - x_j)\mathrm{d}x_i\mathrm{d}x_j = 2\int_0^X (X - r)C_z(r)\mathrm{d}r \tag{2.57}$$

and that the autocovariance function of the integral $\int_0^X z(x)\mathrm{d}x$ as the interval $[0, X]$ is allowed to translate along dimension x is (VanMarcke, 1983):

$$C_{\int_0^X z(x)\mathrm{d}x}(r) = Cov\left[\int_0^X z(x)\mathrm{d}x, \int_r^{r+X} z(x)\mathrm{d}x\right]$$

$$= \int_0^X \int_0^X C_z(r + x_i - x_j)\mathrm{d}x_i\mathrm{d}x_j \tag{2.58}$$

The corresponding moments of the spatial mean $M_X\{z(\mathbf{x})\}$ are:

$$E\left[M_X\{z(\mathbf{x})\}\right] = E\left[\frac{1}{X}\int_0^X z(x)\mathrm{d}x\right] = \int_0^X \frac{1}{X}\mu(x)\mathrm{d}x = \mu \qquad (2.59)$$

$$Var\left[M_X\{z(\mathbf{x})\}\right] = Var\left[\frac{1}{X}\int_0^X z(x)\mathrm{d}x\right] = \frac{2}{X^2}\int_0^X (X-r)C_z(r)\mathrm{d}r \quad (2.60)$$

$$C_{M_X\{z(\mathbf{x})\}}(r) = Cov\left[\frac{1}{X}\int_0^X z(x)\mathrm{d}x, \frac{1}{X}\int_r^{r+X} z(x)\mathrm{d}x\right]$$

$$= \frac{1}{X^2}\int_0^X\int_0^X C_z(r+x_i-x_j)\mathrm{d}x_i\mathrm{d}x_j \qquad (2.61)$$

The effect of spatial averaging is to smooth the process. The variance of the averaged process is smaller than that of the original process $z(x)$, and the autocorrelation of the averaged process is wider. Indeed, averaging is sometimes referred to as *smoothing* (Gelb and Analytic Sciences Corporation Technical Staff, 1974).

The reduction in variance from $z(x)$ to the averaged process $M_X\{z(x)\}$ can be represented in a *variance reduction function*, $\gamma(X)$:

$$\gamma(X) = \frac{Var\left[M_X\{z(\mathbf{x})\}\right]}{var[z(x)]} \qquad (2.62)$$

The variance reduction function is 1.0 for $X = 0$, and decays to zero as X becomes large. $\gamma(X)$ can be calculated from the autocovariance function of $z(x)$ as:

$$\gamma(X) = \frac{2}{x}\int_0^x \left(1 - \frac{r}{X}\right)R_2(r)\mathrm{d}r \qquad (2.63)$$

in which $R_z(r)$ is the autocorrelation function of $z(x)$. Note that the square root of $\gamma(X)$ gives the corresponding reduction of the standard deviation of $z(x)$. Table 2.4 gives one-dimensional variance reduction functions for common autocovariance functions. It is interesting to note that each of these functions is asymptotically proportional to $1/X$. Based on this observation, VanMarcke (1983) proposed a *scale of fluctuation*, θ, such that:

$$\theta = \lim_{X\to\infty} X\ \gamma(X) \qquad (2.64)$$

or $\gamma(X) = \theta/X$, as $X \to \infty$; that is, θ/X is the asymptote of the variance reduction function as the averaging window expands. The function $\gamma(X)$

Table 2.4 Variance reduction functions for common 1D autocovariances (after VanMarcke, 1983).

Model	Autocorrelation	Variance reduction function	Scale of fluctuation
White noise	$R_x(\delta) = \begin{cases} 1 & \text{if } \delta=0 \\ 0 & \text{otherwise} \end{cases}$	$\gamma(X) = \begin{cases} 1 & \text{if } X = 0 \\ 0 & \text{otherwise} \end{cases}$	0
Linear	$R_x(\delta) = \begin{cases} 1 - |\delta|/\delta_n & \text{if } \delta \le \delta_0 \\ 0 & \text{otherwise} \end{cases}$	$\gamma(X) = \begin{cases} 1 - X/3\delta_0 & \text{if } X \le \delta_0 \\ (\delta_0/X)\left[1 - \delta_0/3X\right] & \text{otherwise} \end{cases}$	δ_0
Exponential	$R_x(\delta) = \exp(-\delta/\delta_0)$	$\gamma(X) = 2(\delta_0/X)^2 \left(\frac{X}{\delta_0} - 1 + \exp^2(-X/\delta_0) \right)$	$4\delta_0$
Squared exponential (Gaussian)	$R_x(\delta) = \exp^2(-|\delta|/\delta_0)$	$\gamma(X) = (\delta_0/X)^2 \left[\sqrt{\pi}\frac{X}{\delta_0}\Phi(-X/\delta_0) + \exp^2(-X/\delta_0) - 1 \right]$ in which Φ is the error function	$\sqrt{\pi}\delta_0$

converges rapidly to this asymptote as X increases. For θ to exist, it is necessary that $R_z(r) \to 0$ as $r \to \infty$, that is, that the autocorrelation function decreases faster than $1/r$. In this case, θ can be found from the integral of the autocorrelation function (the moment of $R_z(r)$ about the origin):

$$\theta = 2 \int_0^\infty R_z(r)\mathrm{d}r = \int_{-\infty}^\infty R_z(r)\mathrm{d}r \tag{2.65}$$

This concept of summarizing the spatial or temporal scale of autocorrelation in a single number, typically the first moment of $R_z(r)$, is used by a variety of other workers, and in many fields. Taylor (1921) in hydrodynamics called it the *diffusion constant* (Papoulis and Pillai, 2002); Christakos (1992) in geoscience calls $\theta/2$ the *correlation radius*; Gelhar (1993) in groundwater hydrology calls θ the *integral scale*.

In two dimensions, the equivalent expressions for the mean and variance of the planar integral, $\int_0^X \int_0^X z(x)\mathrm{d}x$, are:

$$E\left[\int_0^X z(x)\mathrm{d}x\right] = \int_0^X \mu(x)\mathrm{d}x = \mu X \tag{2.66}$$

$$Var\left[\int_0^X z(x)\mathrm{d}x\right] = \int_0^X \int_0^X C_z(x_i - x_j)\mathrm{d}x_i\mathrm{d}x_j = 2\int_0^X (X - r)C_z(r)\mathrm{d}r \tag{2.67}$$

Papoulis and Pillai (2002) discuss averaging in higher dimensions, as do Elishakoff (1999) and VanMarcke (1983).

2.4.5 Stochastic differentiation

The continuity and differentiability of a random field depend on the convergence of sequences of random variables $\{z(\mathbf{x}_a), z(\mathbf{x}_b)\}$, in which $\mathbf{x}_a$, $\mathbf{x}_b$ are two locations, with (vector) separation $\mathbf{r} = |\mathbf{x}_a - \mathbf{x}_b|$. The random field is said to be *continuous in mean square* at $\mathbf{x}_a$, if for every sequence $\{z(\mathbf{x}_a), z(\mathbf{x}_b)\}$, $E^2[z(\mathbf{x}_a) - z(\mathbf{x}_b)] \to 0$, as $\mathbf{r} \to 0$. The random field is said to be *continuous in mean square* throughout, if it is continuous in mean square at every $\mathbf{x}_a$. Given this condition, the random field $\mathbf{z}(\mathbf{x})$ is *mean square differentiable*, with partial derivative,

$$\frac{\partial \mathbf{z}(\mathbf{x})}{\partial \mathbf{x}_i} = \lim_{|r| \to 0} \frac{\mathbf{z}(\mathbf{x} + \mathbf{r}\delta_i) - \mathbf{z}(\mathbf{x})}{r} \tag{2.68}$$

in which the delta function is a vector of all zeros, except the ith term, which is unity. While stronger, or at least different, convergence properties could be invoked, mean square convergence is often the most natural form in practice, because we usually wish to use a second-moment representation of the autocovariance function as the vehicle for determining differentiability.

A random field is mean square continuous if and only if its autocovariance function, $C_z(\mathbf{r})$, is continuous at $|\mathbf{r}| = 0$. For this to be true, the first derivatives of the autocovariance function at $|\mathbf{r}|=0$ must vanish:

$$\frac{\partial C_z(\mathbf{r})}{\partial x_i} = 0, \quad \text{for all } i \tag{2.69}$$

If the second derivative of the autocovariance function exists and is finite at $|\mathbf{r}| = 0$, then the field is mean square differentiable, and the autocovariance function of the derivative field is:

$$C_{\partial z/\partial x_i}(\mathbf{r}) = \partial^2 C_z(\mathbf{r})/\partial x_i^2 \tag{2.70}$$

The variance of the derivative field can then be found by evaluating the autocovariance $C_{\partial z/\partial x_i}(\mathbf{r})$ at $|\mathbf{r}| = 0$. Similarly, the autocovariance of the second derivative field is:

$$C_{\partial^2 z/\partial x_i \partial x_j}(\mathbf{r}) = \partial^4 C_z(\mathbf{r})/\partial x_i^2 \partial x_j^2 \tag{2.71}$$

The cross covariance function of the derivatives with respect to x_i and x_j in separate directions is:

$$C_{\partial z/\partial x_i, \partial z/\partial x_j}(\mathbf{r}) = -\partial^2 C_z(\mathbf{r})/\partial x_i \partial x_j \tag{2.72}$$

Importantly, for the case of homogeneous random fields, the field itself, $z(x)$, and its derivative field are uncorrelated (VanMarcke, 1983).

So, the behavior of the autocovariance function in the neighborhood of the origin is the determining factor for mean-square local properties of the field, such as continuity and differentiability (Cramér and Leadbetter, 1967). Unfortunately, the properties of the derivative fields are sensitive to this behavior of $C_z(\mathbf{r})$ near the origin, which in turn is sensitive to the choice of autocovariance model. Empirical verification of the behavior of $C_z(\mathbf{r})$ near the origin is exceptionally difficult. Soong and Grigoriu (1993) discuss the mean square calculus of stochastic processes.

2.4.6 Linear functions of random fields

Assume that the random field, $z(x)$, is transformed by a deterministic function $g(.)$, such that:

$$y(x) = g[z(x)] \tag{2.73}$$

In this equation, $g[z(x_0)]$ is a function of z alone, that is, not of x_0, and not of the value of $z(x)$ at any x other than x_0. Also, we assume that the transformation does not depend on the value of x; that is, the transformation is space- or time-invariant, $y(z + \delta) = g[z(x + \delta)]$. Thus, the random variable $y(x)$ is a deterministic transformation of the random variable $z(x)$, and its probability distribution can be obtained from derived distribution methods. Similarly, the joint distribution of the sequence of random variables $\{y(x_1), \ldots, y(x_n)\}$ can be determined from the joint distribution of the sequence of random variables $\{x(x_1) \ldots, x(x_n)\}$. The mean of $y(x)$ is then:

$$E[y(x)] = \int_{-\infty}^{\infty} g(z)f_z(z(x))dz \tag{2.74}$$

and the autocorrelation function is:

$$R_y(y_1, y_2) = E[y(x_1)y(x_2)] = \int_{-\infty}^{\infty} \int_{-\infty}^{\infty} g(z_1)g(z_2)f_z(z(x_1)z(x_2))dz_1 dz_2$$

$$\tag{2.75}$$

Papoulis and Pillai(2002) show that the process $y(x)$ is (strictly) stationary if $z(x)$ is (strictly) stationary. Phoon (2006b) discusses limitations and practical methods of solving this equation. Among the limitations is that such non-Gaussian fields may not have positive definite covariance matrices.

The solutions for nonlinear transformations are difficult, but for linear functions general results are available. The mean of $y(x)$ for linear $g(z)$ is found by transforming the expected value of $z(x)$ through the function:

$$E[y(x)] = g(E[z(x)]) \tag{2.76}$$

The autocorrelation of $y(x)$ is found in a two-step process:

$$R_{yy}(x_1, x_2) = L_{x_1}[L_{x_2}[R_{zz}(x_1, x_2)]] \tag{2.77}$$

in which L_{x_1} is the transformation applied with respect to the first variable $z(x_1)$ with the second variable treated as a parameter, and L_{x_2} is the transformation applied with respect to the second variable $z(x_2)$ with the first variable treated as a parameter.

2.4.7 Excursions (level crossings)

A number of applications arise in geotechnical practice for which one is interested not in the integrals (averages) or differentials of a stochastic process, but in the probability that the process exceeds some threshold, either positive or negative. For example, we might be interested in the probability that a stochastically varying water inflow into a reservoir exceeds some rate or in the properties of the weakest interval or seam in a spatially varying soil mass. Such problems are said to involve *excursions* or *level crossings* of a stochastic process. The following discussion follows the work of Cramér (1967), Parzen (1964), and Papoulis and Pillai (2002).

To begin, consider the *zero-crossings* of a random process: the points x_i at which $z(x_i) = 0$. For the general case, this turns out to be a surprisingly difficult problem. Yet, for the continuous Normal case, a number of statements or approximations are possible. Consider a process $z(x)$ with zero mean and variance σ^2. For the interval $[x, x+\delta]$, if the product:

$$z(x)z(x+\delta) < 0, \tag{2.78}$$

then there must be an odd number of zero-crossings within the interval, for if this product is negative, one of the values must lie above zero and the other beneath. Papoulis and Pillai(2002) demonstrate that, if the two (zero-mean) variables $z(x)$ and $z(x+\delta)$ are jointly normal with correlation coefficient:

$$r = \frac{E[z(x)z(x+\delta)]}{\sigma_x \sigma_{x+\delta}}, \tag{2.79}$$

then

$$p(z(x)z(x+\delta) < 0) = \frac{1}{2} - \frac{\arcsin(r)}{\pi} = \frac{\arccos(r)}{\pi}$$

$$p(z(x)z(x+\delta) > 0) = \frac{1}{2} + \frac{\arcsin(r)}{\pi} = \frac{\pi - \arccos(r)}{\pi} \tag{2.80}$$

The correlation coefficient, of course, can be taken from the autocorrelation function, $R_z(\delta)$. Thus:

$$\cos[\pi p(z(x)z(x+\delta) < 0)] = \frac{R_z(\delta)}{R_z(0)} \tag{2.81}$$

and the probability that the number of zero-crossings is positive is just the complement of this result.

The probability of exactly one zero-crossing, $p_1(\delta)$, is approximately $p_1(\delta) \approx p_0(\delta)$, and expanding the cosine in a Fourier series and truncating to two terms:

$$1 - \frac{\pi^2 p_1^2(\delta)}{2} = \frac{R_z(\delta)}{R_z(0)} \tag{2.82}$$

or,

$$p_1(\delta) \approx \frac{1}{\pi} \sqrt{\frac{2[R_z(0) - R_z(\delta)]}{R_z(0)}} \tag{2.83}$$

In the case of a regular autocorrelation function, for which the derivative $dR_z(0)/d\delta$ exists and is zero at the origin, the probability of a zero-crossing is approximately:

$$p_1(\delta) \approx \frac{\delta}{\pi}\sqrt{-\frac{d^2R_z(0)/d\delta^2}{R_z(0)}} \tag{2.84}$$

The non-regular case, for which the derivative at the origin is not zero (e.g. $dR_z(\delta) = \exp(\delta/\delta_0)$), is discussed by Parzen (1964). Elishakoff (1999) and VanMarcke (1983) treat higher dimensional results. The related probability of the process crossing an arbitrary level, z^*, can be approximated by noting that, for small δ and thus $r \rightarrow 1$,

$$P\big[\{z(x) - z^*\}\{z(x+\delta) - z^*\} < 0\big] \approx P\big[\{z(x)\}\{z(x+\delta)\} < 0\big]e^{\frac{-\arcsin^2(r)}{2\sigma^2}} \tag{2.85}$$

For small δ, the correlation coefficient $R_z(\delta)$ is approximately 1, and the variances of $z(x)$ and $z(x + \delta)$ are approximately $R_z(0)$, thus:

$$p_{1,z^*}(\delta) \approx p_{1,z^*}(\delta) < 0]e^{\frac{-\arcsin^2(r)}{2R_z(0)}} \tag{2.86}$$

and for the regular case:

$$p_1(\delta) \approx \frac{\delta}{\pi}\sqrt{\frac{-d^2R_z(0)d\delta^2}{R_z(0)}}e^{\frac{-\arcsin^2(r)}{2R_z(0)}} \tag{2.87}$$

Many other results can be found for continuous Normal processes, e.g. the average density of the number of crossings within an interval, the probability of no crossings (i.e. drought) within an interval, and so on. A rich literature is available of these and related results (Yaglom, 1962; Parzen, 1964; Cramér and Leadbetter, 1967; Gelb and Analytic Sciences Corporation Technical Staff, 1974; Adler, 1981; Cliff and Ord, 1981; Cressie, 1991; Christakos, 1992, 2000; Christakos and Hristopulos, 1998).

2.4.8 Example: New Orleans hurricane protection system, Louisiana (USA)

In the aftermath of Hurricane Katrina, reliability analyses were conducted on the reconstructed New Orleans hurricane protection system (HPS) to understand the risks faced in future storms. A first-excursion or level crossing methodology was used to calculate the probability of failure in long embankment sections, following the approach proposed by VanMarcke (1977). This resulted in fragility curves for a reach of levee. The fragility curve gives the conditional probability of failure for known hurricane loads (i.e. surge and wave heights). Uncertainties in the hurricane loads were convolved with these fragility curves in a systems risk model to generate unconditional probabilities and subsequently risk when consequences were included.

As a first approximation, engineering performance models and calculations were adapted from the US Army Corps of Engineers' Design Memoranda describing the original design of individual levee reaches (USACE, 1972). Engineering parameter and model uncertainties were propagated through those calculations to obtain approximate fragility curves as a function of surge and wave loads. These results were later calibrated against analyses which applied more sophisticated stability models, and the risk assessments were updated.

A typical design profile of the levee system is shown in Figure 2.24. Four categories of uncertainty were included in the reliability analysis: geological and geotechnical uncertainties, involving the spatial distribution of soils and soil properties within and beneath the HPS; geotechnical stability modeling of levee performance; erosion uncertainties, involving the performance of levees and fills during overtopping; and mechanical equipment uncertainties, including gates, pumps, and other operating systems, and human operator factors affecting the performance of mechanical equipment.

The principal uncertainty contributing to probability of failure of the levee sections in the reliability analysis was soil engineering properties, specifically undrained strength, S_u, measured in Q-tests (UU tests). Uncertainties in soil engineering properties was presumed to be structured as in Figure 2.16,

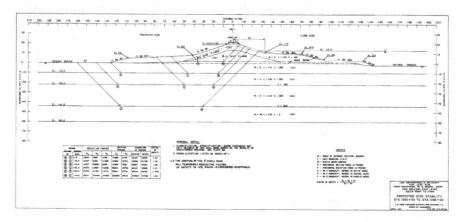

Figure 2.24 Typical design section from the USACE Design Memoranda for the New Orleans Hurricane Protection System (USACE, 1972).

and the variance of the uncertainty in soil properties was divided into four terms:

$$Var(S_u) = Var(x) + Var(e) + Var(m) + Var(b) \qquad (2.88)$$

in which $Var(.)$ is variance, S_u is measured undrained strength, x is the soil property *in situ*, e is measurement error (noise), m is the spatial mean (which has some error due to the statistical fluctuations of small sample sizes), and b is a model bias or calibration term caused by systematic errors in measuring the soil properties. Measured undrained strength for one reach, the New Orleans East lakefront levees, are shown as histograms in Figure 2.25. Test values larger than 750 PCF (36 kPa) were assumed to be local effects and removed from the statistics. The spatial pattern of soil variability was characterized by autocovariance functions in each region of the system and for each soil stratum (Figure 2.26). From the autocovariance analyses two conclusions were drawn: The measurement noise (or fine-scale variation) in the undrained strength data was estimated to be roughly 3/4 the total variance of the data (which was judged not unreasonable given the Q-test methods), and the autocovariance distance in the horizontal direction for both the clay and marsh was estimated to be on the order of 500 feet or more.

The reliability analysis was based on limiting equilibrium calculations. For levees, the analysis was based on General Design Memorandum (GDM) calculations of factor of safety against wedge instability (USACE, 1972)

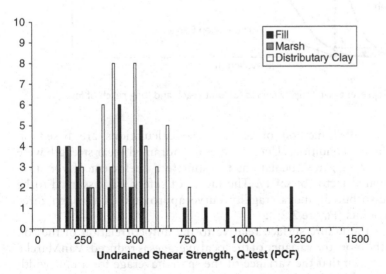

Figure 2.25 Histogram of Q-test (UU) undrained soil strengths, New Orleans East lakefront.

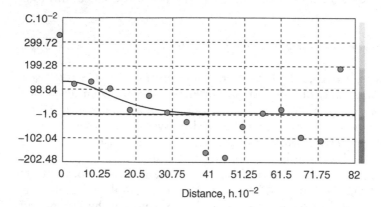

Figure 2.26 Representative autocovariance function for inter-distributary clay undrained strength (Q test), Orleans Parish, Louisiana.

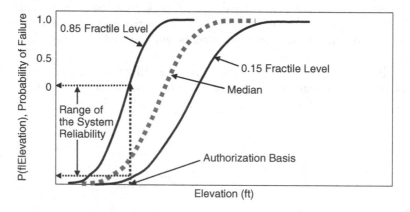

Figure 2.27 Representative fragility curves for unit reach and long reach of levee.

using the so-called method of planes. The calculations are based on undrained failure conditions. Uncertainties in undrained shear strength were propagated through the calculations to estimate a coefficient of variation in the calculated factor of safety. The factor of safety was assumed to be Normally distributed, and a fragility curve approximated through three calculation points (Figure 2.27).

The larger the failure surface relative to the autocorrelation of the soil properties, the more the variance of the local averages is reduced. VanMarcke (1977) has shown that the variance of the spatial average for a unit-width plain strain cross-section decreases approximately in proportion to (L/r_L), for $L > r_L$, in which L is the cross-sectional length of the failure surface, and

r_L is an equivalent autocovariance distance of the soil properties across the failure surface weighted for the relative proportion of horizontal and vertical segments of the surface. For the wedge failure modes this is approximately the vertical autocovariance distance. The variance across the full failure surface of width b along the axis of the levee is further reduced by averaging in the horizontal direction by an additional factor (b/r_H), for $b > r_H$, in which r_H is the horizontal autocovariance distance. At the same time that the variance of the average strength on the failure surface is reduced by the averaging process, so, too, the autocovariance function of this averaged process stretches out from that of the point-to-point variation.

For a failure length of approximately 500 feet along the levee axis and 30 feet deep, typical of those actually observed, with horizontal and vertical autocovariance distances of 500 feet and 10 feet, respectively, the corresponding variance reduction factors are approximately 0.75 for averaging over the cross-sectional length L, and between 0.73 and 0.85 for averaging over the failure length b, assuming either an exponential or squared-exponential (Gaussian) autocovariance. The corresponding reduction to the COV of soil strength based on averaging over the failure plane is the root of the product of these two factors, or between 0.74 and 0.8.

For a long levee, the chance of at least one failure is equivalent to the chance that the variations of the mean soil strength across the failure surface drop below that required for stability at least once along the length. VanMarcke demonstrated that this can be determined by considering the first crossings of a random process. The approximation to the probability of at least one failure as provided by VanMarcke was used in the present calculations to obtain probability of failure as a function of levee length.

2.5 Concluding comments

In this chapter we have described the importance of spatial variation in geotechnical properties and how such variation can be dealt with in a probabilistic analysis. Spatial variation consists essentially of two parts: an underlying trend and a random variation superimposed on it. The distribution of the variability between the trend and the random variation is a decision made by the analyst and is not an invariant function of nature.

The second-moment method is widely used to describe the spatial variation of random variation. Although not as commonly used in geotechnical practice, Bayesian estimation has many advantages over moment-based estimation. One of them is that it yields an estimate of the probabilities associated with the distribution parameters rather than a confidence interval that the data would be observed if the process were repeated.

Spatially varying properties are generally described by random fields. Although these can become extremely complicated, relatively simple models such as Gaussian random field have wide application. The last portion of the chapter demonstrates how they can be manipulated by differentiation, by transformation through linear processes, and by evaluation of excursions beyond specified levels.

Notes

1 It is true that not all soil or rock mass behaviors of engineering importance are governed by averages, for example, block failures in rock slopes are governed by the least favorably positioned and oriented joint. They are extreme values processes. Nevertheless, averaging soil or rock properties does reduce variance.
2 In early applications of geotechnical reliability, a great deal of work was focused on appropriate distributional forms for soil property variability, but this no longer seems a major topic. First, the number of measurements typical of site characterization programs is usually too few to decide confidently whether one distributional form provides a better fit than another, and, second, much of practical geotechnical reliability work uses second-moment characterizations rather than full distributional analysis, so distributional assumptions only come into play at the end of the work. Second-moment characterizations use only means, variances, and covariances to characterize variability, not full distributions.
3 In addition to aleatory and epistemic uncertainties, there are also uncertainties that have little to do with engineering properties and performance, yet which affect decisions. Among these is the set of objectives and attributes considered important in reaching a decision, the value or utility function defined over these attributes, and discounting for outcomes distributed in time. These are outside the present scope.

References

Adler, R. J. (1981). *The Geometry of Random Fields*. Wiley, Chichester.

Ang, A. H.-S. and Tang, W. H. (1975). *Probability Concepts in Engineering Planning and Design*. Wiley, New York.

Azzouz, A., Baligh, M. M. and Ladd, C. C. (1983). Corrected field vane strength for embankment design. *Journal of Geotechnical Engineering, ASCE*, 109(3), 730–4.

Baecher, G. B.(1980). Progressively censored sampling of rock joint traces. *Journal of the International Association of Mathematical Geologists*, 12(1), 33–40.

Baecher, G. B. (1987). Statistical quality control for engineered fills. *Engineering Guide*, U.S. Army Corps of Engineers, Waterways Experiment Station, GL-87-2.

Baecher, G. B., Ladd, C. C., Noiray, L. and Christian, J. T. (1997) Formal observational approach to staged loading. Transportation Research Board Annual Meeting, Washington, D.C.

Barnett, V. (1982). *Comparative Statistical Inference*. John Wiley and Sons, London.

Berger, J. O. (1993). *Statistical Decision Theory and Bayesian Analysis*. Springer, New York.

Berger, J. O., De Oliveira, V. and Sanso, B. (2001). Objective Bayesian analysis of spatially correlated data. *Journal of the American Statistical Association*, 96(456), 1361–74.

Bjerrum, L. (1972). Embankments on soft ground. In *ASCE Conference on Performance of Earth and Earth-Supported Structures*, Purdue, pp. 1–54.

Bjerrum, L. (1973). Problems of soil mechanics and construction on soft clays. In *Eighth International Conference on Soil Mechanics and Foundation Engineering*, Moscow. Spinger, New York, pp. 111–59.

Chiasson, P., Lafleur, J., Soulie, M. and Law, K. T. (1995). Characterizing spatial variability of a clay by geostatistics. *Canadian Geotechnical Journal*, 32, 1–10.

Christakos, G. (1992). *Random Field Models in Earth Sciences*. Academic Press, San Diego.

Christakos, G. (2000). *Modern Spatiotemporal Geostatistics*. Oxford University Press, Oxford.

Christakos, G. and Hristopulos, D. T. (1998). *Spatiotemporal Environmental Health Modelling : A Tractatus Stochasticus*. Kluwer Academic, Boston, MA.

Christian, J. T., Ladd, C. C. and Baecher, G. B. (1994). Reliability applied to slope stability analysis. *Journal of Geotechnical Engineering*, ASCE, 120(12), 2180–207.

Cliff, A. D. and Ord, J. K. (1981). *Spatial Processes: Models & Applications*. Pion, London.

Cramér, H. and Leadbetter, M. R. (1967). *Stationary and Related Stochastic Processes; Sample Function Properties and their Applications*. Wiley, New York.

Cressie, N. A. C. (1991). *Statistics for Spatial Data*. Wiley, New York.

Davis, J. C. (1986). *Statistics and Data Analysis in Geology*. Wiley, New York.

DeGroot, D. J. and Baecher, G. B. (1993). Estimating autocovariance of in situ soil properties. *Journal Geotechnical Engineering Division*, ASCE, 119(1), 147–66.

Elishakoff, I. (1999). *Probabilistic Theory of Structures*. Dover Publications, Mineola, NY.

Fuleihan, N. F. and Ladd, C. C. (1976). Design and performance of Atchafalaya flood control levees. *Research Report R76-24*, Department of Civil Engineering, Massachusetts Institute of Technology, Cambridge, MA.

Gelb, A. and Analytic Sciences Corporation Technical Staff (1974). *Applied Optimal Estimation*. M.I.T. Press, Cambridge, MA.

Gelhar, L. W. (1993). *Stochastic Subsurface Hydrology*. Prentice-Hall, Englewood Cliffs, NJ.

Hartford, D. N. D. (1995). How safe is your dam? Is it safe enough? *MEP11-5*, BC Hydro, Burnaby, BC.

Hartford, D. N. D. and Baecher, G. B. (2004). *Risk and Uncertainty in Dam Safety*. Thomas Telford, London.

Hilldale, C. (1971). A probabilistic approach to estimating differential settlement. MS thesis, Civil Engineering, Massachusetts Institute of Technology, Cambridge, MA.

Javete, D. F. (1983). A simple statistical approach to differential settlements on clay. Ph. D. thesis, Civil Engineering, University of California, Berkeley.

Journel, A. G. and Huijbregts, C. (1978). *Mining Geostatistics*. Academic Press, London.

Jowett, G. H. (1952). The accuracy of systematic simple from conveyer belts. *Applied Statistics*, 1, 50–9.

Kitanidis, P. K. (1985). Parameter uncertainty in estimation of spatial functions: Bayesian analysis. *Water Resources Research*, 22(4), 499–507.

Kitanidis, P. K. (1997). *Introduction to Geostatistics: Applications to Hydrogeology*. Cambridge University Press, Cambridge.

Kolmogorov, A. N. (1941). The local structure of turbulence in an incompressible fluid at very large Reynolds number. *Doklady Adademii Nauk SSSR*, 30, 301–5.

Lacasse, S. and Nadim, F. (1996). Uncertainties in characteristic soil properties. In *Uncertainty in the Geological Environment*, ASCE specialty conference, Madison, WI. ASCE, Reston, VA, pp. 40–75.

Ladd, C.C. and Foott, R. (1974). A new design procedure for stability of soft clays. *Journal of Geotechnical Engineering*, 100(7), 763–86.

Ladd, C. C. (1975). Foundation design of embankments constructed on connecticut valley varved clays. *Report R75-7*, Department of Civil Engineering, Massachusetts Institute of Technology, Cambridge, MA.

Ladd, C. C., Dascal, O., Law, K. T., Lefebrve, G., Lessard, G., Mesri, G. and Tavenas, F. (1983). *Report of the subcommittee on embankment stability— annexe II*, Committe of specialists on Sensitive Clays on the NBR Complex. Societé d'Energie de la Baie James, Montreal.

Lambe, T.W. and Associates (1982). Earthquake risk to patio 4 and site 400. *Report to Tonen Oil Corporation*, Longboat Key, FL.

Lambe, T. W. and Whitman, R. V. (1979). *Soil Mechanics, SI Version*. Wiley, New York.

Lee, I. K., White, W. and Ingles, O. G. (1983). *Geotechnical Engineering*. Pitman, Boston.

Lumb, P. (1966). The variability of natural soils. *Canadian Geotechnical Journal*, 3, 74–97.

Lumb, P. (1974). Application of statistics in soil mechanics. In *Soil Mechanics: New Horizons*, Ed. I. K. Lee. Newnes-Butterworth, London, pp. 44–112.

Mardia, K. V. and Marshall, R. J. (1984). Maximum likelihood estimation of models for residual covariance in spatial regression. *Biometrika*, 71(1), 135–46.

Matérn, B. (1960). Spatial Variation. *49(5)*, Meddelanden fran Statens Skogsforskningsinstitut.

Matérn, B. (1986). *Spatial Variation*. Springer, Berlin.

Matheron, G. (1971). *The Theory of Regionalized Variables and Its Application – Spatial Variabilities of Soil and Landforms*. Les Cahiers du Centre de Morphologie mathematique 8, Fontainebleau.

Papoulis, A. and Pillai, S. U. (2002). *Probability, Random Variables, and Stochastic Processes*, 4th ed. McGraw-Hill, Boston, MA.

Parzen, E. (1964). *Stochastic Processes*. Holden-Day, San Francisco, CA.

Parzen, E. (1992). *Modern Probability Theory and its Applications*. Wiley, New York.

Phoon, K. K. (2006a). Bootstrap estimation of sample autocorrelation functions. In *GeoCongress*, Atlanta, ASCE.

Phoon, K. K. (2006b). Modeling and simulation of stochastic data. In *GeoCongress*, Atlanta, ASCE.

Phoon, K. K. and Fenton, G. A. (2004). Estimating sample autocorrelation functions using bootstrap. In *Proceedings, Ninth ASCE Specialty Conference on Probabilistic Mechanics and Structural Reliability*, Albuquerque.

Phoon, K. K. and Kulhawy, F. H. (1996). On quantifying inherent soil variability. *Uncertainty in the Geologic Environment*, ASCE specialty conference, Madison, WI. ASCE, Reston, VA, pp. 326–40.

Soong, T. T. and Grigoriu, M. (1993). *Random Vibration of Mechanical and Structural Systems*. Prentice Hall, Englewood Cliffs, NJ.

Soulie, M. and Favre, M. (1983). Analyse geostatistique d'un noyau de barrage tel que construit. *Canadian Geotechnical Journal*, 20, 453–67.

Soulie, M., Montes, P. and Silvestri, V. (1990). Modeling spatial variability of soil parameters. *Canadian Geotechnical Journal*, 27, 617–30.

Switzer, P. (1995). Spatial interpolation errors for monitoring data. *Journal of the American Statistical Association*, 90(431), 853–61.

Tang, W. H. (1979). Probabilistic evaluation of penetration resistance. *Journal of the Geotechnical Engineering Division, ASCE*, 105(10): 1173–91.

Taylor, G. I. (1921). Diffusion by continuous movements. *Proceeding of the London Mathematical. Society (2)*, 20, 196–211.

USACE (1972). New Orleans East Lakefront Levee Paris Road to South Point Lake Pontchartrain. *Barrier Plan DM 2 Supplement 5B*, USACE New Orleans District, New Orleans.

VanMarcke, E. (1983). *Random Fields, Analysis and Synthesis*. MIT Press, Cambridge, MA.

VanMarcke, E. H. (1977). Reliability of earth slopes. *Journal of the Geotechnical Engineering Division, ASCE*, 103(GT11), 1247–65.

Weinstock, H. (1963). The description of stationary random rate processes. *E-1377*, Massachusetts Institute of Technology, Cambridge, MA.

Wu, T. H. (1974). Uncertainty, safety, and decision in soil engineering. *Journal Geotechnical Engineering Division, ASCE*, 100(3), 329–48.

Yaglom, A. M. (1962). *An Introduction to the Theory of Random Functions*. Prentice-Hall, Englewood Cliffs, NJ.

Zellner, A. (1971). *An Introduction to Bayesian Inference in Econometrics*. Wiley, New York.

Chapter 3

Practical reliability approach using spreadsheet

Bak Kong Low

3.1 Introduction

In a review on first-order second-moment reliability methods, USACE (1999) rightly noted that a potential problem with both the Taylor's series method and the point estimate method is their lack of invariance for nonlinear performance functions. The document suggested the more general Hasofer–Lind reliability index (Hasofer and Lind, 1974) as a better alternative, but conceded that "many published analyses of geotechnical problems have not used the Hasofer–Lind method, probably due to its complexity, especially for implicit functions such as those in slope stability analysis," and that "the most common method used in Corps practice is the Taylor's series method, based on a Taylor's series expansion of the performance function about the expected values."

A survey of recent papers on geotechnical reliability analysis reinforces the above USACE observation that although the Hasofer–Lind index is perceived to be more consistent than the Taylor's series mean value method, the latter is more often used.

This chapter aims to overcome the computational and conceptual barriers of the Hasofer–Lind index, for correlated normal random variables, and the first-order reliability method (FORM), for correlated nonnormals, in the context of three conventional geotechnical design problems. Specifically, the conventional bearing capacity model involving two random variables is first illustrated, to elucidate the procedures and concepts. This is followed by a reliability-based design of an anchored sheet pile wall involving six random variables, which are first treated as correlated normals, then as correlated nonnormals. Finally, reliability analysis with search for critical noncircular slip surface based on a reformulated Spencer method is presented. This probabilistic slope stability example includes testing the robustness of search for noncircular critical slip surface, modeling lognormal random variables, deriving probability density functions from reliability indices, and comparing results inferred from reliability indices with Monte Carlo simulations.

The expanding ellipsoidal perspective of the Hasofer–Lind reliability index and the practical reliability approach using object-oriented constrained optimization in the ubiquitous spreadsheet platform were described in Low and Tang (1997a, 1997b), and extended substantially in Low and Tang (2004) by testing robustness for various nonnormal distributions and complicated performance functions, and by providing enhanced operational convenience and versatility.

Reasonable statistical properties are assumed for the illustrative cases presented in this chapter; actual determination of the statistical properties is not covered. Only parametric uncertainty is considered and model uncertainty is not dealt with. Hence this chapter is concerned with reliability method and perspectives, and not reliability in its widest sense. The focus is on introducing an efficient and rational design approach using the ubiquitous spreadsheet platform.

The spreadsheet reliability procedures described herein can be applied to stand-alone numerical (e.g. finite element) packages via the established response surface method (which itself is straightforward to implement in the ubiquitous spreadsheet platform). Hence the applicability of the reliability approach is not confined to models which can be formulated in the spreadsheet environment.

3.2 Reliability procedure in spreadsheet and expanding ellipsoidal perspective

3.2.1 A simple hands-on reliability analysis

The proposed spreadsheet reliability evaluation approach will be illustrated first for a case with two random variables. Readers who want a better understanding of the procedure and deeper appreciation of the ellipsoidal perspective are encouraged to go through the procedure from scratch on a blank Excel worksheet. After that, some Excel files for hands-on and deeper appreciation can be downloaded from http://alum.mit.edu/www/bklow.

The example concerns the bearing capacity of a strip footing sustaining non-eccentric vertical load. Extensions to higher dimensions and more complicated scenarios are straightforward.

With respect to bearing capacity failure, the performance function (*PerFn*) for a strip footing, in its simplest form, is:

$$PerFn = q_{\mathrm{u}} - q \tag{3.1a}$$

$$\text{where} \quad q_{\mathrm{u}} = cN_c + p_{\mathrm{o}}N_q + \frac{B}{2}\gamma N_\gamma \tag{3.1b}$$

in which q_{u} is the ultimate bearing capacity, q the applied bearing pressure, c the cohesion of soil, p_{o} the effective overburden pressure at foundation level, B the foundation width, γ the unit weight of soil below the base of foundation, and N_c, N_q, and N_γ are bearing capacity factors, which are established functions of the friction angle (ϕ) of soil:

$$N_q = e^{\pi \tan\phi} \tan^2 \left(45 + \frac{\phi}{2}\right) \tag{3.2a}$$

$$N_c = \left(N_q - 1\right) \cot(\phi) \tag{3.2b}$$

$$N_\gamma = 2\left(N_q + 1\right) \tan\phi \tag{3.2c}$$

Several expressions for N_γ exist. The above N_γ is attributed to Vesic in Bowles (1996).

The statistical parameters and correlation matrix of c and ϕ are shown in Figure 3.1. The other parameters in Equations (3.1a) and (3.1b) are assumed known with values $q = Q_v/B = (200 \text{ kN/m})/B$, $p_o = 18 \text{ kPa}$, $B = 1.2 \text{ m}$, and $\gamma = 20 \text{ kN/m}^3$. The parameters c and ϕ in Equations (3.1) and (3.2) read their values from the column labeled x^*, which were initially set equal to the mean values. These x^* values, and the functions dependent on them,

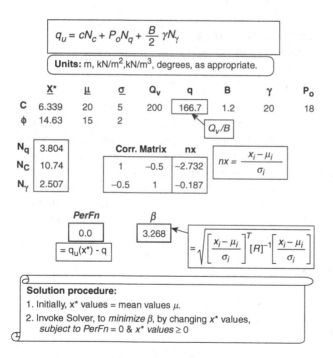

Figure 3.1 A simple illustration of reliability analysis involving two correlated random variables which are normally distributed.

change during the optimization search for the most probable failure point. Subsequent steps are:

1 The formula of the cell labeled β in Figure 3.1 is Equation (3.5b) in Section 3.2.3: "=sqrt(mmult(transpose(nx), mmult(minverse(crmat), nx)))." The arguments *nx* and *crmat* are entered by selecting the corresponding numerical cells of the column vector $(x_i - \mu_i)/\sigma_i$ and the correlation matrix, respectively. This array formula is then entered by pressing "Enter" while holding down the "Ctrl"and "Shift" keys. Microsoft Excel's built-in matrix functions *mmult*, *transpose*, and *minverse* have been used in this step. Each of these functions contains program codes for matrix operations.

2 The formula of the performance function is $g(\mathbf{x}) = q_u - q$, where the equation for q_u is Equation (3.1b) and depends on the x^* values.

3 Microsoft Excel's built-in constrained optimization program Solver is invoked (via Tools\Solver), to *Minimize* β, *By Changing* the x^* values, *Subject To PerFn* ≤ 0, and x^* values ≥ 0. (If used for the first time, Solver needs to be activated once via Tools\Add-ins\Solver Add-in.)

The β value obtained is 3.268. The spreadsheet approach is simple and intuitive because it works in the original space of the variables. It does not involve the orthogonal transformation of the correlation matrix, and iterative numerical partial derivatives are done automatically on spreadsheet objects which may be implicit or contain codes.

The following paragraphs briefly compares lumped factor of safety approach, partial factors approach, and FORM approach, and provide insights on the meaning of reliability index in the original space of the random variables. More details can be found in Low (1996, 2005a), Low and Tang (1997a, 2004), and other documents at http://alum.mit.edu/www/bklow.

3.2.2 Comparing lumped safety factor and partial factors approaches with reliability approach

For the bearing capacity problem of Figure 3.1, a long-established deterministic approach evaluates the lumped factor of safety (F_s) as:

$$F_s = \frac{q_u - p_o}{q - p_o} = f(c, \phi, \ldots) \tag{3.3}$$

where the symbols are as defined earlier. If $c = 20$ kPa and $\phi = 15°$, and with the values of Q_v, B, γ and p_o as shown in Figure 3.1, then the factor of

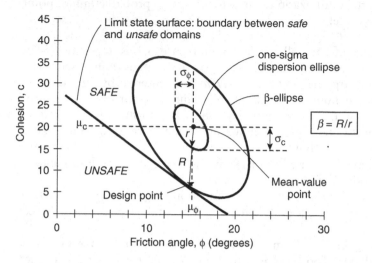

Figure 3.2 The design point, the mean-value point, and expanding ellipsoidal perspective of the reliability index in the original space of the random variables.

safety is $F_s \approx 2.0$, by Equation (3.3). In the two-dimensional space of c and ϕ (Figure 3.2) one can plot the F_s contours for different combinations of c and ϕ, including the $F_s = 1.0$ curve which separates the unsafe combinations from the safe combinations of c and ϕ. The average point ($c = 20$ kPa and $\phi = 15°$) is situated on the contour (not plotted) of $F_s = 2.0$. Design, by the lumped factor of safety approach, is considered satisfactory with respect to bearing capacity failure if the factor of safety by Equation (3.3) is not smaller than a certain value (e.g. when $F_s \geq 2.0$).

A more recent and logical approach (e.g. Eurocode 7) applies partial factors to the parameters in the evaluation of resistance and loadings. Design is acceptable if:

Bearing capacity (based on reduced c and ϕ) $\geq$ Applied pressure

(amplified)   (3.4)

A third approach is reliability-based design, where the uncertainties and correlation structure of the parameters are represented by a one-standard-deviation dispersion ellipsoid (Figure 3.2) centered at the mean-value point, and safety is gaged by a reliability index which is the shortest distance (measured in units of directional standard deviations, R/r) from the safe mean-value point to the most probable failure combination of parameters ("the design point") on the limit state surface (defined by $F_s = 1.0$, for the problem in hand). Furthermore, the probability of failure (P_f)

can be estimated from the reliability index β using the established equation $P_f = 1 - \Phi(\beta) = \Phi(-\beta)$, where Φ is the cumulative distribution (CDF) of the standard normal variate. The relationship is exact when the limit state surface is planar and the parameters follow normal distributions, and approximate otherwise.

The merits of a reliability-based design over the lumped factor-of-safety design is illustrated in Figure 3.3a, in which case A and case B (with different average values of effective shear strength parameters c' and ϕ') show the same values of lumped factor of safety, yet case A is clearly safer than case B. The higher reliability of case A over case B will correctly be revealed when the reliability indices are computed. On the other hand, a slope may have a computed lumped factor of safety of 1.5, and a particular foundation (with certain geometry and loadings) in the same soil may have a computed lumped factor of safety of 2.5, as in case C of Figure 3.3(b). Yet a reliability analysis may show that they both have similar levels of reliability.

The design point (Figure 3.2) is the most probable failure combination of parametric values. The ratios of the respective parametric values at the center of the dispersion ellipsoid (corresponding to the mean values) to those at the design point are similar to the partial factors in limit state design, except that these factored values at the design point are arrived at automatically (and as by-products) via spreadsheet-based constrained optimization. The reliability-based approach is thus able to reflect varying parametric sensitivities from case to case in the same design problem (Figure 3.3a) and across different design realms (Figure 3.3b).

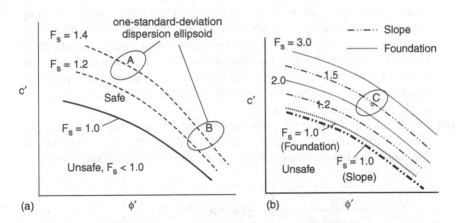

Figure 3.3 Schematic scenarios showing possible limitations of lumped factor of safety: (a) Cases A and B have the same lumped $F_s = 1.4$, but Case A is clearly more reliable than Case B; (b) Case C may have $F_s = 1.5$ for a slope and $F_s = 2.5$ for a foundation, and yet have similar levels of reliability.

3.2.3 Hasofer–Lind index reinterpreted via expanding ellipsoid perspective

The matrix formulation (Veneziano, 1974; Ditlevsen, 1981) of the Hasofer–Lind index β is:

$$\beta = \min_{\mathbf{x} \in F} \sqrt{(\mathbf{x} - \boldsymbol{\mu})^T \mathbf{C}^{-1} (\mathbf{x} - \boldsymbol{\mu})} \tag{3.5a}$$

or, equivalently:

$$\beta = \min_{\mathbf{x} \in F} \sqrt{\left[\frac{x_i - \mu_i}{\sigma_i}\right]^T [\mathbf{R}]^{-1} \left[\frac{x_i - \mu_i}{\sigma_i}\right]} \tag{3.5b}$$

where $\mathbf{x}$ is a vector representing the set of random variables x_i, $\boldsymbol{\mu}$ the vector of mean values μ_i, $\mathbf{C}$ the covariance matrix, $\mathbf{R}$ the correlation matrix, σ_i the standard deviation, and F the failure domain. Low and Tang (1997b; 2004) used Equation (3.5b) in preference to Equation (3.5a) because the correlation matrix $\mathbf{R}$ is easier to set up, and conveys the correlation structure more explicitly than the covariance matrix $\mathbf{C}$. Equation (3.5b) was entered in step (1) above.

The "x^*" values obtained in Figure 3.1 represent the most probable failure point on the limit state surface. It is the point of tangency (Figure 3.2) of the expanding dispersion ellipsoid with the bearing capacity limit state surface. The following may be noted:

(a) The x^* values shown in Figure 3.1 render Equation (3.1a) (*PerFn*) equal to zero. Hence the point represented by these x^* values lies on the bearing capacity limit state surface, which separates the safe domain from the unsafe domain. The one-standard-deviation ellipse and the β-ellipse in Figure 3.2 are tilted because the correlation coefficient between c and ϕ is -0.5 in Figure 3.1. The *design point* in Figure 3.2 is where the expanding dispersion ellipse touches the limit state surface, at the point represented by the x^* values of Figure 3.1.

(b) As a multivariate normal dispersion ellipsoid expands, its expanding surfaces are contours of decreasing probability values, according to the established probability density function of the multivariate normal distribution:

$$f(\mathbf{x}) = \frac{1}{(2\pi)^{\frac{n}{2}} |C|^{0.5}} \exp\left[-\frac{1}{2}(\mathbf{x} - \boldsymbol{\mu})^T \mathbf{C}^{-1} (\mathbf{x} - \boldsymbol{\mu})\right] \tag{3.6a}$$

$$= \frac{1}{(2\pi)^{\frac{n}{2}} |C|^{0.5}} \exp\left[-\frac{1}{2}\beta^2\right] \tag{3.6b}$$

where β is defined by Equation (3.5a) or (3.5b), without the "min." Hence, to minimize β (or β^2 in the above multivariate normal distribution) is to maximize the value of the multivariate normal probability density function, and to find the smallest ellipsoid tangent to the limit state surface is equivalent to finding the most probable failure point (the *design point*). This intuitive and visual understanding of the *design point* is consistent with the more mathematical approach in Shinozuka (1983, equations 4, 41, and associated figure), in which all variables were transformed into their standardized forms and the limit state equation had also to be written in terms of the standardized variables. The differences between the present original space versus Shinozuka's standardized space of variables will be further discussed in (h) below.

(c) Therefore the design point, being the first point of contact between the expanding ellipsoid and the limit state surface in Figure 3.2, is the most probable failure point with respect to the safe mean-value point at the centre of the expanding ellipsoid, where $F_s \approx 2.0$ against bearing capacity failure. The reliability index β is the axis ratio (R/r) of the ellipse that touches the limit state surface and the one-standard-deviation dispersion ellipse. By geometrical properties of ellipses, this co-directional axis ratio is the same along any "radial" direction.

(d) For each parameter, the ratio of the mean value to the x^* value is similar in nature to the partial factors in limit state design (e.g. Eurocode 7). However, in a reliability-based design one does not specify the partial factors. The design point values (x^*) are determined automatically and reflect sensitivities, standard deviations, correlation structure, and probability distributions in a way that prescribed partial factors cannot reflect.

(e) In Figure 3.1, the mean value point, at 20 kPa and 15°, is safe against bearing capacity failure; but bearing capacity failure occurs when the c and ϕ values are decreased to the values shown: (6.339, 14.63). The distance from the safe mean-value point to this most probable failure combination of parameters, in units of directional standard deviations, is the reliability index β, equal to 3.268 in this case.

(f) The probability of failure (P_f) can be estimated from the reliability index β. Microsoft Excel's built-in function NormSDist(.) can be used to compute $\Phi(.)$ and hence P_f. Thus for the bearing capacity problem of Figure 3.1, $P_f = \text{NormSDist}(-3.268) = 0.054\%$. This value compares remarkably well with the range of values $0.051-0.060\%$ obtained from several Monte Carlo simulations each with 800,000 trials using the commercial simulation software @RISK (http://www.palisade.com). The correlation matrix was accounted for in the simulation. The excellent agreement between 0.054% from reliability index and the range

0.051−0.060% from Monte Carlo simulation is hardly surprising given the almost linear limit state surface and normal variates shown in Figure 3.2. However, for the anchored wall shown in the next section, where six random variables are involved and nonnormal distributions are used, the six-dimensional equivalent hyperellispoid and the limit state hypersurface can only be perceived in the mind's eye. Nevertheless, the probabilities of failure inferred from reliability indices are again in close agreement with Monte Carlo simulations. Computing the reliability index and $P_f = \Phi(-\beta)$ by the present approach takes only a few seconds. In contrast, the time needed to obtain the probability of failure by Monte Carlo simulation is several orders of magnitude longer, particularly when the probability of failure is small and many trials are needed. It is also a simple matter to investigate sensitivities by re-computing the reliability index β (and P_f) for different mean values and standard deviations in numerous what-if scenarios.

(g) The probability of failure as used here means the probability that, in the presence of parametric uncertainties in c and ϕ, the factor of safety, Equation (3.3), will be ≤ 1.0, or, equivalently, the probability that the performance function, Equation (3.1a), will be ≤ 0.

(h) Figure 3.2 defines the reliability index β as the dimensionless ratio R/r, in the direction from the mean-value point to the design point. This is the axis ratio of the β-ellipsoid (tangential to the limit state surface) to the one-standard-deviation dispersion ellipsoid. This axis ratio is dimensionless and independent of orientation, when R and r are co-directional. This axis-ratio interpretation in the original space of the variables overcomes a drawback in Shinozuka's (1983) standardized variable space that "the interpretation of β as the shortest distance between the origin (of the standardized space) and the (transformed) limit state surface is no longer valid" if the random variables are correlated. A further advantage of the original space, apart from its intuitive transparency, is that it renders feasible and efficient the two computational approaches involving nonnormals as presented in Low & Tang (2004) and (2007), respectively.

3.3 Reliability-based design of an anchored wall

This section illustrates reliability-based design of anchored sheet pile wall, drawing material from Low (2005a, 2005b). The analytical formulations in a deterministic anchored wall design are the basis of the performance function in a probabilistic-based design. Hence it is appropriate to briefly describe the deterministic approach, prior to extending it to a probabilistic-based design. An alternative design approach is described in BS8002 (1994).

3.3.1 Deterministic anchored wall design based on lumped factor and partial factors

The deterministic geotechnical design of anchored walls based on the free earth support analytical model was lucidly presented in Craig (1997). An example is the case in Figure 3.4, where the relevant soil properties are the effective angle of shearing resistance ϕ', and the interface friction angle δ between the retained soil and the wall. The characteristic values are $c' = 0$, $\phi' = 36°$ and $\delta = \frac{1}{2}\phi'$. The water table is the same on both sides of the wall. The bulk unit weight of the soil is 17 kN/m³ above the water table and 20 kN/m³ below the water table. A surcharge pressure $q_s = 10$ kN/m² acts at the top of the retained soil. The tie rods act horizontally at a depth 1.5 m below the top of the wall.

In Figure 3.4, the active earth pressure coefficient K_a is based on the Coulomb-wedge closed-form equation, which is practically the same as the Kerisel–Absi active earth pressure coefficient (Kerisel and Absi, 1990). The passive earth pressure coefficient K_p is based on polynomial equations

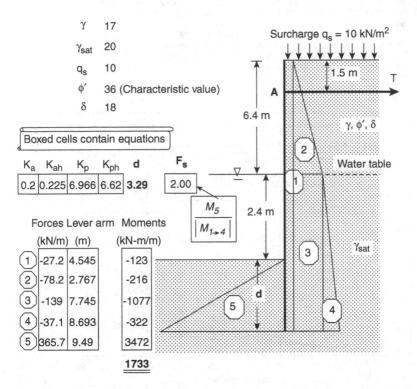

Figure 3.4 Deterministic design of embedment depth d based on a lumped factor of safety of 2.0.

(Figure 3.5) fitted to the values of Kerisel–Absi (1990), for a wall with a vertical back and a horizontal retained soil surface.

The required embedment depth d of 3.29 m in Figure 3.4 – for a lumped factor of safety of 2.0 against rotational failure around anchor point A – agrees with Example 6.9 of Craig (1997).

If one were to use the limit state design approach, with a partial factor of 1.2 for the characteristic shear strength, one enters $\tan^{-1}(\tan\phi'/1.2) = 31°$ in the cell of ϕ', and changes the embedment depth d until the summation of moments is zero. A required embedment depth d of 2.83 m is obtained, in agreement with Craig (1997).

```
Function KpKeriselAbsi(phi, del)
'Passive pressure coefficient Kp, for vertical wall back and horizontal retained fill
'Based on Tables in Kerisel & Absi (1990), for beta = 0, lamda = 0, and
' del/phi = 0,0.33, 0.5, 0.66, 1.00
x = del / phi
Kp100=0.00007776 * phi ^ 4 - 0.006608 * phi ^ 3 + 0.2107 * phi ^ 2 - 2.714 * phi + 13.63
Kp66=0.00002611 * phi ^ 4 - 0.002113 * phi ^ 3 + 0.06843 * phi ^ 2 - 0.8512 * phi + 5.142
Kp50 = 0.00001559 * phi ^ 4 - 0.001215 * phi ^ 3 + 0.03886 * phi ^ 2 - 0.4473 * phi + 3.208
Kp33 = 0.000007318 * phi ^ 4 - 0.0005195 * phi ^ 3 + 0.0164 * phi ^ 2 - 0.1483 * phi + 1.798
Kp0 = 0.000002636 * phi ^ 4 - 0.0002201 * phi ^ 3 + 0.008267 * phi ^ 2 - 0.0714 * phi + 1.507
Select Case x
Case 0.66 To 1: Kp = Kp66 + (x - 0.66) / 1 - 0.66) * (Kp100 - Kp66)
Case 0.5 To 0.66: Kp = Kp50 + (x - 0.5) / (0.66 - 0.5) * (Kp66 - Kp50)
Case 0.33 To 0.5: Kp = Kp33 + (x - 0.33) / (0.5 - 0.33) * (Kp50 - Kp33)
Case 0 To 0.33: Kp = Kp0 + x / 0.33 * (Kp33 - Kp0)
End Select
KpKeriselAbsi = Kp
End Function
```

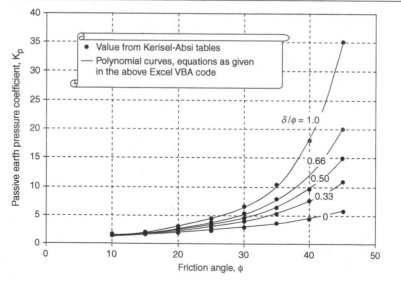

Figure 3.5 User-created Excel VBA function for K_p.

The partial factors in limit state design are applied to the characteristic values, which are themselves conservative estimates and not the most probable or average values. Hence there is a two-tier nested safety: first during the conservative estimate of the characteristic values, and then when the partial factors are applied to the characteristic values. This is evident in Eurocode 7, where Section 2.4.3 clause (5) states that the characteristic value of a soil or rock parameter shall be selected as a cautious estimate of the value affecting the occurrence of the limit state. Clause (7) further states that characteristic values may be lower values, which are less than the most probable values, or upper values, which are greater, and that for each calculation, the most unfavorable combination of lower and upper values for independent parameters shall be used.

The above Eurocode 7 recommendations imply that the characteristic value of ϕ' (36°) in Figure 3.4 is lower than the mean value of ϕ'. Hence in the reliability-based design of the next section, the mean value of ϕ' adopted is higher than the characteristic value of Figure 3.4.

While characteristic values and partial factors are used in limit state design, mean values (not characteristic values) are used with standard deviations and correlation matrix in a reliability-based design.

3.3.2 From deterministic to reliability-based anchored wall design

The anchored sheet pile wall will be designed based on reliability analysis (Figure 3.6). As mentioned earlier, the *mean value* of ϕ' in Figure 3.6 is larger – 38° is assumed – than the *characteristic value* of Figure 3.4. In total there are six normally distributed random variables, with mean and standard deviations as shown. Some correlations among parameters are assumed, as shown in the correlation matrix. For example, it is judged logical that the unit weights γ and γ_{sat} should be positively correlated, and that each is also positively correlated to the angle of friction ϕ', since $\gamma' = \gamma_{sat} - \gamma_w$.

The analytical formulations based on force and moment equilibrium in the deterministic analysis of Figure 3.4 are also required in a reliability analysis, but are expressed as limit state functions or performance functions: "= Sum (Moments$_{1 \to 5}$)."

The array formula in cell β of Figure 3.6 is as described in step 1 of the bearing capacity example earlier in this chapter.

Given the uncertainties and correlation structure in Figure 3.6, we wish to find the required total wall height H so as to achieve a reliability index of 3.0 against rotational failure about point "A." Initially the column x^* was given the mean values. Microsoft Excel's built-in constrained optimization tool Solver was then used to minimize β, by changing (automatically) the x^* column, subject to the constraint that

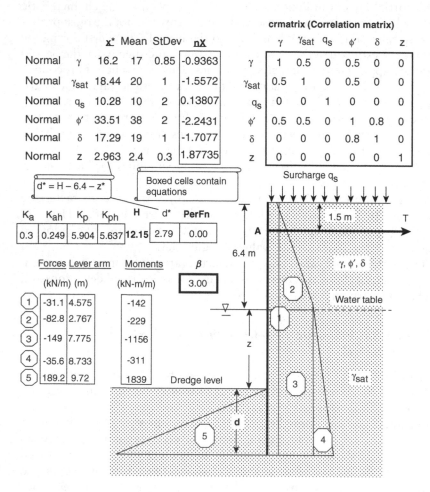

Figure 3.6 Design total wall height for a reliability index of 3.0 against rotational failure. Dredge level and hence z and d are random variables.

the cell *PerFn* be equal to zero. The solution (Figure 3.6) indicates that a total height of 12.15 m would give a reliability index of 3.0 against rotational failure. With this wall height, the mean-value point is safe against rotational failure, but rotational failure occurs when the mean values descend/ascend to the values indicated under the *x** column. These *x** values denote the *design point* on the limit state surface, and represent the most likely combination of parametric values that will cause failure. The distance between the mean-value point and the design point, in units of directional standard deviations, is the Hasofer–Lind reliability index.

As noted in Simpson and Driscoll (1998: 81, 158), clause 8.3.2.1 of Eurocode 7 requires that an "overdig" allowance shall be made for walls which rely on passive resistance. This is an allowance "for the unforeseen activities of nature or humans who have no technical appreciation of the stability requirements of the wall." For the case in hand, the reliability analysis in Figure 3.6 accounts for uncertainty in z, requiring only the mean value and the standard deviation of z (and its distribution type, if not normal) to be specified. The expected embedment depth is $d = 12.15 - 6.4 - \mu_z = 3.35$ m. At the failure combination of parametric values the design value of z is $z^* = 2.9632$, and $d^* = 12.15 - 6.4 - z^* = 2.79$ m. This corresponds to an "overdig" allowance of 0.56 m. Unlike Eurocode 7, this "overdig" is determined automatically, and reflects uncertainties and sensitivities from case to case in a way that specified "overdig" cannot. Low (2005a) illustrates and discusses this automatic probabilistic overdig allowance in a reliability-based design.

The nx column indicates that, for the given mean values and uncertainties, rotational stability is, not surprisingly, most sensitive to ϕ' and the dredge level (which affects z and d and hence the passive resistance). It is least sensitive to uncertainties in the surcharge q_s, because the average value of surcharge (10 kN/m^2) is relatively small when compared with the over 10 m thick retained fill. Under a different scenario where the surcharge is a significant player, its sensitivity scale could conceivably be different. It is also interesting to note that at the design point where the six-dimensional dispersion ellipsoid touches the limit state surface, both unit weights γ and γ_{sat} (16.20 and 18.44, respectively) are lower than their corresponding mean values, contrary to the expectation that higher unit weights will increase active pressure and hence greater instability. This apparent paradox is resolved if one notes that smaller γ_{sat} will (via smaller γ') reduce passive resistance, smaller ϕ' will cause greater active pressure and smaller passive pressure, and that γ, γ_{sat}, and ϕ' are logically positively correlated.

In a reliability-based design (such as the case in Figure 3.6) one does not prescribe the ratios *mean/x** – such ratios, or ratios of (*characteristic values*)/*x**, are prescribed in limit state design – but leave it to the expanding dispersion ellipsoid to seek the most probable failure point on the limit state surface, a process which automatically reflects the sensitivities of the parameters. The ability to seek the most-probable design point without presuming any partial factors and to automatically reflect sensitivities from case to case is a desirable feature of the reliability-based design approach. The sensitivity measures of parameters may not always be obvious from a priori reasoning. A case in point is the strut with complex supports analyzed in Low and Tang (2004: 85), where the mid-span spring stiffness k_3 and the rotational stiffness λ_1 at the other end both turn out to have surprisingly negligible sensitivity weights; this sensitivity conclusion was confirmed by previous elaborate deterministic parametric plots. In contrast, reliability analysis achieved the same conclusion relatively effortlessly.

The spreadsheet-based reliability-based design approach illustrated in Figure 3.6 is a more practical and relatively transparent intuitive approach that obtains the same solution as the classical Hasofer–Lind method for correlated normals and FORM for correlated nonnormals (shown below). Unlike the classical computational approaches, the present approach does not need to rotate the frame of reference or to transform the coordinate space.

3.3.3 Positive reliability index only if mean-value point is in safe domain

In Figure 3.6, if a trial H value of 10 m is used, and the entire "x^*" column given the values equal to the "mean" column values, the performance function $PerFn$ exhibits a value of -448.5, meaning that the mean value point is already inside the unsafe domain. Upon Solver optimization with constraint $PerFn = 0$, a β index of 1.34 is obtained, which should be regarded as a negative index, i.e. -1.34, meaning that the *unsafe* mean value point is at some distance from the nearest safe point on the limit state surface that separates the safe and unsafe domains. In other words, the computed β index can be regarded as positive only if the $PerFn$ value is positive at the mean value point. For the case in Figure 3.6, the mean value point (prior to Solver optimization) yields a positive $PerFn$ for $H > 10.6$ m. The computed β index increases from 0 (equivalent to a lumped factor of safety equal to 1.0, i.e. on the verge of failure) when H is 10.6 m to 3.0 when H is 12.15 m, as shown in Figure 3.7.

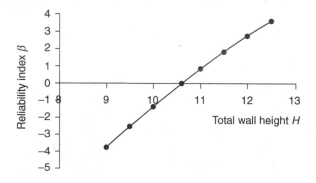

Figure 3.7 Reliability index is 3.00 when $H = 12.15$ m. For H smaller than 10.6 m, the mean-value point is in the unsafe domain, for which the reliability indices are negative.

3.3.4 Reliability-based design involving correlated nonnormals

The two-parameter normal distribution is symmetrical and, theoretically, has a range from $-\infty$ to $+\infty$. For a parameter that admits only positive values, the probability of encroaching into the negative realm is extremely remote if the coefficient of variation (Standard deviation/Mean) of the parameter is 0.20 or smaller, as for the case in hand. Alternatively, the lognormal distribution has often been suggested in lieu of the normal distribution, since it excludes negative values and affords some convenience in mathematical derivations. Figure 3.8 shows an efficient reliability-based design when the random variables are correlated and follow lognormal distributions. The two columns labeled μ^N and σ^N contain the formulae "=EqvN(..., 1)" and "EqvN(..., 2),", respectively, which invoke the user-created functions shown in Figure 3.8 to perform the equivalent normal transformation (when variates are lognormals) based on the following Rackwitz–Fiessler two-parameter equivalent normal transformation (Rackwitz and Fiessler, 1978):

$$\text{Equivalent normal standard deviation: } \sigma^N = \frac{\phi\left\{\Phi^{-1}\left[F(x)\right]\right\}}{f(x)} \qquad (3.7a)$$

$$\text{Equivalent normal mean: } \mu^N = x - \sigma^N \times \Phi^{-1}\left[F(x)\right] \qquad (3.7b)$$

where x is the original nonnormal variate, $\Phi^{-1}[.]$ is the inverse of the cumulative probability (CDF) of a standard normal distribution, $F(x)$ is the original nonnormal CDF evaluated at x, $\phi\{.\}$ is the probability density function (PDF) of the standard normal distribution, and $f(x)$ is the original nonnormal probability density ordinates at x.

For lognormals, closed form equivalent normal transformation is available and has been used in the VBA code of Figure 3.8. Efficient Excel VBA codes for equivalent normal transformations of other nonnormal distributions (including Gumbel, uniform, exponential, gamma, Weibull, triangular, and beta) are given in Low and Tang (2004, 2007), where it is shown that reliability analysis can be performed with various distributions merely by entering "Normal," "Lognormal," "Exponential," "Gamma," ..., in the first column of Figure 3.8, and distribution parameters in the columns to the left of the x^* column. In this way the required Rackwitz–Fiessler equivalent normal evaluations (for μ^N and σ^N) are conveniently relegated to functions created in the VBA programming environment of Microsoft Excel.

Therefore, the single spreadsheet cell object β in Figure 3.8 contains several Excel function objects and substantial program codes.

For correlated nonnormals, the ellipsoid perspective (Figure 3.2) and the constrained optimization approach still apply in the original coordinate

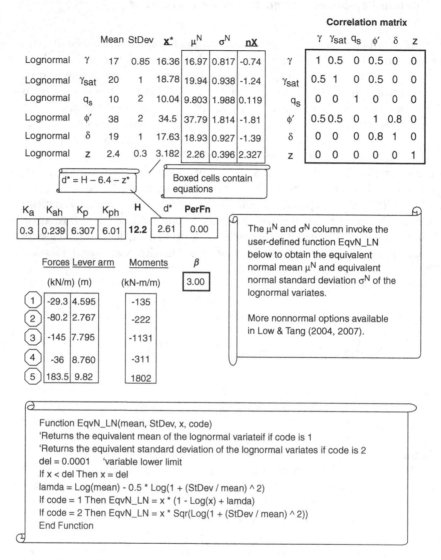

Correlation matrix

		Mean	StDev	$\underline{x^*}$	μ^N	σ^N	$\underline{nX}$		γ	γ_{sat}	q_s	ϕ'	δ	z
Lognormal	γ	17	0.85	16.36	16.97	0.817	-0.74	γ	1	0.5	0	0.5	0	0
Lognormal	γ_{sat}	20	1	18.78	19.94	0.938	-1.24	γ_{sat}	0.5	1	0	0.5	0	0
Lognormal	q_s	10	2	10.04	9.803	1.988	0.119	q_s	0	0	1	0	0	0
Lognormal	ϕ'	38	2	34.5	37.79	1.814	-1.81	ϕ'	0.5	0.5	0	1	0.8	0
Lognormal	δ	19	1	17.63	18.93	0.927	-1.39	δ	0	0	0	0.8	1	0
Lognormal	z	2.4	0.3	3.182	2.26	0.396	2.327	z	0	0	0	0	0	1

$d^* = H - 6.4 - z^*$

Boxed cells contain equations

K_a	K_{ah}	K_p	K_{ph}	**H**	d^*	**PerFn**
0.3	0.239	6.307	6.01	**12.2**	2.61	0.00

The μ^N and σ^N column invoke the user-defined function EqvN_LN below to obtain the equivalent normal mean μ^N and equivalent normal standard deviation σ^N of the lognormal variates.

More nonnormal options available in Low & Tang (2004, 2007).

	Forces	Lever arm	Moments	β
	(kN/m)	(m)	(kN-m/m)	**3.00**
1	-29.3	4.595	-135	
2	-80.2	2.767	-222	
3	-145	7.795	-1131	
4	-36	8.760	-311	
5	183.5	9.82	1802	

```
Function EqvN_LN(mean, StDev, x, code)
'Returns the equivalent mean of the lognormal variateif if code is 1
'Returns the equivalent standard deviation of the lognormal variates if code is 2
del = 0.0001    'variable lower limit
If x < del Then x = del
lamda = Log(mean) - 0.5 * Log(1 + (StDev / mean) ^ 2)
If code = 1 Then EqvN_LN = x * (1 - Log(x) + lamda)
If code = 2 Then EqvN_LN = x * Sqr(Log(1 + (StDev / mean) ^ 2))
End Function
```

Figure 3.8 Reliability-based design of anchored wall; correlated lognormals.

system, except that the nonnormal distributions are replaced by an equivalent normal ellipsoid, centered not at the original mean of the nonnormal distributions, but at an equivalent normal mean μ^N:

$$\beta = \min_{x \in F} \sqrt{\left[\frac{x_i - \mu_i^N}{\sigma_i^N} \right]^T [\mathbf{R}]^{-1} \left[\frac{x_i - \mu_i^N}{\sigma_i^N} \right]} \tag{3.8}$$

as explained in Low and Tang (2004, 2007). One Excel file associated with Low (2005a) for reliability analysis of anchored sheet pile wall involving correlated nonnormal variates is available for download at http://alum.mit.edu/www/bklow.

For the case in hand, the required total wall height H is practically the same whether the random variables are normally distributed (Figure 3.6) or lognormally distributed (Figure 3.8). Such insensitivity of the design to the underlying probability distributions may not always be expected, particularly when the coefficient of variation (standard deviation/mean) or the skewness of the probability distribution is large.

If desired, the original correlation matrix (ρ_{ij}) of the nonnormals can be modified to ρ'_{ij} in line with the equivalent normal transformation, as suggested in Der Kiureghian and Liu (1986). Some tables of the ratio ρ'_{ij}/ρ_{ij} are given in Appendix B2 of Melchers (1999), including a closed-form solution for the special case of lognormals. For the cases illustrated herein, the correlation matrix thus modified differs only slightly from the original correlation matrix. Hence, for simplicity, the examples of this chapter retain their original unmodified correlation matrices.

This section has illustrated an efficient reliability-based design approach for an anchored wall. The correlation structure of the six variables was defined in a correlation matrix. Normal distributions and lognormal distributions were considered in turn (Figures 3.6 and 3.8), to investigate the implication of different probability distributions. The procedure is able to incorporate and reflect the uncertainty of the passive soil surface elevation. Reliability index is the shortest distance between the mean-value point and the limit state surface – the boundary separating safe and unsafe combinations of parameters – measured in units of directional standard deviations. It is important to check whether the mean-value point is in the safe domain or unsafe domain before performing reliability analysis. This is done by noting the sign of the performance function (*PerFn*) in Figures 3.6 and 3.8 when the x^* columns were initially assigned the mean values. If the mean value point is safe, the computed reliability index is positive; if the mean-value point is already in the unsafe domain, the computed reliability index should be considered a negative entity, as illustrated in Figure 3.7.

The differences between reliability-based design and design based on specified partial factors were briefly discussed. The merits of reliability-based design are thought to lie in its ability to explicitly reflect correlation structure, standard deviations, probability distributions and sensitivities, and to automatically seek the most probable failure combination of parametric values case by case without relying on fixed partial factors. Corresponding to each desired value of reliability index, there is also a reasonably accurate simple estimate of the probability of failure.

3.4 Practical probabilistic slope stability analysis based on reformulated Spencer equations

This section presents a practical procedure for implementing Spencer method reformulated for a computer age, first deterministically, then probabilistically, in the ubiquitous spreadsheet platform. The material is drawn from Low (2001, 2003, both available at the author's website) and includes testing the robustness of search for noncircular critical slip surface, modeling lognormal random variables, deriving probability density functions from reliability indices, and comparing results inferred from reliability indices with Monte Carlo simulations. The deterministic modeling is described first, as it underlies the limit state function (i.e. performance function) of the reliability analysis.

3.4.1 Deterministic Spencer method, reformulated

Using the notations in Nash (1987), the sketch at the top of Figure 3.9 (below columns I and J) shows the forces acting on a slice (slice i) that forms part of the potential sliding soil mass. The notations are: weight W_i, base length l_i, base inclination angle α_i, total normal force P_i at the base of slice i, mobilized shearing resistance T_i at the base of slice i, horizontal and vertical components (E_i, E_{i-1}, $\lambda_i E_i$, $\lambda_{i-1} E_{i-1}$) of side force resultants at the left and right vertical interfaces of slice i, where λ_{i-1} and λ_i are the tangents of the side force inclination angles (with respect to horizontal) at the vertical interfaces. Adopting the same assumptions as Spencer (1973), but reformulated for spreadsheet-based constrained optimization approach, one can derive the following from Mohr–Coulomb criterion and equilibrium considerations:

$$T_i = \left[c'_i l_i + (P_i - u_i l_i) \tan \phi'_i \right] / F \quad \text{(Mohr–Coulomb criteria)} \tag{3.9}$$

$$P_i \cos \alpha_i = W_i - \lambda_i E_i + \lambda_{i-1} E_{i-1} - T_i \sin \alpha_i \quad \text{(vertical equilibrium)} \tag{3.10}$$

$$E_i = E_{i-1} + P_i \sin \alpha_i - T_i \cos \alpha_i \quad \text{(horizontal equilibrium)} \tag{3.11}$$

$$P_i = \frac{\left[\begin{array}{l} W_i - (\lambda_i - \lambda_{i-1}) E_{i-1} \\ -\dfrac{1}{F} (c'_i l_i - u_i l_i \tan \phi'_i)(\sin \alpha_i - \lambda_i \cos \alpha_i) \end{array} \right]}{\left[\begin{array}{l} \lambda_i \sin \alpha_i + \cos \alpha_i \\ +\dfrac{1}{F} \tan \phi'_i (\sin \alpha_i - \lambda_i \cos \alpha_i) \end{array} \right]}$$

$$\text{(from above three equations)} \tag{3.12}$$

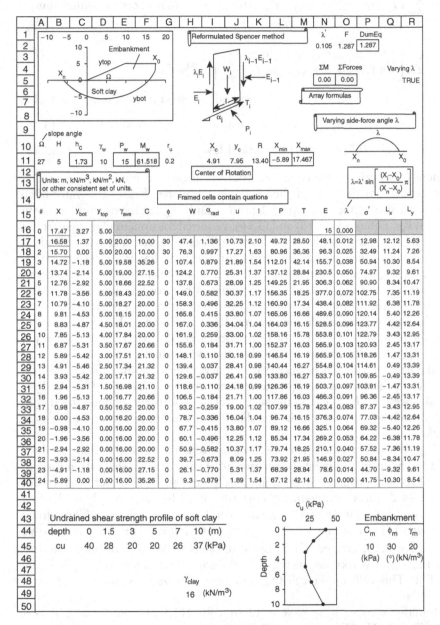

Figure 3.9 Deterministic analysis of a 5 m high embankment on soft ground with depth-dependent undrained shear strength. The limit equilibrium method of slices is based on reformulated Spencer method, with half-sine variation of side force inclination.

$$\sum [T_i \cos\alpha_i - P_i \sin\alpha_i] - P_w = 0 \qquad \text{(overall horizontal equilibrium)}$$

$$(3.13)$$

$$\sum \left[\begin{array}{c} (T_i \sin\alpha_i + P_i \cos\alpha_i - W_i) * L_{xi} \\ + (T_i \cos\alpha_i - P_i \sin\alpha_i) * L_{yi} \end{array} \right] - M_w = 0$$

(overall moment equilibrium)   (3.14)

$$L_{xi} = 0.5 (x_i + x_{i-1}) - x_c \quad \text{(horizontal lever arm of slice } i\text{)} \qquad (3.15)$$

$$L_{yi} = y_c - 0.5 (y_i + y_{i-1}) \quad \text{(vertical lever arm of slice } i\text{)} \qquad (3.16)$$

where c_i, ϕ_i and u_i are cohesion, friction angle and pore water pressure, respectively, at the base of slice i, P_w is the water thrust in a water-filled vertical tension crack (at x_0) of depth h_c, and M_w the overturning moment due to P_w. Equations (3.15) and (3.16), required for noncircular slip surface, give the lever arms with respect to an arbitrary center. The use of both λ_i and λ_{i-1} in Equation (3.12) allows for the fact that the right-most slice (slice #1) has a side that is adjacent to a water-filled tension crack, hence $\lambda_0 = 0$ (i.e. the direction of water thrust is horizontal), and for different λ values (either constant or varying) on the other vertical interfaces.

The algebraic manipulation that results in Equation (3.12) involves opening the term $(P_i - u_i l_i)\tan\phi'$ of Equation (3.9), an action legitimate only if $(P_i - u_i l_i) \geq 0$, or, equivalently, if the effective normal stress $\sigma_i' (= P_i/l_i - u_i)$ at the base of a slice is nonnegative. Hence, after obtaining the critical slip surface in the section to follow, one needs to check that $\sigma_i' \geq 0$ at the base of all slices and $E_i \geq 0$ at all the slice interfaces. Otherwise, one should consider modeling tension cracks for slices near the upper exit end of the slip surface.

Figure 3.9 shows the spreadsheet set-up for deterministic stability analysis of a 5 m high embankment on soft ground. The undrained shear strength profile of the soft ground is defined in rows 44 and 45. The subscript m in cells P44:R44 denotes *embankment*. Formulas need be entered only in the first or second cell (row 16 or 17) of each column, followed by autofilling down to row 40. The columns labeled y_{top}, γ_{ave} and c invoke the functions shown in Figure 3.10, created via Tools/Macro/VisualBasicEditor/Insert/Module on the Excel worksheet menu. The dummy equation in cell P2 is equal to F*1. This cell, unlike cell O2, can be minimized because it contains a formula.

Initially $x_c = 6$, $y_c = 8$, $R = 12$ in cells I11:K11, and $\lambda' = 0$, F = 1 in cells N2:O2. Microsoft Excel's built-in Solver was then invoked to set target and constraints as shown in Figure 3.11. The Solver option "Use Automatic Scaling" was also activated. The critical slip circle and factor of safety $F = 1.287$ shown in Figure 3.9 were obtained automatically within seconds by Solver via cell-object oriented constrained optimization.

```
Function Slice_c(ybmid, dmax, dv, cuv, cm)
'comment: dv = depth vector,
'cuv = cu vector
If ybmid > 0 Then
  Slice_c = cm
  Exit Function
End If
ybmid = Abs(ybmid)
If ybmid > dmax Then      'undefined domain,
  Slice_c = 300    'hence assume hard stratum.
  Exit Function
End If
For j = 2 T o dv.Count    'array size=dv.Count
  If dv(j) >= ybmid Then
    interp = (ybmid - dv(j - 1)) / (dv(j) - dv(j - 1))
    Slice_c = cuv(j - 1) + (cuv(j) - cuv(j - 1)) * interp
    Exit For
  End If
Next j
End Function


Function ytop(x, omega, H)
grad = Tan(omega * 3.14159 / 180)
If x < 0 Then ytop = 0
If x >=0 And x < H / grad Then ytop = x * grad
If x >=H / grad Then ytop = H
End Function


Function AveGamma(ytmid, ybmid, gm, gclay)
If ybmid < 0 Then
    Sum = (ytmid * gm + Abs(ybmid) * gclay)
    AveGamma = Sum / (ytmid - ybmid)
    Else: AveGamma = gm
End If
End Function
```

Figure 3.10 User-defined VBA functions, called by columns y_{top}, γ_{ave}, and c of Figure 3.9.

Noncircular critical slip surface can also be searched using Solver as in Figure 3.11, except that "By Changing Cells" are N2:O2, B16, B18, B40, C17, and C19:C39, and with the following additional cell constraints: B16 ≥ B11/tan(radians(A11)), B16 ≥ B18, B40 ≤ 0, C19:C39 ≤ D19:D39, O2 ≥ 0.1, and P17:P40 ≥ 0.

Figure 3.12 tests the robustness of the search for noncircular critical surface. Starting from four arbitrary initial circles, the final noncircular critical surfaces (solid curves, each with 25 degrees of freedom) are close enough to each other, though not identical. Perhaps more pertinent, their factors of safety vary narrowly within 1.253 – 1.257. This compares with the minimum factor of safety 1.287 of the critical circular surface of Figure 3.9.

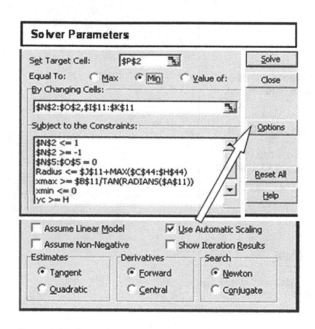

Figure 3.11 Excel Solver settings to obtain the solution of Figure 3.9.

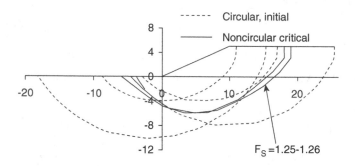

Figure 3.12 Testing the robustness of search for the deterministic critical noncircular slip surface.

3.4.2 Reformulated Spencer method extended probabilistically

Reliability analyses were performed for the embankment of Figure 3.9, as shown in Figure 3.13. The coupling between Figures 3.9 and 3.13 is brought about simply by entering formulas in cells C45:H45 and P45:R45 of Figure 3.9 to read values from column v_i of Figure 3.13. The matrix form of the Hasofer–Lind index, Equation (3.5b), will be used, except that the symbol v_i is used to denote random variables, to distinguish it from the symbol x_i (for x-coordinate values) used in Figure 3.9.

Spatial correlation in the soft ground is modeled by assuming an autocorrelation distance (δ) of 3 m in the following established negative exponential model:

$$\rho_{ij} = e^{-\dfrac{|\text{Depth}(i) - \text{Depth}(j)|}{\delta}} \tag{3.17}$$

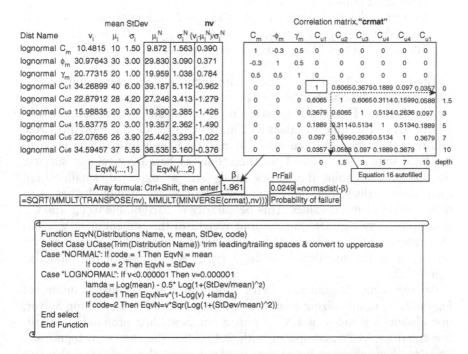

Figure 3.13 Reliability analysis of the case in Figure 3.9, accounting for spatial variation of undrained shear strength. The two templates are easily coupled by replacing the values c_m, ϕ_m, γ_m and the c_u values of Figure 3.9 with formulas that point to the values of the v_i column of this figure.

(Reliability analysis involving horizontal spatial variations of undrained shear strengths and unit weights is illustrated in Low et al. 2007, for a clay slope in Norway.)

Only normals and lognormals are illustrated. Unlike a widely documented classical approach, there is no need to diagonalize the correlation matrix (which is equivalent to rotating the frame of reference). Also, the iterative computations of the equivalent normal mean (μ^N) and equivalent normal standard deviation (σ^N) for each trial design point are automatic during the constrained optimization search.

Starting with a deterministic critical noncircular slip surface of Figure 3.12, the λ' and F values of Figure 3.9 were reset to 0 and 1, respectively. The v_i values in Figure 3.13 were initially assigned their respective mean values. The Solver routine in Microsoft Excel was then invoked to:

- *Minimize* the quadratic form (a 9-dimensional hyperellipsoid in original space), i.e. cell "β."
- *By changing* the nine random variables v_i, the λ' in Figure 3.9, and the 25 coordinate values $x_0, x_2, x_{24}, y_{b1}, y_{b3} : y_{b23}$ of the slip surface. (F remains at 1, i.e. at failure, or at limit state.)
- *Subject to the constraints* $-1 \leq \lambda' \leq 1$, $x_0 \geq H/\tan(\text{radians}(\Omega))$, $x_0 \geq x_2$, $x_{24} \leq 0$, $y_{b3} : y_{b23} \leq y_{t3} : y_{t23}$, $\sum \text{Forces} = 0$, $\sum M = 0$, and $\sigma'_1 : \sigma'_{24} \geq 0$. The Solver option "Use automatic scaling" was also activated.

The β index is 1.961 for the case with lognormal variates (Figure 3.13). The corresponding probability of failure based on the hyperplane assumption is 2.49%. The reliability-based noncircular slip surface is shown in Figure 3.14. The nine v_i values in Figure 3.13 define the most probable failure point, where the equivalent dispersion hyperellipsoid is tangent to the limit state surface $(F = 1)$ of the reliability-based critical slip surface. At this tangent point the values of c_u are, as expected, smaller than their mean values, but the c_m and ϕ_m values are slightly higher than their respective mean values. This peculiarity is attributable to c_m and ϕ_m being positively correlated to the unit weight of the embankment γ_m, and also reflects the dependency of tension crack depth h_c (an adverse effect) on c_m, since the equation $h_c = 2c_m/(\gamma_m \sqrt{K_a})$ is part of the model in Figure 3.9.

By replacing "lognormal" with "normal" in the first column of Figure 3.13, re-initializing column v_i to mean values, and invoking Solver, one obtains a β index of 1.857, with a corresponding probability of failure equal to 3.16%, compared with 2.49% for the case with lognormal variates. The reliability-based noncircular critical slip surface of the case with normal variates is practically indistinguishable from the case with lognormal variates. Both are, however, somewhat different (Figure 3.14) from the deterministic critical noncircular surface from which they evolved via

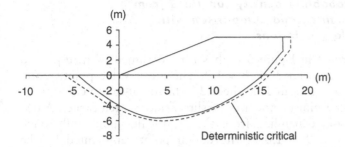

Figure 3.14 Comparison of reliability-based critical noncircular slip surfaces (the two upper curves, for normal variates and lognormal variates) with the deterministic critical noncircular slip surface (the lower dotted curve).

Table 3.1 Reliability indices for the example case.

	[Lognormal variates]	[Normal variates]
Reliability index	1.961	1.857
	(1.971)*	(1.872)*
Prob. of failure	2.49%	3.16%
	(2.44%)*	(3.06%)*

* at deterministic critical noncircular surface

Table 3.2 Solver's computing time on a computer.

F, for specified slip surface	≈ 0.3 s
F, search for noncircular surface	15 s
β, for specified slip surface	2 s
β, search for noncircular surface	20 s

25 degrees of freedom during Solver's constrained optimization search. This difference in slip surface geometry matters little, however, for the following reason. If the deterministic critical noncircular slip surface is used for reliability analysis, the β index obtained is 1.971 for the case with lognormals, and 1.872 for the case with normals. These β values are only slightly higher (<1%) than the β values of 1.961 and 1.857 obtained earlier with a "floating" surface. Hence, for the case in hand, performing reliability analysis based on the fixed deterministic critical noncircular slip surface will yield practically the same reliability index as the reliability-based critical slip surface. Table 3.1 summarizes the reliability indices for the case in hand. Table 3.2 compares the computing time.

3.4.3 Deriving probability density functions from reliability indices, and comparison with Monte Carlo simulations

The reliability index β in Figure 3.13 has been obtained with respect to the limit state surface defined by $F = 1.0$. The mean value point, represented by column μ_i, is located on the 1.253 factor-of-safety contour, as calculated in the deterministic noncircular slip surface search earlier. A dispersion hyperellipsoid expanding from the mean-value point will touch the limit state surface $F = 1.0$ at the design point represented by the values in the column labeled v_i. The probability of failure 2.49% approximates the integration of probability density in the failure domain $F \leq 1.0$. If one defines another limit state surface $F = 0.8$, a higher value of β is obtained, with a correspondingly smaller probability that $F \leq 0.8$, reflecting the fact that the limit state surface defined by $F = 0.8$ is further away from the mean-value point (which is on the factor-of-safety contour 1.253). In this way, 42 values of reliability indices (from positive to negative) corresponding to different specified limit states (from $F = 0.8$ to $F = 1.8$) were obtained promptly using a short VBA code that automatically invokes Solver for each specified limit state F. The series of β indices yields the cumulative distribution function (CDF) of the factor of safety, based on $\Pr[F \leq F_{\text{LimitState}}] \approx \Phi(-\beta)$, where $\Phi(.)$ is the standard normal cumulative distribution function. The probability density function (PDF) plots in Figure 3.15 (solid curves) were then obtained readily by applying cubic spline interpolation (e.g. Kreyszig, 1988) to the CDF, during which process the derivatives (namely the PDF) emerged from the tridiagonal spline matrix. The whole process is achieved easily using standard spreadsheet matrix functions. Another alternative was given in Lacasse and Nadim (1996), which examined pile axial capacity, and approximated the CDF by appropriate probability functions.

For comparison, direct Monte Carlo simulations with 20,000 realizations were performed using the commercial software @RISK with Latin Hypercube sampling. The random variables were first assumed to be normally distributed, then lognormally distributed. The mean values, standard deviations and correlation structure were as in Figure 3.13.

The solid PDF curves derived from the β indices agree remarkably well with the Monte Carlo PDF curves (the dashed curves in Figure 3.15).

Using existing state-of-the-art personal computer, the time taken by Solver to determine either the factor of safety F or the reliability index β of the embankment in hand is as shown in Table 3.2.

The computation time for 20,000 realizations in Monte Carlo simulations would be prohibitive ($20,000 \times 15$ s, or 83 h) if a search for the critical noncircular slip surface is carried out for each random set (c_m, ϕ_m, γ_m, and the c_u values) generated. Hence the PDF plots of Monte Carlo

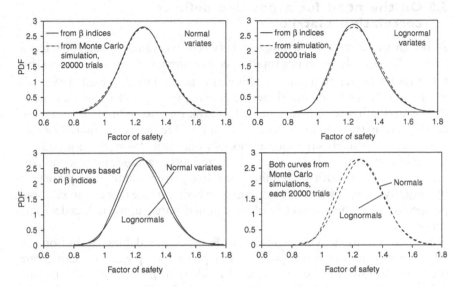

Figure 3.15 Comparing the probability density functions (PDF) obtained from reliability indices with those obtained from Monte Carlo simulations. "Normals" means the nine input variables are correlated normal variates, and "Lognormals" means they are correlated lognormal variates.

simulations shown in Figure 3.15 have been obtained by staying put at the reliability-based critical noncircular slip surface found in the previous section. In contrast, computing 42 values of β indices (with fresh noncircular surface search for each limit state) will take only about 14 min. Nevertheless, for consistency of comparison with the Monte Carlo PDF plots, the 42 β values underlying each solid PDF curve of Figure 3.15 have also been computed based on the same reliability-based critical noncircular slip surface shown in Figure 3.14. Thus it took about 1.4 min (42×2 s) to produce each reliability indices-based PDF curve, but about 70 times as long ($\approx 20,000 \times 0.3$ s) to generate each simulation-based PDF curve. What-if scenarios can be investigated more efficiently using the present approach of first-order reliability method than using Monte Carlo simulations.

The probabilities of failure ($F \leq 1$) from Monte Carlo simulations with 20,000 realizations are 2.41%$\pm$ for lognormal variates, and 3.28%$\pm$ for normal variates. These compare well with the 2.49% and 3.16% shown in Table 3.1. Note that only one β index value is needed to compute each probability of failure in Table 3.1. Only for PDF does one need about 30 or 40 β values of different limit states, in order to define the CDF, from which the PDF are obtained.

3.5 On the need for a positive definite correlation matrix

Although the correlation coefficient between two random variables has a range $-1 \leq \rho_{ij} \leq 1$, one is not totally free in assigning any values within this range for the correlation matrix. This was explained in Ditlevsen (1981: 85) where a 3×3 correlation matrix with $\rho_{12} = \rho_{13} = \rho_{23} = -1$ implies a contradiction and is hence inconsistent. A full-rank (i.e. no redundant variables) correlation matrix has to be positive definite. One may note that Hasofer–Lind index is defined as the square-root of a quadratic form (Equation (3.5a) or (3.5b)), and unless the quadratic form and its covariance matrix and correlation matrix are positive-definite, one is likely to encounter the square root of a negative entity, which is meaningless whether in the dispersion ellipsoid perspective or in the perspective of minimum distance in the reduced variate space.

Mathematically, a symmetric matrix $\mathbf{R}$ is positive definite if and only if all the eigenvalues of $\mathbf{R}$ are positive. For present purposes, one can create the equation $\text{QForm} = \mathbf{u}^{\mathsf{T}}\mathbf{R}^{-1}\mathbf{u}$ side by side with $\beta = \sqrt{(\text{QForm})}$. During Solver's searches, if the cell Qform shows negative value, it means that the quadratic form is not positive definite and hence the correlation matrix is inconsistent.

The Monte Carlo simulation program @RISK (http://www.palisade.com) will also display a warning of "Invalid correlation matrix" and offer to correct the matrix if simulation is run involving an inconsistent correlation matrix.

That a full-rank correlation or covariance matrix should be positive definite is an established theoretical requirement, although not widely known. This positive-definiteness requirement is therefore not a limitation of the paper's proposed procedure.

The joint PDF of an n-dimensional multivariate normal distribution is given by Equation (3.6a). The PDF has a peak at mean-value point and decreasing ellipsoidal probability contours surrounding μ only if the quadratic form and hence the covariance matrix and its inverse are positive definite. (By lemmas of matrix theory: $\mathbf{C}$ positive definite implies $\mathbf{C}^{-1}$ is positive definite, and also means the correlation matrix $\mathbf{R}$ and its inverse are positive definite.)

Without assuming any particular probability distributions, it is documented in books on multivariate statistical analysis that a covariance matrix (and hence a correlation matrix) cannot be allowed to be indefinite, because this would imply that a certain weighted sum of its variables has a negative variance, which is inadmissible, since all real-valued variables must have nonnegative variances. (The classical proof can be viewed as *reduction ad absurdum*, i.e. a method of proving the falsity of a premise by showing that the logical consequence is absurd.)

An internet search using the Google search engine with "positive definite correlation matrix" as search terms also revealed thousands of relevant web pages, showing that many statistical computer programs (in addition to the @RISK simulation software) will check for positive definiteness of correlation matrix and offer to generate the closest valid matrix to replace the entered invalid one. For example, one such web site titled "Not Positive Definite Matrices – Causes and Cures" at http://www.gsu. edu/~mkteer/npdmatri.html refers to the work of Wothke (1993).

3.6 Finite element reliability analysis via response surface methodology

Programs can be written in spreadsheet to handle implicit limit state functions (e.g. Low et al., 1998; Low, 2003; and Low and Tang, 2004: 87). However, there are situations where serviceability limit states can only be evaluated using stand-alone finite element or finite difference programs. In these circumstances, reliability analysis and reliability-based design by the present approach can still be performed, provided one first obtains a response surface function (via the established response surface methodology) which closely approximates the outcome of the stand-alone finite element or finite difference programs. Once the closed-form response functions have been obtained, performing reliability-based design for a target reliability index is straightforward and fast. Performing Monte-Carlo simulation on the closed form approximate response surface function also takes little time. Xu and Low (2006) illustrate the finite element reliability analysis of embankments on soft ground via the response surface methodology.

3.7 Limitations and flexibilities of object-oriented constrained optimization in spreadsheet

If equivalent normal transformation is used for first-order reliability method involving nonnormal distributions other than the lognormal, it is necessary to compute the inverse of the standard normal cumulative distributions (Equations (3.7a) and (3.7b)). There is no closed form equation for this purpose. Numerical procedures such as that incorporated in Excel's NormSInv is necessary. In Excel 2000 and earlier versions, NormSInv(CDF) returns correct values only for CDF between about 0.00000035 and 0.99999965, i.e. within about ±5 standard deviations from the mean. There are undesirable and idiosyncratic sharp kinks at both ends of the validity range. In EXCEL XP (2002) and later versions, the NormSInv (CDF) function has been refined, and returns correct values for a much broader band of CDF: $10^{-16} < \text{CDF} < (1 - 10^{-16})$, or about ±8 standard deviations from the mean. In general, robustness and computation efficiency will be better with the broader validity band of NormSInv(CDF) in Excel XP. This is because, during Solver's trial solutions using the generalized reduced gradient method, the trial x^* values may occasionally stray beyond 5 standard deviations from the mean, as dictated by the directional search, before returning to the real design point, which could be fewer than 5 standard deviations from the mean.

Despite its simple appearance, the single spreadsheet cell that contains the formula for reliability index β is in fact a conglomerate of program routines. It contains Excel's built-in matrix operating routines "*mmult*," "*transpose*," and "*minverse*," which can operate on up to 50×50 matrices. The members of the "**nx**" vector in the formula for β are each functions of their respective VBA codes in the user-created function EqvN (Figure 3.13). Similarly, the single spreadsheet cell for the performance function $g(x)$ can also contain many nested functions (built-in, or user-created). In addition, up to 100 constraints can be specified in the standard Excel Solver. Regardless of their complexities and interconnections, they (β index, $g(x)$, constraints) appear as individual cell objects on which Excel's built-in Solver optimization program performs numerical derivatives and iterative directional search using the generalized reduced gradient method as implemented in Lasdon and Waren's GRG2 code (http://www.solver.com/technology4.htm).

The examples of Low *et al.* (1998, 2001) with implicit performance functions and numerical methods coded in Excel's programming environment involved correlated normals only, and have no need for the Excel NormSInv function, and hence are not affected by the limitations of NormSInv in Excel 2000 and earlier versions.

3.8 Summary

This chapter has elaborated on a practical object-oriented constrained optimization approach in the ubiquitous spreadsheet platform, based on the work of Low and Tang (1997a, 1997b, 2004). Three common geotechnical problems have been used as illustrations; namely, a bearing capacity problem, an anchored sheet pile wall embedment depth design, and a reformulated Spencer method. Slope reliability analyses involving spatially correlated normal and lognormal variates were demonstrated, with search for the critical noncircular slip surface. The Hasofer–Lind reliability index was re-interpreted using the perspective of an expanding equivalent dispersion ellipsoid centered at the mean in the original space of the random variables. When nonnormals are involved, the perspective is one of equivalent ellipsoids. The probabilities of failure inferred from reliability indices are in good agreement with those from Monte Carlo simulations for the examples in hand. The probability density functions (of the factor of safety) were also derived from reliability indices using simple spreadsheet-based cubic spline interpolation, and found to agree well with those generated by the far more time-consuming Monte Carlo simulation method.

The coupling of spreadsheet-automated cell-object oriented constrained optimization and minimal macro programming in the ubiquitous spreadsheet platform, together with the intuitive expanding ellipsoid perspective, renders a hitherto complicated reliability analysis problem conceptually more transparent and operationally more accessible to practicing engineers. The computational procedure presented herein achieves the same reliability index as the classical first order reliability method (FORM) – well-documented in Ang and Tang (1984), Madsen *et al.* (1986), Haldar and Mahadevan (1999), Baecher and Christian (2003), for example – but without involving the concepts of eigenvalues, eigenvectors, and transformed space, which are required in a classical approach. It also affords possibilities yet unexplored or difficult hitherto. Although bearing capacity, anchored wall and slope stability are dealt with in this chapter, the perspectives and techniques presented could be useful in many other engineering problems.

Although only correlated normals and lognormals are illustrated, the VBA code shown in Figure 3.13 can be extended (Low and Tang, 2004 and 2007) to deal with the triangular, the exponential, the gamma, the Gumbel, and the beta distributions, for example.

In reliability-based design, it is important to note whether the mean-value point is in the safe domain or in the unsafe domain. When the mean-value point is in the safe domain, the distance from the *safe mean-value point* to the most probable failure combination of parametric values (the design point) on the limit state surface, in units of directional standard deviations, is a positive reliability index. When the mean-value point is in the unsafe domain (due to insufficient embedment depth of sheet pile wall, for example), the distance from the *unsafe mean-value point* to the most probable safe combination of parametric values on the limit state surface, in units of directional standard deviations, is a negative reliability index. This was shown in Figure 3.7 for the anchored sheet pile wall example. It is also important to appreciate that the correlation matrix, to be consistent, must be positive definite.

The meaning of the computed reliability index and the inferred probability of failure is only as good as the analytical model underlying the performance function. Nevertheless, even this restrictive sense of *reliability* or *failure* is much more useful than the lumped factor of safety approach or the partial factors approach, both of which are also only as good as their analytical models.

In a reliability-based design, the design point reflects sensitivities, standard deviations, correlation structure, and probability distributions in a way that prescribed partial factors cannot.

The spreadsheet-based reliability approach presented in this chapter can operate on stand-alone numerical packages (e.g. finite element) via the response surface method, which is itself readily implementable in spreadsheet.

The present chapter does not constitute a comprehensive risk assessment approach, but it may contribute component blocks necessary for such a final edifice. It may also help to overcome a language barrier that hampers wider adoption of the more consistent Hasofer–Lind reliability index and reliability-based design. Among the issues not covered in this chapter are model uncertainty, human uncertainty, and estimation of statistical parameters. These and other important issues were discussed in VanMarcke (1977), Christian *et al.* (1994), Whitman (1984, 1996), Morgenstern (1995), Baecher and Christian (2003), among others.

Acknowledgment

This chapter was written at the beginning of the author's four-month sabbatical leave at NGI, Norway, summer 2006.

References

Ang, H. S. and Tang, W. H. (1984). *Probability Concepts in Engineering Planning and Design, Vol. 2 – Decision, Risk, and Reliability.* John Wiley, New York.

Baecher, G. B. and Christian, J. T. (2003). *Reliability and Statistics in Geotechnical Engineering.* John Wiley, Chichester, UK.

Bowles, J. E. (1996). *Foundation Analysis and Design,* 5th ed. McGraw-Hill, New York.

BS 8002 (1994). *Code of Practice for Earth Retaining Structures.* British Standards Institution, London.

Christian, J. T., Ladd, C. C., and Baecher, G. B. (1994). Reliability applied to slope stability analysis. *Journal of Geotechnical Engineering, ASCE,* 120(12), 2180–207.

Craig, R. F. (1997). *Soil Mechanics,* 6th ed. Chapman & Hall, London.

Der Kiureghian, A. and Liu, P. L. (1986). Structural reliability under incomplete probability information. *Journal of Engineering Mechanics, ASCE,* 112(1), 85–104.

Ditlevsen, O. (1981). *Uncertainty Modeling: With Applications to Multidimensional Civil Engineering Systems.* McGraw-Hill, New York.

ENV 1997-1 (1994). *Eurocode 7: Geotechnical Design. Part 1: General Rules.* CEN, European Committee for Standardization, Brussels.

Haldar, A. and Mahadevan, S. (1999). *Probability, Reliability and Statistical Methods in Engineering Design.* John Wiley, New York.

Hasofer, A. M. and Lind, N. C. (1974). Exact and invariant second-moment code format. *Journal of Engineering Mechanics, ASCE,* 100, 111–21.

Kerisel, J. and Absi, E. (1990). *Active and Passive Earth Pressure Tables,* 3rd ed. A.A. Balkema, Rotterdam.

Kreyszig, E. (1988). *Advanced Engineering Mathematics,* 6th ed. John Wiley & Sons, New York, pp. 972–3.

Lacasse, S. and Nadim, F. (1996). Model uncertainty in pile axial capacity calculations. *Proceedings of the 28th Offshore Technology Conference,* Texas, pp. 369–80.

Low, B. K. (1996). Practical probabilistic approach using spreadsheet. Geotechnical Special Publication No. 58, *Proceedings, Uncertainty in the Geologic Environment: From Theory to Practice*, ASCE, USA, Vol. 2, pp. 1284–302.

Low, B. K. (2001). Probabilistic slope analysis involving generalized slip surface in stochastic soil medium. *Proceedings, 14th Southeast Asian Geotechnical Conference*, December 2001, Hong Kong. A.A. Balkema, Rotterdam, pp. 825–30.

Low, B. K. (2003). Practical probabilistic slope stability analysis. In *Proceedings, Soil and Rock America*, MIT, Cambridge, MA, June 2003, Verlag Glückauf GmbH Essen, Germany, Vol. 2, pp. 2777–84.

Low, B. K. (2005a). Reliability-based design applied to retaining walls. *Geotechnique*, 55(1), 63–75.

Low, B. K. (2005b). Probabilistic design of anchored sheet pile wall. *Proceedings 16th International Conference on Soil Mechanics and Geotechnical Engineering*, 12–16 September 2005, Osaka, Japan, Millpress, pp. 2825–8.

Low, B. K. and Tang, W. H. (1997a). Efficient reliability evaluation using spreadsheet. *Journal of Engineering Mechanics*, ASCE, 123(7), 749–52.

Low, B. K. and Tang, W. H. (1997b). Reliability analysis of reinforced embankments on soft ground. *Canadian Geotechnical Journal*, 34(5), 672–85.

Low, B. K. and Tang, W. H. (2004). Reliability analysis using object-oriented constrained optimization. *Structural Safety*, 26(1), 69–89.

Low, B. K. and Tang, W. H. (2007). Efficient spreadsheet algorithm for first-order reliability method. *Journal of Engineering Mechanics, ASCE*, 133(12), 1378–87.

Low, B. K., Lacasse, S. and Nadim, F. (2007). Slope reliability analysis accounting for spatial variation. *Georisk: Assesment and Management of Risk for Engineering Systems and Geohazards*. Taylor & Francis, 1(4), 177–89.

Low, B. K., Gilbert, R. B. and Wright, S. G. (1998). Slope reliability analysis using generalized method of slices. *Journal of Geotechnical and Geoenvironmental Engineering*, ASCE, 124(4), 350–62.

Low, B. K., Teh, C. I. and Tang, W. H. (2001). Stochastic nonlinear p–y analysis of laterally loaded piles. *Proceedings of the Eight International Conference on Structural Safety and Reliability, ICOSSAR '01*, Newport Beach, California, 17–22 June 2001, 8 pages. A.A. Balkema, Rotterdam.

Madsen, H. O., Krenk, S. and Lind, N. C. (1986). *Methods of Structural Safety*. Prentice Hall, Englewood Cliffs, NJ.

Melchers, R. E. (1999). *Structural Reliability Analysis and Prediction*, 2nd ed. John Wiley, New York.

Morgenstern, N. R. (1995). Managing risk in geotechnical engineering. Proceedings, Pan American Conference, ISSMFE.

Nash, D. (1987). A comparative review of limit equilibrium methods of stability analysis. In *Slope Stability*, Eds. M. G. Anderson and K. S. Richards. Wiley, New York, pp. 11–75.

Rackwitz, R. and Fiessler, B. (1978). Structural reliability under combined random load sequences. *Computers and Structures*, 9, 484–94.

Shinozuka, M. (1983). Basic analysis of structural safety. *Journal of Structural Engineering, ASCE*, 109(3), 721–40.

Simpson, B. and Driscoll, R. (1998). *Eurocode 7, A Commentary*. ARUP/BRE, Construction Research Communications Ltd, London.

Spencer, E. (1973). Thrust line criterion in embankment stability analysis. *Geotechnique*, 23, 85–100.

U.S. Army Corps of Engineers (1999) ETL 1110-2-556, Risk-based analysis in geotechnical engineering for support of planning studies, Appendix A, pages A11 and A12. (http://www.usace.army.mil/publications/eng-tech-ltrs/etl1110-2-556/toc.html)

VanMarcke, E. H. (1977). Reliability of earth slopes. *Journal of Geotechnical Engineering, ASCE*, 103(11), 1247–66.

Veneziano, D. (1974). *Contributions to Second Moment Reliability*. Research Report No. R74–33. Department of Civil Engineering. MIT, Cambridge, MA.

Whitman, R. V. (1984). Evaluating calculated risk in geotechnical engineering. *Journal of Geotechnical Engineering, ASCE*, 110(2), 145–88.

Whitman, R. V. (1996). Organizing and evaluating uncertainnty in geotechnical engineering. *Proceedings, Uncertainty in Geologic Environment: From Theory to Practice*, ASCE Geotechnical Special Publication #58, ASCE, New York, V1, pp. 1–28.

Wothke, W. (1993). Nonpositive definite matrices in structural modeling, In *Testing Structural Equation Models*, Eds. K.A. Bollen and J. S. Long. Sage, Newbury Park, CA, pp. 257–93.

Xu, B. and Low, B. K. (2006). Probabilistic stability analyses of embankments based on finite-element method, *Journal of Geotechnical and Geoenvironmental Engineering, ASCE*, 132(11), 1444–54.

Chapter 4

Monte Carlo simulation in reliability analysis

Yusuke Honjo

4.1 Introduction

In this chapter, Monte Carlo Simulation (MCS) techniques are described in the context of reliability analysis of structures. Special emphasis is placed on the recent development of MCS for reliability analysis. In this context, generation of random numbers by low-discrepancy sequences (LDS) and subset Markov Chain Monte Carlo (MCMC) techniques are explained in some detail. The background to the introduction of such new techniques is also described in order for the readers to make understanding of the contents easier.

4.2 Random number generation

In Monte Carlo Simulation (MCS), it is necessary to generate random numbers that follow arbitrary probability density functions (PDF). It is, however, relatively easy to generate such random numbers once a sequence of uniform random numbers in (0.0, 1.0) is given (this will be described in section 2.3).

Because of this, the generation of a sequence of random numbers in (0.0, 1.0) is discussed first in this section. Based on the historical developments, the methods can be classified to pseudo random number generation and low-discrepancy sequences (or quasi random number) generation.

4.2.1 Pseudo random numbers

Pseudo random numbers are a sequence of numbers that look like random numbers by deterministic calculations. This is a suitable method for a computer to generate a large number of reproducible random numbers and is used commonly today (Rubinstein, 1981; Tsuda, 1995).

Pseudo random numbers are usually assumed to satisfy the following conditions:

1   A great number of random numbers can be generated instantaneously.
2   If there is a cycle in the generation (which usually exists), it should be long enough to be able to generate a large number of random numbers.
3   The random numbers should be reproducible.
4   The random numbers should have appropriate statistical properties, which can usually be examined by statistical testings, for example Chi square goodness of fit test.

The typical methods of pseudo random number generation include the middle-square method and linear recurrence relations (sometimes called the congruential method).

The origin of the middle-square method goes back to Von Neumann, who proposed generating a random number of $2a$ digits from x_n of $2a$ digits by squaring x_n to obtain $4a$ digits number, and then cut the first and last a digits of this number to obtain x_{n+1}. This method, however, was found to have relatively short cycle and also does not satisfy satisfactory statistical condition. This method is rarely used today.

On the other hand, the general form of linear recurrence relations can be given by the following equation:

$$x_{n+1} = a_0 x_n + a_1 x_{n-1} + \ldots + a_j x_{n-j} + b \quad (mod\ P) \tag{4.1}$$

where $mod\ P$ implies that $x_{n+1} = a_0 x_n + a_1 x_{n-1} + \ldots + a_j x_{n-j} + b - P k_n$, and $k_n = [(x_{n+1} = a_0 x_n + a_1 x_{n-1} + \ldots + a_j x_{n-j} + b)/P]$ denotes the largest positive integer in $(x_{n+1} = a_0 x_n + a_1 x_{n-1} + \ldots + a_j x_{n-j} + b)/P$.

The simplest form of Equation (4.1) may be given as

$$x_{n+1} = x_n + x_{n-1} \quad (mod\ P) \tag{4.2}$$

This is called the Fibonacci method. If the generated random numbers are 2 digits integer and suppose the initial two numbers are $x_1 = 11$ and $x_2 = 36$, and $P = 100$, the generated numbers are $47, 83, 30, 14, 43, \ldots$.

The modified versions of Equation (4.1) are multiplicative congruential method and mixed congruential method, which are still popularly used, and programs are easily available (see for example Koyanagi, 1989).

4.2.2 Low-discrepancy sequences (quasi random numbers)

One of the most important applications of MCS is to solve multi-dimensional integration problems, which includes reliability analysis. The problem can be generally described as a multi-dimensional integration in a unit hyper cube:

$$I = \int_0^1 \cdots \int_0^1 f(\mathbf{x})d\mathbf{x} \qquad [\mathbf{x} = (x_1, x_2, \ldots x_k)] \tag{4.3}$$

The problem solved by MCS can be described as the following equation:

$$S(N) = \frac{1}{N} \sum_{i=1}^{N} f(\mathbf{x}_i) \tag{4.4}$$

where, $\mathbf{x}_i$'s are N randomly generated points in the unit hyper cube.

The error of $|S(N) - I|$ is proportional to $O(N^{-\frac{1}{2}})$, which is related to the sample number N. Therefore, if one wants to improve the accuracy of the integration by one digit, one needs to increase the sample point number 100 times.

Around 1960, some mathematicians showed that, if one can generate the points sequences with some special conditions, the accuracy can be improved by $O(N^{-1})$, or even $O(N^{-2})$. Such points sequences were termed quasi random numbers (Tsuda, 1995).

However, it was found later that quasi random numbers are not so effective, as the dimensions of the integration increase to more than 50. On the other hand, there was much demand, especially in financial engineering, to solve integrations of more than 1000 dimensions, which expedited research in this area in 1990s.

One of the triggers that accelerated the research in this area was the Koksma–Hlawka theorem, which related the convergence of the integration and discrepancy of numbers generated in the unit hyper cube. According to this theorem, the convergence of integration is faster as the discrepancy of the points sequences is smaller. Based on this theorem, the main objective of the research in the 1990s was to generate point sequences in the very high dimensional hyper cube with as low discrepancy as possible. In other words, how to generate point sequences with maximum possible uniformity became the main target of the research. Such point sequences were termed low-discrepancy sequences (LDS). In fact, LDS are point sequences that allow the accuracy of the integration in k dimensions to improve in proportion to $O\left(\frac{(\log N)^k}{N}\right)$ or less (Tezuka, 1995, 2003).

Based on this development, the term LDS is more often used than quasi random numbers in order to avoid confusion between pseudo and quasi random numbers.

The discrepancy is defined as follows (Tezuka, 1995, 2003):

$$D_N^{(k)} = \sup_{y \in [0,1]^k} \left| \frac{\#([0, \mathbf{y}); N)}{N} - \prod_{i=1}^{k} y_i \right| \tag{4.5}$$

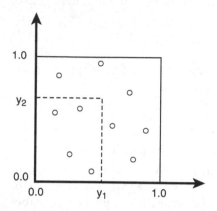

Figure 4.1 Concept of discrepancy in 2-dimensional space.

where $\#([0,\mathbf{y}];N)$ denotes number of points in $[0,\mathbf{y})$. Therefore, $D_N^{(k)}$ indicates discrepancy between the distribution of N points $\mathbf{x_n}$ in the k dimensional hyper cube $[0,1]^k$ from ideally uniform points distributions.

In Figure 4.1, discrepancy in 2-dimensional space is illustrated conceptually where $N = 10$. If the number of the points in a certain area is always in proportion to the area rate, $D_N^{(k)}$ is 0, otherwise $D_N^{(k)}$ becomes larger as points distribution is biased.

A points sequences $\mathbf{x_i}$ is said to be LDS when it satisfies the conditions below:

$$|S(N) - I| \le V_f D_N^{(k)} \qquad (4.6)$$

$$D_N^{(k)} = c(k)\frac{(\log N)^k}{N}$$

where V_f indicates the fluctuation of function $f(\mathbf{x})$ within the domain of the integration, and $c(k)$ is a constant which only depends on dimension k.

Halton, Faure and Sobol' sequences are known to be LSD. Various algorithms are developed to generate LSD (Tezuka, 1995, 2003).

In Figure 4.2, 1000 samples generated by the mixed congruential method and LSD are compared. It is understood intuitively from the figure that points generated by LSD are more uniformly distributed.

4.2.3 Random number generation following arbitrary PDF

In this section, generation of random numbers following arbitrary PDF is discussed. The most popular method to carry out this is to employ inverse transformation method.

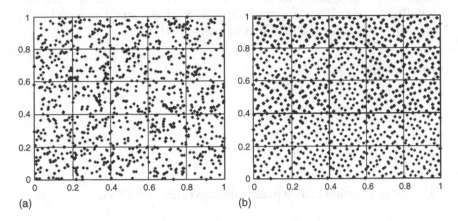

Figure 4.2 (a) Pseudo random numbers by mixed congruential method and (b) point sequences by LDS. 1000 generated sample points in 2-dimensional unit space.

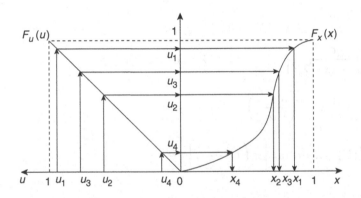

Figure 4.3 Concept of generation of random number x_i by uniform random number u_i based on the inverse transformation.

Suppose X is a random variable whose CDF is $F_X(x) = p$.

$F_X(x)$ is a non-decreasing function, and $0 \le p \le 1.0$. Thus, the following inverse transformation always exists for any p as (Rubinstein, 1981):

$$F_X^{-1}(p) = \min\left[x : F_X(x) \ge p\right] \qquad 0 \le p \le 1.0 \qquad (4.7)$$

By using this inverse transformation, any random variable X that follows CDF $F_X(x)$ can be generated based on uniform random numbers in $(0, 1)$ (Figure 4.3).

$$X = F_X^{-1}(U) \qquad (4.8)$$

Some of the frequently used inverse transformations are summarized below (Rubinstein, 1981; Ang and Tang, 1985; Reiss and Thomas, 1997). Note that U is a uniform random variable in $(0, 1)$.

Exponential distribution:

$$F_X(x) = 1 - \exp[-\lambda x] \quad (x \geq 0)$$

$$X = F_X^{-1}(U) = -\frac{1}{\lambda} \ln(1 - U)$$

The above equation can be practically rewritten as follows:

$$X = F_X^{-1}(U) = -\frac{1}{\lambda} \ln(U)$$

Cauchy distribution:

$$f_X(x) = \frac{\alpha}{\pi \{\alpha^2 + (x - \lambda)^2\}} \qquad \alpha > 0, \lambda > 0, -\infty < x < \infty$$

$$F_X(x) = \frac{1}{2} + \pi^{-1} \tan^{-1}\left(\frac{x - \lambda}{\alpha}\right)$$

$$X = F_X^{-1}(U) = \lambda + \alpha \tan\left[\pi\left(U - \frac{1}{2}\right)\right]$$

Gumbel distribution:

$$F_X(x) = \exp\left[-\exp\{-a(x - b)\}\right] \qquad (-\infty < x < \infty)$$

$$X = F_X^{-1}(U) = -\frac{1}{a} \ln\left(-\ln(U)\right) + b$$

Frechet distribution:

$$F_X(x) = \exp\left[-\left(\frac{v}{x - \varepsilon}\right)^k\right] \qquad (\varepsilon < x < \infty)$$

$$X = F_X^{-1}(U) = v\left(-\ln(U)\right)^{-1/k} + \varepsilon$$

Weibull distribution:

$$F_X(x) = \exp\left[-\left(\frac{\omega - x}{\omega - v}\right)^k\right] \qquad (-\infty < x < \omega)$$

$$X = F_X^{-1}(U) = (\omega - v)\left(\ln(U)\right)^{1/k} + \omega$$

Generalized Pareto distribution:

$$F_X(x) = 1 - \left(1 + \gamma \frac{x - \mu}{\sigma}\right)^{-1/\gamma} \qquad (\mu \le x)$$

$$X = F_X^{-1}(U) = \frac{\sigma}{\gamma}\left[(1 - U)^{-\gamma} - 1\right] + \mu$$

The normal random numbers, which are considered to be used most frequently next to the uniform random numbers, do not have inverse transformation in an explicit form. One of the most common ways to generate the normal random numbers is to employ the Box and Muller method, which is described below (see Rubinstein, 1981: 86–7 for more details).

U_1 and U_2 are two independent uniform random variables. Based on these variables, two independent standard normal random numbers, Z_1 and Z_2 can be generated by using following equations:

$$Z_1 = (-2 \ln(U_1))^{1/2} \cos(2\pi U_2) \tag{4.9}$$

$$Z_2 = (-2 \ln(U_1))^{1/2} \sin(2\pi U_2)$$

A program based on this method can be obtained easily from standard subroutine libraries (e.g. Koyanagi, 1989).

4.2.4 Applications of LDS

In Yoshida and Sato (2005a), reliability analyses by MCS are presented, where results by LDS and by OMCS (ordinary Monte Carlo simulation) are compared. The performance function they employed is:

$$g(x, S) = \sum_{i=1}^{5} x_i - \left(1 + \frac{x_1}{2500}\right) S \tag{4.10}$$

where x_i are independent random variables which follow identically a log normal distribution with mean 250 and COV 1.0.

The convergence of the failure probability with the number of simulation runs are presented in Figure 4.4 for LSD and 5 OMCS runs (which are

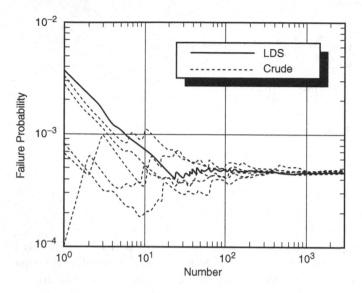

Figure 4.4 Relationship between calculated P_f and number of simulation runs (Yoshida and Sato, 2005a).

indicated by "Crude" in the figure). It is observed that LDS has faster convergence than OMCS.

4.3 Some methods to improve the efficiency of MCS

4.3.1 Accuracy of ordinary Monte Carlo Simulation (OMCS)

We consider the integration of the following type in this section:

$$P_f = \int \cdots \int I[g(\mathbf{x}) \leq 0] f_X(\mathbf{x}) d\mathbf{x} \tag{4.11}$$

where f_X is the PDF of basic variables $\mathbf{X}$, and I is an indicator function defined below:

$$I(\mathbf{x}) = 1 : g(\mathbf{x}) \leq 0$$
$$0 : g(\mathbf{x}) > 0$$

The indicator function identifies the domain of integration. Function $g(\mathbf{x})$ is a performance function in reliability analysis, thus the integration provides the failure probability, P_f. Note that the integration is carried out on all the domains where basic variables are defined.

When this integration is evaluated by MCS, P_f can be evaluated as follows:

$$P_f \approx \frac{1}{N} \sum_{j=1}^{N} I\left[g\left(\mathbf{x}_j\right) \le 0\right] \tag{4.12}$$

where $\mathbf{x}_j$ is j-th sample point generated based on PDF $f_{\mathbf{x}}()$.

One of the important aspects of OMCS is to find the necessary number of simulation runs to evaluate P_f of required accuracy. This can be generally done in the following way (Rubinstein, 1981: 115–18).

Let N be the total number of runs, and N_f be the ones which indicated the failure. Then the failure probability P_f is estimated as:

$$\hat{P}_f = \frac{N_f}{N} \tag{4.13}$$

The mean and variance of P_f can be evaluated as:

$$E\left[\hat{P}_f\right] = \frac{1}{N} E[N_F] = \frac{NP_f}{N} = P_f \tag{4.14}$$

$$Var\left[\hat{P}_f\right] = Var\left[\frac{N_f}{N}\right] = \frac{1}{N^2} Var[N_f]$$

$$= \frac{1}{N^2} NP_f\left(1 - P_f\right) = \frac{1}{N} P_f\left(1 - P_f\right) \tag{4.15}$$

Note that, here, these values are obtained by recognizing the fact that N_f follows binominal distribution of N trials, where the mean and the variance are given as NP_f and $NP_f(1 - P_f)$, respectively.

Suppose one wants to obtain a necessary number of simulation runs for probability $\left|P_f - \hat{P}_f\right| \le \varepsilon$ to be more than $100\alpha\%$:

$$\text{Prob}\left[\left|P_f - \hat{P}_f\right| \le \varepsilon\right] \ge \alpha \tag{4.16}$$

On the other hand, the following relationship is given based on Chebyshev's inequality formula.

$$\text{Prob}\left[\left|P_f - \hat{P}_f\right| < \varepsilon\right] \ge 1 - \frac{Var\left[\hat{P}_f\right]}{\varepsilon^2} \tag{4.17}$$

From Equations (4.16) and (4.17), and using Equation (4.15), the following relationship is obtained:

$$\alpha \le 1 - \frac{\text{Var}\left[\hat{P}_f\right]}{\varepsilon^2} = 1 - \frac{1}{N\varepsilon^2}P_f\left(1 - P_f\right)$$

$$1 - \alpha \ge \frac{1}{N\varepsilon^2}P_f\left(1 - P_f\right)$$

$$N \ge \frac{P_f\left(1 - P_f\right)}{\left(1 - \alpha\right)\varepsilon^2} \tag{4.18}$$

Based on Equation (4.18), the necessary number of simulation runs, N, can be obtained for any given ε and α. Let $\varepsilon = \delta P_f$, and one obtains for N,

$$N \ge \frac{P_f\left(1 - P_f\right)}{\left(1 - \alpha\right)\delta^2 P_f^2} = \frac{\left(1/P_f\right) - 1}{\left(1 - \alpha\right)\delta^2} \approx \frac{1}{\left(1 - \alpha\right)\delta^2 P_f} \tag{4.19}$$

The necessary numbers of simulation runs, N, are calculated in Table 4.1 for $\varepsilon = 0.1P_f$, i.e. $\delta = 0.1$ and $\alpha = 0.95$. In other words, N obtained in this table ensures the error of P_f to be within $\pm 10\%$ for 95% confidence probability.

The evaluation shown in Table 4.1 is considered to be very conservative, because it is based on the very general Chebyshev's inequality relationship. Broding (Broding *et al.*, 1964) gives necessary simulation runs as below (Melchers, 1999):

$$N > \frac{-\ln(1 - \alpha)}{P_f} \tag{4.20}$$

Table 4.1 Necessary number of simulation runs to evaluate P_f of required accuracy.

δ	α	P_f	N by Chebyshev	N by Bording
0.1	0.95	1.00E-02	200,000	300
		1.00E-04	20,000,000	29,957
		1.00E-06	2,000,000,000	2,995,732
0.5	0.95	1.00E-02	8,000	300
		1.00E-04	800,000	29,957
		1.00E-06	80,000,000	2,995,732
1.0	0.95	1.00E-02	2,000	300
		1.00E-04	200,000	29,957
		1.00E-06	20,000,000	2,995,732
5.0	0.95	1.00E-02	80	300
		1.00E-04	8,000	29,957
		1.00E-06	800,000	2,995,732

where N is necessary number of simulation runs for one variable, α is given confidence level and P_f is the failure probability. For example, if $\alpha = 0.95$ and $P_f = 10^{-3}$, the necessary N is, by Equation (4.20), about 3000. In the case of more than one independent random variable, one should multiply this number by the number of variables. N evaluated by Equation (4.20) is also presented in Table 4.1.

4.3.2 Importance sampling (IMS)

As can be understood from Table 4.1, the required number of simulation runs in OMCS to obtain P_f of reasonable confidence level is not small, especially for cases of smaller failure probability. In most practical reliability analyses, the order of P_f is of 10^{-4} or less, thus this problem becomes quite serious. One of the MCS methods to overcome this difficulty is the Importance Sampling (IMS) technique, which is discussed in this section.

Theory

Equation (4.11) to calculate P_f can be rewritten as follows:

$$P_f = \int \cdots \int I[g(\mathbf{x}) \leq 0] \frac{f_X(\mathbf{x})}{h_V(\mathbf{x})} h_V(\mathbf{x}) d\mathbf{x} \tag{4.21}$$

where $\mathbf{v}$ is a random variable following PDF h_V.

$$P_f \approx \frac{1}{N} \left\{ \sum_{j=1}^{N} I[g(\mathbf{v}_j) \leq 0] \frac{f_X(\mathbf{v}_j)}{h_V(\mathbf{v}_j)} \right\} \tag{4.22}$$

where h_V is termed the Importance Sampling function, and acts as a controlling function for sampling. Furthermore, the optimum h_V is known to be the function below (Rubinstein, 1981: 122–4):

$$h_V(\mathbf{v}) = \frac{I[g(\mathbf{v}) \leq 0] f_X(\mathbf{v})}{P_f} \tag{4.23}$$

This function $h_V(\mathbf{v})$, in the case of a single random variable, is illustrated in Figure 4.5.

It can be shown that the estimation variance of P_f is null for the sampling function $h_V(\mathbf{v})$ given in Equation (4.23). In other words, P_f can be obtained

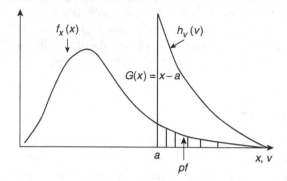

Figure 4.5 An illustration of the optimum Importance Function in the case of a single variable.

exactly by using this sample function as demonstrated below:

$$P_f = \iint \cdots \int I[g(\mathbf{x}) \leq 0] \frac{f_X(\mathbf{x})}{h_V(\mathbf{x})} h_V(\mathbf{x}) d\mathbf{x}$$

$$= \iint \cdots \int I[g(\mathbf{x}) \leq 0] f_X(\mathbf{x}) \left[\frac{I[g(\mathbf{x}) \leq 0] f_X(\mathbf{x})}{P_f} \right]^{-1} h_V(\mathbf{x}) d\mathbf{x}$$

$$= P_f \iint \cdots \int h_V(\mathbf{x}) d\mathbf{x} = P_f$$

Therefore, P_f can be obtained without carrying out MCS if the importance function of Equation (4.23) is employed.

However, P_f is unknown in practical situations, and the result obtained here does not have an actual practical implication. Nevertheless, Equation (4.23) provides a guideline on how to select effective important sampling functions; a sampling function that has a closer form to Equation (4.23) may be more effective. It is also known that if an inappropriate sampling function is chosen, the evaluation becomes ineffective and estimation variance increases.

Selection of importance sampling functions in practice

The important region in reliability analyses is the domain where $g(\mathbf{x}) \leq 0$. The point which has the maximum $f_X(\mathbf{x})$ in this domain is known as the design point, which is denoted by $\mathbf{x}^*$. One of the most popular ways to set h_V is to set the mean value of this function at this design point (Melcher, 1999). Figure 4.6 conceptually illustrates this situation.

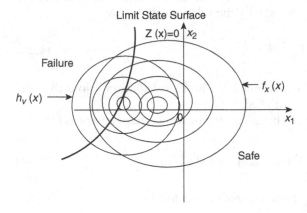

Figure 4.6 A conceptual illustration of importance sampling function set at the design point.

If $g(\mathbf{x})$ is a linear function and the space is standardized normal space, the sample points generated by this h_V fall in the failure region by about 50%, which is considered to give much higher convergence compared to OMCS with a smaller number of sample points.

However, if the characteristic of a performance function becomes more complex and also if the number of basic variables increases, it is not a simple task to select an appropriate sample function. Various so-called adaptive methods have been proposed, but most of them are problem-dependent. It is true to say that there is no comprehensive way to select an appropriate sample function at present that can guarantee better results in all cases.

4.3.3 Subset MCMC (Markov Chain Monte Carlo) method

Introduction to the subset MCMC method

The subset MCMC method is a reliability analysis method by MCS proposed by Au and Beck (2003). They combined the subset concept with the MCMC method, which had been known as a very flexible MCS technique to generate random numbers following any arbitrary PDF (see, for example, Gilks *et al.*, 1996). They combined these two concepts to develop a very effective MCS technique for structural reliability analysis.

The basic concept of this technique is explained in this subsection, whereas a more detailed calculation procedure is presented in the next subsection. Some complementary mathematical proofs are presented in the appendix of this chapter, and some simple numerical examples in subsection 4.3.3.

Let F be a set indicating the failure region. Also, let F_0 be the total set, and $F_i(I = 1, \cdots m)$ be subsets that satisfy the relationship below:

$$F_0 \supset F_1 \supset F_2 \supset \ldots \supset F_m \supset F \qquad (4.24)$$

where F_m is the last subset assumed in the P_f calculation. Note that the condition $F_m \supset F$ is always satisfied.

The failure probability can be calculated by using these subsets as follows:

$$P_f = P(F) = P(F \mid F_m)P(F_m \mid F_{m-1})\ldots P(F_1 \mid F_0) \qquad (4.25)$$

Procedure to calculate P_f by subset MCMC method

(1) Renewal of subsets

The actual procedure to set a subset is as follows, where $z_i = g(x_i)$ is the calculated value of a given performance function at sample point x_i, N_t is the total number of generated points in each subset, N_s are number of points selected among N_t points from the smaller z_i values, and N_f is number of failure points (i.e. z_i is negative.) in N_t. Also note that x used in this section can be a random variable or a random vector that follows a specified PDF.

Step 1   In the first cycle, N_t points are generated by MCS that follow the given PDF.

Step 2   Order the generated samples in ascending order by the value of z_i and select smaller $N_s (< N_t)$ points. A new subset F_{k+1} is defined by the equation below:

$$F_{k+1} = \left\{ x \mid z(x) \leq \frac{z_s + z_{s+1}}{2} \right\} \qquad (4.26)$$

Note that z_i's in this equation is assembled in ascending order. The probability that a point generated in subset F_{k+1} for points in subset F_k is calculated to be $P(F_{k+1} \mid F_k) = N_s/N_t$.

Step 3   By repeating **Step 2** for a range of subsets, Equation (4.25) can be evaluated.

Note that x_i $(i = 1,\ldots N_s)$ are used as seed points in MCMC to generate the next N_t samples, which implies that N_t/N_s samples are generated from each of these seed points. In other words, N_t/N_s points are generated from each seed point by MCMC.

Furthermore, the calculation is ceased when N_f reaches certain number with respect to N_t.

(2) Samples generation by MCMC

By MCMC method, N_t samples are generated in subset F_{k+1}. This methodology is described in this subsection.

Let $\pi(x)$ be a PDF of x, that follows PDF $f_X(x)$ and yet conditioned to be within the subset F_k. PDF $\pi(x)$ is defined as follows:

$$\pi(x) = c \cdot f_X(x) \cdot I_D(x) \tag{4.27}$$

where

$$I_D(x) = \begin{cases} 1 \text{ if } (x \subset F_k) \\ 0 \text{ otherwise} \end{cases}$$

where c is a constant for PDF π to satisfy the condition of a PDF.

Points following this PDF $\pi(x)$ can be generated by MCMC by the following procedure (Gilks et al., 1996; Yoshida and Sato, 2005):

Step 1   Determine any arbitrary PDF $q(x' \mid x_i)$, which is termed proposal density. In this research, a uniform distribution having mean x_i and appropriate variance is employed. (The author tends to use the variance of the original PDF to this uniform distribution. The smaller variance tends to give higher acceptance rate (i.e. Equation (4.29)), but requires more runs due to slow movement of the subset boundary.)

Step 2   The acceptance probability $\alpha(x', x_i)$ is calculated. Then set $x_{i+1} = x'$ with probability α, or $x_{i+1} = x_i$ with probability $1 - \alpha$. The acceptance probability α is defined as the following equation:

$$\alpha(x', x_i) = \min\left\{1.0, \ \frac{q(x_i \mid x') \cdot \pi(x')}{q(x' \mid x_i)\pi(x_i)}\right\}$$

$$= \min\left\{1.0, \ \frac{q(x_i \mid x') \cdot f_X(x') \cdot I_D(x')}{q(x' \mid x_i)f_X(x_i)}\right\} \tag{4.28}$$

Equation (4.28) can be rewritten as below considering the fact that $q(x' \mid x_i)$ is a uniform distribution and point x_i is certainly in the subset under consideration:

$$\alpha = \min\left\{1.0, \ \frac{f_X(x') \cdot I_D(x')}{f_X(x_i)}\right\} \tag{4.29}$$

Step 3   Repeat **Step 2** for necessary cycles.

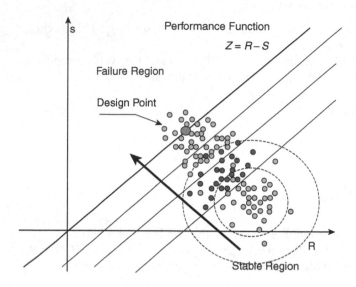

Figure 4.7 A conceptual illustration of the subset MCMC method.

(3) Evaluation of the failure probability P_f

In this study, the calculation is stopped when N_f reaches certain number with respect to N_t, otherwise the calculation in **Step 2** is repeated for a renewed subset which is smaller than the previous one.

Some of more application examples of the subset MCMC to geotechnical reliability problems can be seen in Yoshida and Sato (2005b).

The failure probability can be calculated as below:

$$P_f = P(z < 0) = \left\{ \frac{N_s}{N_t} \right\}^{m-1} \frac{N_f}{N_t} \tag{4.30}$$

The mathematical proof of this algorithm is presented in the next section.

The conceptual illustration of this procedure is presented in Figure 4.7. It is understood that by setting the subsets, more samples are generated to the closer region to the performance function, which is supposed to improve the efficiency of MCS considerably compared to OMCS.

Examples of subset MCMC

Two simple reliability analyses are presented here to demonstrate the nature of subset MCMC method.

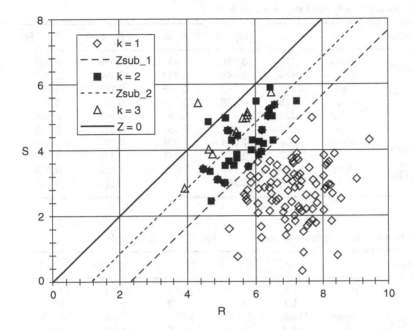

Figure 4.8 A calculation example of subset MCMC method when $Z = R - S$.

(1) An example on a linear performance function

The first example employs the simplest performance function $Z = R - S$. The resistance R follows a normal distribution, $N(7.0, 1.0)$, whereas the external force S, a normal distribution, $N(3.0, 1.0)$. The true value of the failure probability of this problem is $P_f = 2.34E - 03$.

Figure 4.8 exhibits a calculation example of this reliability analysis. In this calculation, $N_t = 100$, and two subsets are formed before the calculation is ceased. It is observed that more samples are generated closer to the limit state line as the calculation proceeds.

In order to examine the performance of subset MCMC method, the number of selected sample points to proceed to the next subset, N_s, and stopping criteria of the calculation are altered. Actually, N_s is set to 2, 5, 10, 20 or 50. The stopping criteria are set based on the rate of N_f to N_t as $N_f \geq 0.50, 0.20, 0.10$ or $0.05N_t$. Also a criterion $N_f \geq N_s$ is added. N_t is set to 100 in all cases.

Table 4.2 lists the results of the calculation. 1000 calculations are made for each case, where mean and COV of $\log_{10}(\text{calculated } P_f)/\log_{10}(\text{true } P_f)$ are presented.

In almost all cases, the means are close to 1.0, which exhibit very little bias in the present calculation for evaluating P_f. COV seems to be somewhat smaller, for case N_s is larger.

Table 4.2 Accuracy of P_f estimation for $Z = R - S$.

Cutoff criterion		$N_s = 2$	$N_s = 5$	$N_s = 10$	$N_s = 20$	$N_s = 50$
①$N_f \geq 0.02N$	Mean	0.992	0.995	0.997	0.976	0.943
	Cov	0.426	0.448	0.443	0.390	0.351
②$N_f \geq 0.05N$	Mean	0.995	0.993	0.989	0.989	0.971
	Cov	0.428	0.378	0.349	0.371	0.292
③$N_f \geq 0.10N$	Mean	1.033	0.994	0.991	0.985	0.985
	Cov	0.508	0.384	0.367	0.312	0.242
④$N_f \geq 0.20N$	Mean	1.056	1.074	1.006	0.993	0.990
	Cov	0.587	0.489	0.348	0.305	0.165
⑤$N_f \geq 0.50N$	Mean	1.268	1.111	1.056	1.020	0.996
	Cov	0.948	0.243	0.462	0.319	0.162

Table 4.3 Number of performance function calls for $Z = R - S$ example.

Cutoff criterion		$N_s = 2$	$N_s = 5$	$N_s = 10$	$N_s = 20$	$N_s = 50$
①$N_f \geq 0.02N$	Mean	171.9	188.1	207.7	232.9	343.2
	s.d.	25.1	37.1	43.8	56.1	109.2
②$N_f \geq 0.05N$	Mean	185.8	209.0	229.1	271.6	440.0
	s.d.	31.3	56.2	44.5	54.0	88.2
③$N_f \geq 0.10N$	Mean	202.9	221.8	251.8	299.0	506.6
	s.d.	32.4	61.6	39.6	53.2	75.4
④$N_f \geq 0.20N$	Mean	212.9	240.6	270.6	326.5	569.4
	s.d.	31.2	48.6	38.9	48.0	68.1
⑤$N_f \geq 0.50N$	Mean	226.0	258.0	293.0	361.1	654.7
	s.d.	25.7	37.1	42.6	48.4	62.7

Table 4.3 presents a number of performance function calls in each case. The mean and s.d. (standard deviation) of these numbers are shown. It is obvious from this table that the numbers of the function calls increase as N_s increases. It is understandable because as N_s increases, the speed of subsets converging to the limit state line decreases, which makes the number of calls larger.

As far as the present calculation is concerned, the better combination of N_s and the stopping criterion seems to be around $N_s = 20 \sim 50$ and $N_f \geq 0.20 \sim 0.50\ N_t$. However, the number of performance function calls also increases considerably for these cases.

(2) An example on a parabolic performance function

The second example employs a parabolic performance function $Z = (R - 11.0)^2 - (S - 6)$. The resistance R follows a normal distribution,

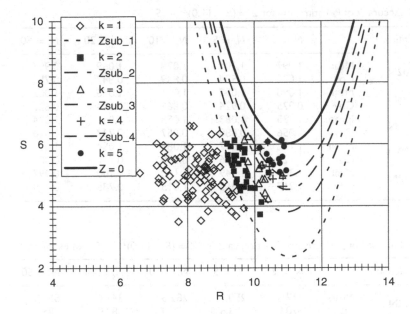

Figure 4.9 A calculation example of subset MCMC method when $Z = (R - 11.0)^2 - (S - 6)$.

$N(8.5, 0.707^2)$, whereas the external force S follows a normal distribution, $N(5.0, 0.707^2)$. The true value of the failure probability of this problem, which is obtained by one million runs of OMCS, is $P_f = 3.20E - 04$.

Figure 4.9 exhibits a calculation example of this reliability analysis. In this calculation, $N_t = 100$, and four subsets were formed before the calculation was ceased. It is observed that more samples are generated closer to the limit state line as the calculation proceeds, which is the same as for the first example.

All the cases that have been calculated in the previous example are repeated in this example as well, where $N_t = 100$.

Table 4.4 exhibits the same information as in the previous example. The mean value is practically 1.0 in most of the cases, which suggests there is little bias in the evaluation. On the other hand, COV tends to be smaller for larger N_s.

Table 4.5 also provides the same information as the previous examples. The number of function calls increases as N_s is set larger. It is difficult to identify the optimum N_s and the stopping criterion.

4.4 Final remarks

Some of the recent developments in Monte Carlo simulation (MCS) techniques, namely low-discrepancy sequences (LDS) and the subset MCMC

Table 4.4 Accuracy of P_f estimation for $Z = (R - 11.0)^2 - (S - 6)$.

Cutoff criterion		$N_s = 2$	$N_s = 5$	$N_s = 10$	$N_s = 20$	$N_s = 50$
①$N_f \geq 0.02N$	Mean	1.097	1.101	1.078	1.069	0.997
	Cov	1.027	0.984	0.839	0.847	0.755
②$N_f \geq 0.05N$	Mean	1.090	1.092	1.071	1.056	1.002
	Cov	0.975	0.924	0.865	0.844	0.622
③$N_f \geq 0.10N$	Mean	1.135	1.083	1.055	1.051	1.024
	Cov	1.056	0.888	0.857	0.737	0.520
④$N_f \geq 0.20N$	Mean	1.172	1.077	1.089	1.054	1.013
	Cov	1.129	0.706	0.930	0.777	0.387
⑤$N_f \geq 0.50N$	Mean	1.309	1.149	1.160	1.100	0.997
	Cov	1.217	0.862	1.015	0.786	0.366

Table 4.5 Number of performance function calls for $Z = (R - 11.0)^2 - (S - 6)$ example.

Cutoff criterion		$N_s = 2$	$N_s = 5$	$N_s = 10$	$N_s = 20$	$N_s = 50$
①$N_f \geq 0.02N$	Mean	217.3	250.1	282.5	341.6	551.5
	s.d.	70.9	73.5	78.9	83.0	125.4
②$N_f \geq 0.05N$	Mean	234.3	270.3	309.1	370.7	638.8
	s.d.	51.9	62.8	79.4	67.8	111.3
③$N_f \geq 0.10N$	Mean	245.7	286.6	323.5	399.9	712.4
	s.d.	36.1	77.5	69.6	65.6	92.6
④$N_f \geq 0.20N$	Mean	262.7	299.7	347.5	430.8	772.7
	s.d.	62.1	84.8	73.2	90.8	88.0
⑤$N_f \geq 0.50N$	Mean	267.9	318.7	370.3	467.3	842.7
	s.d.	39.2	81.6	71.6	85.1	92.3

method, that are considered to be useful in structural reliability problems, are highlighted in this chapter.

There are many other topics that need to be introduced for MCS in structural reliability analysis. These include generation of a three-dimensional random field, MCS technique for code calibration and partial factors determination.

It is the author's observation that classic reliability analysis tools, such as FORM, are now facing difficulties when applied to recent design methods where the calculations become more and more complex and sophisticated. On the other hand, our computational capabilities have progressed tremendously compared to the time when these classic methods were developed. We need to develop more flexible and more user-friendly reliability analysis tools, and MCMS is one of the strong candidate methods to be used for this purpose. More development, however, is still necessary to fulfill these requirements.

Appendix: Mathematical proof of MCMC by Metropolis–Hastings algorithm

(1) Generation of random numbers by MCMC based on Metropolis–Hastings algorithm

Let us generate x that follows PDF $\pi(x)$. Note that x used in this section can be a random variable or a random vector that follows the specified PDF $\pi(x)$.

Let us consider a Markov chain process $x^{(t)} \to x^{(t+1)}$. The procedure to generate this chain is described as follows (Gilks *et al.*, 1996):

Step 1   Fix an appropriate proposal density, which can generally be described as:

$$q\left(x' \mid x^{(t)}\right) \tag{4.31}$$

Step 2   Calculate the acceptance probability which is defined by the equation below:

$$\alpha\left(x'; x^{(t)}\right) = \min\left\{1, \frac{q\left(x^{(t)} \mid x'\right) \pi\left(x'\right)}{q\left(x' \mid x^{(t)}\right) \pi\left(x^{(t)}\right)}\right\} \tag{4.32}$$

α is the acceptance probability in $(0.0, 1.0)$.

Step 3 $x^{(t)}$ is advanced to $x^{(t+1)}$ by the condition below based on the acceptance probability calculated in the previous step:

$$x^{(t+1)} = \begin{cases} x' & \text{with probability } \alpha \\ x^{(t)} & \text{with probability } 1 - \alpha \end{cases}$$

$x^{(t+1)}$ generated by this procedure surely depends on the previous step value $x^{(t)}$.

(2) Proof of the Metropolis–Hastings algorithm

It is proved in this subsection that x generated following the procedure described above actually follows PDF $\pi(x)$. The proof is given in the three steps:

1   First, the transition density, which is to show the probability to move form x to x' in this Markov chain, is defined:

$$p\left(x' \mid x\right) = q\left(x' \mid x\right) \alpha\left(x'; x\right) \tag{4.33}$$

2   It is shown next that the transition density used in the Metropolis–Hastings algorithm satisfies so-called detailed balance. This implies that $p(x' \mid x)$ is reversible with respect to time:

$$p\left(x'|x\right)\pi\left(x\right) = p\left(x|x'\right)\pi\left(x'\right) \tag{4.34}$$

3   Finally, it is proved that generated sequences of numbers by this Markov chain follow a stationary distribution $\pi(x)$, which is expressed by the following relationship:

$$\int p\left(x'|x\right)\pi\left(x\right)\mathrm{d}x = \pi\left(x'\right) \tag{4.35}$$

It is proved here that the transition density used in Metropolis–Hastings algorism satisfies the detailed balance:

It is obvious that Equation (4.34) is satisfied when $x = x'$.

When $x \neq x'$,

$$p\left(x'|x\right)\pi\left(x\right) = q\left(x'|x\right)\alpha\left(x';x\right)\pi\left(x\right)$$

$$= q\left(x'|x\right)\min\left\{1, \frac{q(x|x')\pi(x')}{q(x'|x)\pi(x)}\right\}\pi\left(x\right)$$

$$= \min\left\{q\left(x'|x\right)\pi\left(x\right), q\left(x|x'\right)\pi\left(x'\right)\right\} \tag{4.36}$$

Therefore, $p\left(x'|x\right)\pi\left(x\right)$ is symmetric with respect to x and x'. Thus,

$$p\left(x'|x\right)\pi\left(x\right) = q\left(x'|x\right)\pi\left(x'\right) \tag{4.37}$$

From this relationship, the detailed balance of Equation (4.34) is proved.

Finally, the stationarity of generated x and x' are proved based on the detailed balance of Equation (4.34). This can be proved if the results below are obtained.

$$x \sim \pi\left(x\right)$$

$$x' \sim \pi\left(x'\right) \tag{4.38}$$

This is equivalent to prove the stationarity of the Markov chain:

$$\int p\left(x' \mid x\right)\pi\left(x\right)\mathrm{d}x = \pi\left(x'\right) \tag{4.39}$$

which is obtained as follows:

$$\int p\left(x'|x\right)\pi\left(x\right)\mathrm{d}x = \int p\left(x|x'\right)\pi\left(x'\right)\mathrm{d}x$$

$$= \pi\left(x'\right)\int p\left(x|x'\right)\mathrm{d}x$$

$$= \pi\left(x'\right)$$

Thus, a sequence of numbers generated by the Metropolis–Hastings algorithm follow PDF $\pi(x)$.

References

Ang, A. and Tang, W. H. (1984). *Probability Concepts in Engineering Planning and Design, Vol. II: Decision, Risk and Reliability*. John Wiley & Sons.

Au, S.-K. and Beck, J. L. (2003). Subset Simulation and its Appplication to Seismic Risk Based Dynamic Analysis. *Journal of Engineering Mechanics, ASCE*, 129 (8), 901–17.

Broding, W. C., Diederich, F. W. and Parker, P. S. (1964) Structural optimization and design based on a reliability design criterion, *J. Spacecraft*, 1(1), 56–61

Gilks, W. R., Richardson, S. and Spiegelhalter, D. J. (1996). *Markov Chain Monte Carlo in Practice*. Chapman & Hall/CRC, London.

Koyanagi, Y. (1989). *Random Numbers, Numerical Calculation Softwares by FORTRAN 77*, Eds. T. Watabe, M. Natori and T. Oguni. Maruzen Co., Tokyo, pp. 313–22 (in Japanese).

Melcher, R. E. (1999). *Structure Reliability Analysis and Prediction*, 2nd ed. John Wiley & Sons Ltd, England.

Reiss, R. D. and Thomas, M. (1997). *Statistical Analysis of Extreme Values*. Birkhauser, Basel.

Rubinstein, R. Y. (1981). *Simulation and the Monte Carlo Method*. John Wiley & Sons Ltd, New York.

Tezuka, S. (1995). *Uniform Random Numbers: Theory and Practice*. Kluwer, Boston.

Tezuka, S. (2003). *Mathematics in Low Discrepancy Sequences. In Computational Statistics I*. Iwanami Publisher, Tokyo, pp. 65–120 (in Japanese).

Tsuda, T. (1995). *Monte Carlo Methods and Simulation*, 3rd ed. Baifukan, Tokyo (in Japanese).

Yoshida, I. and Sato, T. (2005a). *A Fragility Analysis by Low Discrepancy Sequences, Journal of Structural Engineering*, 51A, 351–6 (in Japanese).

Yoshida, I. and Sato, T. (2005b). *Effective Estimation Method of Low Failure Probability by using Markov Chain Monte Carlo, Journal of Structural Engineering*, No.794/I-72m. 43–53 (in Japanese).

Chapter 5

Practical application of reliability-based design in decision-making

Robert B. Gilbert, Shadi S. Najjar, Young-Jae Choi and Samuel J. Gambino

5.1 Introduction

A significant advantage of reliability-based design (RBD) approaches is to facilitate and improve decision-making. While the attention in RBD has generally been focused on the format and calibration of design-checking equations, the basic premise of a reliability-based design is that the target of the design should achieve an acceptable level of reliability. With this premise in mind, designers and stakeholders can utilize a range of possible alternatives in order to achieve the desired reliability. The objective of this chapter is to present practical methods and information that can be used to this end.

This chapter begins with a discussion of decision trees and decision criteria. Next, the methods for conducting reliability analyses and the results from reliability analyses are considered in the context of decision-making. Finally, the calibration of mathematical models with data is addressed for the purposes of reliability analyses and, ultimately, decision-making.

The emphasis in presenting the material in this chapter is on illustrating basic concepts with practical examples. These examples are all based on actual applications where the concepts were used in solving real-world problems to achieve more reliable and/or more cost-effective designs.

In order to provide consistency throughout the chapter, the examples are intentionally narrowed in scope to the design of foundations for offshore structures. For context, two types of structures are considered herein: fixed jacket platforms and floating production systems. Fixed jackets consist of a steel frame with legs supported at the sea floor with driven pipe piles (Figure 5.1). Fixed jackets are used in water depths up to several hundred meters with driven piles that are approximately 1 m in diameter and 100 m long. Floating production systems consist of a steel hull moored to the sea floor with 8–16 mooring lines anchored by steel caissons that are jacked into the soil using under-pressure and are called suction caissons. Floating systems are used in water depths of 1000 m or more with suction caissons that are approximately 5 m in diameter and 30 m long. Both structures

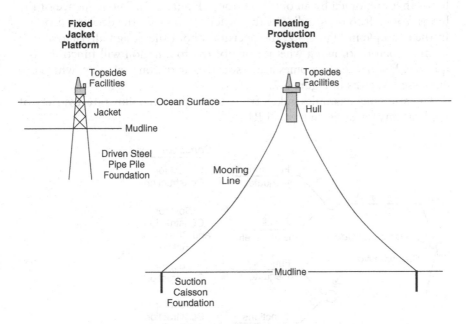

Figure 5.1 Schematic of typical offshore structures.

provide a platform for the production and processing of oil and gas offshore. The structures typically cost between $100 million for jackets up to more than $1 billion for floating systems, and the foundation can cost anywhere from 5 to 50% of the cost of the structure.

5.2 Decision trees

Decision trees are a useful tool for structuring and making decisions (Benjamin and Cornell, 1970; Ang and Tang, 1984). They organize the information that is needed in making a decision, provide for a repeatable and consistent process, allow for an explicit consideration of uncertainty and risk, and facilitate communication both in eliciting input from stakeholders to the decision and in presenting and explaining the basis for a decision.

There are two basic components in a decision tree: alternatives and outcomes. A basic decision tree in reliability-based design is shown in Figure 5.2 to illustrate these components. There are two alternatives, e.g. two different pile lengths. For each alternative, there are two possible outcomes: the design functions adequately, e.g. the settlements in the foundation do not distress the structure; or it functions inadequately, e.g. the settlements are excessive. In a decision tree, the alternatives are represented by limbs that are joined by a decision node, while the outcomes are represented by limbs that are joined by an outcome node (Figure 5.2). The sequence of the limbs from left to right indicates the order of events in the decision process. In the example in Figure 5.2, the decision about the design alternative will be made before knowing whether or not the foundation will function adequately. Finally, the outcomes are associated with consequences, which are expressed as costs in Figure 5.2.

The following examples, taken from actual projects, illustrate a variety of applications for decision trees in RBD.

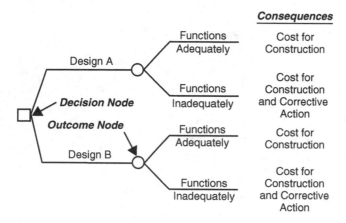

Figure 5.2 Basic decision tree.

Box 5.1 Example 1 – Decision tree for design of new facility

The combination of load and resistance factors to be used for the design of a caisson foundation for a floating offshore structure was called into question due to the unique nature of the facility: the loads on the foundation were dominated by the sustained buoyant load of the structure versus transient environmental loads, and the capacity of the foundation was dominated by its weight versus the shear resistance of the soil. The design-build contractor proposed to use a relatively small safety margin, defined as the ratio of the load factor divided by the resistance factor, due to the relatively small uncertainty in both the load and the capacity. The owner wanted to consider this proposed safety margin together with a typical value for more conventional offshore structures, which would be higher, and an intermediate value. The two considerations in this decision were the cost of the foundation, which was primarily affected by needing to use larger vessels for installation as the weight and size of the caisson increased, and the cost associated with a pull-out failure of the caisson if its capacity was not adequate. The decision tree in Figure 5.3 shows the structure of decision alternatives and outcomes that were considered.

Box 5.1 cont'd

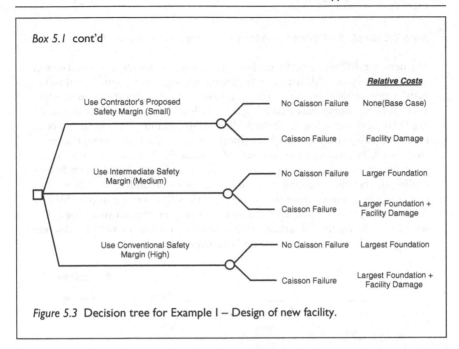

Figure 5.3 Decision tree for Example 1 – Design of new facility.

5.3 Decision criteria

Once the potential alternatives have been established, through use of decision trees or by some other means, the criterion for making a decision among potential alternatives should logically depend on the consequences for possible outcomes and their associated probabilities of happening. A rational and defensible criterion to compare alternatives when the outcomes are uncertain is the value of the consequence that is expected if that alternative is selected (Benjamin and Cornell, 1970; Ang and Tang, 1984). The expected consequence, $E(C)$, is obtained mathematically as follows:

$$E(C) = \sum_{\text{all } c_i} c_i P_C(c_i) \quad \text{or} \quad \int_{\text{all } c} c f_C(c) \, dc \qquad (5.1)$$

where C is a random variable representing the consequence and $P_C(c_i)$ or $f_C(c)$ is the probability distribution for c in discrete or continuous form, respectively.

The consequences of a decision outcome generally include a variety of attributes such as monetary value, human health and safety, environmental impact, social and cultural impact, and public perception. Within the context of expected consequence (Equation (5.1)), a simple approach to accommodate multiple attributes is to assign a single value, such as a

The need for drilling geotechnical borings in order to design new structures in a mature offshore field, where considerable geologic and geotechnical information was already available, was questioned. The trade-off for not drilling a site-specific boring was that an increased level of conservatism would be needed in design to account for the additional uncertainty. The decision tree in Figure 5.4 shows the structure of decision alternatives and outcomes that were considered. In this case, the outcome of drilling the boring, i.e. the geotechnical design properties derived from the boring data, is not known before drilling the boring. In addition, there is a range of possible design properties that could be obtained from the boring, and therefore a range of possible pile designs that would be needed for a given set of geotechnical properties. The semi-circles in Figure 5.4 indicate that there is a continuous versus a discrete set of possibilities at the outcome and decision nodes.

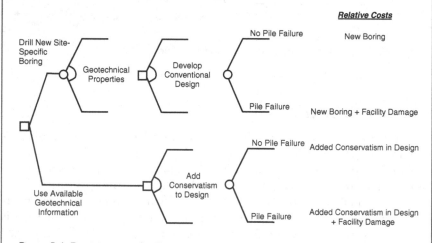

Figure 5.4 Decision tree for Example II – Site investigation in mature field.

The pile foundation for an offshore structure in a frontier area was designed based on a preliminary analysis of the site investigation data. The geotechnical properties of the site were treated in design as if the soil conditions were similar to other offshore areas where the experience base was large. The steel was then ordered. Subsequently, a more detailed analysis of the geotechnical properties showed that the soil conditions were rather unusual, calling into question

Box 5.3 cont'd

how the properties should be used in design and leading to relatively large uncertainty in the estimated pile capacity. The owner was faced with a series of decisions. First, should they stay with the original pile design or change it given the uncertainty in the pile capacity, considering that changing the pile design after the steel had been ordered would substantially impact the cost and schedule of the project? Second, if they decided to stay with the original pile design, should they monitor the installation to confirm the capacity was acceptable, considering that this approach required a flexible contract where the pile design may need to be updated after the pile is installed? The decision tree in Figure 5.5 shows the structure of decision alternatives and outcomes that were considered; all of the possibilities are shown for completeness even though some, such as keeping the design in spite of evidence that the capacity is unacceptable, could intuitively be ruled out.

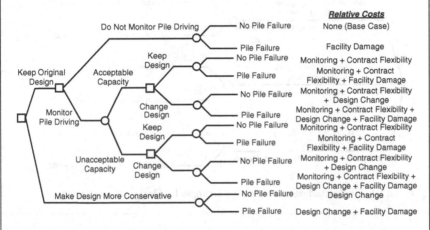

Figure 5.5 Decision tree for Example III – Pile monitoring in frontier field.

Box 5.4 Decision criterion for Example I – Design of new facility

The decision tree for the design example described in Box 5.1 (Figure 5.3) is completed in Figure 5.6 with information for the values and probabilities of the consequences. The probabilities and costs for caisson failure correspond to the event that the caisson will pull out of the mudline at some time during the 20-year design life for the facility. The costs are reported as negative

Continued

Box 5.4 cont'd

values in order to denote monetary loss. It is assumed that a caisson failure will result in a loss of $500 million. The cost of redesigning and installing the caisson to achieve an intermediate safety margin is $1 million, while the cost of redesigning and installing the caisson to achieve a high safety margin is $5 million. The cost of the caisson increases significantly if the highest safety margin is used because a more costly installation vessel will be required. In this example, the maximum expected consequence (or the minimum expected cost) is obtained for the intermediate safety margin. Note that the contribution of the expected cost of failure to the total expected cost becomes relatively insignificant for small probabilities of failure.

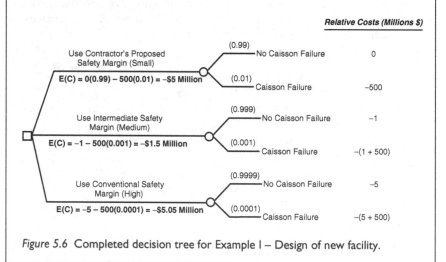

Figure 5.6 Completed decision tree for Example 1 – Design of new facility.

monetary value, that implicitly considers all of the possible attributes for an outcome.

A second approach to consider factors that cannot necessarily be directly related to monetary value, such as human fatalities, is to establish tolerable probabilities of occurrence for an event as a function of the consequences. These tolerable probabilities implicitly account for all of the attributes associated with the consequences. A common method of expressing this information is on a risk tolerance chart, which depicts the annual probability of occurrence for an event versus the consequences expressed as human fatalities. These charts are sometimes referred to as F–N charts, where F stands for probability or frequency of exceedance, typically expressed on an annual basis, and N stands for number of fatalities (Figure 5.7). These charts establish a level of risk, expressed by an envelope or line, which will be tolerated by society in exchange for benefits, such as economical energy.

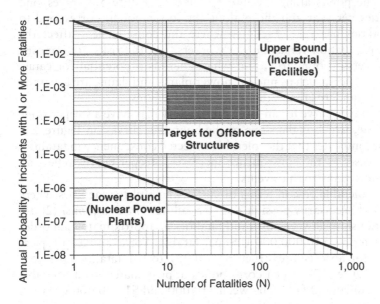

Figure 5.7 Risk tolerance chart for an engineered facility.

The risk is considered acceptable if the combination of probability and consequences falls below the envelope. Excellent discussions concerning the basis for, applications of, and limitations with these types of charts are provided in Fischhoff *et al.* (1981), Whitman (1984), Whipple (1985), ANCOLD (1998), Bowles (2001), USBR (2003) and Christian (2004).

The chart on Figure 5.7 includes an upper and lower bound for risk tolerance based on a range of *F–N* curves that have been published. The lower bound (most stringent criterion) on Figure 5.7 corresponds to a published curve estimating the risk tolerated by society for nuclear power plants (USNRC, 1975). For comparison purposes, government agencies in both the United States and Australia have established risk tolerance curves for dams that are one to two orders of magnitude above the lower bound on Figure 5.7 (ANCOLD, 1998 and USBR, 2003). The upper bound (least stringent criterion) on Figure 5.7 envelopes the risks that are being tolerated for a variety of industrial applications, such as refineries and chemical plants, based on published data (e.g. Whitman, 1984). As an example, the target levels of risk that are used by the oil industry for offshore structures are shown as a shaded rectangular box on Figure 5.7 (Bea, 1991; Stahl *et al.*, 1998; Goodwin *et al.*, 2000).

The level of risk that is deemed acceptable for the same consequences on Figure 5.7 is not a constant and depends on the type of facility. If expected consequences are calculated for an event which results in a given number of fatalities, the points along the upper bound on Figure 5.7 correspond to 0.1 fatalities per year and those along the lower bound correspond to 1×10^{-5} fatalities per year. Facilities where the consequences affect the surrounding population versus on-site workers, i.e., the risks are imposed involuntarily versus voluntarily, and where there is a potential for catastrophic consequences, such as nuclear power plants and dams, are generally held to higher standards than other engineered facilities.

The risk tolerance curve for an individual facility may have a steeper slope than one to one, which is the slope of the bounding curves on Figure 5.7. A steeper slope indicates that the tolerable expected consequence in fatalities per year decreases as the consequence increases. Steeper curves reflect an aversion to more catastrophic events.

Expressing tolerable risk in terms of costs is helpful for facilities where fatalities are not necessarily possible due to a failure. *The Economist* (2004) puts the relationship between costs and fatalities at approximately, in order of magnitude terms, $10,000,000 U.S. per fatality. For example, a typical tolerable risk for an offshore structure is 0.001 to 0.1 fatalities per year (shaded box Figure 5.7), which corresponds approximately to a tolerable expected consequence of failure between $10,000 and $1,000,000 per year. A reasonable starting point for a target value in a preliminary evaluation of a facility is $100,000 per year.

The effect of time is not shown explicitly in a typical risk tolerance curve since the probability of failure is expressed on an annual basis. However, time can be included if the annual probabilities on Figure 5.7 are assumed to apply for failure events that occur randomly with time, such as explosions, hurricanes and earthquakes. Based on the applications for which these curves were developed, this assumption is reasonable. In this way, the tolerability curve can be expressed in terms of probability of failure in the lifetime of the facility, meaning that failure events that may or may not occur randomly with time can be considered on a consistent basis.

Box 5.5 Tolerable risk for Example I – Design of new facility

Revisiting the example for the design of a floating system in Boxes 5.1 and 5.4 (Figures 5.3 and 5.6), the event of "caisson failure" does not lead to fatalities. The primary consequences for the failure of this very large and expensive facility are the economic damage to the operator and the associated negative perception from the public. In addition, the event of "caisson failure" is not an event that occurs randomly with time over the 20-year lifetime of the facility. The maximum load on the foundation occurs when a maintenance operation is performed, which happens once every five years of operation. The probability of "caisson failure" in Figure 5.6 corresponds

Box 5.5 cont'd

to the probability that the caisson will be overloaded in at least one of these operations over its 20-year life. The information on Figure 5.7 has been re-plotted on Figure 5.8 for this example by assuming that incidents occur randomly according to a Poisson process, e.g. an incident with an annual probability of 0.001 per year on Figure 5.7 corresponds to probability of occurrence of $1 - e^{-(0.001/\text{year})(20 \text{ years})} = 0.02$ in 20 years. Figure 5.8 shows that an acceptable probability of failure in the 20-year lifetime for this facility is between 0.002 and 0.02. Therefore, the first design alternative on Figure 5.6 (use contractor's proposed factor of safety (small) with a probability of caisson failure equal to 0.01), would provide a level of risk that is marginal concerning what would generally be tolerated.

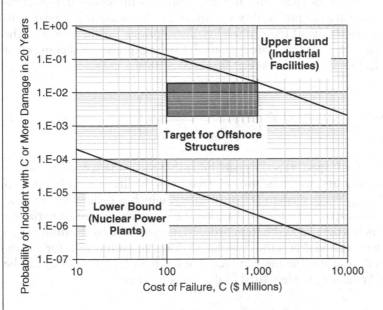

Figure 5.8 Risk tolerance chart for Example I – Design of floating system.

A third approach to consider factors in decision making that cannot necessarily be directly related to monetary value is to use multi-attribute utility theory (Kenney and Raiffa, 1976). This approach provides a rational and systematic means of combining consequences expressed in different measures to be combined together into a single scale of measure, a utility value. While flexible and general, implementation of multi-attribute utility theory is cumbersome and not commonly used in engineering practice.

5.4 Reliability analysis

Reliability analyses are an integral part of the decision-making process because they establish the probabilities for outcomes in the decision trees (e.g. the probability of caisson failure in Figure 5.6). The following sections illustrate practical applications for relating reliability analyses to decision making.

5.4.1 Simplified first-order analyses

The probability of failure for a design can generally be expressed as the probability that a load exceeds a capacity. A useful starting point for a reliability analysis is to assume that the load on the foundation and the capacity of the foundation are independent random variables with lognormal distributions. In this way, the probability of failure can be analytically related to the first two moments of the probability distributions for the load and the capacity. This simplified first-order analysis can be expressed in a convenient mathematical form, as follows (e.g. Wu *et al.*, 1989):

$$P\left(\text{Load} > \text{Capacity}\right) \cong \Phi\left(-\frac{\ln\left(FS_{\text{median}}\right)}{\sqrt{\delta_{\text{load}}^2 + \delta_{\text{capacity}}^2}}\right) \tag{5.2}$$

where $P(\text{Load} > \text{Capacity})$ is the probability that the load exceeds the capacity in the design life, which is also referred to as the lifetime probability of failure; FS_{median} is the median factor of safety, which is defined as the ratio of the median capacity to the median load; δ is the coefficient of variation (c.o.v.), which is defined as the standard deviation divided by the mean value for that variable, and $\Phi()$ is the cumulative distribution function for a standard normal variable. The median factor of safety in Equation (5.2) can be related to the factor of safety used in design as follows:

$$FS_{\text{median}} = FS_{\text{design}} \times \frac{\left(\dfrac{\text{capacity}_{\text{median}}}{\text{capacity}_{\text{design}}}\right)}{\left(\dfrac{\text{load}_{\text{median}}}{\text{load}_{\text{design}}}\right)} \tag{5.3}$$

where the subscript "design" indicates the value used to design the foundation. The ratios of the median to design values represent biases between the median value (or the most likely value since the mode equals the median for a lognormal distribution) in the design life and the value that is used in the design check with the factor of safety or with load and resistance factors. The coefficients of variation in Equation (5.2) represent uncertainty in the load and the capacity. The denominator in Equation (5.2) is referred to as the total coefficient of variation:

$$\delta_{total} = \sqrt{\delta^2_{load} + \delta^2_{capacity}} \qquad (5.4)$$

An exact solution for Equation (5.2) is obtained if the denominator, i.e., the total c.o.v. in Equation (5.4), is replaced by the following expression: $\sqrt{\ln\left(1 + \delta^2_{load}\right) + \ln\left(1 + \delta^2_{capacity}\right)}$; the approximation in Equation (5.2) is reasonable for values of the individual coefficients of variation that are less than about 0.3.

The relationship between the probability of failure and the median factor of safety and total c.o.v. is shown in Figure 5.9. An increase in the median factor of safety and a decrease in the total c.o.v. both reduce the probability of failure.

5.4.2 Physical bounds

In many practical applications of reliability analyses, there are physical bounds on the maximum load that can be applied to the foundation or the minimum capacity that will be available in response to the load. It is

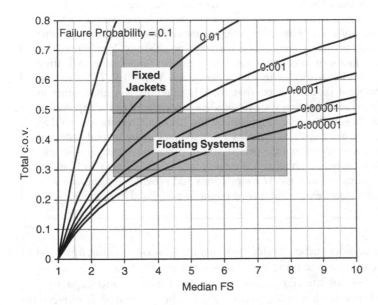

Figure 5.9 Graphical solutions for simplified first-order reliability analysis.

Box 5.6 Reliability benchmarks for Example I – Design of new facility

In evaluating the appropriate level of conservatism for the design of a new floating system, as described in Box 5.1 (Figure 5.3), it is important to consider the reliability levels achieved for similar types of structures. Fixed steel jacket structures have been used for more than 50 years to produce oil and gas offshore in relatively shallow water depths up to about 200 m. In recent years, floating systems have been used in water depths ranging from 1000 to 3000 m. To a large extent, the foundation design methods developed for driven piles in fixed structures have been applied directly to the suction caissons that are used for floating systems.

For a pile in a fixed jacket with a design life of 20 years, the median factor of safety is typically between three and five and the total c.o.v. is typically between 0.5 and 0.7 (Tang and Gilbert, 1993). Hence, the resulting probability of failure in the lifetime is on the order of 0.01 based on Figure 5.9. Note that the event of foundation failure, i.e., axial overload of a single pile in the foundation, does not necessarily lead to collapse of a jacket; failure probabilities for the foundation system are ten to 100 times smaller than those for a single pile (Tang and Gilbert, 1993).

For comparison, typical values were determined for the median factor of safety and the total c.o.v. for a suction caisson foundation in a floating production system with a 20-year design life (Gilbert *et al.*, 2005a). The median factor of safety ranges from three to eight. The median factor of safety tends to be higher for the floating versus fixed systems for two reasons. First, a new source of conservatism was introduced for floating systems in that the foundations are checked for a design case where the structure is damaged (i.e., one line is removed from the mooring system). Second, the factors of safety for suction caissons were generally increased above those for driven piles due to the relatively small experience base with suction caissons. In addition to a higher median factor of safety, the total c.o.v. value for foundations in floating systems tends to be smaller, with values between 0.3 and 0.5. This decrease compared to fixed jackets reflects that there is generally less uncertainty in the load applied to a mooring system foundation compared to that applied to a jacket foundation. The resulting probabilities of foundation failure tend to be smaller for floating systems compared to fixed jackets by several orders of magnitude (Figure 5.9). From the perspective of decision making, this simplified reliability analysis highlights a potential lack of consistency in how the design code is applied to offshore foundations and provides insight into why such an inconsistency exists and how it might be addressed in practice.

---

important to consider these physical bounds in decision making application because they can have a significant effect on the reliability.

One implication of the significant role that a lower-bound capacity can have on reliability is that information about the estimated lower-bound capacity should possibly be included in design-checking equations for

Box 5.7 Lower-bound capacity for Example I – Design of new facility

In evaluating the appropriate level of conservatism for the design of a new floating system, as described in Box 5.1 (Figure 5.3), it is important to consider the range of possible values for the capacity of the caisson foundations. A reasonable estimate of a lower-bound on the capacity for a suction caisson in normally consolidated clay can be obtained using the remolded strength of the clay to calculate side friction and end bearing. Based on an analysis of load-test data for suction caissons, Najjar (2005) found strong statistical evidence for the existence of this physical lower bound and that the ratio of lower-bound capacities to measured capacities ranged from 0.25 to 1.0 with an average value of 0.6.

The effect of a lower-bound capacity on the reliability of a foundation for a floating production system is shown in Figure 5.10. The structure is anchored with suction caisson foundations in a water depth of 2000 m; details of the analysis are provided in Gilbert *et al.* (2005a). Design information for the caissons is as follows: $load_{median}/load_{design} = 0.7$; $\delta_{load} = 0.14$; $capacity_{median}/capacity_{design} = 1.3$; and $\delta_{capacity} = 0.3$. The lognormal distribution for capacity is assumed to be truncated at the lower-bound value using a mixed lognormal distribution, where the probability that the capacity is equal to the lower bound is set equal to the probability that the capacity is

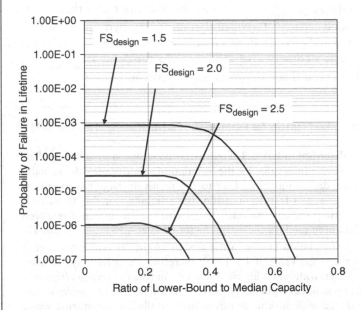

Figure 5.10 Effect of lower-bound capacity on foundation reliability for Example I – Design of floating system.

Continued

Box 5.7 cont'd

less than or equal to the lower bound for the non-truncated distribution. The probability of failure is calculated through numerical integration.

The results on Figure 5.10 show the significant role that a lower-bound capacity can have on the reliability. For a lower-bound capacity that is 0.6 times the median capacity, the probability of failure is more than 1000 times smaller with the lower-bound than without it (when the ratio of the lower-bound to median capacity is zero) for a design factor of safety of 1.5.

RBD codes. For example, an alternative code format would be to have two design-checking equations:

$$\phi_{capacity} capacity_{design} \geq \gamma_{load} load_{design}$$

or   (5.5)

$$\phi_{capacity_{lower\ bound}} capacity_{lower\ bound} \geq \gamma_{load} load_{design}$$

where $\phi_{capacity}$ and γ_{load} are the conventional resistance and load factors, respectively, and $\phi_{capacity_{lower\ bound}}$ is an added resistance factor that is applied to the lower-bound capacity. Provided that one of the two equations is satisfied, the design would provide an acceptable level of reliability. Gilbert et al. (2005b) show that a conservative value of 0.75 for $\phi_{capacity_{lower\ bound}}$ would cover a variety of typical conditions in foundation design.

Box 5.8 Lower-bound capacity for Example III – Pile monitoring in frontier field

The example described in Box 5.3 (Figure 5.5) illustrates a practical application where a lower-bound value for the foundation capacity provides valuable information for decision-making. Due to the relatively large uncertainty in the pile capacity at this frontier location, the reliability for this foundation was marginal using the conventional design check even though a relatively large value had been used for the design factor of safety, $FS_{design} = 2.25$. The probability of foundation failure in the design life of 20 years was slightly greater than the target value of 1×10^{-3}; therefore, the alternative of using the existing design without additional information (the top design alternative on Figure 5.5) was not acceptable. However, the alternative of supplementing the existing design with information from pile installation monitoring would only be economical if the probability that the design would need to be changed after installation was small. Otherwise, the preferred alternative for the owner would be to modify the design before installation.

Box 5.8 cont'd

In general, pile monitoring information for piles driven into normal to slightly overconsolidated marine clays cannot easily be related to the ultimate pile capacity due to the effects of set-up following installation. The capacity measured at the time of installation may only be 20–30% of the capacity after set-up. However, the pile capacity during installation does, arguably, provide a lower-bound on the ultimate pile capacity. In order to establish the probability of needing to change the design after installation, the reliability of the foundation was related to the estimated value for the lower-bound capacity based on pile driving, as shown on Figure 5.11. In this analysis, the pile capacity estimated at the time of driving was considered to be uncertain; it was modeled as a normally distributed random variable with a mean equal to the estimated value and a coefficient of variation of 0.2 to account for errors in the estimated value. The probability of foundation failure in the design life was then calculated by integrating the lower-bound value over the range of possible values for a given estimate. The result is shown in Figure 5.11. Note that even though there is uncertainty in the lower-bound value, it still can have a large effect on the reliability.

The expected value of driving resistance to be encountered and measured during installation is indicated by the arrow labeled "installation" in

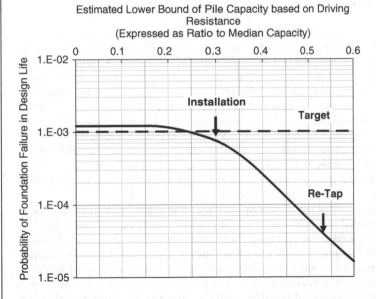

Figure 5.11 Reliability versus lower-bound capacity for Example III – Pile monitoring in frontier field.

Continued

Box 5.8 cont'd

Figure 5.11. While it was expected that the installation information would just barely provide for an acceptable level of reliability (Figure 5.11), the consequence of having to change the design at that point was very costly. In order to provide greater assurance, the owner decided to conduct a re-tap analysis 5 days after driving. The arrow labeled "re-tap" in Figure 5.11 shows that this information was expected to provide the owner with significant confidence that the pile design would be acceptable. The benefits of conducting the re-tap analysis were considered to justify the added cost.

5.4.3 Systems versus components

Nearly all design codes, whether reliability-based or not, treat foundation design on the basis of individual components. However, from a decision-making perspective, it is insightful to consider how the reliability of an individual foundation component is related to the reliability of the overall foundation and structural system.

Box 5.9 System reliability analysis for Example I – Design of new facility

The mooring system for a floating offshore facility, such as that described in Box 5.1 (Figure 5.3), includes multiple lines and foundation anchors. The results from a component reliability analysis are shown in Figure 5.12 for a facility moored in three different water depths (Choi *et al.*, 2006). These results correspond to the most heavily loaded line in the system, where the primary source of loading is from hurricanes. Each mooring line consists of segments of steel chain and wire rope or polyester. The points labeled "rope & chain" in Figure 5.12 correspond to a failure anywhere within the segments of the mooring line, the points labeled "anchor" correspond to a failure at the foundation, and the points labeled "total" correspond to a failure anywhere within the line and foundation. The probability that the foundation fails is about three orders of magnitude smaller than that for the line.

The reliability of the system of lines was also related to that for the most heavily loaded line (Choi *et al.*, 2006). In this analysis, the event that the system fails is related to that for an individual line in two ways. First, the loads will be re-distributed to the remaining lines when a single line fails. Second, the occurrence of a single line failure indicates information about the loads on the system; if a line has failed, it is more likely that the system is being exposed to a severe hurricane (although it is still possible that the loads are relatively small and the line capacity for that particular line was small). The results for this analysis are shown in Figure 5.13, where the probability that the mooring system fails is expressed as a conditional probability for the event that a

Box 5.9 cont'd

single line in the system fails during a hurricane. For the mooring systems in 2000 and 3000 m water depths, the redundancy is significant in that there is less than a 10% chance the system will fail even if the most heavily loaded line fails during a hurricane. Also, the mooring system in the 1000 m water depth has substantially less redundancy than those in deeper water.

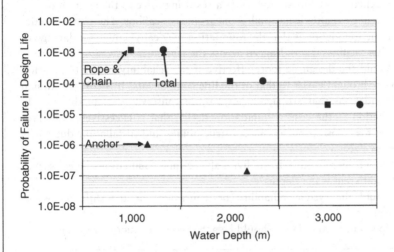

Figure 5.12 Comparison of component reliabilities for most heavily loaded line in offshore mooring system.

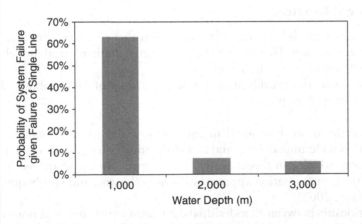

Figure 5.13 System redundancy for offshore mooring system.

Continued

Box 5.9 cont'd

The results in Figures 5.12 and 5.13 raise a series of questions for consideration by the writers of design codes.

(1) Is it preferable to have a system fail in the line or at the foundation? The benefit of a failure in the foundation is that the weight of the foundation attached to the line still provides a restoring force to the entire mooring system even after it has pulled out from the seafloor. The cost of failure in the foundation is that the foundation may cause collateral damage to other facilities if it is dragged across the seafloor.

(2) Would the design be more effective if the reliability of the foundation were brought closer to that of the line components? It is costly to design foundations that are significantly more conservative than the line. In addition to the cost, a foundation design that is excessively conservative may pose problems in constructability and installation due to its large size.

(3) Should the reliability for individual lines and foundations be consistent for different water depths?

(4) Should the system reliability be consistent for the different water depths?

(5) How much redundancy should be achieved in the mooring system?

5.5 Model calibration

The results from a reliability analysis need to be realistic to be of practical value in decision-making. Therefore, the calibration of mathematical models with real-world data is very important.

A general framework for calibrating a model with a set of data is obtained from the following assumptions.

• The variation for an individual measurement about its mean value is described by a Hermite Polynomial transformation of a standard normal distribution, which can theoretically take on any possible shape as the order of the transformation approaches infinity (Journel and Huijbregts, 1978; Wang, 2002).

• The relationship between an individual data point with other data points is described by a linear correlation.

• The shape of the Hermite Polynomial transformation is consistent between data points, i.e., affine correlation as described in Journel and Huijbregts (1978).

In mathematical terms, the probability distribution to describe variations between a single measurement y_i and a model of that measurement, Y_i, is described by a mean value, a standard deviation, correlation coefficients, and a conditional probability distribution:

$$\mu_{Y_i} = g_\mu \left(\mathbf{x}_{Y_i}, \theta_1, \ldots, \theta_{n_\mu} \right) \tag{5.6}$$

$$\sigma_{Y_i} = g_\sigma \left(\mathbf{x}_{Y_i}, \mu_{Y_i}, \theta_{n_\mu+1}, \ldots, \theta_{n_\mu+n_\sigma} \right) \tag{5.7}$$

$$\rho_{Y_i,Y_j} = g_\rho \left(\mathbf{x}_{Y_i}, \mathbf{x}_{Y_j}, \theta_{n_\mu+n_\sigma+1}, \ldots, \theta_{n_\mu+n_\sigma+n_\rho} \right) \tag{5.8}$$

$$F_{Y_i|y_1,\ldots,y_{i-1}} \left(y_i | y_1, \ldots, y_{i-1} \right) = g_F \left(\mathbf{x}_{Y_i}, \mu_{Y_i|y_1,\ldots,y_{i-1}}, \sigma_{Y_i|y_1,\ldots,y_{i-1}}, \right.$$
$$\left. \theta_{n_\mu+n_\sigma+n_\rho+1}, \ldots, \theta_{n_\mu+n_\sigma+n_\rho+n_F} \right) \tag{5.9}$$

where μ_{Y_i} is the mean value and $g_\mu()$ is a model with n_μ model parameters, $\theta_1, \ldots, \theta n_\mu$, that relate the mean to attributes of the measurement, $\mathbf{x}_{Y_i}$; σ_{Y_i} is the standard deviation and $g_\sigma()$ is a model with n_σ model parameters, $\theta_{n_\mu+1}, \ldots, \theta_{n_\mu+n_\sigma}$, that relate the standard deviation to attributes and the mean value of the measurement; ρ_{Y_i,Y_j} is the correlation coefficient between measurements i and j and $g_\rho()$ is a model with n_ρ model parameters, $\theta_{n_\mu+n_\sigma+1}, \ldots, \theta_{n_\mu+n_\sigma+n_\rho}$, that relate the correlation coefficient to the attributes for measurements i and j; and $F_{Y_i|y_1,\ldots,y_{i-1}} \left(y_i | y_1, \ldots, y_{i-1} \right)$ is the cumulative distribution function for measurement i conditioned on the measurements y_1 to y_{i-1} and $g_F()$ is a model with n_F model parameters, $\theta_{n_\mu+n_\sigma+n_\rho+1}, \ldots, \theta_{n_\mu+n_\sigma+n_\rho+n_F}$, that relate the cumulative distribution function to the data attributes and the conditional mean value, $\mu_{Y_i|y_1,\ldots,y_{i-1}}$, and standard deviation, $\sigma_{Y_i|y_1,\ldots,y_{i-1}}$, which are obtained from the mean values, standard deviations and correlations coefficients for measurements 1 to i and the measurements y_1 to y_{i-1}. The n_F model parameters in $g_F()$ are the coefficients in the Hermite Polynomial transformation function, where the order of the transformation is $n_F + 1$:

$$g_F \left(\mathbf{x}_{Y_i}, \mu_{Y_i|y_1,\ldots,y_{i-1}}, \sigma_{Y_i|y_1,\ldots,y_{i-1}}, \theta_{n_\mu+n_\sigma+n_\rho+1}, \ldots, \theta_{n_\mu+n_\sigma+n_\rho+n_F} \right)$$
$$= \int_{\varphi(u) \leq \dfrac{y_i - \mu_{Y_i|y_1,\ldots,y_{i-1}}}{\sigma_{Y_i|y_1,\ldots,y_{i-1}}}} \frac{1}{\sqrt{2\pi}} e^{-\frac{1}{2}u^2} du \tag{5.10}$$

where

$$\varphi(u) = \frac{\left[H_1(u) + \sum_{m=2}^{n_F+1} \frac{\psi_m}{m!} H_m(u) \right]}{-\sqrt{1 + \sum_{m=2}^{n_F+1} \frac{\psi_m^2}{m!}}} = \frac{y - \mu_{Y_i|y_1,\dots,y_{i-1}}}{\sigma_{Y_i|y_1,\dots,y_{i-1}}}$$

(5.11)

in which ψ_m are polynomial coefficients that are expressed as model parameters, $\psi_m = \theta_{n_\mu + n_\sigma + n_\rho + m-1}$, and $H_m(u)$ are Hermite polynomials: $H_{m+1}(u) = -uH_m(u) - mH_{m-1}(u)$ with $H_0(u) = 1$ and $H_1(u) = -u$. The practical implementation of Equation (5.10) involves first finding all of the intervals of u in Equation (5.11) where $\varphi(u) \leq \frac{y_i - \mu_{Y_i|y_1,\dots,y_{i-1}}}{\sigma_{Y_i|y_1,\dots,y_{i-1}}}$ and then integrating the standard normal distribution over those intervals. If the data points are assumed to follow a normal distribution, then the order of the polynomial transformation is one, $n_F = 0$, $\varphi(u) = u = \frac{y - \mu_{Y_i|y_1,\dots,y_{i-1}}}{\sigma_{Y_i|y_1,\dots,y_{i-1}}}$, and $F_{Y_i|y_1,\dots,y_{i-1}}(y_i|y_1,\dots,y_{i-1}) = \Phi(u)$ where $\Phi(u)$ is the standard normal function.

The framework described above contains $n_\theta = n_\mu + n_\sigma + n_\rho + n_F$ model parameters, $\theta_1, \dots, \theta_{n_\mu + n_\sigma + n_\rho + n_F}$, that need to be estimated or calibrated based on available information. It is important to bear in mind that these model parameters have probability distributions of their own since they are not known with certainty. For a given set of measurements, the calibration of the various model parameters follows a Bayesian approach

$$f_{\Theta_1,\dots,\Theta_{n_\theta}|y_1,\dots,y_n}\left(\theta_1,\dots,\theta_{n_\theta}|y_1,\dots,y_n\right)$$

(5.12)

$$= \frac{P\left(y_1,\dots,y_n|\theta_1,\dots,\theta_{n_\theta}\right) f_{\Theta_1,\dots,\Theta_{n_\theta}}\left(\theta_1,\dots,\theta_{n_\theta}\right)}{P\left(y_1,\dots,y_n\right)}$$

where $f_{\Theta_1,\dots,\Theta_{n_\theta}|y_1,\dots,y_n}\left(\theta_1,\dots,\theta_{n_\theta}|y_1,\dots,y_n\right)$ is the calibrated or updated probability distribution for the model parameters, $P\left(y_1,\dots,y_n|\theta_1,\dots,\theta_{n_\theta}\right)$ is the probability of obtaining this specific set of measurements for a given set of model parameters, and $f_{\Theta_1,\dots,\Theta_{n_\theta}}\left(\theta_1,\dots,\theta_{n_\theta}\right)$ is the probability distribution for the model parameters based on any additional information that is independent of the measurements. The probability of obtaining the measurements, $P\left(y_1,\dots,y_n|\theta_1,\dots,\theta_{n_\theta}\right)$, can include both point measurements as well as interval measurements such as proof loads or censored values

(Finley, 2004). While approximations exist for solving Equation (5.12) (e.g. Gilbert, 1999), some form of numerical integration is generally required.

The information from model calibration can be used in decision making in two ways. First, the calibrated distribution from Equation (5.12) provides models that are based on all available information at the time of making a decision; uncertainty in the calibrated parameters can be included directly in the decision since the model parameters are represented by a probability distribution and not deterministic point estimates. Second, the effect and therefore value of obtaining additional information before making a decision (e.g. Figure 5.4) can be assessed.

Box 5.10 Probability distribution of capacity for Example 1 – Design of new facility

The probability distribution for axial pile capacity plays an important role in reliability-based design, such as described in Box 5.1, because it dictates the probability of foundation failure due to overloading. A database of pile load tests with 45 driven piles in clay was analyzed by Najjar (2005) to investigate the shape of the probability distribution. First, a conventional analysis was conducted on the ratio of measured to predicted capacity, where the distribution for this ratio was assumed to be lognormal. Individual data points were assumed to be statistically independent. In terms of the framework presented in Equations (5.6) through (5.11), the attribute for each data point, x_i, is the predicted capacity; the function g_μ () is $\mu_{\ln Y_i} = \ln(\theta_1 x_i)$; the function g_σ () is $\sigma_{\ln Y_i} = \sqrt{\ln(1 + \theta_2^2)}$; the data points are statistically independent, i.e., the correlation coefficients between data points are zero; and the cumulative distribution function for $\ln Y_i$, g_F (), is a normal distribution with mean $\mu_{\ln Y_i}$ and standard deviation $\sigma_{\ln Y_i}$. Therefore, there are two model parameters to be calibrated with the data, θ_1 and θ_2.

In calibrating the model parameters with Equation (5.12), the only information used was the measured data points in the load test database; i.e., $f_{\Theta_1, \Theta_2}(\theta_1, \theta_2)$ is taken as a non-informative diffuse prior distribution. The calibrated expected values for θ_1 and θ_2 from the updated probability distribution obtained with Equation (5.12) are 0.96 and 0.24, respectively. The resulting probability distribution is shown on Figure 5.14 and labeled "conventional lognormal."

In Boxes 5.7 and 5.8, the effect of a lower bound on the pile capacity was shown to be significant. In order to investigate the existence of a lower-bound value, an estimate for a lower-bound capacity was established for each data point based on the site-specific soil properties and pile geometry. Specifically, remolded undrained shear strengths were used to calculate a lower-bound capacity for driven piles in normally consolidated to slightly overconsolidated

Continued

Box 5.10 cont'd

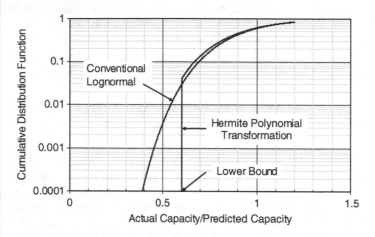

Figure 5.14 Calibrated probability distributions for axial pile capacity.

clays and residual drained shear strengths and at-rest conditions were used to calculate a lower-bound capacity for driven piles in highly overconsolidated clays (Najjar, 2005). For all 45 data points, the measured capacity was above the calculated value for the lower-bound capacity. To put this result in perspective, the probability of obtaining this result if the distribution of capacity really is a lognormal distribution (with a lower bound of zero) is essentially zero, meaning that a conventional lognormal distribution is not plausible.

The model for the probability distribution of the ratio of measured to predicted capacity was therefore refined to account for a more complicated shape than a simple lognormal distribution. The modifications are the following: an additional attribute, the predicted lower-bound capacity, is included for each data point; and a 7th-order Hermite Polynomial transformation is used as a general model for the shape of the probability distribution. In terms of Equations (5.6) through (5.11), the attributes for each data point are the predicted capacity, $x_{1,i}$, and the calculated lower-bound capacity, $x_{2,i}$; the function g_μ () is $\mu_{Y_i} = \theta_1 x_{1,i}$; the function g_σ () is $\sigma_{Y_i} = \theta_2 \mu_{Y_i}$; the correlation coefficients between data points were zero; and the cumulative distribution function for $\ln Y_i$, g_F (), is obtained by combining Equations (5.11) and (5.12) as follows

$$
F_Y(y_i) = \begin{cases} 0 & y_i < x_{2,i} \\ \Phi(u_i)\ \text{where}\ \dfrac{H_1(u_i) + \sum\limits_{m=2}^{7} \dfrac{\theta_{m+1}}{m!} H_m(u_i)}{-\sqrt{1 + \sum\limits_{m=2}^{7} \dfrac{\theta_{m+1}^2}{m!}}} = \dfrac{y_i - \mu_{Y_i}}{\sigma_{Y_i}} & y_i \geq x_{2,i} \end{cases}
$$

$$(5.13)$$

Box 5.10 cont'd

Therefore, there are now eight model parameters to be calibrated with the data, θ_1 to θ_8.

The calibrated expected values for the model parameters from the updated probability distribution obtained with Equation (5.12) are $\theta_1 = 0.96$, $\theta_2 = 0.26$, $\theta_3 = -0.289$, $\theta_4 = 0.049$, $\theta_5 = -0.049$, $\theta_6 = 0.005$, $\theta_7 = 0.05$, and $\theta_8 = 0.169$. The resulting probability distribution is shown on Figure 5.14 and labeled "Hermite polynomial transformation."

There are two practical conclusions from this calibrated model. First, the left-hand tail of the calibrated distribution with the 7th-order Hermite Polynomial transformation is substantially different than that for the conventional lognormal distribution (Figure 5.14). This difference is significant because it is the lower percentiles of the distribution for capacity, say less than 10%, which govern the probability of failure. Second, truncating the conventional lognormal distribution at the calculated value for the lower-bound capacity provides a reasonable and practical fit to the more complicated 7th-order Hermite Polynomial transformation.

Box 5.11 Value of new soil boring for Example II – Site investigation in mature field

A necessary input to the decision about whether or not to drill an additional soil boring in a mature field, as described in Box 5.2 and illustrated in Figure 5.4, is a model describing spatial variability in pile capacity across the field. The geology for the field in this example is relatively uniform marine clays that are normally to slightly overconsolidated. The available data are soil borings where the geotechnical properties were measured and a design capacity was determined. Two types of soil borings are available, modern borings where high-quality soil samples were obtained and older borings where soil samples were of lower quality. Pile capacity for these types of piles is governed by side shear capacity.

The model for the spatial variability in design pile capacity, expressed in terms of the average unit side shear over a length of pile penetration, is given by Equations (5.14) through (5.17):

$$\mu_{Y_i} = \left(\theta_1 + \theta_2 x_{1,i} + \theta_3 x_{1,i}^2 \right) e^{x_{2,i}\theta_4} \tag{5.14}$$

where Y_i is the average unit side shear, which is the total capacity due to side shear divided by the circumference of the pile multiplied by its length, $x_{1,i}$, is the penetration of the pile, and $x_{2,i}$ is 0 for modern borings and 1 for older

Continued

Box 5.11 cont'd

borings (the term $e^{x_{2,i}\theta_4}$ accounts for biases due to the quality of soil samples);

$$\sigma_{Y_i} = e^{\theta_4} e^{x_{2,i}\theta_5} \qquad (5.15)$$

where the term e^{θ_4} accounts for a standard deviation that does not depend on pile penetration and that is non-negative, and the term $e^{x_{2,i}\theta_5}$ accounts for any effects on the spatial variability caused by the quality of the soil samples, including greater or smaller variability;

$$\rho_{Y_i,Y_j} = \left(e^{-\sqrt{(x_{3,i}-x_{3,j})^2 + (x_{4,i}-x_{4,j})^2} \Big/ (e^{\theta_6} + x_{1,i}e^{\theta_7})} \right) e^{-|x_{2,i}-x_{2,j}| \big/ e^{\theta_8}} \qquad (5.16)$$

where $x_{3,i}$ and $x_{4,i}$ are the coordinates describing the horizontal position of boring i, the term $e^{-\sqrt{(x_{3,i}-x_{3,j})^2 + (x_{4,i}-x_{4,j})^2} \big/ (e^{\theta_6} + x_{1,i}e^{\theta_7})}$ accounts for a positive correlation between nearby borings that is described by a positive correlation distance of $e^{\theta_6} + x_{1,i}e^{\theta_7}$, and the term $e^{-|x_{2,i}-x_{2,j}| \big/ e^{\theta_8}}$ accounts for a smaller positive correlation between borings with different qualities of samples than those with the same qualities of samples; and the probability distribution for the average side friction is modeled by a normal distribution

$$F_{Y_i|y_1,\dots y_{i-1}} \left(y_i | y_1, \dots y_{i-1} \right) = \Phi \left(\frac{y_i - \mu_{Y_i|y_1,\dots y_{i-1}}}{\sigma_{Y_i|y_1,\dots y_{i-1}}} \right) \qquad (5.17)$$

Therefore, there are eight parameters to be calibrated with the data, θ_1 to θ_8. The detailed results for this calibration are provided in Gambino and Gilbert (1999).

Before considering the decision at hand, there are several practical conclusions that can be obtained from the calibrated model in its own right. The expected values for the mean and standard deviation of average unit side shear from modern borings are 98 kPa and 7.2 kPa, respectively. The coefficient of variation for pile capacity is then 7.2/98 or 0.07, meaning that the magnitude of spatial variability in pile capacity across this field is relatively small. The expected horizontal correlation distance is shown with respect to pile penetration in Figure 5.15; it is on the order of thousands of meters and it increases with pile penetration. The expected bias due to older soil borings is equal to 0.91 ($e^{x_{2,i}\theta_4}$ in Equation (5.14)), meaning that designs based on older borings tend to underestimate the pile capacity that would be obtained from a modern boring. Also, the expected effect on the spatial variability due to older soil borings is to reduce the standard deviation by 0.86 ($e^{x_{2,i}\theta_5}$) in

Box 5.11 cont'd

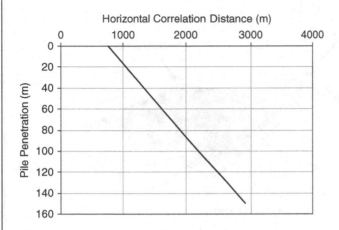

Figure 5.15 Calibrated horizontal correlation distance for pile capacities between borings.

Equation (5.15), meaning that the data from the older soil borings tends to mask some of the naturally occurring spatial variations that are picked up with modern borings, thereby reducing the apparent standard deviation. Furthermore, the expected correlation coefficient between an older boring and a modern boring at the exact same location is 0.58 $(e^{-|x_{2,i}-x_{2,j}|}/e^{\theta_8}$ in Equation (5.16) and notably less than the ideal value of 1.0, meaning that it is not possible to deterministically predict the pile capacity obtained from a modern boring by using information from an older boring.

The results for this calibrated model are illustrated in Figures 5.16a and 5.16b for an example 4000 m × 4000 m block in this field. Within this block, there are two soil borings available: a modern one and an older one at the respective locations shown in Figures 5.16a and 5.16b. The three-dimensional surface shown in Figure 15.16a denotes the expected value for the calculated design capacity at different locations throughout the block. The actual design pile capacities at the two boring locations are both above the average for the field (Figure 5.16a). Therefore, the expected values for the design pile capacity at other locations of an offshore platform within this block are above average; however, as the location moves further away from the existing borings, the expected value tends toward the unconditional mean for the field, approximately 30 MN.

The uncertainty in the pile capacity due to not having a site-specific, modern soil boring is expressed as a coefficient of variation on Figure 5.16b.

Continued

Box 5.11 cont'd

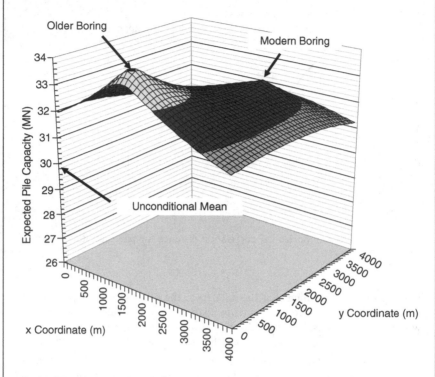

Figure 5.16a Expected value for estimated pile capacity versus location for a 100 m long, 1 m diameter steel pipe pile.

At the location of the modern boring, this c.o.v. is zero since a modern boring is available at that location and the design capacity is known. However, at the location of the older boring, the c.o.v. is greater than zero since the correlation between older and modern borings is not perfect (Figure 5.16b); therefore, even though the design capacity is known at this location, the design capacity that would have been obtained based on a modern soil boring is not known. As the platform location moves away from both borings, the c.o.v. approaches the unconditional value for the field, approximately 0.075.

The final step is to put this information from the calibrated model for spatial variability into the decision tree in Figure 5.4. The added uncertainty in not having a site-specific, modern boring is included in the reliability analysis by adding uncertainty to the capacity; the greater the uncertainty, the lower the reliability for the same set of load and resistance factors in a reliability-based design code. In order to achieve the same reliability as if a modern soil boring

Box 5.11 cont'd

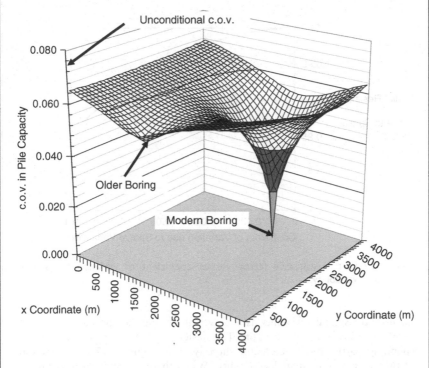

Figure 5.16b Coefficient of variation for estimated pile capacity versus location for a 100 m long, 1 m diameter steel pipe pile.

were available, the resistance factor needs to be smaller, as expressed in the following design check format:

$$\phi_{\text{spatial}} \left(\phi_{\text{capacity}} \text{capacity}_{\text{design}} \right) \geq \gamma_{\text{load}} \text{load}_{\text{design}} \qquad (5.17)$$

where ϕ_{spatial} is a partial resistance factor that depends on the magnitude of spatial variability obtained from Figure 5.16b and the design capacity is obtained from Figure 5.16a. Figure 5.17 shows the relationship between ϕ_{spatial} and the coefficient of variation due to spatial variability for the following assumptions: design capacity is normally distributed with a c.o.v. of 0.3 when a site-specific boring is available; load is normally distributed with a c.o.v. of 0.5; and the target probability of failure for the pile foundation is 6×10^{-3} (Gilbert *et al.*, 1999). To use this chart, a designer would first

Continued

Box 5.11 cont'd

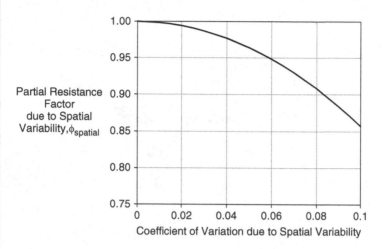

Figure 5.17 Partial resistance factor versus coefficient of variation for spatial variability.

quantify the spatial variability as a c.o.v. value using the approach described above, as shown in Figure 5.16b. The designer would then read the corresponding partial resistance factor directly from Figure 5.17 as a function of this c.o.v. due to spatial variability. When there is not additional uncertainty because a modern soil boring is available at the platform location (i.e., the c.o.v. due to spatial variability is 0), the partial resistance factor is 1.0 and does not affect the design. As the c.o.v. due to spatial variability increases, the partial resistance factor decreases and has a larger effect on the design.

The information in Figures 5.16a, 5.16b, and 5.17 can be combined together to support the decision-making process. For example, consider a location in the block where the c.o.v. due to spatial variability is 0.06 from Figure 5.16b. The required partial resistance factor to account for this spatial variability is 0.95, or 95% of the design capacity could be relied upon in this case compared to what is expected if a site-specific boring is drilled. Since pile capacity is approximately proportional to pile length over a small change in length, this reduction in design capacity is roughly equivalent to a required increase in pile length of about 5%. Therefore, the cost of drilling a site-specific boring could be compared to the cost associated with increasing the pile lengths by 5% in the decision tree in Figure 5.4 to decide whether or not to drill an additional soil boring.

5.6 Summary

There can be considerable value in using the reliability-based design approach to facilitate and improve decision-making. This chapter has demonstrated the practical potential of this concept. Several different levels of decisions were addressed:

- project-specific decisions, such as how long to make a pile foundation or how many soil borings to drill;
- design code decisions, such as what resistance factors should be used; and
- policy decisions, such as what level of reliability should be achieved for components and systems in different types of facilities.

An emphasis has been placed on first establishing what is important in the context of the decisions that need to be made before developing and applying mathematical tools. An emphasis has also been placed on practical methods to capture the realistic features that will influence decision-making. It is hoped that the real-world examples described in this chapter motivate the reader to consider how the role of decision making can be utilized in the context of reliability-based design for a variety of different applications.

Acknowledgments

We wish to acknowledge gratefully the following organizations for supporting the research and consulting upon which this paper is based: American Petroleum Institute, American Society of Civil Engineers, BP, ExxonMobil, National Science Foundation, Offshore Technology Research Center, Shell, State of Texas Advanced Research and Technology Programs, Unocal, and United States Minerals Management Service. In many cases, these organizations provided real-world problems and data, as well as financial support. We would also like to acknowledge the support of our colleagues and students at The University of Texas at Austin. The views and opinions expressed herein are our own and do not reflect any of the organizations, parties or individuals with whom we have worked.

References

ANCOLD (1998). *Guidelines on Risk Assessment*. Working Group on Risk Assessment, Australian National Committee on Large Dams, Sydney, New South Wales, Australia.

Ang, A. A-S. and Tang, W. H. (1984). *Probability Concepts in Engineering Planning and Design, Volume II – Decision, Risk and Reliability*. John Wiley & Sons, New York.

Bea, R. G. (1991). Offshore platform reliability acceptance criteria. *Drilling Engineering*, Society of Petroleum Engineers, June, 131–6.

Benjamin, J. R. and Cornell, C. A. (1970). *Probability, Statistics, and Decision for Civil Engineers*. McGraw-Hill, New York.

Bowles, D. S. (2001). Evaluation and use of risk estimates in dam safety decisionmaking. In *Proceedings, Risk-Based Decision-Making in Water Resources*, ASCE, Santa Barbara, California, 17 pp.

Choi, Y. J., Gilbert, R. B., Ding, Y. and Zhang, J. (2006). Reliability of mooring systems for floating production systems. *Final Report for Minerals Management Service*, Offshore Technology Research Center, College Station, Texas, 90 pp.

Christian, J. T. (2004). Geotechnical engineering reliability: how well do we know what we are doing? *Journal of Geotechnical and Geoenvironmental Engineering*, ASCE, 130 (10), 985–1003.

Finley, C. A. (2004). Designing and analyzing test programs with censored data for civil engineering applications. Ph.D. dissertation, The University of Texas at Austin.

Fischhoff, B., Lichtenstein, S., Slovic, P., Derby, S. L. and Keeney, R. L. (1981). *Acceptable Risk*. Cambridge University Press, Cambridge.

Gambino, S. J. and Gilbert, R.B. (1999). Modeling spatial variability in pile capacity for reliability-based design. *Analysis, Design, Construction and Testing of Deep Foundations*, ASCE Geotechnical Special Publication No. 88, 135–49.

Gilbert, R. B. (1999). First-order, second-moment bayesian method for data analysis in decision making. *Geotechnical Engineering Center Report*, Department of Civil Engineering, The University of Texas at Austin, Austin, Texas, 50 pp.

Gilbert, R. B., Gambino, S. J. and Dupin, R. M. (1999). Reliability-based approach for foundation design without with-specific soil borings. In *Proceedings, Offshore Technology Conference*, Houston, Texas, OTC 10927 Society of Petroleum Engineers, Richardson, pp. 631–40.

Gilbert, R. B., Choi, Y. J., Dangyach, S. and Najjar, S. S. (2005a). Reliability-based design considerations for deepwater mooring system foundations. In *Proceedings, ISFOG 2005, Frontiers in Offshore Geotechnics*, Perth, Western Australia. Taylor & Francis, London, pp. 317–24.

Gilbert, R. B., Najjar, S. S. and Choi, Y. J. (2005b). Incorporating lower-bound capacities into LRFD codes for pile foundations. In *Proceedings, Geo-Frontiers*, Austin, TX, ASCE, Virginia, pp. 361–77.

Goodwin, P., Ahilan, R. V., Kavanagh, K. and Connaire, A. (2000). Integrated mooring and riser design: target reliabilities and safety factors. In *Proceedings, Conference on Offshore Mechanics and Arctic Engineering*, New Orleans, The American Society of Mechanical Engineers (ASME), New York, 785–92.

Journel, A. G. and Huijbregts, Ch. J. (1978). *Mining Geostatistics*. Academic Press, San Diego.

Kenney, R. L. and Raiffa, H. (1976). *Decision with Multiple Objectives: Preferences and Value Tradeoffs*. John Wiley and Sons, New York.

Najjar, S. S. (2005). *The importance of lower-bound capacities in geotechnical reliability assessments*. PhD. dissertation, The University of Texas at Austin.

Stahl, B., Aune, S., Gebara, J. M. and Cornell, C. A. (1998). Acceptance criteria for offshore platforms. In *Proceedings, Conference on Offshore Mechanics and Arctic Engineering*, OMAE98-1463.

Tang, W. H. and Gilbert, R. B. (1993). Case study of offshore pile system reliability. In *Proceedings, Offshore Technology Conference*, OTC 7196, Houston, Society of Petroleum Engineers, pp. 677–83.

The Economist (2004). The price of prudence. *A Survey of Risk*, 24 January, 6–8.

USBR (2003). *Guidelines for Achieving Public Protection in Dam Safety Decision Making*. Dam Safety Office, United States Bureau of Reclamation, Denver, CO.

USNRC (1975). *Reactor safety study: an assessment of accident risks in U.S. commercial nuclear power plants*. United States Nuclear Regulatory Commission, NUREG-75/014, Washington, D. C.

Wang, D. (2002). *Development of a method for model calibration with non-normal data*. PhD dissertation, The University of Texas at Austin.

Whipple, C. (1985). Approaches to acceptable risk. In *Proceedings, Risk-Based Decision Making in Water Resources*, Ed. Y. Y. Haimes, and E.Z. Stakhiv, ASCE, New York, CA, pp. 31–45.

Whitman, R. V. (1984). Evaluating calculated risk in geotechnical engineering. *Journal of Geotechnical Engineering*, ASCE, 110(2), 145–88.

Wu, T. H., Tang, W. H., Sangrey, D. A. and Baecher, G. (1989). Reliability of offshore foundations. *Journal of Geotechnical Engineering*, ASCE, 115(2), 157–78.

Chapter 6

Randomly heterogeneous soils under static and dynamic loads

Radu Popescu, George Deodatis and Jean-Hervé Prévost

6.1 Introduction

The values of soil properties used in geotechnical engineering and geomechanics involve a significant level of uncertainty arising from several sources such as: inherent random heterogeneity (also referred to as spatial variability), measurement errors, statistical errors (due to small sample sizes), and uncertainty in transforming the index soil properties obtained from soil tests into desired geomechanical properties (e.g. Phoon and Kulhawy, 1999). In this chapter, only the first source of uncertainty will be examined (inherent random heterogeneity), as it can have a dramatic effect on soil response under loading (in particular soil failure), as will be demonstrated in the following. As opposed to lithological heterogeneity (which can be described deterministically), the aforementioned random soil heterogeneity refers to the natural spatial variability of soil properties within geologically distinct layers. Spatially varying soil properties are treated probabilistically and are usually modeled as random fields (also referred to as stochastic fields). In the last decade or so, the probabilistic characteristics of the spatial variability of soil properties have been quantified using results of various sets of in-situ tests.

The first effort in accounting for uncertainties in the soil mass in geotechnical engineering and geomechanics problems involved the random variable approach. According to this approach, each soil property is modeled by a random variable following a prescribed probability distribution function (PDF). Consequently, the soil property is constant over the analysis domain. This approach allowed the use of established deterministic analysis methods developed for uniform soils, but neglected any effects of their spatial variability. It therefore reduced the natural spatial variation of soil properties to an uncertainty in their mean values only. This rather simplistic approach was quickly followed by a more sophisticated one modeling the spatial heterogeneity of various soil properties as random fields. This became possible from an abundance of data from in-situ tests that allowed the establishment of various probabilistic characteristics of the spatial variability of soil properties.

In the last decade or so, a series of papers have appeared in the literature dealing with the effect of inherent random soil heterogeneity on the mechanical behavior of various problems in geomechanics and geotechnical engineering using random field theory. A few representative papers are mentioned here grouped by the problem examined: (1) foundation settlements: Brzakala and Pula (1996), Paice et al. (1996), Houy et al. (2005); (2) soil liquefaction: Popescu et al. (1997, 2005a,b), Fenton and Vanmarke (1998), Koutsourelakis et al. (2002), Elkateb et al. (2003), Hicks and Onisiphorou (2005); (3) slope stability: El Ramly et al. (2002, 2005), Griffiths and Fenton (2004); and (4) bearing capacity of shallow foundations: Cherubini (2000), Nobahar and Popescu (2000), Griffiths et al. (2002), Fenton and Griffiths (2003), Popescu et al. (2005c). The methodology used in essentially all of these studies was Monte Carlo Simulation (MCS).

First-order reliability analysis approaches (see e.g. Christian (2004) for a review), as well as perturbation/expansion techniques, postulate the existence of an "average response" that depends on the average values of the soil properties. This "average response" is similar to that obtained from a corresponding uniform soil having properties equal to the average properties of the randomly variable soil. The inherent soil variability (random heterogeneity) is considered to induce uncertainty in the computed response only as a random fluctuation around the "average response." For example, in problems involving a failure surface, it is quite common to assume that the failure mechanism is deterministic, depending only on the average values of the soil properties. More specifically, in most reliability analyses of slopes involving spatially variable soil, deterministic analysis methods are first used to estimate the critical slip surface (corresponding to a minimum factor of safety and calculated assuming uniform soil properties). After that, the random variability of soil properties along this pre-determined slip surface is used in the stochastic analysis (e.g. El Ramly et al., 2002, 2005).

In contrast to the type of work mentioned in the previous paragraph, a number of recent MCS-based studies that did not impose any restrictions on the type and geometry of the failure mechanism (e.g. Paice et al., 1996, for foundation settlements; Popescu et al., 1997, 2005a,b, for soil liquefaction; Griffiths and Fenton, 2004, for slope stability; and Popescu et al., 2005c, for bearing capacity of shallow foundations) observed that the inherent spatial variability of soil properties can significantly modify the failure mechanism of spatially variable soils, compared to the corresponding mechanism of uniform (deterministic) soils. For example, Focht and Focht (2001) state that "the actual failure surface can deviate from its theoretical position to pass through weaker material so that the average mobilized strength is less than the apparent average strength." The deviation of the failure surface from its deterministic configuration can be dramatic, as there can be a complete change in the topology of the surface. Furthermore, the average response of spatially variable soils (obtained from MCS) was observed to be significantly different from the response of the corresponding uniform soil. For example, Paice *et al.* (1996) predicted an up to 12% increase in average settlements for an elastic heterogeneous soil with coefficient of variation CV = 42%, compared to corresponding settlements of a uniform soil deposit with the same mean soil properties. Nobahar and Popescu (2000) and Griffiths *et al.* (2002) found a 20–30% reduction in the mean bearing capacity of heterogeneous soils with CV = 50%, compared to the corresponding bearing capacity of a uniform soil with the same average properties. Popescu *et al.* (1997) predicted an increase of about 20% in the amount of pore-water pressure build-up for a heterogeneous soil deposit with CV = 40%, compared to the corresponding results of uniform soil with the same mean properties. It should be noted that the effects of spatial variability were generally stronger for phenomena governed by highly nonlinear constitutive laws (such as bearing capacity and soil liquefaction) than for phenomena governed by linear laws (e.g. settlements of foundations on elastic soil).

This chapter discusses the effects of inherent spatial variability (random heterogeneity) of soil properties on two very different phenomena: bearing capacity failure of shallow foundations under static loads and seismically induced soil liquefaction. The first implies a limit equilibrium type of failure involving a large volume of soil, while the second is based on a micro-scale (soil grain level) mechanism. Based on results of numerical work previously published by the authors, it is shown that the end effects of spatial variability on limit loads are qualitatively similar for both phenomena.

The probabilistic characteristics of the spatial variability of soils are briefly discussed in the first part of this chapter. A general MCS approach for geotechnical systems exhibiting random variation of their properties is also presented. The second part focuses on the effects of inherent soil spatial variability. Based on examples dealing with the bearing capacity of shallow foundations and seismically induced soil liquefaction, it is shown that the spatial variability of soil properties can change the soil behavior from the well known theoretical results obtained for uniform (deterministic) soils dramatically. The mechanisms of this intriguing behavior are analyzed in some detail and interpreted using experimental and numerical evidence. It is also shown that, in some situations, the resulting factors of safety and failure loads for randomly heterogeneous soils can be considerably lower than the corresponding ones determined from some current reliability approaches used in geotechnical engineering, with potentially significant implications for design. The concept of characteristic values or "equivalent uniform" soil properties is discussed in the last part of the chapter as a means of accounting for the effects of soil variability in practical applications.

6.2 Representing the random spatial variability of soil properties

The natural heterogeneity in a supposedly homogeneous soil layer may be due to small-scale variations in mineral composition, environmental conditions during deposition, past stress history, variations in moisture content, etc. (e.g. Tang, 1984; Lacasse and Nadim, 1996). From a geotechnical design viewpoint, this small-scale heterogeneity is manifested as spatial variation of relevant soil properties (such as relative density, shear strength, hydraulic conductivity, etc.).

An example of soil spatial variability is presented in Figure 6.1 in terms of recorded cone tip resistance values in a number of piezocone (CPTu) profiles, performed for core verification at one of the artificial islands (namely: Tarsiut P-45) constructed in the Canadian Beaufort Shelf and used as drilling platforms for oil exploration in shallow waters (Gulf Canada Resources, 1984). The cone tip resistance q_c is directly related to shear strength and to liquefaction potential of sandy soils. The profiles shown in Figure 6.1 were located on a straight line at 9 m center-to-center distances. Those records are part of a more extended soil investigation program consisting of 32 piezocone tests, providing an almost uniform coverage of the area of interest (72 m × 72 m). The soil deposit at Tarsiut P-45 consists of two distinct

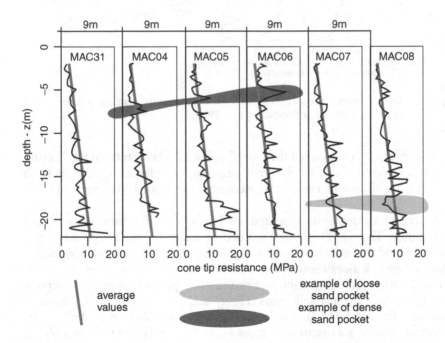

Figure 6.1 Cone tip resistance recorded in the sand core at Tarsiut P-45 (after Popescu et al., 1997).

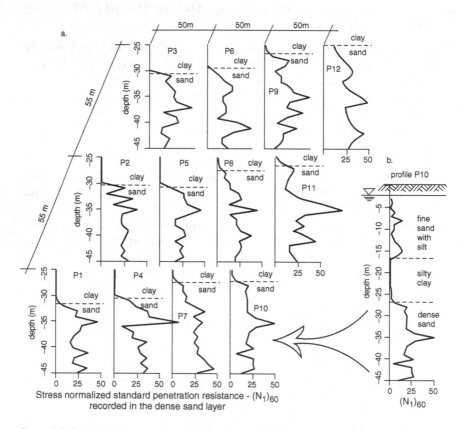

Figure 6.2 Standard Penetration Test results from a 2D measurement array in the Tokyo Bay area, Japan (after Popescu *et al.*, 1998a).

layers: an upper layer termed the "core" and a base layer termed the "berm" (see e.g. Popescu *et al.*, 1998a, for more details). Selected q_c records from the core are shown in Figure 6.1. Records from both layers are shown in Figure 6.3a.

A second example is presented in Figure 6.2 in terms of Standard Penetration Test (SPT) results recorded at a site in the Tokyo Bay area, Japan, where an extended in situ soil test program had been performed for liquefaction risk assessment. A two-dimensional measurement array consisting of 24 standard penetration test profiles was performed in a soil deposit formed of three distinct soil layers (Figure 6.2b): a fine sand with silt inclusions, a silty clay layer, and a dense to very dense sand layer. The results shown in Figure 6.2a represent stress normalized SPT blowcounts $(N_1)_{60}$, recorded in the dense sand layer in 12 of the profiles. Results from all soil layers are shown in Figure 6.3b.

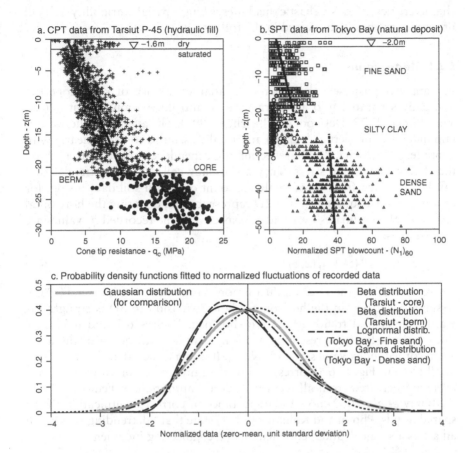

Figure 6.3 In-situ penetration test results at two sites (after Popescu *et al.*, 1998a) and corresponding fitted probability distribution functions for the normalized data (zero mean, unit standard deviation): (a) cone penetration resistance from 8 CPT profiles taken at an artificial island in the Beaufort Sea (Gulf Canada Resources, 1984); (b) normalized standard penetration data from 24 borings at a site in the Tokyo Bay area involving a natural soil deposit; (c) fitted PDFs to normalized fluctuations of field data. Thick straight lines in (a) and (b) represent average values (spatial trends).

It can be easily observed from Figures 6.1 and 6.2 that soil shear strength – which is proportional to q_c and $(N_1)_{60}$ – is not uniform over the test area but varies from one location to another (both in the horizontal and vertical directions). As deterministic descriptions of this spatial variability are practically impossible – owing to the prohibitive cost of sampling and to uncertainties induced by measurement errors (e.g. VanMarcke, 1989) – it has become standard practice today in the academic community to model this small scale random heterogeneity using stochastic field theory. The probabilistic characteristics of a stochastic field describing spatial variability of soil properties are briefly discussed in the following.

6.2.1 Mean value

For analysis purposes, the inherent spatial variability of soil properties is usually separated into an average value and fluctuations about it (e.g. VanMarcke, 1977; DeGroot and Baecher, 1993). The average values of certain index soil properties (e.g. q_c) are usually assumed to vary linearly with depth (e.g. Rackwitz, 2000), reflecting the dependence of shear strength on the effective confining stress. They are usually called "spatial trends" and can be estimated from a reasonable amount of in-situ soil data (using, for example, a least square fit). In a 2D representation, denoting the horizontal and vertical coordinates by x and z, respectively, the recorded q_c values are expressed as:

$$q_c(x, z) = q_c^{av}(z) + u_f(x, z) \tag{6.1}$$

where $u_f(x, z)$ are the random fluctuations around the spatial trend $q_c^{av}(z)$. Assuming isotropy in the horizontal plane, extension to 3D is straightforward. The spatial trends $q_c^{av}(z)$ are shown in Figures 6.1 and 6.3a,b by thick straight lines, representing a linear increase with depth of the average recorded index values q_c and $(N_1)_{60}$. It is mentioned that the thick lines in all plots in Figure 6.1 correspond to the same trend for all profiles (one average calculated from all records). Several authors (e.g. Fenton, 1999b; El-Ramly et al., 2002) note the importance of correct identification of the spatial trends. Phoon and Kulhawy (2001) state that "detrending is as much an art as a science" and recommend using engineering judgment.

6.2.2 Degree of variability

The degree of variability – or equivalently the magnitude of the random fluctuations around the mean values – is quantified by the variance or the standard deviation. As the systematic trends can be identified and expressed deterministically, they are usually removed from the subsequent analysis and it's only the local fluctuations that are modeled using random field theory. For mathematical convenience, the fluctuations $u_f(x, z)$ in Equation (6.1) are normalized by the standard deviation $\sigma(z)$:

$$u(x, z) = [q_c(x, z) - q_c^{av}(z)]/\sigma(z) \tag{6.2}$$

so that the resulting normalized fluctuations $u(x, z)$ can be described by a zero-mean, unit standard deviation random field.

As implied by Equation (6.2), the standard deviation $\sigma(z)$ may be a function of the depth z. A depth-dependent standard deviation can be inferred by evaluating it in smaller intervals, over which it is assumed to be constant, rather than using the entire soil profile at one time. For example, the standard deviation of the field data shown in Figure 6.1 was calculated over 1 m long moving intervals and was found to vary between 1 and 4 MPa (Popescu, 1995).

The most commonly used parameter for quantifying the magnitude of fluctuations in soil properties around their mean values is the coefficient of variation (CV). For the data in Figure 6.1 $CV(z) = \sigma(z)/q_c^{av}(z)$ was found to vary between 20% and 60%. The average CV of SPT data from the dense sand layer in Figure 6.2 was about 40%.

It should be mentioned at this point that the CPT and SPT records shown in Figures 6.1 and 6.2 include measurement errors, and therefore the degrees of variability discussed here may be larger than the actual variability of soil properties. The measurement errors are relatively small for the case of CPT records (e.g. for the electrical cone, the average standard deviation of measurement errors is about 5%; American Society for Testing and Materials, 1989), but they can be significant for SPT records, which are affected by many sources of errors (discussed by Schmertmann, 1978, among others).

6.2.3 Correlation structure

The fluctuations of soil properties around their spatial trends exhibit in general some degree of coherence/correlation as a function of depth, as can be observed in Figures 6.1 and 6.2. For example, areas with q_c values consistently larger than average indicate the presence of dense (stronger) pockets in the soil mass, while areas with q_c values consistently smaller than average indicate the presence of loose (less-resistant) soil. Examples of dense and loose soil zones inferred from the fluctuations of q_c are identified in Figure 6.1. Similarly, the presence of a stronger soil zone at a depth of about 35 m and spanning profiles P5 and P8–P12 can be easily observed from the SPT results presented in Figure 6.2.

The similarity between fluctuations recorded at two points as a function of the distance between those two points is quantified by the correlation structure. The correlation between values of the same material property measured at different locations is described by the auto-correlation function. A significant parameter associated with the auto-correlation function is called the scale of fluctuation (or correlation distance) and represents a length over which significant coherence is still manifested. Owing to the geological soil formation processes for most natural soil deposits, the correlation distance in the horizontal direction is significantly larger than the one in the vertical direction. This can be observed in Figure 6.1 by the shape of the dense and loose soil pockets. In this respect, separable correlation structure models (e.g. VanMarcke, 1983) seem appropriate to model different spatial variability characteristics in the vertical direction (normal to soil strata) and in the horizontal one.

There are several one-dimensional theoretical auto-correlation function models that can be combined into separable correlation structures and used to describe natural soil variability (e.g. VanMarcke, 1983; Shinozuka and Deodatis, 1988; Popescu, 1995; Rackwitz, 2000). Inferring the parameters of a theoretical correlation model from field records is straightforward for the vertical (depth) direction where a large number of closely spaced data are usually available (e.g. from CPT records). However, it is much more challenging for the horizontal direction, where sampling distances are much larger. DeGroot and Baecher (1993) and Fenton (1999a) review the most common methods for estimating the auto-correlation function of soil variability in the vertical direction. Przewlocki (2000) and Jaksa (2007) present methods for estimating the two-dimensional correlation of soil spatial variability based on close-spaced field measurements in both vertical and horizontal directions. As close-spaced records in horizontal direction are seldom available in practice, Popescu (1995) and Popescu et al. (1998a) introduce a method for estimating the horizontal auto-correlation function based on a limited number of vertical profiles.

As a note of caution, it is mentioned here that various different definitions for the correlation distance exist in the literature, yielding different values for this parameter (see e.g. Popescu et al., 2005c, for a brief discussion). Another problem stems from the fact that when estimating correlation distances from discrete data, the resulting values are dependent on the sampling distance (e.g. Der Kiureghian and Ke, 1988) and the length scale, or the distance over which the measurements are taken (e.g. Fenton, 1999b).

When two or more different soil properties can be recorded simultaneously, it is interesting to assess the spatial interdependence between them and how it affects soil behavior. For example, from piezocone test results one can infer another index property – besides q_c – called soil classification index, which is related to grain size and soil type (see e.g. Jefferies and Davies, 1993; Robertson and Wride, 1998). The spatial dependence between two different properties measured at different locations is described by the cross-correlation function. In this case, the variability of multiple soil properties is modeled by a multivariate random field (also known as a vector field), and the ensemble of auto-correlation and cross-correlation functions form the cross-correlation matrix.

As the field data $u(x, z)$ employed to estimate the correlation structure have been detrended and normalized (refer to Equation (6.2)), the resulting field is stationary. It is always possible to model directly non-stationary characteristics in spatial correlation, but this is very challenging because of limited data and modeling difficulties (e.g. Rackwitz, 2000). Consequently, stationarity is almost always assumed (at least in a piecewise manner).

6.2.4 Probability distribution

Based on several studies reported in the literature, soil properties can follow different probability distribution functions (PDFs) for different types of soils and sites, but due to physical reasons, they always have to follow distributions defined only for nonnegative values of the soil properties (excluding thus the Gaussian distribution). Unlike Gaussian stochastic fields where the first two moments provide complete probabilistic information, non-Gaussian fields require knowledge of moments of all orders. As it is extremely difficult to estimate moments higher than order two from actual data (e.g. Lumb, 1966; Phoon, 2006), the modeling and subsequent simulation of soil properties that are represented as homogeneous non-Gaussian stochastic fields are usually done using their cross-correlation matrix (or equivalently cross-spectral density matrix) and non-Gaussian marginal probability distribution functions.

While there is no clear evidence pointing to any specific non-Gaussian model for the PDF of soil properties, one condition that has to be satisfied is for the PDF to have a lower (nonnegative) bound. The beta, gamma and lognormal PDFs are commonly used for this purpose as they all satisfy this condition. For prescribed values of the mean and standard deviation, the gamma and lognormal PDFs are one-parameter distributions (e.g. the value of the lower bound can be considered as this parameter). They are both skewed to the left. Similarly, for prescribed values of the mean and standard deviation, the beta PDF is a two-parameter distribution, and consequently more flexible in fitting in-situ data. Moreover, it can model data that are symmetrically distributed or skewed to the right. Based on penetration data from an artificial and a natural deposit, Popescu *et al.* (1998a) observed that PDFs of soil strength in shallow layers are skewed to the left, while for deeper soils the corresponding PDF's tend to follow more symmetric distributions (refer to Figure 6.3).

As will be discussed later, the degree of spatial variability (expressed by the coefficient of variation) is a key factor influencing the behavior of heterogeneous soils as compared to uniform ones. It has also been determined that both the correlation structure and the marginal probability distribution functions of soil properties affect the response behavior of heterogeneous soils to significant degrees (e.g. Popescu *et al.*, 1996, 2005a; Fenton and Griffiths, 2003).

6.3 Analysis method

6.3.1 Monte Carlo Simulation approach

Though expensive computationally, Monte Carlo Simulation (MCS) is the only currently available universal methodology for accurately solving problems in stochastic mechanics involving strong nonlinearities and large variations of non-Gaussian uncertain system parameters, as is the case of spatially variable soils. As it will be shown later in this chapter, failure mechanisms of heterogeneous soils may be different from the theoretical ones (derived for assumed deterministic uniform soil) and can change significantly from one realization (sample function) of the random soil properties to another. This type of behavior prevents the use of stochastic analysis methods such as the perturbation/expansion ones that are based on the assumption of an average failure mechanism and variations (fluctuations) around it.

MCS is a means of solving numerically problems in mathematics, physics and other sciences through sampling experiments, and basically consists of three steps (e.g. Elishakoff, 1983): (1) simulation of the random quantities (variables, fields), (2) solution of the resulting deterministic problem for a large number of realizations in step 1, and (3) statistical analysis of results. For the case of spatially variable soils, the realizations of the random quantities in step 1 consist of simulated sample functions of stochastic fields with probabilistic characteristics estimated from field data, as discussed in Section 6.2. Each such sample function represents a possible realization of the relevant soil index properties over the analysis domain. The deterministic problem in step 2 is usually solved by finite element analysis. One such analysis is performed for each realization of the (spatially variable) soil properties.

6.3.2 Simulation of homogeneous non-Gaussian stochastic fields

Stochastic fields modeling spatially variable soil properties are non-Gaussian, multi-dimensional (2D or 3D), and can be univariate (describing the variability of a single soil property) or multivariate (referring to the variability of several soil properties and their interdependence). Among the various methods that have been developed to simulate homogeneous non-Gaussian stochastic fields, the following representative ones are mentioned here: Yamazaki and Shinozuka (1988), Grigoriu (1995, 1998), Gurley and Kareem (1998), Deodatis and Micaletti (2001), Puig et al. (2002), Sakamoto and Ghanem (2002), Masters and Gurley (2003), Phoon et al. (2005).

As discussed in Section 6.2.4, it is practically impossible to estimate non-Gaussian joint PDFs from actual soil data. Therefore, a full description of the random field representing the spatial variability of soil properties at a given site is not achievable. Under these circumstances, the simulation methods considered generate sample functions matching a target cross-correlation structure (CCS) – or equivalently cross-spectral density matrix (CSDM) in the wave number domain – and target marginal PDFs that can be inferred from the available field information. It is mentioned here that certain simulation methods match only lower order moments (such as mean, variance, skewness and kurtosis) instead of marginal PDFs (the reader is referred to Deodatis and Micaletti, 2001, for a review and discussion).

In most methods for non-Gaussian random field simulation, a Gaussian sample function is first generated starting from the target CCS (or target CSDM), and then mapped into the desired non-Gaussian sample function using a transformation procedure that involves nonlinear mapping (e.g. Grigoriu, 1995; Phoon, 2006). While the target marginal PDFs are matched after such a mapping, the nonlinearity of this transformation modifies the CCS and the resulting non-Gaussian sample function is no longer compatible with the target CCS. Cases involving small CVs and non-Gaussian PDFs that are not too far from Gaussianity result in general in changes in the CCS that are small compared to the uncertainties involved in estimating the target CCS from field data. However, for highly skewed marginal PDFs, the differences between target and resulting CCSs may become significant. For example, Popescu (2004) discussed such an example where mapping from a Gaussian to a lognormal random field for very large CVs of soil variability produced differences of up to 150% between target and resulting correlation distances.

There are several procedures for correcting this problem and generating sample functions that are compatible with both the target CCS and marginal PDFs. These methods range from iterative correction of the target CCS (or target CSDM) used to generate the Gaussian sample function (e.g. Yamazaki and Shinozuka, 1988), to analytical transformation methods that are applicable to certain combinations of target CCS – target PDF (e.g. Grigoriu, 1995, 1998). A review of these methods is presented by Deodatis and Micaletti (2001) together with an improved algorithm for simulating highly skewed non-Gaussian fields.

The methodology used for the numerical examples presented in this study combines work done on the spectral representation method by Yamazaki and Shinozuka (1988), Shinozuka and Deodatis (1996) and Deodatis (1996), and extends it to simulation of multi-variate, multi-dimensional (mV–nD), homogeneous, non-Gaussian stochastic fields. According to this methodology, a sample function of an mV–nD Gaussian vector field is first generated using the classic spectral representation method. Then, this Gaussian vector field is transformed into a non-Gaussian one that is compatible with a prescribed cross-spectral density matrix and with prescribed (non-Gaussian) marginal PDFs assumed for the soil properties. This is achieved through the classic memoryless nonlinear transformation of "translation fields" (Grigoriu, 1995) and the iterative scheme proposed by Yamazaki and Shinozuka (1988). For a detailed presentation of this simulation algorithm and a discussion on the convergence of the iterative scheme, the reader is referred to Popescu et al. (1998b). The reason that a more advanced methodology, such as that of Deodatis and Micaletti (2001), is not used in this chapter is the fact that the random fields describing various soil properties are usually only slightly skewed and, consequently, the iterative correction proposed by Yamazaki and Shinozuka (1988) gives sufficiently accurate results.

6.3.3 Deterministic finite element analyses with generated soil properties as input

Solutions of boundary value problems can be readily obtained through finite element analysis (FEA). Along the lines of the MCS approach, a deterministic FEA is performed for every generated sample function representing a possible realization of soil index properties over the analysis domain. In each such analysis, the relevant material (soil) properties are obtained at each spatial location (finite element centroid) as a function of the simulated soil index properties. Two important issues related to such analyses are worth mentioning:

1   The size of finite elements should be selected to accurately reproduce the simulated boundary value problem, and at the same time, to adequately capture the essential features of the stochastic spatial variability of soil properties. Regarding modeling the spatial variability, the optimum mesh refinement depends on both correlation distance and type of correlation structure. For example, Der Kireghian and Ke (1988) recommend the following upper bounds for the size of finite elements to be used with an exponential correlation structure: $\Delta x_k \leq (0.25 - 0.5)\theta_k$, where Δx_k is the finite element size in the spatial direction "k" and θ_k is the correlation distance in the same spatial direction.

2   Quite often, the mesh used for generating sample functions of the random field representing the soil properties is different from the finite element mesh. Therefore, the generated random soil properties have to be transferred from one mesh to another. There are several methods for accomplishing this data transfer (see e.g. Brenner, 1991, for a review). Two of them are mentioned here (see Popescu, 1995, for a comparison study): (1) the spatial averaging method proposed by VanMarcke (1977), which assigns to each element a value obtained as an average of stochastic field values over the element domain; and (2) the midpoint method (e.g. Shinozuka and Dasgupta, 1986; Der Kiureghian and Ke, 1988; Deodatis, 1989), in which the random field is represented by its values at the centroid of each finite element. The midpoint method is used in all numerical examples presented here, as it better preserves the non-Gaussian characteristics describing the variability of different soil properties.

6.4 Understanding mechanisms and effects

6.4.1 A simple example

To provide a first insight on how spatial variability of material properties can modify the mechanical response and which are the main factors affecting structural behavior, a simple example is presented in Figure 6.4. The response of an axially loaded linear elastic bar with initial length L_0, unit area $A_0 = 1$ and uniform Young's modulus $E = 1$, is compared to that of another linear elastic with the same L_0 and A_0, but made of two different materials with Young's moduli equal to $E_1 = 1.2$ and $E_2 = 0.8$. The uniform bar and the non-uniform one are shown in Figures 6.4a and 6.4b, respectively. Note that the non-uniform bar in Figure 6.4b has an "average" modulus $(E_1 + E_2)/2 = 1$, equal to the Young's modulus of the uniform bar. The linear stress–strain relationships of the three materials E, E_1 and E_2 are shown in Figure 6.4c. The two bars are now subjected to an increasing tensile axial stress σ. The ratio of the resulting axial strains ε-variable/ε-uniform (variable denoting the non-uniform bar) is plotted versus a range of values for σ in Figure 6.4e. In this case involving linear elastic materials, only a 4% difference is observed between ε-variable and ε-uniform for all values of σ (continuous line in Figure 6.4e). This small difference is mainly

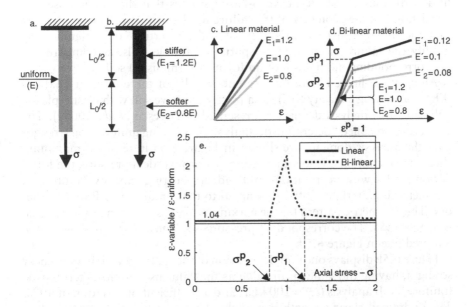

Figure 6.4 Simple example illustrating the effects of material heterogeneity on mechanical response: (a) uniform bar, (b) non-uniform bar, (c) linear stress–strain relationships, (d) bi-linear stress–strain relationships, (e) resulting strain ratios for a range of axial stresses.

induced by the type of averaging considered for Young's modulus and it would vanish if geometric mean were used for the elastic moduli instead of arithmetic mean.

The situation is significantly different for the same bars when the material behavior becomes nonlinear. Assuming bilinear material behavior for all three materials as shown in Figure 6.4d, the resulting differences between the responses in terms of axial strains are now significant. As indicated by the dotted line in Figure 6.4e, the ratio ε-variable/ε-uniform now exceeds 200% for some values of σ. The reason for this different behavior is that for a certain range of axial stresses, the response of the non-uniform bar is controlled by the softer material that reaches the plastic state (with resulting very large strains) before the material of the uniform bar. From this simple example it can be concluded that: (1) mechanical effects induced by material heterogeneity are more pronounced for phenomena governed by highly nonlinear laws, (2) loose zones control the deformation mechanisms, and (3) the effects of material heterogeneity are stronger for certain values of the load intensity. All these observations will be confirmed later based on more realistic examples of heterogeneous soil behavior.

6.4.2 Problems involving a failure surface

In a small number of recent MCS-based studies that did not impose any restrictions on the geometry of the failure mechanism, it has been observed that the inherent spatial variability of soil properties could significantly modify the failure mechanism of spatially variable soils, as compared to the corresponding mechanism of uniform (deterministic) soils.

Along these lines, Figure 6.5 presents results of finite element analyses (FEA) of bearing capacity (BC) for a purely cohesive soil with elastic-plastic behavior (the analysis details are presented in Popescu et al., 2005c). The results of a so-called "deterministic analysis," assuming uniform soil strength over the analysis domain, are shown in Figure 6.5a in terms of maximum shear strain contours. Under increasing uniform vertical pressure, the foundation settles with no rotations and induces bearing capacity failure in a symmetrical pattern, essentially identical to that predicted by Prandtl's theory. The symmetric pattern for the maximum shear strain can be clearly seen in Figure 6.5a. The corresponding pressure–settlement curve is shown with a dotted line in Figure 6.5d.

Figure 6.5b displays one sample function of the spatially variable soil shear strength, having the same mean value as the value used in the deterministic (uniform soil) analysis ($c_u^{av} = 100$ kPa) and a coefficient of variation of 40%. The undrained shear strength is modeled here as a homogeneous random field with a symmetric beta probability distribution function and a separable correlation structure with correlation distances $\theta_H = 5$ m in the horizontal direction and $\theta_V = 1$ m in the vertical direction. Lighter areas in Figure 6.5b

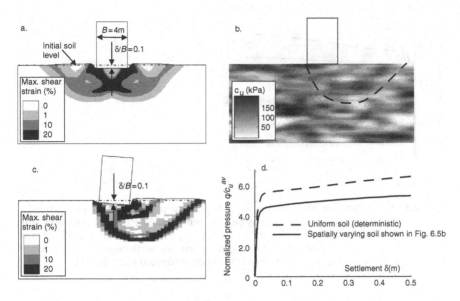

Figure 6.5 Comparison between finite element computations of bearing capacity for a uniform vs. a heterogeneous soil deposit: (a) contours of maximum shear strain for a uniform soil deposit with undrained shear strength, $c_u = 100$ kPa; (b) contours of undrained shear strength (c_u) obtained from a generated sample function of a random field modeling the spatial variability of soil; (c) contours of maximum shear strain for the variable soil shown in Figure 5(b); (d) computed normalized pressure vs. settlement curves. Note that the settlement δ is measured at the midpoint of the width B of the foundation (after Popescu *et al.*, 2005c).

indicate weaker zones of soil, while darker areas indicate stronger zones. Figure 6.5c displays results for the maximum shear strain corresponding to the spatially variable soil deposit shown in Figure 6.5b. An unsymmetric failure surface develops, passing mainly through weaker soil zones, as illustrated by the dotted line in Figure 6.5b. The resulting pressure–settlement curve (continuous line in Figure 6.5d) indicates a lower ultimate BC for the spatially variable soil than that for the uniform one (dotted line in Figure 6.5d). It is important to emphasize here that the failure surface for the spatially variable soil passes mainly through weak soil zones indicated by lighter patches in Figure 6.5b. It should be noted that other sample functions of the spatially variable soil shear strength, different from the one shown in Figure 6.5b, will produce different failure surfaces from the one shown in Figure 6.5c (and different from the deterministic failure pattern shown in Figure 6.5a).

Figure 6.6, also taken from Popescu *et al.* (2005c), presents results involving 100 sample functions in a Monte Carlo simulation type of analysis. The problem configuration here is essentially identical to that in Figure 6.5, the

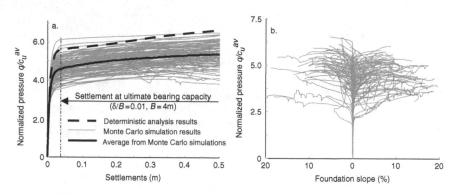

Figure 6.6 Monte Carlo simulation results involving 100 sample functions for a strip foundation on an overconsolidated clay deposit with variable undrained shear strength.
Problem configuration identical to that in Figure 6.5 except marginal PDF of soil
properties: (a) normalized bearing pressures vs. settlements, (b) normalized
bearing pressures vs. footing rotations (after Popescu *et al.*, 2005c).

only difference being that the marginal PDF of soil properties is now modeled
by a Gamma distribution (compared to a beta distribution in Figure 6.5).
The significant variation between different sample functions becomes immediately obvious, as well as the fact that, on the average, the ultimate BC of
spatially variable soils is considerably lower than the corresponding value
of uniform (deterministic) soils. Figure 6.6b indicates also that the spatial
variability of soils induces footing rotations (differential settlements) for
centrally loaded symmetric foundations.

The aforementioned results suggest the following behavior for problems
involving the presence of a failure surface: (1) the consideration of the spatial
variability of soil properties leads to different failure mechanisms (surfaces)
for different realizations of the soil properties. These failure surfaces can
become dramatically different from the classic ones predicted by existing
theories for uniform (deterministic) soils (e.g. Prandtl–Reisner solution for
BC, critical slip surface resulting from limit equilibrium analysis for slope
stability); (2) an immediate consequence is that the failure loads/factors of
safety for the case of spatially variable soils can become significantly lower
(on the average) than the corresponding values for uniform (deterministic)
soils.

6.4.3 Problems involving seismically induced soil
liquefaction

Marcuson (1978) defines liquefaction as the transformation of a granular
material from a solid to a liquefied state as a consequence of increased pore

water pressure and reduced effective stress. This phenomenon occurs most readily in loose to medium dense granular soils that have a tendency to compact when sheared. In saturated soils, pore water pressure drainage may be prevented due to the presence of silty or clayey seam inclusions, or may not have time to occur due to rapid loading – such as in the case of seismic events. In this situation, the tendency to compact is translated into an increase in pore water pressure. This leads to a reduction in effective stress, and a corresponding decrease of the frictional shear strength. If the excess pore water pressure (EPWP) generated at a certain location in a purely frictional soil (e.g. sand) reaches the initial value of the effective vertical stress, then, theoretically, all shear strength is lost at that location and the soil liquefies and behaves like a viscous fluid. Liquefaction-induced large ground deformations are a leading cause of disasters during earthquakes.

Regarding the effects of soil spatial variability on seismically induced liquefaction, both experimental (e.g. Budiman *et al.*, 1995; Konrad and Dubeau, 2002) and numerical results indicate that more EPWP is generated in a heterogeneous soil than in the corresponding uniform soil having geomechanical properties equal to the average properties of the variable soil.

To illustrate the effects of soil heterogeneity on seismically induced EPWP build-up, some of the results obtained by Popescu *et al.* (1997) for a loose to medium dense saturated soil deposit subjected to seismic loading are reproduced in Figure 6.7. The geomechanical properties and spatial variability characteristics were estimated based on the piezocone test results shown in Figure 6.1. The results in Figure 6.7b show the computed contours of the EPWP ratio with respect to the initial effective vertical stress for six sample functions of a stochastic field representing six possible realizations of soil properties over the analysis domain (see Popescu *et al.*, 1997, for more details on the soil properties). Soil liquefaction (EPWP ratio larger than approximately 0.9) was predicted for most sample functions shown in Figure 6.7b. Analysis of an assumed uniform (deterministic) soil deposit, with strength characteristics corresponding to the average strength of the soil samples used in MCS, resulted in no soil liquefaction (the maximum predicted EPWP ratio was 0.44 as shown in the upper-left plot of Figure 6.7c). It can be concluded from these results that both the pattern and the amount of dynamically induced EPWP build-up are strongly affected by the spatial variability of soil properties. For the same average values of the soil parameters, more EPWP build-up was predicted in the stochastic analysis (MCS) accounting for spatial variability than in the deterministic analysis considering uniform soil properties.

It was postulated by Popescu *et al.* (1997) that the presence of loose soil pockets leads to earlier initiation of EPWP and to local liquefaction, compared to the corresponding uniform soil. After that, the pressure gradient between loose and dense zones would induce water migration into

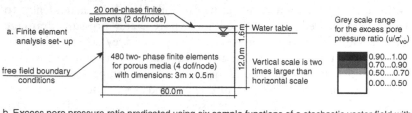

a. Finite element analysis set-up

free field boundary conditions

20 one-phase finite elements (2 dof/node)

480 two-phase finite elements for porous media (4 dof/node) with dimensions: 3m x 0.5m

60.0m

12.0m

1.6m

Water table

Vertical scale is two times larger than horizontal scale

Grey scale range for the excess pore pressure ratio (u/σ'$_{vo}$)

0.90...1.00
0.70...0.90
0.50....0.70
0.00...0.50

b. Excess pore pressure ratio predicated using six sample functions of a stochastic vector field with cross-correlation structure and probability distribution functions estimated from piezocone test results

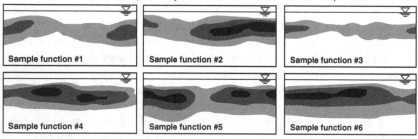

Sample function #1

Sample function #2

Sample function #3

Sample function #4

Sample function #5

Sample function #6

c. Excess pore pressure ratio predicted using **deterministic** input soil parameters

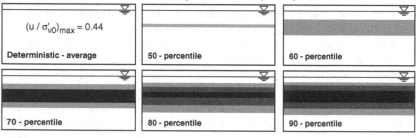

(u / σ'$_{v0}$)$_{max}$ = 0.44

Deterministic - average

50 - percentile

60 - percentile

70 - percentile

80 - percentile

90 - percentile

Figure 6.7 Monte Carlo simulations of seismically induced liquefaction in a saturated soil deposit, accounting for natural variability of the soil properties (after Popescu *et al.*, 1997): (a) finite element analysis setup; (b) contours of EPWP ratio for six sample functions used in the Monte Carlo simulations; (c) contours of EPWP ratio using deterministic (i.e. uniform) soil parameters and various percentiles of soil strength.

neighboring denser soil zones, followed by softening and liquefaction of the dense sand. This assumption – mainly that the presence of loose soil pockets is responsible for lower liquefaction resistance – was further verified by two sets of numerical results. First, for a given average value of the soil strength and a given earthquake intensity, the resulting amount of EPWP build-up in a variable soil increased with the degree of variability of soil properties (expressed by the coefficient of variation). This degree of soil variability directly controls the amount of loose pockets in the soil mass (e.g. Popescu *et al.*, 1998c). Second, for the same coefficient of variation of spatial variability, more EPWP was predicted when the soil strength fluctuations followed a probability distribution function (PDF) with a fatter left tail that yields directly a larger amount of loose soil pockets (e.g. Popescu *et al.*, 1996).

There are very few experimental studies dealing with liquefaction of spatially variable soil. Particularly interesting are the results obtained by Konrad and Dubeau (2002). They performed a series of undrained cyclic triaxial tests with various cyclic stress ratios on uniform dense sand, uniform silt, and layered soil (a silt layer sandwiched between two sand layers). The soils in layered samples were prepared at the same void ratios as the corresponding soils in the uniform samples. It was concluded from the results that the cyclic strength (expressed as number of cycles to liquefaction) of the layered (non-homogeneous) samples was considerably lower than the one of the uniform silt and the uniform sand samples. For example, at a cyclic stress ratio (CSR) $= 0.166$, the numbers of cycles to liquefaction were: $N_L = 150$ for uniform dense sand at relative density 77%, $N_L = 90$ for silt prepared at void ratio 0.78, and $N_L = 42$ for the layered soil. Chakrabortty *et al.* (2004) reproduced the cyclic undrained triaxial tests on uniform and layered samples made of dense sand and silt layers described by Konrad and Dubeau (2002), and studied the mechanism by which a sample made of two different soils liquefies faster than each of the soils tested separately in uniform samples. Their explanation – resulting from a detailed analysis of the numerical results – was that water was *squeezed* from the more deformable silt layer and injected into the neighboring sand, leading to liquefaction of the dense sand.

Regarding liquefaction mechanisms of soil deposits involving the same material, but with spatially variable strength, Ghosh and Madabhushi (2003) performed a series of centrifuge experiments to analyze the effects of localized loose patches in a dense sand deposit subjected to seismic loads. They observed that EPWP is first generated in the loose sand patches, and then water migrates into the neighboring dense sand, reducing the effective stress and loosening the dense soil that can subsequently liquefy. As discussed before, it is believed that a similar phenomenon is responsible for the lower liquefaction resistance of continuously heterogeneous soils compared to that of corresponding uniform soils. To further verify this assumed mechanism, Popescu *et al.* (2006) calibrated a numerical finite element model for liquefaction analysis (Prévost, 1985, 2002) to reproduce the centrifuge experimental results of Ghosh and Madabhushi (2003), and then used it for a detailed analysis of a structure founded on a hypothetical chess board-like heterogeneous soil deposit subjected to earthquake loading. From careful analysis of the EPWP calculated at the border between loose and dense zones, it was clearly observed how pore water pressures built up first in the loose areas, and then transferred into the dense sand zones that eventually experienced the same increase in EPWP. Water "injection" into dense sand loosened the strong pockets and the overall liquefaction resistance of the heterogeneous soil deposit became much lower than that of a corresponding uniform soil, almost as low as the liquefaction resistance of a deposit made entirely of loose sand.

6.5 Quantifying the effects of random soil heterogeneity

6.5.1 General considerations

Analyzing the effects of inherent soil variability by MCS involves the solution of a large number of nonlinear boundary value problems (BVPs) as was described earlier in this chapter. Such an approach is too expensive computationally and too complex conceptually to be used in current design applications. The role of academic research is to eventually provide the geotechnical practice with easy to understand/easy to use guidelines in the form of charts and characteristic values that account for the effects of soil variability. In this respect, MCS results contain a wealth of information that can be processed in various ways to provide insight into the practical aspects of the problem at hand. Two such design guidelines estimated from MCS results are presented here:

1   "Equivalent uniform" soil properties representing values of certain soil properties which – when used in a deterministic analysis assuming uniform soil properties over the analysis domain – would provide a response "equivalent" to the average response resulting from computationally expensive MCS accounting for soil spatial variability. The response resulting from deterministic analysis is termed "equivalent" to the MCS response since only certain components of it can be similar. For example, when studying BC failure of symmetrically loaded foundations, one can obtain in a deterministic analysis an ultimate BC equal to the average one from MCS, but cannot simulate footing rotations. Similarly, for the case of soil liquefaction, a deterministic analysis can reproduce the average amount of EPWP resulting from MCS, but cannot predict any kind of spatially variable liquefaction pattern (see e.g. Figures 6.7b and c). In conclusion, an equivalent uniform analysis would only match the average response provided by a set of MCS accounting for spatial variability of soil properties, and nothing more. The equivalent uniform properties are usually expressed as percentiles of field test recorded soil index properties and are termed "characteristic percentiles."

2   Fragility curves, which are an illustrative and practical way of expressing the probability of exceeding a prescribed threshold in the response (or damage level) as a function of load intensity. They can be used directly by practicing engineers without having to perform any complex

computations and form the basis for all risk analysis/loss estimation/risk reduction calculations of civil infrastructure systems performed by emergency management agencies and insurance companies. Fragility curves can include effects of multiple sources of uncertainty related to material resistance or load characteristics. The fragility curves are usually represented as shifted lognormal cumulative distribution functions. For a specific threshold in the response, each MCS case analyzed is treated as a realization of a Bernoulli experiment. The Bernoulli random variable resulting from each MCS case is assigned to the unity probability level if the selected threshold is exceeded, or to zero if the computed response is less than the threshold. Next, the two parameters of the shifted lognormal distribution are estimated using the maximum likelihood method and the results of all the MCS cases. For a detailed description of this methodology, the reader is referred to Shinozuka *et al.* (2000) and Deodatis *et al.* (2000).

Such guidelines have obvious qualitative value, indicating the most important effects of soil heterogeneity for a particular BVP and type of response, which probabilistic characteristics of spatial variability have major effects, and which are the directions where more investigations would provide maximum benefit. However, several other aspects have to be carefully considered as will be demonstrated in the following, when such guidelines are used in quantitative analyses.

6.5.2 Liquefaction of level ground random heterogeneous soil

Regarding liquefaction analysis of natural soil deposits exhibiting a certain degree of spatial variability in their properties, Popescu *et al.* (1997) suggested that in order to predict more accurate values of EPWP build-up in a deterministic analysis assuming uniform soil properties, one has to use a modified (or equivalent uniform) soil strength. The results of deterministic analyses presented in Figure 6.7c illustrate this idea of an equivalent uniform soil strength for liquefaction analysis of randomly heterogeneous soils. For every such analysis in Figure 6.7c, the uniform (deterministic) soil properties are determined based on a certain percentile of the in-situ recorded q_c (as liquefaction strength is proportional to q_c). From visual examination and comparison of Figures 6.7b and 6.7c, one can select a percentile value somewhere between 70% and 80% that would lead to an equivalent amount of EPWP in a deterministic analysis to the one predicted by multiple MCS. This value is termed "characteristic percentile." A comparison between the results presented in Figures 6.7b and 6.7c also shows clearly the differences in the predicted liquefaction pattern of natural heterogeneous soil versus that of assumed uniform soil. This fact somehow invalidates the use of an equivalent uniform soil strength estimated solely on the basis of average EPWP build-up, as liquefaction effects may be controlled also by the presence of a so called "maximal plane," defined by Fenton and VanMarcke (1988) as the horizontal plane having the highest value of average EPWP ratio. This aspect is discussed in detail by Popescu et al. (2005a).

It should be mentioned that establishing the characteristic percentile by "visual examination and comparison" of results is exemplified here only for illustrative purposes. Popescu et al. (1997) compared the range of MCS predictions with deterministic analysis results in terms of several indices characterizing the severity of liquefaction and in terms of horizontal displacements. Furthermore, these comparisons were made for seismic inputs with various ranges of maximum spectral amplitudes. For the type of soil deposit analyzed and for the probabilistic characteristics of spatial variability considered in that study, it was determined that the 80-percentile of soil strength was a good equivalent uniform value to be used in deterministic analyses.

The computed characteristic percentile of soil strength is valid only for the specific situation analyzed and can be affected by a series of factors, such as: soil type, probabilistic characteristics of spatial variability, seismic motion intensity and frequency content, etc.

Popescu et al. (1998c) studied the effect of the "intensity" of spatial variability on the characteristic percentile of soil strength. A fine sand deposit with probabilistic characteristics of its spatial variability derived from the field data from Tokyo Bay area (upper soil layer in Figures 6.2b and 6.3b) has been analyzed in a parametric study involving a range of values for the CV of soil strength. Some of these results are shown in Figure 6.8 in the form of range of average EPWP ratios predicted by MCS as a function of CV (shaded area). The EPWP ratio was averaged over the entire analysis domain (12 m deep × 60 m long). As only 10 sample functions of spatial variability have been used for each value of CV, these results have only qualitative value. It can be observed that for the situation analyzed in this example, the effects of spatial variability are insignificant for CV<0.2 and are strongly dependent on the degree of soil variability for CV between 0.2 and 0.8. It can be therefore concluded that, for the soil type and seismic intensity considered in this example, CV=0.2 is a threshold beyond which spatial variability has an important effect on liquefaction. The range of average EPWP ratios predicted by MCS is compared in Figure 6.8 with average EPWP ratios computed from deterministic analyses assuming uniform soil properties and using various percentiles of in-situ recorded soil strength. This comparison offers some insight about the relation between the characteristic percentile of soil strength and the CV.

Intensity of seismic ground motion is another important factor. Popescu et al. (2005a) studied the effect of soil heterogeneity on liquefaction for a soil deposit subjected to a series of earthquake ground accelerations

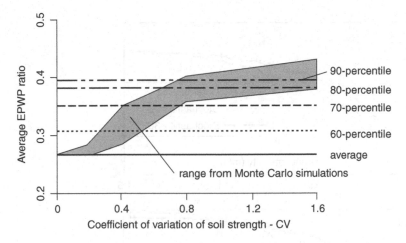

Figure 6.8 Range of average excess pore water pressure (EPWP) ratio from Monte Carlo simulations as a function of the coefficient of variation of soil strength (after Popescu *et al.*, 1998c). Horizontal lines represent average EPWP ratios computed from deterministic analyses assuming uniform soil strength equal to the average and four percentiles of the values used in MCS.

corresponding to a wide range of seismic intensities. A saturated soil deposit with randomly varying soil strength was considered in a range of full 3D analyses (accounting for soil variability in all three spatial directions), as well as in a set of corresponding 2D analyses under a plane strain assumption (and therefore assuming infinite correlation distance in the third direction). The soil variability was modeled by a three-dimensional, two-variate random field, having as components the cone tip resistance q_c and the soil classification index. A value of CV=0.5 was assumed for q_c. For all other probabilistic characteristics considered in that study, the reader is referred to Popescu *et al.* (2005a). Deterministic analyses, assuming average uniform soil properties, were also performed for comparison.

The resulting EPWP ratios were compared in terms of their averages over the entire analysis domain (Figure 6.9a) and averages over the "maximal plane" (Figure 6.9b). A point on each line in Figure 6.9 represents the computed average EPWP ratio for a specific input earthquake intensity, expressed by the Arias Intensity (e.g. Arias, 1970) which is a measure of the total energy delivered per unit mass of soil during an earthquake. The corresponding approximate peak ground accelerations (PGA) are also indicated on Figure 6.9. Similar to results of previous studies, it can be observed that generally higher EPWP ratios were predicted (on average) by MCS than by deterministic analysis. It is important to remark, however, that for the type of soil and input accelerations used in that study, significant differences between

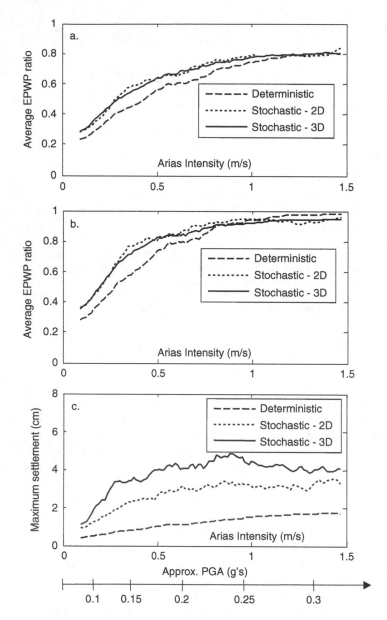

Figure 6.9 Comparison between stochastic (2D and 3D) and deterministic analysis results in terms of predicted: (a) EPWP ratios averaged over the entire analysis domain; (b) EPWP ratios averaged over the maximal plane; (c) maximum settlements (after Popescu *et al.*, 2005a).

MCS and deterministic analyses occurred only for seismic inputs with Arias Intensity less than 1 m/s (or PGA less than 0.25 g), while for stronger earthquakes the results were very similar. It is obvious, therefore, that the "equivalent uniform" soil strengths or characteristic percentiles developed for liquefaction of heterogeneous soils are dependent on the seismic load intensity.

As discussed in Section 6.2, another important probabilistic characteristic of the spatial variability is the correlation distance. Figure 6.9 indicates that MCS using full 3D analysis yielded practically the same results in terms of average EPWP ratios as MCS using 2D analysis in conjunction with the plane strain assumption. While the 3D analyses account for soil spatial variability in both horizontal directions, in 2D plane strain calculations the soil is assumed uniform in one horizontal direction (normal to the plane of analysis), and the correlation distance in that direction is, therefore, considered to be infinite. Therefore, it can be concluded from these results that the value of horizontal correlation distance does not affect the average amount of EPWP build-up in this problem. In contrast, the horizontal correlation distance affects other factors of the response in this same problem, as shown in Figure 6.9c displaying settlements. Maximum settlements predicted by MCS using 3D analyses are about 40% larger than those resulting from MCS using 2D analyses. A similar conclusion was also reached by Popescu (1995) based on a limited study on the effects of horizontal correlation distance on liquefaction of heterogeneous soils: a fivefold increase in horizontal correlation distance did not change the predicted average EPWP build-up, but resulted in significant changes in the pattern of liquefaction.

6.5.3 Seismic response of structures on liquefiable random heterogeneous soil

Some of the effects of inherently spatially variable soil properties on the seismic response of structures founded on potentially liquefiable soils have been studied by Popescu et al. (2005b), who considered a tall structure on a saturated sand deposit. The structure was modeled as a single degree-of-freedom oscillator with a characteristic frequency of 1.4 Hz corresponding to a seven-storey building. The surrounding soil (and consequently the structure too) was subjected to a series of 100 earthquake acceleration time histories, scaled according to their Arias Intensities, with the objective of establishing fragility curves for two response thresholds: exceeding average settlements of 20 cm and differential settlements of 5 cm. These fragility curves are plotted in Figures 6.10a and 6.10b for two different assumptions related to soil properties: (1) variable soils with properties modeled by a random field with coefficient of variation CV = 0.5, correlation distances $\theta_H = 8$ m in the horizontal direction and $\theta_V = 2$ m in the vertical direction, and gamma PDF

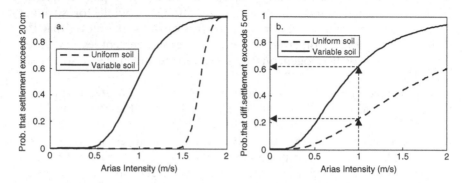

Figure 6.10 Fragility curves for a tall structure on a saturated sand deposit. Comparison between results for variable soil and equivalent uniform soil in terms of: (a) average settlements; (b) differential settlements.

(see Popescu *et al.*, 2005b, for more details), and (2) corresponding uniform soils.

As can be easily observed in Figure 6.10, the fragility curves calculated for variable soil are shifted to the left compared to the ones for uniform soil, indicating that – for a given seismic intensity – there is a significantly higher probability of exceeding the response threshold when soil variability is accounted for than for an assumed corresponding uniform soil. For example, for a seismic ground motion with Arias Intensity of 1 m/s (or PGA of about 0.25 g), a 62% exceedance probability of a 5 cm differential settlement is estimated for spatially variable soils, versus only a 22% exceedance probability for corresponding uniform soils (Figure 6.10b).

The behavior at very large seismic intensities is very interesting. As far as average settlements are concerned (Figure 6.10a), both fragility curves converge to the same probability (approximately unity), indicating that the degree of liquefaction of the soil below structure is about the same for both uniform and variable soils. This is in agreement with the results presented in Figures 6.9a and 6.9b for ground level soil. In contrast, the two fragility curves for differential settlements in Figure 6.10b are still far apart at very large Arias Intensities. This suggests that – even if the computed degree of liquefaction is about the same at the end of the seismic event for both uniform and variable soils – the initiation of liquefaction in variable soils takes place in isolated patches that strongly affect subsequent local soil deformations and therefore differential settlements of foundations. This is believed to be due to lateral migration of pore water from loose soil zones that liquefy first towards denser soil pockets (as discussed in Section 6.4.4), leading eventually to local settlements that cannot be captured in an analysis assuming uniform soil properties.

6.5.4 Static bearing capacity of shallow foundations on random heterogeneous soil

The effects of various probabilistic characteristics of soil variability on the BC of shallow foundations were studied by Popescu *et al.* (2005c) in a parametric study involving the CV, the PDF and the horizontal correlation distance of the random field modeling the spatial variability of soil strength. The layout of the foundation under consideration is the one shown in Figure 6.5 and the main assumptions made are mentioned in Section 6.4.2. The finite element analyses involved in the MCS scheme were performed using ABAQUS/Standard (Hibbitt *et al.*, 1998). Some of the results are presented in Figure 6.11 in terms of fragility curves expressing the probability of BC failure of the strip foundation on randomly heterogeneous soil as a function of the bearing pressure. For comparison with results of corresponding deterministic analyses, the bearing pressures q in Figure 6.11 are normalized with respect to the ultimate BC obtained for

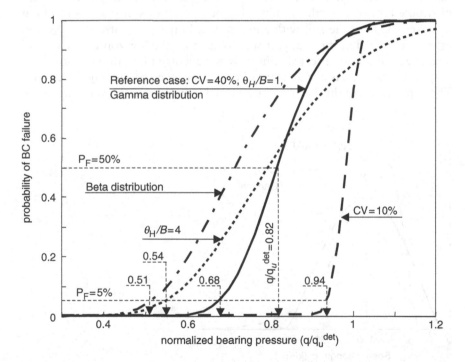

Figure 6.11 Fragility curves illustrating the probability of bearing capacity failure of a strip foundation of width B on randomly heterogeneous soil, as a function of the bearing pressure q normalized with respect to the ultimate bearing capacity q_u^{det}, corresponding to uniform soil with undrained shear strength $c_u = 100$ kPa.

the corresponding uniform soil: q_u^{det}. Each fragility curve represents the results of MCS involving 100 realizations of the spatial variability of soil strength, generated using a set of probabilistic characteristics describing soil variability. One of the sets – corresponding to CV = 40%, $\theta_H/B = 1$ and gamma-distributed soil strength – is labeled "reference case." The other three cases shown in the figure are obtained by varying one parameter of the reference case at a time. It is mentioned that the method for building the fragility curves shown in Figure 6.11 is different from the traditional procedure using Bernoulli trials (e.g. Shinozuka *et al.*, 2000; Deodatis *et al.*, 2000) and is described in detail in Popescu *et al.* (2005c).

Figure 6.11 indicates that the two most important probabilistic characteristics of inherent soil heterogeneity – with respect to their effects on BC failure – are the coefficient of variation (CV) and marginal PDF of soil strength. Both the CV and the PDF control the amount of loose pockets in the soil mass. Figure 6.12 demonstrates that the symmetrical beta distribution used in this study has a thicker left tail than the gamma distribution that is positively skewed. Therefore, for the same CV, the soil with beta-distributed shear strength exhibits a larger amount of soft pockets resulting in lower BC than the soil with gamma-distributed shear strength. For the range considered in this study, it was found that the horizontal correlation distance θ_H does not affect significantly the average ultimate BC. However, it appears that θ_H has a significant effect on the slope of the fragility curves. The explanation of this behavior is believed to be related to averaging of the

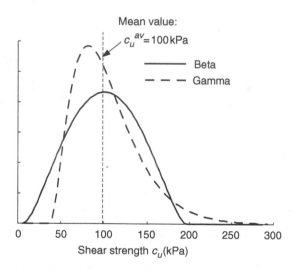

Figure 6.12 The two models selected for the marginal PDF of the undrained shear strength of soil. Both PDFs have the same mean value ($c_u^{av} = 100$ kPa). The coefficient of variation used for this plot is CV = 40% for both PDFs (after Popescu *et al.*, 2005c).

shear strength over the length of the failure surface: larger horizontal correlation distances tend to diminish the averaging effect on the bearing capacity failure mechanism, and therefore yield larger variability of the results in terms of ultimate BC.

An equivalent undrained shear strength c_u for uniform soils can be inferred from the fragility curves presented in Figure 6.11 in the following way. Based on the linear relation between ultimate BC and c_u, it is possible to define an approximate equivalent uniform c_u from the normalized pressures q/q_u^{det} corresponding to a probability of BC failure $P_F = 50\%$. For example, for the type of soils and BVP analyzed in this example, the equivalent uniform c_u for the reference case is $c_u^{EQ} = 0.82c_u^{av} = 82$ kPa (refer to Figure 6.11 and remember that $c_u^{av} = 100$ kPa). Similar c_u^{EQ} values can be inferred for the other cases illustrated in Figure 6.11, as listed in Table 6.1.

Instead of an equivalent uniform c_u corresponding to a failure probability of 50%, most modern codes use nominal values of material strength corresponding to much lower failure probabilities (usually 5 or 10%). The determination of such nominal values for the BC of soil – denoted by q_{uN} – is demonstrated graphically in Figure 6.11 for a probability of BC failure $P_F = 5\%$. The resulting normalized nominal values are displayed in Table 6.1. Regarding the earlier discussion about the effect of the horizontal correlation distance θ_H, Table 6.1 indicates that even if θ_H has little effect on the average behavior of spatially variable soils (less than 4% change in c_u^{EQ} when θ_H/B varies from 1 to 4), it affects significantly the resulting nominal values (21% change in q_{uN} when θ_H/B varies from 1 to 4).

More results such as those presented in Figure 6.11 and Table 6.1, including the effects of soil deformability on the BC, and also the effects of various probabilistic characteristics on differential settlements, are presented in Popescu et al. (2005c).

6.6 Conclusions

Some distinctive features of spatially variable random heterogeneous soil and the main geomechanical aspects governing its behavior under load have been

Table 6.1 Normalized equivalent uniform undrained shear strength and normalized nominal values for the cases illustrated in Figure 6.11.

	$CV = 40\%$, $\theta_H/B = 1$, gamma PDF (reference case)	$CV = 10\%$, $\theta_H/B = 1$, gamma PDF	$CV = 40\%$, $\theta_H/B = 1$, beta PDF	$CV = 40\%$, $\theta_H/B = 4$, gamma PDF
c_u^{EQ}/c_u^{av}	0.82	0.98	0.71	0.79
q_{uN}/q_u^{det}	0.68	0.94	0.51	0.54

analyzed and discussed for two types of soil mechanics problems: seismically induced soil liquefaction, and bearing capacity of shallow foundations. For the case of liquefaction, it was shown that pore water pressures build up first in the loose soil areas, and then, due to pressure gradients, water is "injected" into neighboring dense sand. In this way the strong pockets are softened and the overall liquefaction resistance is reduced, becoming considerably lower than that of the corresponding uniform soil. For the case of geotechnical problems involving presence of a failure surface, the spatial variability of soil properties can change dramatically the failure mechanism from the well known theoretical result obtained for uniform soils (e.g. the Prandtl – Reisner solution for bearing capacity) to a failure surface that seeks the weakest path following predominantly the weak zones (pockets) in the soil mass. Such a surface will have a more or less irregular shape and could differ significantly from one possible realization of the spatial distribution of soil strength to another.

It was demonstrated that random soil heterogeneity has similar overall effects (specifically inducing lower mechanical resistance) on phenomena that are fundamentally different: limit equilibrium type of failures involving a large volume of soil – for BC, and micro-scale pore pressure build-up mechanisms – for liquefaction. Moreover, for both phenomena addressed here, the loose zones (pockets) in the soil mass have a crucial and critical effect on the failure mechanisms (e.g. for soil liquefaction they are the initial sources of EPWP build-up, while for BC the failure surface seeks the weakest path). It also appears that random soil heterogeneity has a more pronounced effect for phenomena governed by highly nonlinear laws.

An outcome of major importance of both experimental investigations and numerical simulations discussed in this chapter is that the inherent spatial variability of soil properties affects the fundamental nature of soil behavior, leading to lower failure loads as compared to those for corresponding uniform soils. Therefore, besides variability in the computed response, spatial variability of soil properties will also produce a response (e.g. failure mechanism or liquefaction pattern) that is, in general, different from the theoretical response of the corresponding uniform soil. On the other hand, other uncertainties considered in reliability analyses, such as random measurement errors and transformation errors, only induce variability in the computed response. Consequently, it is imperative that the effects of inherent soil heterogeneity be estimated separately from those of other uncertainties, and only after that combined for reliability assessments.

For both problems discussed here, it was found that the CV and marginal PDF of soil strength are the two most important parameters of spatial variability in reducing liquefaction resistance or bearing capacity and producing substantial differential settlements. This is due to the fact that both CV and PDF (more specifically the shape of its left tail) control the amount of loose pockets in the soil mass. For the range of correlation distances considered in this study, the horizontal correlation distance was found to have only a minor influence on overall liquefaction resistance or bearing capacity. It was found, however, to significantly affect the variability of the response, and therefore the low fractiles and the nominal values.

References

American Society for Testing and Materials (1989). Standard test method for deep, quasi-static, cone and friction-cone penetration tests of soil (D3441–86). In *Annual Book of Standards*, Vol. 4.08. ASTM, Philadelphia, pp. 414–19.

Arias, A. (1970). A measure of earthquake intensity. In *Seismic Design for Nuclear Power Plants*, Ed. R. J. Hansen. MIT Press, Cambridge, MA, pp. 438–83.

Brenner, C. E. (1991). Stochastic finite element methods: literature review. *Technical Report 35–91*, Institute of Engineering Mechanics, University of Innsbruck, Austria.

Brzakala, W. and Pula, W. (1996). A probabilistic analysis of foundation settlements. *Computers and Geotechnics*, 18(4); 291–309.

Budiman, J. S., Mohammadi, J. and Bandi, S. (1995). Effect of large inclusions on liquefaction of sand. In *Proceedings, Conference on Geotechnical Engr. Div.*, Geotechnical Special Publication 56, ASCE, pp. 48–63.

Chakrabortty, P., Jafari-Mehrabadi, A. and Popescu, R. (2004). Effects of low permeability soil layers on seismic stability of submarine slopes. In *Proceedings of the 57th Canadian Geotechnical Conference*, Quebec City, PQ, on CD-ROM.

Cherubini, C. (2000). Reliability evaluation of shallow foundation bearing capacity on c', ϕ' soils. *Canadian Geotechnical Journal*, 37; 264–9.

Christian, J. T. (2004). Geotechnical engineering reliability: how well do we know what we are doing? *Journal of Geotechnical and Geoenvironmental Engineering, ASCE*, 130(10); 985–1003.

DeGroot, D. J. and Baecher, G. B. (1993). Estimating autocovariance of in-situ soil properties. *Journal of Geotechnical Engineering, ASCE*, 119(1); 147–66.

Deodatis, G. (1989). Stochastic FEM sensitivity analysis of nonlinear dynamic problems. *Probabilistic Engineering Mechanics*, 4(3); 135–41.

Deodatis, G. (1996). Simulation of ergodic multi-variate stochastic processes. *Journal of Engineering Mechanics, ASCE*, 122(8); 778–87.

Deodatis, G. and Micaletti, R. (2001). Simulation of highly skewed non-Gaussian stochastic processes. *Journal of Engineering Mechanics, ASCE*, 127(12); 1284–95.

Deodatis, G., Saxena, V. and Shinozuka, M. (2000). Effect of spatial variability of ground motion on bridge fragility curves. In *Proceedings of the 8th ASCE Specialty Conference on Probabilistic Mechanics and Structural Reliability*, University of Notre Dame, IL, on CD-ROM.

Der Kiureghian, A. and Ke, J. B. (1988). The stochastic finite element method in structural reliability. *Probabilistic Engineering Mechanics*, 3(2); 83–91.

Elishakoff, I. (1983). *Probabilistic Methods in the Theory of Structures*. Willey.

Elkateb, T., Chalaturnyk, R. and Robertson, P. K. (2003). Simplified geostatistical analysis of earthquake-induced ground response at the Wildlife Site, California, USA. *Canadian Geotechnical Journal*, 40; 16–35.

El-Ramly, H., Morgenstern, N. R. and Cruden, D. (2002). Probabilistic slope stability analysis for practice. *Canadian Geotechnical Journal*, 39; 665–83.

El-Ramly, H., Morgenstern, N. R. and Cruden, D. (2005). Probabilistic assessment of a cut slope in residual soil. *Geotechnique*, 55(2); 77–84.

Fenton, G. A. (1999a). Estimation for stochastic soil models. *Journal of Geotechnical and Geoenvironmental Engineering*, 125(6); 470–85.

Fenton, G. A. (1999b). Random field modeling of CPT data. *Journal of Geotechnical and Geoenvironmental Engineering*, 125(6); 486–98.

Fenton, G. A. and Griffiths, D. V. (2003). Bearing capacity prediction of spatially random $c - \phi$ soils. *Canadian Geotechnical Journal*, 40; 54–65.

Fenton, G. A. and VanMarcke, E. H. (1998). Spatial variation in liquefaction risk. *Geotechnique*, 48(6); 819–31.

Focht, J. A. and Focht III, J. A. (2001). Factor of safety and reliability in geotechnical engineering (Discussion). *Journal of Geotechnical and Geoenvironmental Engineering, ASCE*, 127(8); 704–6.

Ghosh, B. and Madabhushi, S. P. G. (2003). Effects of localized soil inhomogeneity in modifying seismic soil–structure interaction. In *Proceeding of the 16th ASCE Engineering Mechanics Conference*, Seattle, WA, July.

Griffths, D. V. and Fenton, G. A. (2004). Probabilistic slope stability analysis by finite elements. *Journal of Geotechnical and Geoenvironmental Engineering, ASCE*, 130(5); 507–18.

Griffths, D. V., Fenton, G. A. and Manoharan, N. (2002). Bearing capacity of rough rigid strip footing on cohesive soil: probabilistic study. *Journal of Geotechnical and Geoenvironmental Engineering, ASCE*, 128(9); 743–55.

Grigoriu, M. (1995). *Applied non-Gaussian Processes*. Prentice-Hall, Englewood Cliffs, NJ.

Grigoriu, M. (1998). Simulation of stationary non-Gaussian translation processes. *Journal of Engineering Mechanics, ASCE*, 124(2); 121–6.

Gulf Canada Resources Inc. (1984). Frontier development, Molikpaq, Tarsiut delineation 1984–85 season. Technical Report 84F012.

Gurley, K. and Kareem, A. (1998). Simulation of correlated non-Gaussian pressure fields. *Meccanica-International, the Journal of the Italian Association of Theoretical and Applied Mechanics*, 33; 309–17.

Hibbitt, Karlsson & Sorensen, Inc. (1998). *ABAQUS version 5.8 – Theory Manual*. Hibbitt, Karlsson & Sorensen, Inc., Pawtucket, RI.

Hicks, M. A. and Onisiphorou, C. (2005). Stochastic evaluation of static liquefaction in a predominantly dilative sand fill. *Geotechnique*, 55(2); 123–33.

Huoy, L., Breysse, D. and Denis, A. (2005). Influence of soil heterogeneity on load redistribution and settlement of a hyperstatic three-support frame. *Geotechnique*, 55(2); 163–70.

Jaksa, M. B. (2007). Modeling the natural variability of over-consolidated clay in Adelaide, South Australia. In *Characterisation and Natural Properties of Natural Soils*, Eds Tan, Phoon, Hight and Leroueil. Taylor & Francis, London, pp. 2721–51.

Jefferies, M. G. and Davies, M. P. (1993). Use of CPTu to estimate equivalent SPT N60. *Geotechnical Testing Journal*, 16(4); 458–68.

Konrad, J-M. and Dubeau, S. (2002). Cyclic strength of stratified soil samples. In *Proceedings of the 55th Canadian Geotechnical Conference: Ground and*

Water: Theory to Practice. Canadian Geotechnical Society, Alliston, ON, October, pp. 89–94.

Koutsourelakis, S., Prevost, J. H. and Deodatis, G. (2002). Risk assessment of an interacting structure–soil system due to liquefaction. *Earthquake Engineering and Structural Dynamics*, 31; 851–79.

Lacasse, S. and Nadim, F. (1996). Uncertainties in characteristic soil properties. In *Proceedings Uncertainty in the Geologic Environment: From Theory to Practice*, ASCE, New York, NY, pp. 49–75.

Lumb, P. (1966). The variability of natural soils. *Canadian Geotechnical Journal*, 3; 74–97.

Marcuson, W. F. III. (1978). Definition of terms related to liquefaction. *Journal of the Geotechnical Engineering Division, ASCE*, 104(9); 1197–200.

Masters, F. and Gurley, K. (2003). Non-Gaussian simulation: CDF map-based spectral correction. *Journal of Engineering Mechanics*, ASCE, 129(12); 1418–1428.

Nobahar, A. and Popescu, R. (2000). Spatial variability of soil properties – effects on foundation design. In *Proceedings of the 53rd Canadian Geotechnical Conference*, Montreal, Québec, 2; pp. 1139–44.

Paice, G. M., Griffiths, G. V. and Fenton, G. A. (1996). Finite element modeling of settlement on spatially random soil. *Journal of Geotechnical Engineering, ASCE*, 122; 777–80.

Phoon, K-K. (2006). Modeling and simulation of stochastic data. In *Proceedings, GeoCongress 2006: Geotechnical Engineering in the Information Technology Age*, on CD-ROM.

Phoon K-K. and Kulhawy, F. H. (1999). Characterization of geotechnical variability. *Canadian Geotechnical Journal*, 36; 612–24.

Phoon K-K. and Kulhawy, F. H. (2001). Characterization of geotechnical variability *and* Evaluation of geotechnical property variability (Reply). *Canadian Geotechnical Journal*, 38; 214–15.

Phoon, K. K., Huang, H. W. and Quek, S. T. (2005). Simulation of strongly non-Gaussian processes using Karhunen–Loeve expansion. *Probabilistic Engineering Mechanics*, 20(2); 188–98.

Popescu, R. (1995). Stochastic variability of soil properties: data analysis, digital simulation, effects on system behavior. PhD Thesis, Princeton University, Princeton, NJ. http://cee.princeton.edu/~radu/papers/phd/.

Popescu, R. (2004). Bearing capacity prediction of spatially random $c - \phi$ soils (Discussion). *Canadian Geotechnical Journal*, 41; 366–7.

Popescu, R., Chakrabortty, P. and Prevost, J. H. (2005b). Fragility curves for tall structures on stochastically variable soil. In *Proceedings of the 9th International Conference on Structural Safety and Reliability (ICOSSAR)*, Eds G. Augusti, G. I. Schueller and M. Ciampoli. Millpress, Rotterdam, pp. 977–84.

Popescu, R., Deodatis, R. and Nobahar, A. (2005c). Effects of soil heterogeneity on bearing capacity. *Probabilistic Engineering Mechanics*, 20(4); 324–41.

Popescu, R., Deodatis, G. and Prevost, J. H. (1998b). Simulation of non-Gaussian homogeneous stochastic vector fields. *Probabilistic Engineering Mechanics*, 13(1); 1–13.

Popescu, R., Prevost, J. H. and Deodatis, G. (1996). Influence of spatial variability of soil properties on seismically induced soil liquefaction. In *Proceedings,*

Uncertainty in the Geologic Environment: From Theory to Practice, ASCE. Madison, Wisconsin, pp. 1098–112.

Popescu, R., Prevost, J. H. and Deodatis, G. (1997). Effects of spatial variability on soil liquefaction: some design recommendations. *Geotechnique*, 47(5); 1019–36.

Popescu, R., Prevost, J. H. and Deodatis, G. (1998a). Spatial variability of soil properties: two case studies. In *Geotechnical Earthquake Engineering and Soil Dynamics*, Geotechnical Special Publication No. 75, ASCE, Seattle, pp. 568–79.

Popescu, R., Prevost, J. H. and Deodatis, G. (1998c). Characteristic percentile of soil strength for dynamic analyses. In *Geotechnical Earthquake Engineering and Soil Dynamics*, Geotechnical Special Publication No. 75, ASCE, Seattle, pp. 1461–71.

Popescu, R., Prevost, J. H. and Deodatis, G. (2005a). 3D effects in seismic liquefaction of stochastically variable soil deposits. *Geotechnique*, 55(1); 21–32.

Popescu, R., Prevost, J. H., Deodatis, G. and Chakrabortty, P. (2006). Dynamics of porous media with applications to soil liquefaction. *Soil Dynamics and Earthquake Engineering*, 26(6–7); 648–65.

Prevost, J. H. (1985). A simple plasticity theory for frictional cohesionless soils. *Soil Dynamics and Earthquake Engineering*, 4(1); 9–17.

Prevost, J. H. (2002). DYNAFLOW – A nonlinear transient finite element analysis program, Version 02, Technical Report, Department of Civil and Environmental Engineering, Princeton University, Princeton, NJ. http://www.princeton.edu/~dynaflow/.

Przewlocki, J. (2000). Two-dimensional random field of mechanical properties. *Journal of Geotechnical and Geoenvironmental Engineering*, ASCE, 126(4); 373–7.

Puig, B., Poirion, F. and Soize, C. (2002). Non-Gaussian simulation using hermite polynomial expansion: convergences and algorithms. *Probabilistic Engineering Mechanics*, 17(3); 253–64.

Rackwitz, R. (2000). Reviewing probabilistic soil modeling. *Computers and Geotechnics*, 26; 199–223.

Robertson, P. K. and Wride, C. E. (1998). Evaluating cyclic liquefaction potential using the cone penetration test. *Canadian Geotechnical Journal*, 35(3); 442–59.

Sakamoto, S. and Ghanem, R. (2002). Polynomial chaos decomposition for the simulation of non-Gaussian non-stationary stochastic processes. *Journal of Engineering Mechanics*, ASCE, 128(2); 190–201.

Schmertmann, J. H. (1978). Use the SPT to measure soil properties? – Yes, but...!, In *Dynamic Geotechnical Testing (STP 654)*. ASTM, Philadelphia, pp. 341–355.

Shinozuka, M. and Dasgupta, G. (1986). Stochastic finite element methods in dynamics. In *Proceedings of the 3rd Conference on Dynamic Response of Structures*, ASCE, UCLA, CA, pp. 44–54.

Shinozuka, M. and Deodatis, G. (1988). Simulation of multi-dimensional Gaussian stochastic fields by spectral representation. *Applied Mechanics Reviews, ASME*, 49(1); 29–53.

Shinozuka, M. and Deodatis, G. (1996). Response variability of stochastic finite element systems. *Journal Engineering Mechanics, ASCE*, 114(3); 499–519.

Shinozuka, M., Feng, M. Q., Lee, J. and Naganuma, T. (2000). Statistical analysis of fragility curves. *Journal of Engineering Mechanics, ASCE*, 126(12); 1224–31.

Tang, W. H. (1984). Principles of probabilistic characterization of soil properties. In *Probabilistic Characterization of Soil Properties: Bridge Between Theory and Practice*. ASCE, Atlanta, pp. 74–89.

VanMarcke, E. H. (1977). Probabilistic modeling of soil profiles. *Journal of Geotechnical Engineering Division, ASCE*, 109(5); 1203–14.

VanMarcke, E. H. (1983). *Random Fields: Analysis and Synthesis*. The MIT Press, Cambridge, MA.

VanMarcke, E. H. (1989). Stochastic finite elements and experimental measurements. In *International Conference on Computational Methods and Experimental Measurements*, Capri, Italy, pp. 137–155. Springer.

Yamazaki, F. and Shinozuka, M. (1988). Digital generation of non-Gaussian stochastic fields. *Journal of Engineering Mechanics, ASCE*, 114(7); 1183–97.

Chapter 7

Stochastic finite element methods in geotechnical engineering

Bruno Sudret and Marc Berveiller

7.1 Introduction

Soil and rock masses naturally present heterogeneity at various scales of description. This heterogeneity may be of two kinds:

- the soil properties can be considered piecewise homogeneous once regions (e.g. layers) have been identified;
- no specific regions can be identified, meaning that the spatial variability of the properties is smooth.

In both cases, the use of deterministic values for representing the soil characteristics is poor, since it ignores the natural randomness of the medium. Alternatively, this randomness may be modeled properly using probability theory.

In the first of the two cases identified above, the material properties may be modeled in each region as random variables whose distribution (and possibly mutual correlation) have to be specified. In the second case, the introduction of random fields is necessary. Probabilistic soil modeling is a long-term story, see for example VanMarcke (1977); DeGroot and Baecher (1993); Fenton (1999a,b); Rackwitz (2000); Popescu *et al.* (2005).

Usually soil characteristics are investigated in order to feed models of geotechnical structures that are in project. Examples of such structures are dams, embankments, pile or raft foundations, tunnels, etc. The design then consists in choosing characterictics of the structure (dimensions, material properties) so that it fulfills some requirements (e.g. retain water, support a building, etc.) under a given set of environmental actions that we will call "loading." The design is carried out practically by satisfying some *design criteria* which usually apply onto model response quantities (e.g. displacements, settlements, strains, stresses, etc.). The conservatism of the design according to codes of practice is ensured first by introducing safety coefficients, and second by using penalized values of the model parameters. In this approach, the natural spatial variability of the soil is completely hidden.

From another point of view, when the uncertainties and variability of the soil properties have been identified, methods that allow *propagation* of these uncertainties throughout the model have to be used. Classically, the methods can be classified according to the type of information on the (random) response quantities they provide.

- The *perturbation method* allows computation of the mean value and variance of the mechanical response of the system (Baecher and Ingra, 1981; Phoon *et al.*, 1990). This gives a feeling on the central part of the response probability density function (PDF).
- *Structural reliability methods* allow investigation of the tails of the response PDF by computing the probability of exceedance of a pre-scribed threshold (Ditlevsen and Madsen, 1996). Among these methods, FOSM (first-order second-moment methods) have been used in geotechnical engineering (Phoon *et al.*, 1990; Mellah *et al.*, 2000; Eloseily *et al.*, 2002). FORM/SORM and importance sampling are applied less in this context, and have proven successful in engineering mechanics, both in academia and more recently in industry.
- *Stochastic finite element (SFE) methods*, named after the pioneering work by Ghanem and Spanos (1991), aim at representing the complete response PDF in an intrinsic way. This is done by expanding the response (which, after proper discretization of the problem, is a random vector of unknown joint PDF) onto a particular basis of the space of random vectors of finite variance called the *polynomial chaos* (PC). Applications to geotechnical problems can be found in Ghanem and Brzkala (1996); Sudret and Der Kiureghian (2000); Ghiocel and Ghanem (2002); Clouteau and Lafargue (2003); Sudret *et al.* (2004, 2006); Berveiller *et al.* (2006).

In the following, we will concentrate on this last class of methods. Indeed, once the coefficients of the expansion have been computed, a straightforward post-processing of these quantities gives the statistical moments of the response under consideration, the probability of exceeding a threshold or the full PDF.

The chapter is organized as follows. Section 7.2 presents methods for representing random fields that are applicable for describing the spatial variability of soils. Section 7.3 presents the principles of polynomial chaos expansions for representing both the (random) model response and possibly the non-Gaussian input. Section 7.4 and (respectively, 7.5) reviews the so-called *intrusive* (respectively, *non-intrusive*) solving scheme in stochastic finite element problems. Section 7.6 is devoted to the practical post-processing of polynomial chaos expansions. Finally, Section 7.7 presents some application examples.

7.2 Representation of spatial variability

7.2.1 Basics of probability theory and notation

Probability theory gives a sound mathematical framework to the representation of uncertainty and randomness. When a random phenomenon is observed, the set of all possible outcomes defines the sample space denoted by Θ. An event E is defined as a subset of Θ containing outcomes $\theta \in \Theta$. The set of events defines the σ-algebra $\mathcal{F}$ associated with Θ. A *probability measure* allows one to associate numbers to events, i.e. their *probability of occurrence*. Finally the probability space constructed by means of these notions is denoted by $(\Theta, \mathcal{F}, P)$.

A real *random variable* X is a mapping X: $(\Theta, \mathcal{F}, P) \longrightarrow \mathbb{R}$. For continuous random variables, the PDF and cumulative distribution function (CDF) are denoted by $f_X(x)$ and $F_X(x)$, respectively:

$$F_X(x) = P(X \le x) \qquad f_X(x) = \frac{\mathrm{d}F_X(x)}{\mathrm{d}x} \tag{7.1}$$

The mathematical expectation will be denoted by $\mathrm{E}[\cdot]$. The mean, variance and nth moment of X are:

$$\mu \equiv \mathrm{E}[X] = \int_{-\infty}^{\infty} x f_X(x)\,\mathrm{d}x \tag{7.2}$$

$$\sigma^2 = \mathrm{E}\left[(X - \mu)^2\right] = \int_{-\infty}^{\infty} (x - \mu)^2 f_X(x)\,\mathrm{d}x \tag{7.3}$$

$$\mu'_n = \mathrm{E}[X^n] = \int_{-\infty}^{\infty} x^n f_X(x)\,\mathrm{d}x \tag{7.4}$$

A *random vector* $\mathbf{X}$ is a collection of random variables whose probabilistic description is contained in its joint PDF denoted by $f_{\mathbf{X}}(\mathbf{x})$. The covariance of two random variables X and Y (e.g. two components of a random vector) is:

$$\mathrm{Cov}[X, Y] = \mathrm{E}\left[(X - \mu_X)(Y - \mu_Y)\right] \tag{7.5}$$

Introducing the joint distribution $f_{X,Y}(x, y)$ of these variables, Equation (7.5) can be rewritten as:

$$\mathrm{Cov}[X, Y] = \int_{-\infty}^{\infty} \int_{-\infty}^{\infty} (x - \mu_X)(y - \mu_Y) f_{X,Y}(x, y)\,\mathrm{d}x\,\mathrm{d}y \tag{7.6}$$

The vectorial space of real random variables with finite second moment $(\mathrm{E}[X^2] < \infty)$ is denoted by $\mathcal{L}^2(\Theta, \mathcal{F}, P)$. The expectation operator defines an inner product on this space:

$$\langle X, Y \rangle = \mathrm{E}[XY] \tag{7.7}$$

This allows in particuler definition of *orthogonal* random variables (when their inner product is zero).

7.2.2 Random fields

A unidimensional random field $H(\mathbf{x}, \theta)$ is a collection of random variables associated with a continuous index $\mathbf{x} \in \Omega \subset \mathbb{R}^n$, where $\theta \in \Theta$ is the coordinate in the outcome space. Using this notation, $H(\mathbf{x}, \theta_0)$ denotes a particular realization of the field (i.e. a usual function mapping Ω into $\mathbb{R}$) whereas $H(\mathbf{x}_0, \theta)$ is *the* random variable associated with point $\mathbf{x}_0$. Gaussian random fields are of practical interest because they are completely described by a mean function $\mu(\mathbf{x})$, a variance function $\sigma^2(\mathbf{x})$ and an autocovariance function $C_{HH}(\mathbf{x}, \mathbf{x}')$:

$$C_{HH}(\mathbf{x}, \mathbf{x}') = \mathrm{Cov}[H(\mathbf{x}), H(\mathbf{x}')] \tag{7.8}$$

Alternatively, the correlation structure of the field may be prescribed through the autocorrelation coefficient function $\rho(\mathbf{x}, \mathbf{x}')$ defined as:

$$\rho(\mathbf{x}, \mathbf{x}') = \frac{C_{HH}(\mathbf{x}, \mathbf{x}')}{\sigma(\mathbf{x})\sigma(\mathbf{x}')} \tag{7.9}$$

Random fields are non-numerable infinite sets of correlated random variables, which is computationally intractable. Discretizing the random field $H(\mathbf{x})$ consists in approximating it by $\hat{H}(\mathbf{x})$, which is defined by means of a *finite set* of random variables $\{\chi_i, i = 1, \dots n\}$, gathered in a random vector denoted by χ:

$$H(\mathbf{x}, \theta) \overset{\text{Discretization}}{\longrightarrow} \hat{H}(\mathbf{x}, \theta) = \mathcal{G}[\mathbf{x}, \chi(\theta)] \tag{7.10}$$

Several methods have been developed since the 1980s to carry out this task, such as the *spatial average method* (VanMarcke and Grigoriu, 1983), the *midpoint method* (Der Kiureghian and Ke, 1988) and the *shape function method* (W. Liu et al., 1986a,b). A comprehensive review and comparison of these methods is presented in Li and Der Kiureghian (1993). These early methods are relatively inefficient, in the sense that a large number of random variables is required to achieve a good approximation of the field.

More efficient approaches for discretization of random fields using series expansion methods have been introduced in the past 15 years, including the *Karhunen–Loève Expansion* (KL) (Ghanem and Spanos, 1991), the *Expansion Optimal Linear Estimation* (EOLE) method (Li and Der Kiureghian, 1993) and the *Orthogonal Series Expansion* (OSE) (Zhang and Ellingwood, 1994). Reviews of these methods have been presented in Sudret and Der Kiureghian (2000), see also Grigoriu (2006). The KL and EOLE methods are now briefly presented.

7.2.3 Karhunen–Loève expansion

Let us consider a Gaussian random field $H(\mathbf{x})$ defined by its mean value $\mu(\mathbf{x})$ and autocovariance function $C_{HH}(\mathbf{x}, \mathbf{x}') = \sigma(\mathbf{x})\sigma(\mathbf{x}')\rho(\mathbf{x}, \mathbf{x}')$. The Karhunen–Loève expansion of $H(\mathbf{x})$ reads:

$$H(\mathbf{x}, \theta) = \mu(\mathbf{x}) + \sum_{i=1}^{\infty} \sqrt{\lambda_i}\,\xi_i(\theta)\,\varphi_i(\mathbf{x}) \tag{7.11}$$

where $\{\xi_i(\theta),\ i = 1, ...\}$ are zero-mean orthogonal variables and $\{\lambda_i, \varphi_i(\mathbf{x})\}$ are solutions of the eigenvalue problem:

$$\int_{\Omega} C_{HH}(\mathbf{x}, \mathbf{x}')\varphi_i(\mathbf{x}')\,\mathrm{d}\Omega_{\mathbf{x}'} = \lambda_i\varphi_i(\mathbf{x}) \tag{7.12}$$

Equation (7.12) is a Fredholm integral equation. Since *kernel* $C_{HH}(\cdot, \cdot)$ is an autocovariance function, it is bounded, symmetric and positive definite. Thus the set of $\{\varphi_i\}$ forms a *complete orthogonal basis*. The set of eigenvalues (spectrum) is moreover real, positive and numerable. In a sense, Equation (7.11) corresponds to a *separation* of the space and randomness variables in $H(\mathbf{x}, \theta)$.

The Karhunen–Loève expansion possesses other interesting properties (Ghanem and Spanos, 1991).

- It is possible to order the eigenvalues λ_i in a descending series converging to zero. Truncating the ordered series (7.11) after the Mth term gives the KL approximated field:

$$\hat{H}(\mathbf{x}, \theta) = \mu(\mathbf{x}) + \sum_{i=1}^{M} \sqrt{\lambda_i}\,\xi_i(\theta)\,\varphi_i(\mathbf{x}) \tag{7.13}$$

- The covariance eigenfunction basis $\{\varphi_i(\mathbf{x})\}$ is optimal in the sense that the mean square error (integrated over Ω) resulting from a truncation after the Mth term is minimized (with respect to the value it would take when any other complete basis $\{h_i(\mathbf{x})\}$ is chosen).
- The set of random variables appearing in (7.11) is orthonormal if and only if the basis functions $\{h_i(\mathbf{x})\}$ and the constants λ_i are solutions of the eigenvalue problem (7.12).

- As the random field is Gaussian, the set of $\{\xi_i\}$ are *independent* standard normal variables. Furthermore, it can be shown (Loève, 1977) that the Karhunen–Loève expansion of Gaussian fields is almost surely convergent. For non-Gaussian fields, the KL expansion also exists; however, the random variables appearing in the series are of unknown law and may be correlated (Phoon *et al.*, 2002b, 2005; Li *et al.*, 2007).

- From Equation (7.13), the error variance obtained when truncating the expansion after M terms turns out to be, after basic algebra:

$$
\text{Var}\left[H(\mathbf{x}) - \hat{H}(\mathbf{x})\right] = \sigma^2(\mathbf{x}) - \sum_{i=1}^{M} \lambda_i \varphi_i^2(\mathbf{x})
$$

$$
= \text{Var}\left[H(\mathbf{x})\right] - \text{Var}\left[\hat{H}(\mathbf{x})\right] \geq 0 \qquad (7.14)
$$

The right-hand side of the above equation is always positive because it is the variance of some quantity. This means that the Karhunen–Loève expansion always *underrepresents* the true variance of the field. The accuracy of the truncated expansion has been investigated in details in Huang *et al.* (2001).

Equation (7.12) can be solved analytically only for few autocovariance functions and geometries of Ω. Detailed closed form solutions for triangular and exponential covariance functions for one-dimensional homogeneous fields can be found in Ghanem and Spanos (1991). Otherwise, a numerical solution to the eigenvalue problem (7.12) can be obtained (same reference, chapter 2). Wavelet techniques have been recently applied for this purpose in Phoon *et al.* (2002a), leading to a fairly efficient approximation scheme.

7.2.4 The EOLE method

The *expansion optimal linear estimation* method (EOLE) was proposed by Li and Der Kiureghian (1993). It is based on the pointwise regression of the original random field with respect to selected values of the field, and a compaction of the data by spectral analysis.

Let us consider a Gaussian random field as defined above and a grid of points $\{\mathbf{x}_1, \ldots \mathbf{x}_N\}$ in the domain Ω. Let us denote by χ the random vector $\{H(\mathbf{x}_1), \ldots H(\mathbf{x}_N)\}$. By construction, χ is a Gaussian vector whose mean value $\boldsymbol{\mu}_\chi$ and covariance matrix $\boldsymbol{\Sigma}_{\chi\chi}$ read:

$$
\mu_\chi^i = \mu(\mathbf{x}_i) \qquad (7.15)
$$

$$
\left(\boldsymbol{\Sigma}_{\chi\chi}\right)_{i,j} = \text{Cov}\left[H(\mathbf{x}_i), H(\mathbf{x}_j)\right] = \sigma(\mathbf{x}_i)\sigma(\mathbf{x}_j)\rho(\mathbf{x}_i, \mathbf{x}_j) \qquad (7.16)
$$

The *optimal linear estimation* (OLE) of random variable $H(\mathbf{x})$ onto the random vector χ reads:

$$H(\mathbf{x}) \approx \mu(\mathbf{x}) + \boldsymbol{\Sigma}'_{H\chi}(\mathbf{x}) \cdot \boldsymbol{\Sigma}_{\chi\chi}^{-1} \cdot \left(\chi - \mu_\chi\right) \tag{7.17}$$

where $(.)'$ denotes the transposed matrix and $\boldsymbol{\Sigma}_{H\chi}(\mathbf{x})$ is a vector whose components are given by:

$$\boldsymbol{\Sigma}^j_{H\chi}(\mathbf{x}) = \text{Cov}\left[H(\mathbf{x}), \chi_j\right] = \text{Cov}\left[H(\mathbf{x}), H(\mathbf{x}_j)\right] \tag{7.18}$$

Let us now consider the spectral decomposition of the covariance matrix $\boldsymbol{\Sigma}_{\chi\chi}$:

$$\boldsymbol{\Sigma}_{\chi\chi}\,\boldsymbol{\phi}_i = \lambda_i\boldsymbol{\phi}_i \qquad i = 1, \ldots, N \tag{7.19}$$

This allows one to linearly transform the original vector χ:

$$\chi(\theta) = \mu_\chi + \sum_{i=1}^{N} \sqrt{\lambda_i}\,\xi_i(\theta)\,\boldsymbol{\phi}_i \tag{7.20}$$

where $\{\xi_i, i = 1, \ldots N\}$ are *independent* standard normal variables. Substituting for (7.20) in (7.17) and using (7.19) yields the EOLE representation of the field :

$$\hat{H}(\mathbf{x}, \theta) = \mu(\mathbf{x}) + \sum_{i=1}^{N} \frac{\xi_i(\theta)}{\sqrt{\lambda_i}}\,\boldsymbol{\phi}_i^T\,\boldsymbol{\Sigma}_{H(\mathbf{x})\chi} \tag{7.21}$$

As in the Karhunen–Loève expansion, the series can be truncated after $r \leq N$ terms, the eigenvalues λ_i being sorted first in descending order. The variance of the error for EOLE is:

$$\text{Var}\left[H(\mathbf{x}) - \hat{H}(\mathbf{x})\right] = \sigma^2(\mathbf{x}) - \sum_{i=1}^{r} \frac{1}{\lambda_i}\left(\boldsymbol{\phi}_i^T\,\boldsymbol{\Sigma}_{H(\mathbf{x})\chi}\right)^2 \tag{7.22}$$

As in KL, the second term in the above equation is identical to the variance of $\hat{H}(\mathbf{x})$. Thus EOLE also always *underrepresents* the true variance. Due to the form of (7.22), the error decreases monotonically with r, the minimal error being obtained when no truncation is made ($r = N$). This allows one to define automatically the cut-off value r for a given tolerance in the variance error.

7.3 Polynomial chaos expansions

7.3.1 Expansion of the model response

Having recognized that the input parameters such as the soil properties can be modeled as random fields (which are discretized using standard normal random variables), it is clear that the response of the system is a nonlinear function of these variables. After a discretization procedure (e.g. finite element or finite difference scheme), the response may be considered as a random vector S, whose probabilistic properties are yet to be determined.

Due to the above representation of the input, it is possible to expand the response S onto the so-called *polynomial chaos basis*, which is a basis of the space of random variables with finite variance (Malliavin, 1997):

$$S = \sum_{j=0}^{\infty} S_j \Psi_j \left(\{\xi_n\}_{n=1}^{\infty}\right) \tag{7.23}$$

In this expression, the Ψ_j's are the multivariate Hermite polynomials defined by means of the ξ_n's. This basis is orthogonal with respect to the Gaussian measure, i.e. the expectation of products of two different such polynomials is zero (see details in Appendix A).

Computationnally speaking, the input parameters are represented using M independent standard normal variables, see Equations (7.13) and (7.21). Considering all M-dimensional Hermite polynomials of degree not exceeding p, the response may be approximated as follows:

$$S \approx \sum_{j=0}^{P-1} S_j \Psi_j (\xi), \quad \xi = \{\xi_1, \ldots, \xi_M\} \tag{7.24}$$

The number of unknown (vector) coefficients in this summation is:

$$P = \binom{M+p}{p} = \frac{(M+p)!}{M!\, p!} \tag{7.25}$$

The practical construction of a polynomial chaos of order M and degree p is described in Appendix A. The problem is now recast as computing the expansion coefficients $\{S_j, j = 0, \ldots, P-1\}$. Two classes of methods are presented below in Sections 7.4 and 7.5.

7.3.2 Representation of non-Gaussian input

In Section 7.2, the representation of the spatial variability through Gaussian random fields has been shown. It is important to note that many soil properties should not be modeled as Gaussian random variables or fields. For instance, the Poisson ratio is a bounded quantity, whereas Gaussian variables are defined on $\mathbb{R}$. Parameters such as the Young's modulus or the cohesion are positive in nature: modeling them as Gaussian introduces an approximation that should be monitored carefully. Indeed, when a large dispersion of the parameter is considered, choosing a Gaussian representation can easily lead to negative realizations of the parameter, which have no physical meaning (lognormal variables or fields are often appropriate).

As a consequence, if the parameter under consideration is modeled by a non-Gaussian random field, it is not possible to expand it as a linear expression in standard normal variables as in Equations (7.13) and (7.21). Easy-to-define non-Gaussian random fields $H(\mathbf{x})$ are obtained by *translation* of a Gaussian field $N(\mathbf{x})$ using a nonlinear transform $\mathfrak{h}(.)$:

$$H(\mathbf{x}) = \mathfrak{h}(N(\mathbf{x})) \tag{7.26}$$

The discretization of this kind of field is straightforward: the nonlinear transform $\mathfrak{h}(.)$ is directly applied to the discretized underlying Gaussian field $\hat{N}(\mathbf{x})$ (see e.g. Ghanem, 1999, for lognormal fields).

$$\hat{H}(\mathbf{x}) = \mathfrak{h}(\hat{N}(\mathbf{x})) \tag{7.27}$$

From another point of view, the description of the spatial variability of parameters is in some cases beyond the scope of the analysis. For instance, soil properties may be considered homogeneous in some domains. These parameters are not well known though, and it may be relevant to model them as (usually non-Gaussian) random variables.

It is possible to transform any continuous random variable with finite variance into a standard normal variable using the iso-probabilistic transform: denoting by $F_X(x)$ (respectively, $\Phi(x)$) the CDF of X (respectively, a standard normal variable ξ), the direct and inverse transform read:

$$\xi = \Phi^{-1} \circ F_X(x) \qquad X = F_X^{-1} \circ \Phi(\xi) \tag{7.28}$$

If the input parameters are modeled by a random vector with independent components, it is possible to represent it using a standard normal random vector of the same size by applying the above transform componentwise. If the input random vector has a prescribed joint PDF, it is generally *not* possible to transform it exactly into a standard normal random vector. However, when only marginal PDF and correlations are known, an approximate representation may be obtained by the Nataf transform (Liu and Der Kiureghian, 1986).

As a conclusion, the input parameters of the model, which do or do not exhibit spatial variability, may always be represented after some discretization process, mapping, or combination thereof, as functionals of standard normal random variables:

- for non-Gaussian independent random variables, see Equation (7.28);
- for Gaussian random fields, see Equations(7.13) and (7.21);
- for non-Gaussian random fields obtained by translation, see Equation (7.27).

Note that Equation (7.28) is an exact representation, whereas the field discretization techniques provide only approximations (which converge to the original field if the number of standard normal variables tends to infinity).

In the sequel, we consider that the discretized input fields and non Gaussian random variables are represented through a set of independent standard normal variables $\boldsymbol{\xi}$ of size M and we denote by $\mathbf{X}(\boldsymbol{\xi})$ the functional that yields the original variables and fields.

7.4 Intrusive SFE method for static problems

The historical SFE approach consists in computing the polynomial chaos coefficients of the vector of nodal displacements $\mathbf{U}(\theta)$ (Equation (7.24)). It is based on the minimization of the residual in the balance equation in the Galerkin sense (Ghanem and Spanos, 1991). To illustrate this method, let us consider a linear mechanical problem, whose finite element discretization leads to the following linear system (in the deterministic case):

$$\mathbf{K} \cdot \mathbf{U} = \mathbf{F} \tag{7.29}$$

Let us denote by N_{ddl} the number of degrees of freedom of the structure, i.e. the size of the above linear system. If the material parameters are described by random variables and fields, the stiffness matrix $\mathbf{K}$ in the above equation becomes random. Similarly, the load vector $\mathbf{F}$ may be random. These quantities can be expanded onto the polynomial chaos basis:

$$\mathbf{K} = \sum_{j=0}^{\infty} \mathbf{K}_j \Psi_j \tag{7.30}$$

$$\mathbf{F} = \sum_{j=0}^{\infty} \mathbf{F}_j \Psi_j \tag{7.31}$$

In these equations, $\mathbf{K}_j$ are deterministic matrices whose complete description can be found elsewhere (e.g. Ghanem and Spanos (1991) in the case when the input Young's modulus is a random field, and Sudret *et al.* (2004) when the Young's modulus and the Poisson ratio are non-Gaussian random variables).

In the same manner, $\mathbf{F}_j$ are deterministic load vectors obtained from the data (Sudret *et al.*, 2004; Sudret *et al.*, 2006).

As a consequence, the vector of nodal displacements $\mathbf{U}$ is random and may be represented on the same basis:

$$U = \sum_{j=0}^{\infty} \mathbf{U}_j \Psi_j \qquad (7.32)$$

When the three expansions (7.30)–(7.32) are truncated after P terms and substituted for in Equation (7.29), the residual ϵ_P in the stochastic balance equation reads:

$$\epsilon_P = \left(\sum_{i=0}^{P-1} \mathbf{K}_i \Psi_i \right) \cdot \left(\sum_{j=0}^{P-1} \mathbf{U}_j \Psi_j \right) - \sum_{j=0}^{P-1} \mathbf{F}_j \Psi_j \qquad (7.33)$$

Coefficients $\{\mathbf{U}_0, \dots, \mathbf{U}_{P-1}\}$ are obtained by minimizing the residual using a Galerkin technique. This minimization is equivalent to requiring the residual be orthogonal to the subspace of $\mathcal{L}^2(\Theta, F, P)$ spanned by $\{\Psi_j\}_{j=0}^{P-1}$:

$$\mathrm{E}\left[\epsilon_P \Psi_k\right] = 0, \, k = \{0, \dots, P-1\} \qquad (7.34)$$

After some algebra, this leads to the following linear system, whose size is equal to $N_{ddl} \times P$:

$$\begin{pmatrix} \mathbf{K}_{0,0} & \cdots & \mathbf{K}_{0,P-1} \\ \mathbf{K}_{1,0} & \cdots & \mathbf{K}_{1,P-1} \\ \vdots & & \vdots \\ \mathbf{K}_{P-1,0} & \cdots & \mathbf{K}_{P-1,P-1} \end{pmatrix} \cdot \begin{pmatrix} \mathbf{U}_0 \\ \mathbf{U}_1 \\ \vdots \\ \mathbf{U}_{P-1} \end{pmatrix} = \begin{pmatrix} \mathbf{F}_0 \\ \mathbf{F}_1 \\ \vdots \\ \mathbf{F}_{P-1} \end{pmatrix} \qquad (7.35)$$

where $\mathbf{K}_{j,k} = \sum_{i=0}^{P-1} d_{ijk} \mathbf{K}_i$ and $d_{ijk} = \mathrm{E}[\Psi_i \Psi_j \Psi_k]$.

Once the system has been solved, the coefficients $\mathbf{U}_j$ may be post-processed in order to represent the response PDF (e.g. by Monte Carlo simulation), to compute the mean value, standard deviation and higher moments or to evaluate the probability of exceeding a given threshold. The post-processing techniques are detailed in Section 7.6. It is important to note already that the set of $\mathbf{U}_j$'s contains all the probabilistic information on the response, meaning that post-processing is carried out without additional computation on the mechanical model.

The above approach is easy to apply when the mechanical model is linear. Although nonlinear problems have been recently addressed (Matthies and Keese, 2005; Nouy, 2007), their treatment is still not completely mature. Moreover, this approach naturally yields the expansion of the basic response quantities (such as the nodal displacements in mechanics). When derived quantities such as strains or stresses are of interest, additional work (and approximations) is needed. Note that in the case of non-Gaussian input random variables, expansion of these variables onto the PC basis is needed in order to apply the method, which introduces an approximation of the input. Finally, the implementation of the historical method as described in this section has to be carried out for each class of problem: this is why it has been qualified as *intrusive* in the literature. All these reasons have led to the development of so-called *non-intrusive methods* that in some sense provide an answer to the above drawbacks.

7.5 Non-intrusive SFE methods

7.5.1 Introduction

Let us consider a scalar response quantity S of the model under consideration, e.g. a nodal displacement, strain or stress component in a finite element model:

$$S = \mathfrak{h}(\mathbf{X}) \tag{7.36}$$

Contrary to Section 7.4, each response quantity of interest is directly expanded onto the polynomial chaos as follows:

$$S = \sum_{j=0}^{\infty} S_j \Psi_j \tag{7.37}$$

The P-term approximation reads:

$$\tilde{S} = \sum_{j=0}^{P-1} S_j \Psi_j \tag{7.38}$$

Two methods are now proposed to compute the coefficients in this expansion from a series of *deterministic* finite element analysis.

7.5.2 Projection method

The *projection* method is based on the orthogonality of the polynomial chaos (Le Maître *et al.*, 2002; Ghiocel and Ghanem, 2002). By pre-multiplying

Equation (7.38) by Ψ_j and taking the expectation of both members, it comes:

$$E\left[S\Psi_j\right] = E\left[\sum_{i=0}^{\infty} S_i \Psi_i \Psi_j\right] \tag{7.39}$$

Due to the orthogonality of the basis, $E\left[\Psi_i \Psi_j\right] = 0$ for any $i \neq j$. Thus:

$$S_j = \frac{E\left[S\Psi_j\right]}{E\left[\Psi_j^2\right]} \tag{7.40}$$

In this expression, the denominator is known analytically (see Appendix A) and the numerator may be cast as a multidimensional integral:

$$E\left[S\Psi_j\right] = \int_{\mathbb{R}^M} \mathfrak{h}(\mathbf{X}(\boldsymbol{\xi})) \, \Psi_j(\boldsymbol{\xi}) \, \varphi_M(\boldsymbol{\xi}) \, d\boldsymbol{\xi} \tag{7.41}$$

where φ_M is the M-dimensional multinormal PDF, and where the dependency of S in $\boldsymbol{\xi}$ through the iso-probabilistic transform of the input parameters $\mathbf{X}(\boldsymbol{\xi})$ has been given for the sake of clarity.

This integral may be computed by crude Monte Carlo simulation (Field, 2002) or Latin Hypercube Sampling (Le Maître *et al.*, 2002). However the number of samples required in this case should be large enough, say 10,000–100,000, to obtain a sufficient accuracy. In cases when the response S is obtained by a computationally demanding finite element model, this approach is practically not applicable. Alternatively, the use of quasi-random numbers instead of Monte Carlo (Niederreiter, 1992) simulation has been recently investigated in Sudret *et al.* (2007), and appears promising.

An alternative approach presented in Berveiller *et al.* (2004) and Matthies and Keese (2005) is the use of a Gaussian quadrature scheme to evaluate the integral. Equation (7.41) is computed as a weighted summation of the integrands evaluated at selected points (the so-called integration points):

$$E\left[S\Psi_j\right] \approx \sum_{i_1=1}^{K} \cdots \sum_{i_M=1}^{K} \omega_{i_1} \cdots \omega_{i_M} \, \mathfrak{h}\left(\mathbf{X}\left(\xi_{i_1}, \ldots, \xi_{i_M}\right)\right) \Psi_j\left(\xi_{i_1}, \ldots, \xi_{i_M}\right) \tag{7.42}$$

In this expression, the integration points $\{\xi_{i_j}, \quad 1 \leq i_1 \leq \cdots \leq i_M \leq K\}$ and weights $\{\omega_{i_j}, \quad 1 \leq i_1 \leq \cdots \leq i_M \leq K\}$ in each dimension are computed using the theory of orthogonal polynomials with respect to the Gaussian measure. For a Kth order scheme, the integration points are the roots of the Kth order Hermite polynomial (Abramowitz and Stegun, 1970).

The proper order of the integration scheme K is selected as follows: if the response S in Equation (7.37) was polynomial of order p (i.e. $S_j = 0$ for $j \geq P$), the terms in the integral (7.41) would be of degree less than or equal to $2p$. Thus an integration scheme of order $K = p + 1$ would give the *exact* value of the expansion coefficients. We take this as a rule in the general case, where the result now is only an approximation of the true value of S_j.

As seen from Equations (7.40),(7.42), the projection method allows one to compute the expansion coefficients from selected evaluations of the model. Thus the method is qualified as *non-intrusive* since the deterministic computation scheme (i.e. a finite element code) is used without any additional implementation or modification.

Note that in finite element analysis, the response is usually a vector (e.g. of nodal displacements, nodal stresses, etc.). The above derivations are strictly valid for a vector response **S**, the expectation in Equation (7.42) being computed component by component.

7.5.3 Regression method

The *regression* method is another approach for computing the response expansion coefficients. It is nothing but the regression of the exact solution S with respect to the polynomial chaos basis $\{\Psi_j(\boldsymbol{\xi}), j = 1, \ldots, P - 1\}$. Let us assume the following expression for a scalar response quantity S:

$$S = \mathfrak{h}(\mathbf{X}) = \tilde{S}(\boldsymbol{\xi}) + \varepsilon \qquad \tilde{S}(\boldsymbol{\xi}) = \sum_{j=0}^{P-1} S_j \Psi_j(\boldsymbol{\xi}) \tag{7.43}$$

where the residual ε is supposed to be a zero-mean random variable, and $\mathcal{S} = \{S_j, j = 0, \ldots, P - 1\}$ are the unknown coefficients. The minimization of the variance of the residual with respect to the unknown coefficients leads to:

$$\mathcal{S} = \operatorname{Argmin} \mathrm{E}\left[(\mathfrak{h}(\mathbf{X}(\boldsymbol{\xi})) - \tilde{S}(\boldsymbol{\xi}))^2 \right] \tag{7.44}$$

In order to solve Equation (7.44), we choose a set of Q regression points in the standard normal space, say $\{\boldsymbol{\xi}^1, \ldots \boldsymbol{\xi}^Q\}$. From these points, the isoprobabilistic transform (7.28) gives a set of Q realizations of the input vector **X**, say $\{\mathbf{x}^1, \ldots \mathbf{x}^Q\}$. The mean-square minimization (7.44) leads to solve the following problem:

$$\mathcal{S} = \operatorname{Argmin} \frac{1}{Q} \sum_{i=1}^{Q} \left\{ \mathfrak{h}(\mathbf{x}^i) - \sum_{j=0}^{P-1} S_j \Psi_j(\boldsymbol{\xi}^i) \right\}^2 \tag{7.45}$$

Denoting by $\boldsymbol{\Psi}$ the matrix whose coefficients are given by $\boldsymbol{\Psi}_{ij} = \Psi_j(\boldsymbol{\xi}^i)$, $i = 1, \ldots, Q$; $j = 0, \ldots, P-1$, and by $\boldsymbol{\mathcal{S}}_{\text{ex}}$ the vector containing the *exact* response values computed by the model $\boldsymbol{\mathcal{S}}_{\text{ex}} = \{\mathfrak{h}(\mathbf{x}^i), i = 1, \ldots, Q\}$, the solution to Equation (7.45) reads:

$$\boldsymbol{\mathcal{S}} = (\boldsymbol{\Psi}^T \cdot \boldsymbol{\Psi})^{-1} \cdot \boldsymbol{\Psi}^T \cdot \boldsymbol{\mathcal{S}}_{\text{ex}} \tag{7.46}$$

The regression approach detailed above is comparable with the so-called *response surface method* used in many domains of science and engineering. In this context, the set of $\{\mathbf{x}^1, \ldots \mathbf{x}^Q\}$ is the so-called *experimental design*. In Equation (7.46), $\boldsymbol{\Psi}^T \cdot \boldsymbol{\Psi}$ is the *information matrix*. Computationally speaking, it may be ill-conditioned. Thus a particular solver such as the Singular Value Decomposition method should be employed (Press *et al.*, 2001).

It is now necessary to specify the choice of the experimental design. Starting from the early work by Isukapalli (1999), it has been shown in Berveiller (2005) and Sudret (2005) that an efficient design can be built from the roots of the Hermite polynomials as follows.

- If p denotes the maximal degree of the polynomials in the truncated PC expansion, then the $p+1$ roots of the Hermite polynomial of degree $p+1$ (denoted by He_{p+1}) are computed, say $\{r_1, \ldots, r_{p+1}\}$.
- From this set, M-uplets are built using all possible combinations of the roots: $\mathbf{r}^k = (r_{i_1}, \ldots, r_{i_M})$, $1 \leq i_1 \leq \cdots \leq i_M \leq p+1$, $k = 1, \ldots, (p+1)^M$.
- The Q points in the experimental design $\{\boldsymbol{\xi}^1, \ldots, \boldsymbol{\xi}^Q\}$ are selected among the r^j's by retaining those which are closest to the origin of the space, i.e. those with the smallest norm, or equivalently those leading to the largest values of the PDF $\varphi_M(\boldsymbol{\xi}^j)$.

To choose the size Q of the experimental design, the following empirical rule was proposed by Berveiller (2005) based on a large number of numerical experiments:

$$Q = (M-1)P \tag{7.47}$$

A slightly more efficient rule leading to a smaller value of Q has been recently proposed by Sudret (2006, 2008), based on the invertibility of the information matrix.

7.6 Post-processing of the SFE results

7.6.1 Representation of response PDF

Once the coefficients S_j of the expansion of a response quantity are known, the polynomial approximation can be simulated using Monte

Carlo simulation. A sample of standard normal random vector is generated, say $\{\boldsymbol{\xi}^{(1)}, \ldots, \boldsymbol{\xi}^{(n)}\}$. Then the PDF can be plotted using a histogram representation, or better, kernel smoothing techniques (Wand and Jones, 1995).

7.6.2 Computation of the statistical moments

From Equation (7.38), due to the orthogonality of the polynomial chaos basis, it is easy to see that the mean and variance of S is given by:

$$E[S] = S_0 \tag{7.48}$$

$$\text{Var}[S] = \sigma_S^2 = \sum_{j=1}^{P-1} S_j^2 E\left[\Psi_j^2\right] \tag{7.49}$$

where the expectation of Ψ_j^2 is given in Appendix A. Moments of higher order are obtained in a similar manner. Namely the skewness and kurtosis coefficients of response variable S (denoted by δ_S and κ_S, respectively) are obtained as follows:

$$\delta_S \equiv \frac{1}{\sigma_S^3} E\left[(S - E[S])^3\right] = \frac{1}{\sigma_S^3} \sum_{i=1}^{P-1}\sum_{j=1}^{P-1}\sum_{k=1}^{P-1} E[\Psi_i \Psi_j \Psi_k] S_i S_j S_k \tag{7.50}$$

$$\kappa_S \equiv \frac{1}{\sigma_S^4} E\left[(S - E[S])^4\right] = \frac{1}{\sigma_S^4} \sum_{i=1}^{P-1}\sum_{j=1}^{P-1}\sum_{k=1}^{P-1}\sum_{l=1}^{P-1} E[\Psi_i \Psi_j \Psi_k \Psi_l] S_i S_j S_k S_l \tag{7.51}$$

Here again, expectation of products of three (respectively four) Ψ_j's are known analytically; see for example Sudret et al. (2006).

7.6.3 Sensitivity analysis: selection of important variables

The problem of selecting the most "important" input variables of a model is usually known as sensitivity analysis. In a probabilistic context, methods of *global* sensitivity analysis aim at quantifying which input variable (or combination of input variables) influences most the response variability. A state-of-the-art of such techniques is available in Saltelli *et al.* (2000). They include *regression-based* methods such as the computation of standardized regression coefficients (SRC) or partial correlation coefficients (PCC) and *variance-based* methods, also called *ANOVA techniques* for "ANalysis Of VAriance." In this respect, the Sobol' indices (Sobol', 1993; Sobol' and Kucherenko, 2005) are known as the most efficient tool to find out the important variables of a model.

The computation of Sobol' indices is traditionnally carried out by Monte Carlo simulation (Saltelli *et al.*, 2000), which may be computationally unaffordable in case of time-consuming models. In the context of stochastic finite element methods, it has been recently shown in Sudret (2006, 2008) that the Sobol' indices can be derived *analytically* from the coefficients of the polynomial chaos expansion of the response, once the latter have been computed by one of the techniques detailed in Sections 7.4 and 7.5. For instance, the first order Sobol' indices $\{\delta_i, i = 1, ..., M\}$, which quantify what fraction of the response variance is due to each input variable $i = 1, ..., M$:

$$\delta_i = \frac{\mathrm{Var}_{X_i}\left[\mathrm{E}\left[S|X_i\right]\right]}{\mathrm{Var}\left[S\right]} \tag{7.52}$$

can be evaluated from the coefficients of the PC expansion (Equation (7.38)) as follows:

$$\delta_i^{PC} = \sum_{\alpha \in \mathcal{I}_i} S_\alpha^2 \, \mathrm{E}\left[\Psi_\alpha^2\right] / \sigma_S^2 \tag{7.53}$$

In this equation, σ_S^2 is the variance of the model response computed from the PC coefficients (Equation (7.49)) and the summation set (defined using the multi-index notation detailed in Appendix) reads:

$$\mathcal{I}_i = \left\{\boldsymbol{\alpha} : \alpha_i > 0, \alpha_{j \neq i} = 0\right\} \tag{7.54}$$

Higher-order Sobol' indices, which correspond to interactions of the input parameters, can also be computed using this approach; see Sudret (2006) for a detailed presentation and an application to geotechnical engineering.

7.6.4 Reliability analysis

Structural reliability analysis aims at computing the probability of failure of a mechanical system with respect to a prescribed failure criterion by accounting for uncertainties arising in the model description (geometry, material properties) or the environment (loading). It is a general theory whose development began in the mid 1970s. The research on this field is still active – see Rackwitz (2001) for a review.

Surprisingly, the link between structural reliability and the stochastic finite element methods based on polynomial chaos expansions is relatively new (Sudret and Der Kiureghian, 2000, 2002; Berveiller, 2005). For the sake of completeness, three essential techniques for solving reliability problems are reviewed in this section. Then their application, together with (a) a deterministic finite element model and (b) a PC expansion of the model response, is detailed.

Problem statement

Let us denote by $\mathbf{X} = \{X_1, X_2, \ldots, X_M\}$ the set of random variables describing the randomness in the geometry, material properties and loading. This set also includes the variables used in the discretization of random fields, if any. The failure criterion under consideration is mathematically represented by a *limit state function* $g(\mathbf{X})$ defined in the space of parameters as follows:

- $g(\mathbf{X}) > 0$ defines the *safe state* of the structure.
- $g(\mathbf{X}) \leq 0$ defines the *failure state*.
- $g(\mathbf{X}) = 0$ defines the *limit state surface*.

Denoting by $f_{\mathbf{X}}(\mathbf{x})$ the joint PDF of random vector $\mathbf{X}$, the probability of failure of the structure is:

$$P_f = \int_{g(\mathbf{x}) \leq 0} f_{\mathbf{X}}(\mathbf{x})\, d\mathbf{x} \tag{7.55}$$

In all but academic cases, this integral cannot be computed analytically. Indeed, the failure domain is often defined by means of response quantities (e.g. displacements, strains, stresses, etc.), which are computed by means of computer codes (e.g. finite element code) in industrial applications, meaning that the failure domain is implicitly defined as a function of $\mathbf{X}$. Thus numerical methods have to be employed.

Monte Carlo simulation

Monte Carlo simulation (MCS) is a universal method for evaluating integrals such as Equation (7.55). Denoting by $1_{[g(\mathbf{x}) \leq 0]}(x)$ the indicator function of the failure domain (i.e. the function that takes the value 0 in the safe domain and 1 in the failure domain), Equation (7.55) rewrites:

$$P_f = \int_{\mathbb{R}^M} 1_{[g(\mathbf{x}) \leq 0]}(\mathbf{x})\, f_{\mathbf{X}}(\mathbf{x})\, d\mathbf{x} = \mathrm{E}\left[1_{[g(\mathbf{x}) \leq 0]}(x) \right] \tag{7.56}$$

where $\mathrm{E}[.]$ denotes the mathematical expectation. Practically, Equation (7.56) can be evaluated by simulating N_{sim} realizations of the random vector $\mathbf{X}$, say $\{\mathbf{X}^{(1)}, \ldots, \mathbf{X}^{(N_{\text{sim}})}\}$. For each sample, $g\left(\mathbf{X}^{(i)}\right)$ is evaluated. An estimation of P_f is given by the empirical mean:

$$\hat{P}_f = \frac{1}{N_{\text{sim}}} \sum_{i=1}^{N_{\text{sim}}} 1_{[g(\mathbf{x}) \leq 0]}(\mathbf{X}^{(i)}) = \frac{N_{\text{fail}}}{N_{\text{sim}}} \tag{7.57}$$

where N_{fail} denotes the number of samples that are in the failure domain. As mentioned above, MCS is applicable whatever the complexity of the deterministic model. However, the number of samples N_{sim} required to get an accurate estimation of P_f may be dissuasive, especially when the value of P_f is small. Indeed, if the order of magnitude of P_f is about 10^{-k}, a total number $N_{\text{sim}} \approx 4.10^{k+2}$ is necessary to get accurate results when using Equation (7.57). This number corresponds approximately to a coefficient of variation CV equal to 5% for the estimator $\hat{P}_f$. Thus crude MCS is not applicable when small values of P_f are sought and/or when the CPU cost of each run of the model is non-negligible.

FORM method

The first-order reliability method (FORM) has been introduced to get an approximation of the probability of failure at a low cost (in terms of number of evaluations of the limit state function).

The first step consists in recasting the problem in the standard normal space by using a iso-probabilistic transformation $\mathbf{X} \to \boldsymbol{\xi} = T(\mathbf{X})$. The Rosenblatt or Nataf transformations may be used for this purpose. Thus Equation (7.56) rewrites:

$$P_f = \int_{g(\mathbf{x}) \leq 0} f_{\mathbf{X}}(\mathbf{x})\, d\mathbf{x} = \int_{g(T^{-1}(\boldsymbol{\xi})) \leq 0} \varphi_M(\boldsymbol{\xi})\, d\boldsymbol{\xi} \qquad (7.58)$$

where $\varphi_M(\boldsymbol{\xi})$ stands for the standard multinormal PDF:

$$\varphi_M(\boldsymbol{\xi}) = \frac{1}{\left(\sqrt{2\pi}\right)^n} \exp\left(-\frac{1}{2}\left(\xi_1^2 + \cdots + \xi_M^2\right)\right) \qquad (7.59)$$

This PDF is maximal at the origin and decreases exponentially with $\|\boldsymbol{\xi}\|^2$. Thus the points that contribute at most to the integral in Equation (7.58) are those of the failure domain that are closest to the origin of the space.

The second step in FORM thus consists in determining the so-called design point, i.e. the point of the failure domain closest to the origin in the standard normal space. This point P^* is obtained by solving an optimization problem:

$$P^* = \boldsymbol{\xi}^* = \text{Argmin}\left\{\|\boldsymbol{\xi}\|^2 / g\left(T^{-1}(\boldsymbol{\xi})\right) \leq 0\right\} \qquad (7.60)$$

Several algorithms are available to solve the above optimisation problem, e.g. the Abdo–Rackwitz (Abdo and Rackwitz, 1990) or the SQP (sequential quadratic programming) algorithm. The corresponding reliability index is defined as:

$$\beta = \text{sign}\left[g(T^{-1}(0))\right] \cdot \|\boldsymbol{\xi}^*\| \tag{7.61}$$

It corresponds to the algebraic distance of the design point to the origin, counted as positive if the origin is in the safe domain, or negative in the other case.

The third step of FORM consists in replacing the failure domain by the half space $HS(P*)$ defined by means of the hyperplane which is tangent to the limit state surface at the design point (see Figure 7.1). This leads to:

$$P_f = \int\limits_{g(T^{-1}(\boldsymbol{\xi}))\leq 0} \varphi_M(\boldsymbol{\xi}) \, d\boldsymbol{\xi} \approx \int\limits_{HS(P*)} \varphi_M(\boldsymbol{\xi}) \, d\boldsymbol{\xi} \tag{7.62}$$

The latter integral can be evaluated in a closed form and gives the first order approximation of the probability of failure:

$$P_f \approx P_{f,FORM} = \Phi(-\beta) \tag{7.63}$$

where $\Phi(x)$ denotes the standard normal CDF. The unit normal vector $\boldsymbol{\alpha} = \boldsymbol{\xi}^*/\beta$ allows definition of the sensitivity of the reliability index with respect to each variable. Precisely the squared components α_i^2 of $\boldsymbol{\alpha}$ (which sum to one) are a measure of the importance of each variable in the computed reliability index.

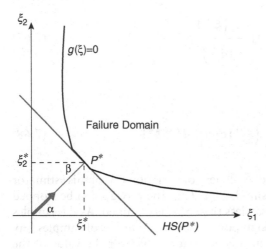

Figure 7.1 Principle of the first-order reliability method (FORM).

Importance sampling

FORM as described above gives an *approximation* of the probability of failure without any measure of its accuracy, contrary to Monte Carlo simulation, which provides an estimator of P_f together with the coefficient of variation thereof. *Importance sampling* (IS) is a technique that allows to combine both approaches. First the expression of the probability of failure is modified as follows:

$$P_f = \int_{\mathbb{R}^M} I_{D_f} (\mathbf{X}(\boldsymbol{\xi})) \varphi_M (\boldsymbol{\xi}) \, d\boldsymbol{\xi} \tag{7.64}$$

where $I_{D_f} (\mathbf{X}(\boldsymbol{\xi}))$ is the indicator function of the failure domain. Let us introduce the *sampling density* $\varrho(\boldsymbol{\xi})$ in the above equation, which may be any valid M-dimensional PDF:

$$P_f = \int_{\mathbb{R}^M} I_{D_f} (\mathbf{X}(\boldsymbol{\xi})) \frac{\varphi_M (\boldsymbol{\xi})}{\varrho (\boldsymbol{\xi})} \varrho (\boldsymbol{\xi}) \, d\boldsymbol{\xi} = E_\varrho \left[I_{D_f} (\mathbf{X}(\boldsymbol{\xi})) \frac{\varphi_M (\boldsymbol{\xi})}{\varrho (\boldsymbol{\xi})} \right] \tag{7.65}$$

where $E_\varrho [.]$ denotes the expectation with respect to the sampling density $\varrho(\boldsymbol{\xi})$. To smartly apply IS after a FORM analysis, the following sampling density is chosen:

$$\varrho (\boldsymbol{\xi}) = (2\pi)^{-M/2} \exp \left(-\frac{1}{2} \| \boldsymbol{\xi} - \boldsymbol{\xi}^* \|^2 \right) \tag{7.66}$$

This allows one to concentrate the sampling around the design point $\boldsymbol{\xi}^*$. Then the following estimator of the probability of failure is computed:

$$\hat{P}_{f,IS} = \frac{1}{N_{sim}} \sum_{i=1}^{N_{sim}} 1_{D_f} \left(\mathbf{X}(\boldsymbol{\xi}^{(i)}) \right) \frac{\varphi_M \left(\boldsymbol{\xi}^{(i)} \right)}{\varrho \left(\boldsymbol{\xi}^{(i)} \right)} \tag{7.67}$$

which may be rewritten as:

$$\hat{P}_{f,IS} = \frac{\exp[-\beta^2/2]}{N_{sim}} \sum_{i=1}^{N_{sim}} 1_{D_f} \left(\boldsymbol{\xi}^{(i)} \right) \exp \left[-\boldsymbol{\xi}^{(i)} \cdot \boldsymbol{\xi}^* \right] \tag{7.68}$$

As in any simulation method, the coefficient of variation CV of this estimator can be monitored all along the simulation. Thus the process can be stopped as soon as a small value of CV, say less than 5%, is obtained. As the samples are concentrated around the design point, a limited number of samples, say 100–1000, is necessary to obtain this accuracy, whatever the value of the probability of failure.

7.6.5 Reliability methods coupled with FE/SFE models

The reliability methods (MCS, FORM and IS) described in the section above are general, i.e. *not* limited to stochastic finite element methods. They can actually be applied whatever the nature of the model, may it be analytical or algorithmic.

- When the model is a finite element model, a coupling between the reliability algorithm and the finite element code is necessary. Each time the algorithm requires the evaluation of the limit state function, the finite element code is called with the current set of input parameters. Then the limit state function is evaluated. This technique is called *direct coupling* in the next section dealing with application examples.
- When an SFE model has been computed first, the response is approximately represented as a polynomial series in standard normal random variables (Equation (7.37)). This is an *analytical* function that can now be used together with any of the reliability methods mentioned above.

In the next section, several examples are presented. In each case when a reliability problem is addressed, the direct coupling and the post-processing of a PC expansion are compared.

7.7 Application examples

The application examples presented in the sequel have been originally published elsewhere, namely in Sudret and Der Kiureghian (2000, 2002) for the first example, Berveiller *et al.* (2006) for the second example and Berveiller *et al.* (2004) for the third example.

7.7.1 Example #1: Foundation problem – spatial variability

Description of the deterministic problem

Consider an elastic soil layer of thickness t lying on a rigid substratum. A superstructure to be founded on this soil mass is idealized as a uniform pressure P applied over a length $2B$ of the free surface (see Figure 7.2). The soil is modeled as an elastic linear isotropic material. A plane strain analysis is carried out.

Due to the symmetry, half of the structure is modeled by finite elements. Strictly speaking, there is no symmetry in the system when random fields of material properties are introduced. However, it is believed that this simplification does not significantly influence the results. The parameters selected for the deterministic model are listed in Table 7.1.

Table 7.1 Example #1 – Parameters of the deterministic model.

Parameter	Symbol	Value
Soil layer thickness	t	30 m
Foundation width	$2B$	10 m
Applied pressure	P	0.2 MPa
Soil Young's modulus	E	50 MPa
Soil Poisson's ratio	v	0.3
Mesh width	L	60 m

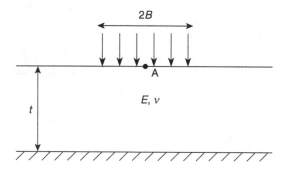

Figure 7.2 Settlement of a foundation – problem definition.

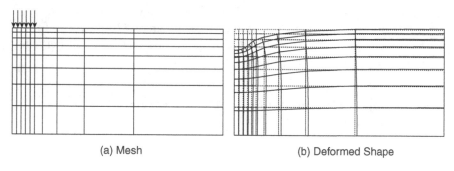

(a) Mesh	(b) Deformed Shape

Figure 7.3 Finite element mesh and deformed shape for mean values of the parameters by a deterministic analysis.

A refined mesh was first used to obtain the "exact" maximum displacement under the foundation (point A in Figure 7.2). Less-refined meshes were then tried in order to design a mesh with as few elements as possible that yielded no more than 1% error in the computed maximum settlement. The mesh displayed in Figure 7.3(a) was eventually chosen. It contains 99 nodes and 80 elements. The maximum settlement computed with this mesh is equal to 5.42 cm.

Description of the probabilistic data

The assessment of the serviceability of the foundation described in the above paragraph is now investigated under the assumption that the Young's modulus of the soil mass is spatially varying.

The Young's modulus of the soil is considered to vary only in the vertical direction, so that it is modeled as a one-dimensional random field along the depth. This is a reasonable model for a layered soil medium. The field is assumed to be lognormal and homogeneous. Its second-moment properties are considered to be the mean $\mu_E = 50$ MPa, the coefficient of variation $\delta_E = \sigma_E/\mu_E = 0.2$. The autocorrelation coefficient function of the underlying Gaussian field (see Equation (7.26)) is $\rho_{EE}(z, z') = \exp(-|z - z'|/\ell)$, where z is the depth coordinate and $\ell = 30$ m is the correlation length.

The accuracy of the discretization of the underlying Gaussian field $N(\mathbf{x})$ depends on the number of terms M retained in the expansion. For each value of M, a global indicator of the accuracy of the discretization, $\bar{\varepsilon}$, is computed from

$$\bar{\varepsilon} = \frac{1}{|\Omega|} \int_\Omega \frac{\text{Var}\left[N(\mathbf{x}) - \hat{N}(\mathbf{x})\right]}{\text{Var}\left[N(\mathbf{x})\right]} \, d\Omega \qquad (7.69)$$

A relative accuracy in the variance of 12% (respectively, 8%, 6%) is obtained when using $M = 2$ (respectively, 3, 4) terms in the KL expansion of $N(\mathbf{x})$. Of course, these values are closely related to the parameters defining the random field, particularly the correlation length ℓ. As ℓ is comparable here to the size of the domain Ω, an accurate discretization is obtained using few terms.

Reliability analysis

The limit state function is defined in terms of the maximum settlement u_A at the center of the foundation:

$$g(\xi) = u_0 - u_A(\xi) \qquad (7.70)$$

where u_0 is an admissible threshold initially set equal to 10 cm and ξ is the vector used for the random field discretization.

Table 7.2 reports the results of the reliability analysis carried out either by direct coupling between the finite element model and the FORM algorithm (column #2), or by the application of FORM after solving the SFE problem (column #6, for various values of p). Both results have been validated using importance sampling (columns #3 and #7, respectively). In the direct coupling approach, 1000 samples (corresponding to 1000 deterministic FE runs) were used, leading to a coefficient of variation of the simulation less than 6%. In the SFE approach, the polynomial chaos expansion of the response is used

Table 7.2 Example #1 – Reliability index β – Influence of the orders of expansion M and p ($u_0 = 10$cm).

M	β_{direct}^{FORM}	β_{direct}^{IS}	p	P	β_{SFE}^{FORM}	β_{SFE}^{IS}
2	3.452	3.433	2	6	3.617	3.613
			3	10	3.474	3.467
3	3.447	3.421	2	10	3.606	3.597
			3	20	3.461	3.461
4	3.447	3.449	2	15	3.603	3.592
			3	35	3.458	3.459

for importance sampling around the design point obtained by FORM (i.e. no additional finite element run is required), and thus 50,000 samples can be used, leading to a coefficient of variation of the simulation less than 1%.

It appears that the solution is not much sensitive to the order of expansion of the input field (when comparing the results for $M = 2$ with respect to those obtained for $M = 4$). This can be understood easily by the fact that the maximum settlement of the foundation is related to the global (i.e. homogenized) behavior of the soil mass. Modeling in a refined manner the spatial variability of the stiffness of the soil mass by adding terms in the KL expansion does not significantly influence the results.

In contrary, it appears that a PC expansion of third degree ($p = 3$) is required in order to get a satisfactory accuracy on the reliability index.

Parametric study

A comprehensive comparison of the two approaches is presented in Sudret and Der Kiureghian (2000), where the influences of various parameters are investigated. Selected results are reported in this section. More precisely, the accuracy of the SFE method combined with FORM is investigated when varying the value of the admissible settlement from 6 to 20 cm, which leads to an increasing reliability index. A two-term ($M = 2$) KL expansion of the underlying Gaussian field is used. The results are reported in Table 7.3. Column #2 shows the values obtained by direct coupling between FORM and the deterministic finite element model. Column #4 shows the values obtained using FORM after the SFE solution of the problem using an intrusive approach.

The results in Table 7.3 show that the "SFE+FORM" procedure obviously converges to the direct coupling results when p is increased. It appears that a third-order expansion is accurate enough to predict reliability indices up to 5. For larger values of β, a fourth-order expansion should be used.

Note that a *single* SFE analysis is carried out to get the reliability indices associated with the various values of the threshold u_0 (once p is chosen).

Table 7.3 Example #1 – Influence of the threshold in the limit state function.

u_0 (cm)	β_{direct}	p	β_{SFE}
6	0.473	2	0.477
		3	0.488
		4	0.488
8	2.152	2	2.195
		3	2.165
		4	2.166
10	3.452	2	3.617
		3	3.474
		4	3.467
12	4.514	2	4.858
		3	4.559
		4	4.534
15	5.810	2	6.494
		3	5.918
		4	5.846
20	7.480	2	8.830
		3	7.737
		4	7.561

In contrary, a FORM analysis has to be restarted for each value of u_0 when direct coupling is used. As a conclusion, if a single value of β (and related $P_f \approx \Phi(-\beta)$) is of interest, direct coupling using FORM is probably the most efficient method. When the evolution of β with respect to a threshold is investigated, the "SFE+FORM" approach may become more efficient.

7.7.2 Example #2: Foundation problem – non Gaussian variables

Deterministic problem statement

Let us consider now an elastic soil mass made of two layers of different isotropic linear elastic materials lying on a rigid substratum. A foundation on this soil mass is modeled by a uniform pressure P_1 applied over a length $2B_1 = 10\ m$ of the free surface. An additional load P_2 is applied over a length $2B_2 = 5\ m$ (Figure 7.4).

Due to the symmetry, half of the structure is modeled by finite element (Figure 7.4). The mesh comprises 80 QUAD4 elements as in the previous section. The finite element code used in this analysis is the open source code Code_Aster.[1] The geometry is considered as deterministic. The elastic material properties of both layers and the applied loads are modeled by random

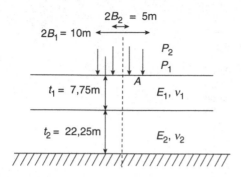

Figure 7.4 Example #2: Foundation on a two-layer soil mass.

Table 7.4 Example #2: Two-layer soil layer mass – Parameters of the model.

Parameter	Notation	Type of PDF	Mean value	Coef. of variation
Upper layer soil thickness	t_1	Deterministic	7.75 m	–
Lower layer soil thickness	t_2	Deterministic	22.25 m	–
Upper layer Young's modulus	E_1	Lognormal	50 MPa	20%
Lower layer Young's modulus	E_2	Lognormal	100 MPa	20%
Upper layer Poisson ratio	ν_1	Uniform	0.3	15%
Lower layer Poisson ratio	ν_2	Uniform	0.3	15%
Load #1	P_1	Gamma	0.2 MPa	20 %
Load #2	P_2	Weibull	0.4 MPa	20 %

variables, whose PDF are specified in Table 7.4. All six random variables are supposed to be independent.

Again the model response under consideration is the maximum vertical displacement at point A (Figure 7.4). The finite element model is thus considered as an algorithmic function $\mathfrak{h}(.)$ that computes the vertical nodal displacement u_A as a function of the six input parameters:

$$u_A = \mathfrak{h}(E_1, E_2, \nu_1, \nu_2, P_1, P_2) \tag{7.71}$$

Reliability analysis

The serviceability of this foundation on a layered soil mass vis-à-vis an admissible settlement is studied. Again, two stategies are compared.

- A direct coupling between the finite element model and the probabilistic code PROBAN (Det Norske Veritas, 2000). The limit state function given in Equation (7.70) is rewritten in this case as:

$$g(X) = u_0 - \mathfrak{h}(E_1, E_2, \nu_1, \nu_2, P_1, P_2) \tag{7.72}$$

where u_0 is the admissible settlement. The failure probability is computed using FORM analysis followed by importance sampling. One thousand samples are used in IS, allowing a coefficient of variation of the simulation less than 5%.

• A SFE analysis using the regression method is carried out, leading to an approximation of the maximal vertical settlement:

$$\tilde{u}_A = \sum_{j=0}^{P} u_j \Psi_j(\{\xi_k\}_{k=1}^6) \tag{7.73}$$

For this purpose, the six input variables $\{E_1, E_2, \nu_1, \nu_2, P_1, P_2\}$ are first transformed into a six-dimensional standard normal gaussian vector $\boldsymbol{\xi} \equiv \{\xi_k\}_{k=1}^6$. Then a third-order ($p = 3$) PC expansion of the response is performed which requires the computation of $P = \binom{6+3}{3} = 84$ coefficients. An approximate limit state function is then considered:

$$\tilde{g}(X) = u_0 - \sum_{j=0}^{P} u_j \Psi_j(\{\xi_k\}_{k=1}^6) \tag{7.74}$$

Then FORM analysis followed by importance sampling is applied (one thousand samples, coefficient of variation less than 1% for the simulation). Note that in this case FORM as well as IS are performed using the *analytical* limit state function Equation (7.74). This computation is almost costless compared to the computation of the PC expansion coefficients $\{u_j\}_{j=0}^{P-1}$ in Equation (7.73).

Table 7.5 shows the probability of failure obtained by direct coupling and by SFE/regression using various numbers of points in the experimental design (see Section 7.5.3). Figure 7.5 shows the evolution of the ratio between the logarithm of the probability of failure (divided by the logarithm of the converged probability of failure) versus the number of regression points for several values of the maximum admissible settlement u_0. Accurate results are obtained when using 420 regression points or more for different values of the failure probability (from 10^{-1} to 10^{-4}). When taking less than 420 points, results are inaccurate. When taking more than 420 points, the accuracy is not improved. Thus this number seems to be the best compromise between accuracy and efficiency. Note that it corresponds to 5×84 points, as pointed out in Equation (7.47).

Table 7.5 Example #2: Foundation on a two-layered soil – probability of failure P_f.

Threshold u_0 (cm)	Direct coupling	Non-intrusive SFE/regression approach				
		84 pts	*168 pts*	*336 pts*	*420 pts*	*4096 pts*
12	$3.09.10^{-1}$	$1.62.10^{-1}$	$2.71.10^{-1}$	$3.31.10^{-1}$	$3.23.10^{-1}$	$3.32.10^{-1}$
15	$6.83.10^{-2}$	$6.77.10^{-2}$	$6.90.10^{-2}$	$8.43.10^{-2}$	$6.73.10^{-2}$	$6.93.10^{-2}$
20	$2.13.10^{-3}$	–	$9.95.10^{-5}$	$8.22.10^{-4}$	$2.01.10^{-3}$	$1.98.10^{-3}$
22	$4.61.10^{-4}$	–	$7.47.10^{-7}$	$1.31.10^{-4}$	$3.80.10^{-4}$	$4.24.10^{-4}$
Number of FE runs required	84	168	336	420	4096	

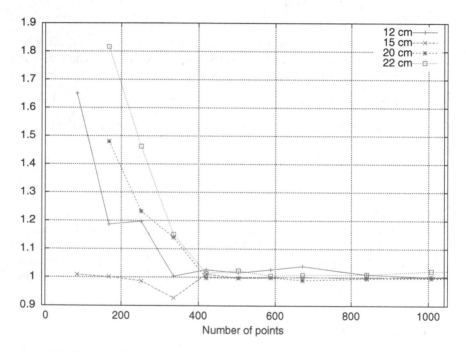

Figure 7.5 Example #2: Evolution of the logarithm of the failure probability divided by the converged value vs. the number of regression points.

7.7.3 Example #3: Deep tunnel problem

Deterministic problem statement and probabilistic model

Let us consider a deep tunnel in an elastic, isotropic homogeneous soil mass. Let us consider a homogeneous initial stress field. The coefficient of earth pressure at rest is defined as $K_0 = \frac{\sigma_{xx}^0}{\sigma_{yy}^0}$. Parameters describing geometry, material properties and loads are given in Table 7.6. The analysis is carried out under plane strain conditions. Due to the symmetry of the problem, only a quarter of the problem is modeled by finite element using appropriate boundary conditions (Figure 7.6). The mesh contains 462 nodes and 420 4-node linear elements, which allow a 1.4%-accuracy evaluation of the radial displacement of the tunnel wall compared to a reference solution.

Moment analysis

One is interested in the radial displacement (convergence) of the tunnel wall, i.e. the vertical displacement of point E denoted by u_E. The value u_E^m obtained for the mean values of the random parameters (see Table 7.6) is 6.24 mm. A third-order ($p = 3$) PC expansion of this nodal displacement is computed. This requires $P = \binom{4+3}{3} = 35$ coefficients. Various SFE

Table 7.6 Example #3 – Parameters of the model.

Parameter	Notation	Type	Mean	Coef. of Var.
Tunnel depth	L	Deterministic	20 m	–
Tunnel radius	R	Deterministic	1 m	–
Vertical initial stress	$-\sigma_{yy}^0$	Lognormal	0.2 MPa	30%
Coefficient of earth pressure at rest	K_0	Lognormal	0.5	10%
Young's modulus	E	Lognormal	50 MPa	20%
Poisson ratio	ν	Uniform [0.1–0.3]	0.2	29%

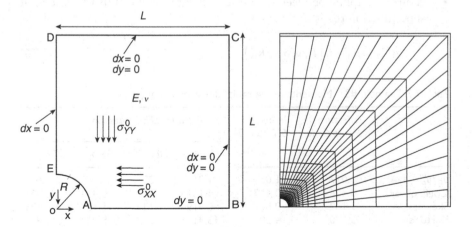

Figure 7.6 Scheme of the tunnel. Mesh of the tunnel.

methods are applied to solve the SFE problem, namely the intrusive method (Section 7.4) and the non intrusive regression method (Section 7.5) using various experimental designs. The statistical moments of u_E (reduced mean value $E[u_E]/u_E^m$, coefficient of variation, skewness and kurtosis coefficients) are reported in Table 7.7.

The reference solution is obtained by Monte Carlo simulation of the deterministic finite element model using 30,000 samples (column #2). The coefficient of variation of this simulation is 0.25%. The regression method gives good results when there are at least 105 regression points (note that this corresponds again to Equation (7.47)). These results are slightly better than those obtained by the intrusive approach (column #3), especially for the skewness and kurtosis coefficients. This is due to the fact that input variables are expanded (i.e. approximated) onto the PC basis when applying the intrusive approach while they are exactly represented through the isoprobabilistic transform in the non intrusive approaches.

Reliability analysis

Let us now consider the reliability of the tunnel with respect to an admissible radial displacement u_0. The deterministic finite element model is considered as an algorithmic function $\mathfrak{h}(.)$ that computes the radial nodal displacement u_E as a function of the four input parameters:

$$u_E = \mathfrak{h}\left(E, v, \sigma_{yy}^0, K_0\right) \qquad (7.75)$$

Two solving stategies are compared.

- A direct coupling between the finite element model PROBAN. The limit state function reads in this case:

$$g(\mathbf{X}) = u_0 - \mathfrak{h}\left(E, v, \sigma_{yy}^0, K_0\right) \qquad (7.76)$$

Table 7.7 Example #3 – Moments of the radial displacement at point E.

	Reference Monte Carlo	Intrusive SFE ($p = 3$)	Non-intrusive SFE/Regression			
			35 pts	70 pts	105 pts	256 pts
u_E/u_E^m	1.017	1.031	1.311	1.021	1.019	1.018
Coeff. of var.	0.426	0.427	1.157	0.431	0.431	0.433
Skewness	−1.182	−0.807	−0.919	−1.133	−1.134	−1.179
Kurtosis	5.670	4.209	13.410	5.312	5.334	5.460
Number of FE runs required	−	35	70	105	256	

where u_0 is the admissible radial displacement. The failure probability is computed using FORM analysis followed by importance sampling. One thousand samples are used in IS, allowing a coefficient of variation of the simulation less than 5%.

• A SFE analysis using the regression method is carried out, leading to an approximation of the radial displacement:

$$\tilde{u}_E = \sum_{j=0}^{P} u_j \Psi_j(\{\xi_k\}_{k=1}^4) \tag{7.77}$$

and the associated limit state function reads:

$$g() = u_0 - \sum_{j=0}^{P} u_j \Psi_j(\{\xi_k\}_{k=1}^4) \tag{7.78}$$

The generalized reliability indices $\beta = -\Phi^{-1}(P_{f,IS})$ associated to limit state functions (7.76) and (7.78) for various values of u_0 are reported in Table 7.8.

The above results show that at least 105 points of regression should be used when a third-order PC expansion is used. Additional points do not improve the accuracy of the results. The intrusive and non-intrusive approaches give very similar results. They are close to the direct coupling results when the obtained reliability index is not too large. For larger values, the third-order PC expansion may not be accurate enough to solve the reliability problem. Anyway, the non-intrusive approach (which does not introduce any approximation of the input variables) is slightly more accurate than the non intrusive method, as explained in Example #2.

Table 7.8 Example #3: Generalized reliability indices $\beta = -\Phi^{-1}(P_{f,IS})$ vs. admissible radial displacement.

Threshold u_0 (cm)	Direct coupling	Intrusive SFE ($p = 3, P = 35$)	Non-intrusive SFE/regression			
			35 pts	70 pts	105 pts	256 pts
7	0.427	0.251	−0.072	0.227	0.413	0.405
8	0.759	0.571	0.038	0.631	0.734	0.752
9	1.046	1.006	0.215	1.034	0.994	1.034
10	1.309	1.309	0.215	1.350	1.327	1.286
12	1.766	1.920	0.538	1.977	1.769	1.747
15	2.328	2.907	0.857	2.766	2.346	2.322
17	2.627	3.425	1.004	3.222	2.663	2.653
20	3.342	4.213	1.244	3.823	3.114	3.192

7.8 Conclusion

Modeling soil material properties properly is of crucial importance in geotechnical engineering. The natural heterogeneity of soil can be fruitfully modeled using probability theory.

- If an accurate description of the spatial variability is required, random fields may be employed. Their use in engineering problems requires their discretization. Two efficient methods have been presented for this purpose, namely the Karhunen–Loève expansion and the EOLE method. These methods should be better known both by researchers and engineers, since they provide a much better accuracy than older methods such as point discretization or local averaging techniques.
- If an homogenized behavior of the soil is sufficient with respect to the geotechnical problem under consideration, the soil characteristics may be modeled as random variables that are usually non Gaussian.

In both cases, *identification* of the parameters of the probabilistic model is necessary. This is beyond the scope of this chapter.

Various methods have been reviewed that predict the impact of input random parameters onto the response of the geotechnical model. Attention has been focused on a class of stochastic finite element methods based on polynomial chaos expansions. It has been shown how the input variables/fields should be first represented using functions of standard normal variables.

Two classes of methods for computing the expansion coefficients have been presented, namely the intrusive and non-intrusive methods. The historical intrusive approach is well-suited to solve linear problems. It has been extended to some particular non linear problems, but proves delicate to apply in these cases. In contrast, the projection and regression methods are easy to apply whatever the physics since they make use only of the deterministic model as available in the finite element code. Several runs of the model for selected values of the input parameters are required. The computed responses are processed in order to get the PC expansion coefficients of the response. Note that the implementation of these non-intrusive methods is done once and for all, and can be applied thereafter with any finite element software at hand, and more generally with any model (possibly analytical). However, the non-intrusive methods may become computationnally expensive when the number of input variables is large, which may be the case when discretized random fields are considered.

Based on a large number of application examples, the authors suggest the use of second-order $(p = 2)$ PC expansions for estimating mean and standard deviation of response quantities. When reliability problems are considered, at least a third-order expansion is necessary to catch the true shape of the response PDF tail.

Acknowledgments

The work of the first author on spatial variability and stochastic finite element methods has been initiated during his post-doctoral stay in the research group of Professor Armen Der Kiureghian (Departement of Civil and Environmental Engineering, University of California at Berkeley), who is gratefully acknowledged. The work on non-intrusive methods (as well as most application examples presented here) is part of the PhD thesis of the second author. This work was co-supervised by Professor Maurice Lemaire (Institut Français de Mécanique Avancée, Clermont-Ferrand, France), who is gratefully acknowledged.

Appendix A

A.1 Hermite polynomials

The Hermite polynomials $He_n(x)$ are solutions of the following differential equation:

$$y'' - xy' + ny = 0 \qquad n \in \mathbb{N} \tag{7.79}$$

They may be generated in practice by the following recurrence relationship:

$$He_0(x) = 1 \tag{7.80}$$

$$He_{n+1}(x) = x\,He_n(x) - n\,He_{n-1}(x) \tag{7.81}$$

They are orthogonal with respect to the Gaussian probability measure:

$$\int_{-\infty}^{\infty} He_m(x)\,He_n(x)\,\varphi(x)\,\mathrm{d}x = n!\,\delta_{mn} \tag{7.82}$$

where $\varphi(x) = 1/\sqrt{2\pi}\,e^{-x^2/2}$ is the standard normal PDF. If ξ is a standard normal random variable, the following relationship holds:

$$E\left[He_m(\xi)\,He_n(\xi)\right] = n!\,\delta_{mn} \tag{7.83}$$

The first three Hermite polynomials are:

$$He_1(x) = x \quad He_2(x) = x^2 - 1 \quad He_3(x) = x^3 - 3x \tag{7.84}$$

A.2 Construction of the polynomial chaos

The Hermite polynomial chaos of order M and degree p is the set of multivariate polynomials obtained by products of univariate polynomials so that the maximal degree is less than or equal to p. Let us define the following integer sequence $\boldsymbol{\alpha}$:

$$\boldsymbol{\alpha} = \{\alpha_i, \, i = 1, \ldots, M\}, \quad \alpha_i \geq 0, \quad \sum_{i=1}^{M} \alpha_i \leq p \tag{7.85}$$

The multivariate polynomial $\Psi_{\boldsymbol{\alpha}}$ is defined by:

$$\Psi_{\boldsymbol{\alpha}}(x_1, \ldots, x_M) = \prod_{i=1}^{M} He_{\alpha_i}(x_i) \tag{7.86}$$

The number of such polynomials of degree not exceeding p is:

$$P = \frac{(M+p)!}{M!p!} \tag{7.87}$$

An original algorithm to determine the set of $\boldsymbol{\alpha}$'s is detailed in Sudret and Der Kiureghian (2000). Let $\mathbf{Z}$ be a standard normal random vector of size M. It is clear that:

$$\mathrm{E}\left[\Psi_{\boldsymbol{\alpha}}(\mathbf{Z})\Psi_{\boldsymbol{\beta}}(\mathbf{Z})\right] = \prod_{i=1}^{M} \mathrm{E}\left[He_{\alpha_i}(Z_i) He_{\beta_i}(Z_i)\right] = \delta_{\boldsymbol{\alpha}\boldsymbol{\beta}} \prod_{i=1}^{M} \mathrm{E}\left[He_{\alpha_i}^2(Z_i)\right] \tag{7.88}$$

The latter equation shows that the polynomial chaos basis is orthogonal.

Note

1 This is an open source finite element code developed by Electricité de France, R&D Division, see http://www.code-aster.org.

References

Abdo, T. and Rackwitz, R. (1990). A new β-point algorithm for large time invariant and time-variant reliability problems. In: A. Der Kiureghian and P. Thoft-Christensen, Editors, Reliability and Optimization of Structural Systems '90. Proc. 3rd WG 7.5 IFIP Conf Berkeley 26–28 March 1990. Springer Verlag, Berlin (1991), pp. 1–12.

Abramowitz, M. and Stegun, I. A. Eds (1970). *Handbook of Mathematical Functions*. Dover Publications, Inc., New York.

Baecher, G.-B. and Ingra, T.-S. (1981). Stochastic finite element method in settlement predictions. *Journal of the Geotechnical Engineering Division, ASCE*, 107(4), 449–63.

Berveiller, M. (2005). Stochastic finite elements: intrusive and non intrusive methods for reliability analysis. PhD thesis, Université Blaise Pascal, Clermont-Ferrand.

Berveiller, M., Sudret, B. and Lemaire, M. (2004). Presentation of two methods for computing the response coefficients in stochastic finite element analysis. In *Proceedings of the 9th ASCE Specialty Conference on Probabilistic Mechanics and Structural Reliability,* Albuquerque, USA.

Berveiller, M., Sudret, B. and Lemaire, M. (2006). Stochastic finite element: a non intrusive approach by regression. *Revue Européenne de Mécanique Numérique,* 15(1–3), 81–92.

Choi, S., Grandhi, R., Canfield, R. and Pettit, C. (2004). Polynomial chaos expansion with latin hypercube sampling for estimating response variability. *AIAA Journal,* 45, 1191–8.

Clouteau, D. and Lafargue, R. (2003). An iterative solver for stochastic soil-structure interaction. In *Computational Stochastic Mechanics (CSM-4),* Eds P. Spanos and G. Deodatis. Millpress, Rotterdam, pp. 119–24.

DeGroot, D. and Baecher, G. (1993). Estimating autocovariance of in-situ soil properties. *Journal of Geotechnical Engineering, ASCE,* 119(1), 147–66.

Der Kiureghian, A. and Ke, J.-B. (1988). The stochastic finite element method in structural reliability. *Probabilistic Engineering Mechanics,* 3(2), 83–91.

Det Norske Veritas (2000). *PROBAN user's manual,* V.4.3.

Ditlevsen, O. and Madsen, H. (1996). *Structural Reliability Methods.* John Wiley and Sons, Chichester.

Eloseily, K., Ayyub, B. and Patev, R. (2002). Reliability assessment of pile groups in sands. *Journal of Structural Engineering, ASCE,* 128(10), 1346–53.

Fenton, G.-A. (1999a). Estimation for stochastic soil models. *Journal of Geotechnical Engineering, ASCE,* 125(6), 470–85.

Fenton, G.-A. (1999b). Random field modeling of CPT data. *Journal of Geotechnical Engineering, ASCE,* 125(6), 486–98.

Field, R. (2002). Numerical methods to estimate the coefficients of the polynomial chaos expansion. In *Proceedings of the 15th ASCE Engineering Mechanics Conference.*

Ghanem, R. (1999). The nonlinear Gaussian spectrum of log-normal stochastic processes and variables. *Journal of Applied Mechanic, ASME,* 66, 964–73.

Ghanem, R. and Brzkala, V. (1996). Stochastic finite element analysis of randomly layered media. *Journal of Engineering Mechanics,* 122(4), 361–9.

Ghanem, R. and Spanos, P. (1991). *Stochastic Finite Elements – A Spectral Approach.* Springer, New York.

Ghiocel, D. and Ghanem, R. (2002). Stochastic finite element analysis of seismic soil-structure interaction. *Journal of Engineering Mechanics, (ASCE),* 128, 66–77.

Grigoriu, M. (2006). Evaluation of Karhunen–Loève, spectral and sampling representations for stochastic processes. *Journal of Engineering Mechanics (ASCE),* 132(2), 179–89.

Huang, S., Quek, S. and Phoon, K. (2001). Convergence study of the truncated Karhunen–Loève expansion for simulation of stochastic processes. *International Journal of Numerical Methods in Engineering,* 52(9), 1029–43.

Isukapalli, S. S. (1999). *Uncertainty Analysis of transport-transportation models.* Ph.D. thesis, The State University of New Jersey, NJ.

Le Maître, O., Reagen, M., Najm, H., Ghanem, R. and Knio, D. (2002). A stochastic projection method for fluid flow-II. Random Process, *Journal of Computational Physics*, 181, 9–44.

Li, C. and Der Kiureghian, A. (1993). Optimal discretization of random fields. *Journal of Engineering Mechanics*, 119(6), 1136–54.

Li, L., Phoon, K. and Quek, S. (2007). Simulation of non-translation processes using non Gaussian Karhunen–Loève expansion. *Computers and Structures*, 85(5–6), pp. 264–276.

Liu, P.-L. and Der Kiureghian, A. (1986). Multivariate distribution models with prescribed marginals and covariances. *Probabilistic Engineering Mechanics*, 1(2), 105–12.

Liu, W., Belytschko, T. and Mani, A. (1986a). Probabilistic finite elements for non linear structural dynamics. *Computer Methods in Applied Mechanics and Engineering*, 56, 61–86.

Liu, W., Belytschko, T. and Mani, A. (1986b). Random field finite elements. *International Journal of Numerical Methods in Engineering*, 23(10), 1831–45.

Loève, M. (1977). *Probability Theory*. Springer, New-York.

Malliavin, P. (1997). *Stochastic Analysis*. Springer, Berlin.

Matthies, H. and Keese, A. (2005). Galerkin methods for linear and nonlinear elliptic stochastic partial differential equations. *Computer Methods in Applied Mechanics and Engineering*, 194, 1295–331.

Mellah, R., Auvinet, G. and Masrouri, F. (2000). Stochastic finite element method applied to non-linear analysis of embankments. *Probabilistic Engineering Mechanics*, 15(3), 251–9.

Niederreiter, H. (1992). *Random Number Generation and Quasi-Monte Carlo Methods*. Society for Industrial and Applied Mathematics, Philadelphia, PA.

Nouy, A. (2007). A generalized spectral decomposition technique to solve a class of linear stochastic partial differential equations, *Computer Methods in Applied Mechanics and Engineering*, 196(45–48), 4521–37.

Phoon, K., Huang, H. and Quek, S. (2005). Simulation of strongly non Gaussian processes using Karhunen-Loève expansion. *Probabilistic Engineering Mechanics*, 20(2), 188–98.

Phoon, K., Huang, S. and Quek, S. (2002a). Implementation of Karhunen–Loève expansion for simulation using a wavelet–Galerkin scheme. *Probabilistic Engineering Mechanics*, 17(3), 293–303.

Phoon, K., Huang, S. and Quek, S. (2002b). Simulation of second-order processes using Karhunen–Loève expansion. *Computers and Structures*, 80(12), 1049–60.

Phoon, K., Quek, S., Chow, Y. and Lee, S. (1990). Reliability analysis of pile settlements. *Journal of Geotechnical Engineering, ASCE*, 116(11), 1717–35.

Popescu, R., Deodatis, G. and Nobahar, A. (2005). Effects of random heterogeneity of soil properties on bearing capacity. *Probabilistic Engineering Mechanics*, 20, 324–41.

Press, W., Vetterling, W., Teukolsky, S.-A. and Flannery, B.-P. (2001). *Numerical Recipes*. Cambridge University Press, Cambridge.

Rackwitz, R. (2000). Reviewing probabilistic soil modeling. *Computers and Geotechnics*, 26, 199–223.

Rackwitz, R. (2001). Reliability analysis – a review and some perspectives. *Structural Safety*, 23, 365–95.

Saltelli, A., Chan, K. and Scott, E. Eds (2000). *Sensitivity Analysis*. John Wiley and Sons, New York.

Sobol', I. (1993). Sensitivity estimates for nonlinear mathematical models. *Mathematical Modeling and Computational Experiment*, 1, 407–14.

Sobol', I. and Kucherenko, S. (2005). Global sensitivity indices for nonlinear mathematical models. Review. *Wilmott Magazine*, 1, 56–61.

Sudret, B. (2005). Des éléments finis stochastiques spectraux aux surfaces de réponse stochastiques: une approche unifiée. In *Proceedings, 17ème Congrès Français de Mécanique – Troyes* (in French).

Sudret, B. (2006). Global sensitivity analysis using polynomial chaos expansions. In *Procedings of the 5th International Conference on Comparative Stochastic Mechanics (CSM5)*, Eds P. Spanos and G. Deodatis, Rhodos.

Sudret, B. (2008). Global sensitivity analysis using polynomial chaos expansions. *Reliability Engineering and System Safety*. In press.

Sudret, B., Berveiller, M. and Lemaire, M. (2004). A stochastic finite element method in linear mechanics. *Comptes Rendus Mécanique*, 332, 531–7 (in French).

Sudret, B., Berveiller, M. and Lemaire, M. (2006). A stochastic finite element procedure for moment and reliability analysis. *Revue Européenne de Mécanique Numérique*, 15(7–8), 1819–35.

Sudret, B., Blatman, G. and Berveiller, M. (2007). Quasi random numbers in stochastic finite element analysis – application to global sensitivity analysis. In *Proceedings of the 10th International Conference on Applications of Statistics and Probability in Civil Engineering (ICASP10)*. Tokyo.

Sudret, B. and Der Kiureghian, A. (2000). Stochastic finite elements and reliability: a state-of-the-art report. Technical Report n° UCB/SEMM-2000/08, University of California, Berkeley.

Sudret, B. and Der Kiureghian, A. (2002). Comparison of finite element reliability methods. *Probabilistic Engineering Mechanics*, 17, 337–48.

VanMarcke, E. (1977). Probabilistic modeling of soil profiles. *Journal of the Geotechnical Engineering Division, ASCE*, 103(GT11), 1227–46.

VanMarcke, E.-H. and Grigoriu, M. (1983). Stochastic finite element analysis of simple beams. *Journal of Engineering Mechanics, ASCE*, 109(5), 1203–14.

Wand, M. and Jones, M. (1995). *Kernel Smoothing*. Chapman and Hall, London.

Zhang, J. and Ellingwood, B. (1994). Orthogonal series expansions of random fields in reliability analysis. *Journal of Engineering Mechanics, ASCE*, 120(12), 2660–77.

Chapter 8

Eurocode 7 and reliability-based design

Trevor L. L. Orr and Denys Breysse

8.1 Introduction

This chapter outlines how reliability-based geotechnical design is being introduced in Europe through the adoption of Eurocode 7, the new European standard for geotechnical design. Eurocode 7 is based on the limit state design concept with partial factors and characteristic values. This chapter traces the development of Eurocode 7; it explains how the overall reliability of geotechnical structures is ensured in Eurocode 7; it shows how the limit state concept and partial factor method are implemented in Eurocode 7 for geotechnical designs; it explains how characteristic values are selected and design values obtained; it presents the partial factors given in Eurocode 7 to obtain the appropriate levels of reliability; and it examines the use of probability-based reliability methods, such as first-order reliability method (FORM) analyses and Monte Carlo simulations, for geotechnical designs and the use of these methods for calibrating the partial factor values. An example is presented of a spread foundation designed to Eurocode 7 using the partial factor method, and this is followed by some examples of the use of probabilistic methods to investigate how uncertainty in the parameter values affects the reliability of geotechnical designs.

8.2 Eurocode program

In the 1975 the Commission of the European Community (CEC) decided to initiate work on the preparation of a program of harmonized codes of practice, known as the Eurocodes, for the design of structures. The purpose and intended benefits of this program were that, by providing common design criteria, it would remove the barriers to trade due to the existence of different codes of practice in the member states of what was then the European Economic Community (EEC) and is now the European Union (EU). The Eurocodes would also serve as reference documents for fulfilling the requirements for mechanical resistance, stability and resistance to fire, including aspects of durability and economy specified in the Construction Products Directive, which have been adopted in each EU member state's building regulations. A further objective of the Eurocodes is to improve the competitiveness of the European construction industry internationally.

The set of Eurocodes consists of 10 codes, which are European standards, i.e. Europäische Norms (ENs). The first Eurocode, EN 1990, sets out the basis of design adopted in the set of Eurocodes; the second, EN 1991, provides the loads on structures, referred to as actions; the codes EN 1992 – EN 1997 and EN 1999 provide the rules for designs involving the different materials; and the code EN 1998 provides the rules for seismic design of structures. Part 1 of EN 1997, called Eurocode 7 and referred to as this throughout this chapter, provides the general rules for geotechnical design. As explained in the next section, EN 1997 is published by CEN (Comité Éuropéen de Normalisation or European Commitee for Standardization), as are the other Eurocodes.

8.3 Development of Eurocode 7

The development of Eurocode 7 started in 1980, when Professor Kevin Nash, Secretary General of the International Society for Soil Mechanics and Foundation Engineering, invited Niels Krebs Ovesen to form a committee, consisting of representatives from eight of the nine EEC member states at that time, which were Belgium, Denmark, France, Germany, Ireland, Italy, Netherlands and the UK (there was no representative from Luxembourg, and Greece joined the committee from the first meeting in 1981 when it joined the EEC), to prepare for the CEC a draft model limit state design code for Eurocode 7, which was published in 1987. Subsequently, the work on all the Eurocodes was transferred from the CEC to CEN, and the pre-standard version of Eurocode 7, which was based on partial material factors, was published in 1994 as ENV 1997-1, *Eurocode 7 Geotechnical design: Part 1 General rules*. Then, taking account of comments received on the ENV version and including partial resistance factors as well as partial material factors, the full standard version of Eurocode 7 – Part 1, EN 1997-1, was published by CEN in November 2004. Since each member state is responsible for the safety levels of structures within its jurisdiction, there was, as specified in Guidance Document Paper L (EC, 2002), a two-year period following publication of the EN, i.e. until November 2006, for each country to prepare a National Annex giving the values of the partial factors and other safety elements so that Eurocode 7 could be used for geotechnical designs in that country.

8.4 Eurocode design requirements and the limit state design concept

The basic Eurocode design requirements, given in EN 1990, are that *a structure shall be designed and executed in such a way that it will, during its intended life, with appropriate degrees of reliability and in an economical way:*

- *sustain all actions and influences likely to occur during execution and use, and*
- *remain fit for the use for which it is required.*

To achieve these basic design requirements, the limit state design concept is adopted in all the Eurocodes, and hence in Eurocode 7. The limit state design concept involves ensuring that, for each design situation, the occurrence of all limit states is sufficiently unlikely, where a limit state is a state beyond which the structure no longer fulfills the design criteria. The limit states that are considered are ultimate limit states (ULSs) and serviceability limit states (SLSs), which are defined as follows.

1. Ultimate limit states are those situations involving safety, such as the collapse of a structure or other forms of failure, including excessive deformation in the ground prior to failure causing failure in the supported structure, or where there is a risk of danger to people or severe economic loss. Ultimate limit states have a low probability of occurrence for well-designed structures, as noted in Section 8.12.
2. Serviceability limit states correspond to those conditions beyond which the specified requirements of the structure or structural element are no longer met. Examples include excessive deformations, settlements, vibrations and local damage of the structure in normal use under working loads such that it ceases to function as intended. Serviceability limit states have a higher probability of occurrence than ultimate limit states, as noted in Section 8.12.

The calculation models used in geotechnical designs to check that the occurrence of a limit state is sufficiently unlikely should describe the behavior of the ground at the limit state under consideration. Thus separate and different calculations should be carried out to check the ultimate and serviceability limit states. In practice, however, it is often known from experience which limit state will govern the design and hence, having designed for this limit state, the avoidance of the other limit states may be verified by a control check. ULS calculations will normally involve analyzing a failure mechanism and using ground strength properties, while SLS calculations will normally involve a deformation analysis and ground stiffness or compressibility properties. The particular limit states to be considered in the case of common geotechnical design situations are listed in the appropriate sections of Eurocode 7.

Geotechnical design differs from structural design since it involves a natural material (soil) which is variable and whose properties need to be determined from geotechnical investigations, rather than a manufactured material (such as concrete or steel) which is made to meet certain specifications. Hence, in geotechnical design, the geotechnical investigations to identify and characterize the relevant ground mass and determine the characteristic values of the ground properties are an important part of the design process. This is reflected in the fact that Eurocode 7 has two parts: *Part 1 – General rules* and *Part 2 – Ground investigation and testing*.

8.5 Reliability-based designs and Eurocode design methods

As noted in the previous section, the aim of designs to the Eurocodes is to provide structures with appropriate degrees of reliability, thus designs to the Eurocodes are reliability-based designs (RBDs). Reliability is defined in EN 1990 as the *ability of a structure or a structural member to fulfill the specified requirements, including the design working life, for which it has been designed,* and it is noted that *reliability is usually expressed in probabilistic terms.* EN 1990 states that the required reliability for a structure shall be achieved by designing in accordance with the appropriate Eurocodes and by the use of appropriate execution and quality management measures.

EN 1990 allows for different levels of reliability which take account of:

- the cause and/or mode of obtaining a limit state (i.e. failure),
- the possible consequences of failure in terms of risk to life, injury or potential economic loss,
- public aversion to failure, and
- the expense and procedures necessary to reduce the risk of failure.

Examples of the use of different levels of reliability are the adoption of a high level of reliability in a situation where a structure poses a high risk to human life, or where the economic, social or environmental consequences of failure are great, as in the case of a nuclear power station, and a low level of reliability where the risk to human life is low and where the economic, social and environmental consequences of failure are small or negligible, as in the case of a farm building.

In geotechnical designs, the required level of reliability is obtained in part through ensuring that the appropriate design measures and control checks are applied to all aspects and at all stages of the design process from conception, through the ground investigation, design calculations and construction, to maintenance. For this purpose, Eurocode 7 provides lists of relevant items and aspects to be considered at each stage. Also, a risk assessment system, known as the Geotechnical Categories, is provided to take account of the different levels of complexity and risk in geotechnical designs. This system is discussed in Section 8.6.

According to EN 1990, designs to ensure that the occurrence of limit state is sufficiently unlikely may be carried out using either of the following design methods:

- the partial factor method, or
- an alternative method based directly on probabilistic methods.

The partial factor method is the normal Eurocode design method and is the method presented in Eurocode 7. This method involves applying appropriate partial factor values, specified in the National Annexes to Eurocode 7, to statistically based characteristic parameter values to obtain the design values for use in relevant calculation models to check that a structure has the required probability that neither an ultimate nor a serviceability limit state will be exceeded during a specified reference period. This design method is referred to as a semi-probabilistic reliability method in EN 1990, as explained in Section 8.12. The selection of characteristic values and the appropriate partial factor values to achieve the required reliability level are discussed in Sections 8.8–8.11. As explained in Section 8.12, there is no procedure in Eurocode 7 to allow for reliability differentiation by modifying the specified partial factor values, and hence the calculated reliability level, in order to take account of the consequences of failure. Instead reliability differentiation may be achieved through using the system of Geotechnical Categories.

Eurocode 7 does not provide any guidance on the direct use of fully probabilistic reliability methods for geotechnical design. However, EN 1990 states that the information provided in Annex C of EN 1990, which includes guidance on the reliability index, β value, may be used as a basis for probabilistic design methods.

8.6 Geotechnical risk and Geotechnical Categories

In Eurocode 7, three Geotechnical Categories, referred to as Geotechnical Categories 1, 2 and 3, have been introduced to take account of the different levels of complexity of a geotechnical design and to establish the minimum requirements for the extent and content of the geotechnical investigations, calculations and construction control checks to achieve the required reliability. The factors affecting the complexity of a geotechnical design include:

1. the nature and size of the structure;
2. the conditions with regard to the surroundings, for example neighboring structures, utilities, traffic, etc.;
3. the ground conditions;
4. the groundwater situation;
5. regional seismicity; and
6. the influence of the environment, e.g. hydrology, surface water, subsidence, seasonal changes of moisture.

The use of the Geotechnical Categories is not a code requirement, as they are presented as an application rule, rather than a principle, and so are optional rather than mandatory. The advantage of the Geotechnical Categories is that they provide a framework for categorizing the different levels of risk in a geotechnical design and for selecting appropriate levels of reliability to account for the different levels of risk. Geotechnical risk is a function of two factors: the geotechnical hazards (i.e. dangers) and the vulnerability of people and the structure to specific hazards. With regard to the design complexity factors listed above, factors 1 and 2 (the structure and its surroundings) relate to vulnerability, and factors 3–6 (ground conditions, groundwater, seismicity and environment) are geotechnical hazards. The various levels of complexity of these factors, in relation to the different Geotechnical Categories and the associated geotechnical risks, are shown in Table 8.1, taken from Orr and Farrell (1999). It is the geotechnical designer's responsibility to ensure, through applying the Eurocode 7 requirements appropriately and by the use of appropriate execution and quality management measures, that structures have the required reliability, i.e. have sufficient safety against failure as a result of any of the potential hazards.

In geotechnical designs to Eurocode 7, the distinction between the Geotechnical Categories lies in the degree of expertise required and in the nature and extent of the geotechnical investigations and calculations to be carried out, as shown in Table 8.2, taken from Orr and Farrell (1999). Some examples of structures in each Geotechnical Category are also shown in this table. As noted in the previous section, Eurocode 7 does not provide for any differentiation in the calculated reliability for the different Geotechnical Categories by allowing variation in the partial factors values. Instead, while satisfying the basic design requirements and using the specified partial factor values, the required reliability is achieved in the higher Geotechnical Categories by greater attention to the quality of the geotechnical investigations, the design calculations, the

Table 8.1 Geotechnical Categories related to geotechnical hazards and vulnerability levels.

Factors to be considered	Geotechnical Categories		
	GC1	GC2	GC3
Geotechnical hazards	Low	Moderate	High
Ground conditions	Known from comparable experience to be straightforward. Not involving soft, loose or compressible soil, loose fill or sloping ground	Ground conditions and properties can be determined from routine investigations and tests	Unusual or exceptionally difficult ground conditions requiring non-routine investigations and tests
Groundwater situation	No excavations below water table, except where experience indicates this will not cause problems	No risk of damage without prior warning to structures due to groundwater lowering or drainage. No exceptional water tightness requirements	High groundwater pressures and exceptional groundwater conditions, e.g. multi-layered strata with variable permeability
Regional seismicity	Areas with no or very low earthquake hazard	Moderate earthquake hazard where seismic design code (EC8) may be used	Areas of high earthquake hazard
Influence of the environment	Negligible risk of problems due to surface water, subsidence, hazardous chemicals, etc.	Environmental factors covered by routine design methods	Complex or difficult environmental factors requiring special design methods
Vulnerability	Low	Moderate	High
Nature and size of the structure and its elements	Small and relatively simple structures or construction. Insensitive structures in seismic areas	Conventional types of structures with no abnormal risks	Very large or unusual structures and structures involving abnormal risks. Very sensitive structures in seismic areas
Surroundings	Negligible risk of damage to or from neighboring structures or services and negligible risk for life	Possible risk of damage to neighboring structures or services due, for example, to excavations or piling	High risk of damage to neighboring structures or services
Geotechnical risk	Low	Moderate	High

monitoring during construction and the maintenance of the completed structure.

The reliability of a geotechnical design is influenced by the expertise of the designer. No specific guidelines are given in Eurocode 7 with regard to the level of expertise required by designers for particular Geotechnical Categories except that, as stated in both EN 1990 and EN 1997-1, it is assumed that structures are designed by appropriately qualified and experienced personnel. Table 8.2 provides an indication of the levels of expertise required by those involved in the different Geotechnical Categories. The main features of the different Geotechnical Categories are summarized in the following paragraphs.

8.6.1 Geotechnical Category 1

Geotechnical Category 1 includes only small and relatively simple structures for which the basic design requirements may be satisfied on the basis of experience and qualitative geotechnical investigations and where there is negligible risk for property and life due to the ground or loading conditions. Geotechnical Category 1 designs involve empirical procedures to ensure the required reliability, without the use of any probability-based analyses or design methods. Apart from the examples listed in Table 8.2, Eurocode 7 does not, and, without local knowledge, cannot provide detailed guidance or specific requirements for Geotechnical Category 1; these must be found elsewhere: for example, in local building regulations, national guidance documents and textbooks. According to Eurocode 7, the design of Geotechnical Category 1 structures requires someone with appropriate comparable experience.

8.6.2 Geotechnical Category 2

Geotechnical Category 2 includes conventional types of structures and foundations with no abnormal risk or unusual or exceptionally difficult ground or loading conditions. Structures in Geotechnical Category 2 require quantitative geotechnical data and analyses to ensure that the basic requirements will be satisfied and require a suitably qualified person, normally a civil engineer with appropriate geotechnical knowledge and experience. Routine procedures may be used for field and laboratory testing and for design and construction. The partial factor method presented in Eurocode 7 is the method normally used for Geotechnical Category 2 designs.

8.6.3 Geotechnical Category 3

Geotechnical Category 3 includes structures or parts of structures that do not fall within the limits of Geotechnical Categories 1 or 2.

Table 8.2 Investigations, designs and structural types related to Geotechnical Categories.

	Geotechnical Categories		
	GC1	*GC2*	*GC3*
Expertise required	Person with appropriate comparable experience	Experienced qualified person	Experienced geotechnical specialist
Geotechnical investigations	Qualitative investigations including trial pits	Routine investigations involving borings, field and laboratory tests	Additional more sophisticated investigations and laboratory tests
Design procedures	Prescriptive measures and simplified design procedures, e.g. design bearing pressures based on experience or published presumed bearing pressures. Stability or deformation calculations may not be necessary	Routine calculations for stability and deformations based on design procedures in EC7	More sophisticated analyses
Examples of structures	• Simple 1 and 2 storey structures and agricultural buildings having maximum design column load of 250 kN and maximum design wall load of 100 kN/m • Retaining walls and excavation supports where ground level difference does not exceed 2 m • Small excavations for drainage and pipes.	Conventional: • Spread and pile foundations • Walls and other retaining structures • Bridge piers and abutments • Embankments and earthworks • Ground anchors and other support systems • Tunnels in hard, non-fractured rock	• Very large buildings • Large bridges • Deep excavations • Embankments on soft ground • Tunnels in soft or highly permeable ground

Geotechnical Category 3 includes very large or unusual structures, structures involving abnormal risks, or unusual or exceptionally difficult ground or loading conditions, and structures in highly seismic areas. Geotechnical Category 3 structures will require the involvement of a specialist, such as a geotechnical engineer. While the requirements in Eurocode 7 are the minimum requirements for Geotechnical Category 3, Eurocode 7 does not provide any special requirements for Geotechnical Category 3. Eurocode 7 states that Geotechnical Category 3 normally includes alternative provisions and rules to those given in Eurocode 7. Hence, in order to take account of the abnormal risk associated with Geotechnical Category 3 structures, it would be appropriate to use probability-based reliability analyses when designing these to Eurocode 7.

8.6.4 Classification into a particular Geotechnical Category

If the Geotechnical Categories are used, the preliminary classification of a structure into a particular category is normally performed prior to any investigation or calculation being carried out. However, this classification may need to be changed during or following the investigation or design as additional information becomes available. Also, when using this system, all parts of a structure do not have to be treated according to the highest Geotechnical Category. Only some parts of a structure may need to be classified in a higher category, and only those parts will need to be treated differently; for example, with regard to the level of investigation or the degree of sophistication of the design. A higher Geotechnical Category may be used to achieve a higher level of reliability or a more economical design.

8.7 The partial factor method and geotechnical design calculations

Geotechnical design calculations involve the following components:

- imposed loads or displacements, referred to as actions, F in the Eurocodes;
- properties of soil, rock and other materials, X or ground resistances, R;
- geometrical data, a;
- partial factors, γ or some other safety elements;
- action effects, E, for example resulting forces or calculated settlements;
- limiting or acceptable values, C of deformations, crack widths, vibrations, etc; and
- a calculation model describing the limit state under consideration.

ULS design calculations, in accordance with Eurocode 7, involve ensuring that the design effect of the actions or action effect, E_d, does not exceed the design resistance, R_d, where the subscript d indicates a design value:

$$E_d \leq R_d \tag{8.1}$$

while SLS calculations involve ensuring that the design action effect, E_d (e.g. settlement), is less than the limiting value of the deformation of the structure, C_d at the SLS:

$$E_d \leq C_d \tag{8.2}$$

In geotechnical design, the action effects and resistances may be functions of loads and material strengths, as well as being functions of geometrical parameters. For example, in the design of a cantilever retaining wall against sliding, due to the frictional nature of soil, both the action effect, E, which is the earth pressure force on the retaining wall, and the sliding resistance, R, are functions of the applied loads as well as being functions of the soil strength and the height of the wall, i.e. in symbolic form they are $E\{F, X, a\}$ and $R\{F, X, a\}$.

When using the partial factor method in a ULS calculation and assuming the geometrical parameters are not factored, the E_d and R_d may be obtained either by applying partial action and partial material factors, γ_F and γ_M, to representative loads, F_{rep} and characteristic soil strengths, X_k or else partial action effect and partial resistance factors, γ_E and γ_R, may be applied to the action effects and the resistances calculated using unfactored representative loads and characteristic soil strengths. This is indicated in the following equations, where the subscripts *rep* and *k* indicate representative and characteristic values, respectively:

$$E_d = E[\gamma_F F_{rep}, X_k/\gamma_M, a_d] \tag{8.3}$$

and $$R_d = R[\gamma_F F_{rep}, X_k/\gamma_M, a_d] \tag{8.4}$$

or $$E_d = \gamma_E E[F_{rep}, X_k, a_d] \tag{8.5}$$

and $$R_d = R[F_{rep}, X_k, a_d]/\gamma_R \tag{8.6}$$

The difference between the Design Approaches presented in Section 8.11 is whether Equations (8.3) and (8.4) or Equations (8.5) and (8.6) are used. The definition of representative and characteristic values and how they are selected, and the choice of partial factor values to obtain E_d and R_d, are discussed in the following sections.

8.8 Characteristic and representative values of actions

EN 1990 defines the characteristic value of an action, F_k, as *its main representative value* and its value is specified *as a mean value, an upper or lower value, or a nominal value.* EN 1990 states that *the variability of permanent loads, G may be neglected if G does not vary significantly during the design working life of the structure and its coefficient of variation, V is small. G_k should then be taken equal to the mean value.* EN 1990 notes that the value of V for G_k *can be in the range 0.05 to 0.10, depending on the type of structure.* For variable loads, Q the characteristic value, Q_k may either be an *upper value with an intended probability of not being exceeded, or a nominal value when a statistical distribution is not known.* In the case of climatic loads, a probability of 0.02 is quoted in EN 1990 for reference period of 1 year, corresponding to a return period of 50 years for the characteristic value. No V value for variable loads is given in EN 1990, but a typical value would be 0.15, which is the value used in the example in Section 8.14. If a load, F, is a normally distributed random variable with a mean value $\mu(F)$, standard deviation $\sigma(F)$ and coefficient of variation $V(F)$, the characteristic value corresponding to the 95% fractile is given by:

$$F_k = \mu(F) + 1.645\sigma(F) = \mu_F(1 + 1.645V(F)) \tag{8.7}$$

The representative values of actions, F_{rep}, are obtained from the equation:

$$F_{rep} = \psi F_k \tag{8.8}$$

where ψ = either 1.0 or ψ_0, ψ_1 or ψ_2.

The factors ψ_0, ψ_1 or ψ_2 are factors that are applied to the characteristic values of the actions to obtain the representative values for different design situations. The factor ψ_0 is the combination factor used to obtain the fundamental combination $\psi_0 Q_k$ of a variable action for persistent and transient design situations and is applied only to the non-leading variable actions. The factor ψ_0 is chosen so that the probability that the effects caused by the combination will be exceeded is approximately the same as the probability of the effects caused by the characteristic value of an individual action. The factor ψ_1 is used to obtain the frequent value, $\psi_1 Q_k$, of a variable action and the factor ψ_2 is used to obtain the quasi-permanent value $\psi_2 Q_k$ of a variable action for accidental or seismic design situations. The factor ψ_1 is chosen so that either the total time, within the reference period, during which it is exceeded is only a small given part of the reference period or so that the probability of it being exceeded is limited to a given value, while the factor ψ_2 is the value determined so that the total period of time for which it will be exceeded is a large fraction of the reference period. Both ψ_1 and ψ_2 are applied to all the variable actions.

The equations for combining the actions for different design situations using the factors ψ_0, ψ_1 and ψ_2 and the values of these factors in the case of imposed, snow and wind loads on buildings are given in EN 1990. An example of a ψ value is $\psi_0 = 0.7$ for an imposed load on a building that is not a load in a storage area. Because the different variable loads are often much less significant than the permanent loads in geotechnical designs compared with the significance of the variable loads to the permanent loads in many structural designs (for example, comparing the permanent and variable loads in the designs of a slope and a roof truss), the combination factors, ψ_0, are used much less in geotechnical design than in structural design. Consequently, in geotechnical design the representative actions are often equal to the characteristic actions and hence F_k is normally used for F_{rep} in design calculations.

8.9 Characteristic values of material properties

EN 1990 defines the characteristic value of a material or product property, X_k, as "the value having a prescribed probability of not being attained in a hypothetical unlimited test series". In structural design, this value generally corresponds to a specified fractile of the assumed statistical distribution of the particular property of the material or product. Where a low value of a material property is unfavorable, EN 1990 states that *the characteristic value should be defined as the 5% fractile*. For this situation and assuming a normal distribution, the characteristic value is given by the equation:

$$X_k = \mu(X) - 1.645\sigma(X) = \mu(X)(1 - 1.645V(X)) \qquad (8.9)$$

where $\mu(X)$ is the mean value, $\sigma(X)$ is the standard deviation and $V(X)$ is the coefficient of variation of the unlimited test series, X, and the coefficient 1.645 provides the 5% fractile of the test results.

The above definition of the characteristic value and Equation (8.9) are applicable in the case of test results that are normally distributed and when the volume of material in the actual structural element being designed is similar to the volume in the test element. This is normally the situation in structural design, when, for example, the failure of a beam is being analyzed and the strength is based on the results of tests on concrete cubes. However, in geotechnical design, the volume of soil involved in a geotechnical failure is usually much greater than the volume of soil involved in a single test; the consequence of this is examined in the following paragraphs.

In geotechnical design, Equation (8.9) may not be applicable because the distribution of soil properties may not be normal and if $V(X)$ is large, the calculated X_k value may violate a physical lower bound, for example a critical state value of ϕ', while if $V(X)$ is greater than $0.6 (= 1/1.645)$, then the calculated X_k value is negative. This problem with Equation (8.9) does not occur in the case of structural materials, such as steel, for which the $V(X)$ values are normally about 0.1, normally distributed, and do not have physical lower bounds.

As noted above, Equation (8.9) is not normally applicable in geotechnical design because the volume of soil involved in a failure is usually much larger than the volume of soil involved in a single field test or in a test on a laboratory specimen. Since soil, even homogeneous soil, is inherently variable, the value affecting the occurrence of a limit state is the mean, i.e. average, value over the relevant slip surface or volume of ground, not a locally measured low value. Hence, in geotechnical designs the characteristic strength is the 5% fractile of the mean strengths measured along the slip surface or in the volume of soil affecting the occurrence of that limit state, which is not normally the 5% fractile of the test results. However, a particular situation when Equation (8.9) is applicable in geotechnical design is when centrifuge tests are used to obtain the mean soil strength of the soil on the entire failure surface because then the characteristic value is the 5% fractile of the soil strengths obtained from the centrifuge tests.

How the volume of soil involved in a failure affects the characteristic value is demonstrated by considering the failures of a slope and a foundation. The strength value governing the stability of the slope is the mean value of a large volume of soil involving the entire slip surface, while the failure mechanism of a spread foundation involves a much smaller volume of ground, and, if the spread foundation is in a weak zone of the same ground as in the slope, the mean value governing the stability of the foundation will be lower than the mean value governing the stability of the slope. Hence the characteristic value chosen to design the spread foundation should be a more cautious (i.e. a lower) estimate of the mean value, corresponding to a local mean value, than the less cautious estimate of the mean value, corresponding to a global mean value, chosen to design the slope in the same soil. Refined slope stability analyses have shown that the existence of weaker zones can change the shape of slip surface crossing these zones. Thus, careful attention needs to be paid to the influence of weaker zones, even if some simplified practical methods have been proposed to account for their influence when deriving characteristic values (Breysse *et al.*, 1999).

A further difference between geotechnical and structural design is that normally only a very limited number of test results are available in geotechnical design. Arising from this, the mean and standard deviation values obtained from the test results may not be the same as the mean and standard deviation values of the soil volume affecting the occurrence of the limit state.

Because of the nature of soil and soil tests, leading to the problems outlined above when using the EN 1990 definition of the characteristic value for soil properties, the definition of the characteristic value of a soil property is given in Eurocode 7 as "a cautious estimate of the value affecting the occurrence of the limit state." There is no mention in the Eurocode 7 definition of the prescribed probability or the unlimited test series that are in the EN 1990 definition. Each of the terms: "cautious," "estimate" and "value affecting the occurrence of the limit state" in the Eurocode 7 definition of the characteristic value is important and how they should be interpreted is explained in the following paragraphs.

The characteristic value is an estimated value and the selection of this value from the very limited number of test results available in geotechnical designs involves taking account of spatial variability and experience of the soil and hence using judgment and caution. The problem in selecting the characteristic value of a ground property is deciding how cautious the estimate of the mean value should be. In an application rule to explain the characteristic value in geotechnical design, Eurocode 7 states that, "if statistical methods are used, the characteristic value should be derived such that the calculated probability of a worst value governing the occurrence of the limit state under consideration is not greater than 5%." It also states that such methods should differentiate between local and regional sampling and should allow the use of a priori knowledge regarding the variability of ground properties. However, Eurocode 7 provides no guidance for the designer on how this "cautious estimate" should be chosen, instead it relies mainly on the designer's professional expertise and relevant experience of the ground conditions.

The following equation for the characteristic value, corresponding to a 95% confidence level that the actual mean value is greater than this value, is given by:

$$X_k = m(X) - (t/\sqrt{N})s(X) \tag{8.10}$$

where $m(X)$ and $s(X)$ are the mean and the standard deviation of the sample of test values and t is known as the Student t value (Student, 1908), the value of which depends on the actual number, N, of test values considered and on the required confidence level.

The following simpler equation for the characteristic value, which has been found to be useful in practice, has been proposed by Schneider (1997):

$$X_k = m(X) - 0.5s(X) \tag{8.11}$$

The selection of the characteristic value is illustrated by the following example. Ten undrained triaxial tests were carried out on samples obtained at various depths from different boreholes through soil in which 10 m long piles are to be founded and the measured undrained shear strength c_u values are plotted in Figure 8.1. It is assumed that the measured c_u values are the correct values at the locations where the samples were taken so that there is

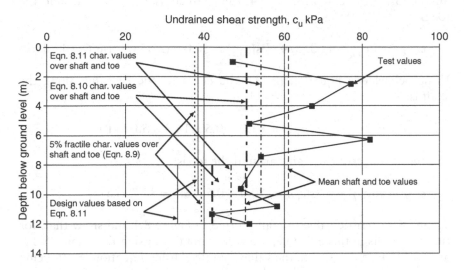

Figure 8.1 Test results and characteristic values obtained using different equations.

no bias or variation in the measured values due to the testing method. The means of the 7 measured values along the shaft and the 4 measured values in the zone 1 m above and 2 m below the toe were calculated as $m(c_{u,s}) = 61$ kPa and $m(c_{u,t}) = 50$ kPa, respectively, and these values are plotted in Figure 8.1. The standard deviations of the measured c_u values along the shaft and around the toe were calculated as $s(c_{u,s}) = 14.3$ kPa and $s(c_{u,t}) = 6.6$ kPa, giving coefficients of variation $V(c_{u,s}) = 0.23$ and $V(c_{u,t}) = 0.13$.

If Equation (8.9) is used and the means and standard deviations of the two sets of test results are assumed to be the same as the means and standard deviations of all the soil along the shaft and around the toe, the characteristic c_u along the shaft is $c_{u,s;k} = m(c_{u,s}) - 1.645\ s(c_{u,s}) = 61.0 - 1.645 \times 14.3 = 37.5$ kPa, while the characteristic c_u around the toe is $c_{u,t;k} = m(c_{u,t}) - 1.64\ s(c_{u,t}) = 50.0 - 1.64 \times 6.6 = 39.2$ kPa. These results are clearly incorrect, as they are both less than the lowest measured values and the calculated characteristic value around the toe is greater than the characteristic value along the shaft. They provide the 5% fractiles of the c_u values measured on the relatively small volume of soil in the samples tested and not the 5% estimates of the mean c_u values of all the soil over the each failure zone.

Using Equation (8.10), and choosing $t = 1.943$ for $N = 7$, the characteristic c_u value along the shaft is:

$$c_{u,sk} = m(c_{u,s}) - (1.943/\sqrt{N})s(c_{u,s}) = 61.0 - (1.943/\sqrt{7})14.3 = 50.5 \text{ kPa}$$

and choosing $t = 2.353$ for $N = 4$, the characteristic c_u value around the toe is:

$$c_{u,tk} = m(c_{u,t}) - (2.353/\sqrt{N})s(c_{u,t}) = 50.0 - (2.353/\sqrt{4})6.6 = 42.3 \text{ kPa}$$

Using Schneider's Equation (8.11), the characteristic c_u value along the shaft is:

$$c_{u,sk} = m(c_{u,s}) - 0.5s(c_{u,s}) = 61.0 - 0.5 \times 14.3 = 53.9 \text{ kPa}$$

and the characteristic c_u value around the toe is:

$$c_{u,tk} = m(c_{u,t}) - 0.5s(c_{u,t}) = 50.0 - 0.5 \times 6.6 = 46.7 \text{ kPa}$$

These characteristic values are all plotted in Figure 8.1 and show that, for the test results in this example, the values obtained using Equations (8.10) and (8.11) are similar, with the values obtained using Equation (8.10) being slightly more conservative than those obtained using Equation (8.11). From the results obtained in this example, it can be seen that the characteristic value of a ground property is similar to the value that has conventionally been selected for use in geotechnical designs, as noted by Orr (1994).

When only very few test results are available, interpretation of the characteristic value using the classical statistical approach outlined above is not possible. However, using prior knowledge in the form of local or regional experience of the particular soil conditions, the characteristic value may be estimated using a Bayesian statistical procedure as shown, for example, by Ovesen and Denver (1994) and Cherubini and Orr (1999). An example of prior knowledge is an estimate of coefficient of variation of the strength for a particular soil deposit. Typical ranges of V for soil parameters are given in Table 8.3. The value $V = 0.10$ for the coefficient of variation of $\tan\phi'$ is used in the example in Section 8.15.

Table 8.3 Typical range of V for soil parameters.

Soil parameter	Typical range of V values	Recommended V value if limited test results available
$\tan\phi'$	0.05–0.15	0.10
c'	0.20–0.40	0.40
c_u	0.20–0.40	0.30
m_v	0.20–0.40	0.40
γ (weight density)	0.01–0.10	0

Duncan (2000) has proposed a practical rule to help the engineer in assessing V. This rule is based on the "3-sigma assumption," which is equivalent to stating that nearly all the values of a random property lie within the $[X_{min}, X_{max}]$ interval, where X_{min} and X_{max} are equal to $X_{mean} - 3\sigma$ and $X_{mean} + 3\sigma$, respectively. Thus, if the engineer is able to estimate the "worst" value X_{min} and the "best" value X_{max} that X can take, then σ can be calculated as $\sigma = [X_{max} - X_{min}]/6$ and Equation (8.9) used to determine the X_k value. However, Duncan added that "a conscious effort should be made to make the range between X_{min} and X_{max} as wide as seemingly possible, or even wider, to overcome the natural tendency to make the range too small."

8.10 Characteristic values of pile resistances and ξ values

When the compressive resistance of a pile is measured in a series of static pile load tests, these tests provide the values of the measured mean compressive resistance, $R_{c;m}$, controlling the settlement and ultimate resistance of the pile at the locations of the test piles. In this situation and when the structure does not have the capacity to transfer load from "weak" to "strong" piles, Eurocode 7 states that the characteristic compressive pile resistance, $R_{c;k}$, should be obtained by applying the correlation factors, ξ, in Table 8.4 to the mean and lowest $R_{c;m}$ values so as to satisfy the following equation:

$$R_{c,k} = \text{Min}\left\{ \frac{(R_{c,m})_{mean}}{\xi_1}, \frac{(R_{c,m})_{min}}{\xi_2} \right\} \tag{8.12}$$

Using Equation (8.12) and the ξ values in Table 8.4, it is found that, when only one static pile load test is carried out, the characteristic pile resistance is the measured resistance divided by 1.4, whereas if five or more tests are carried out, the characteristic resistance is equal to lowest measured value. When pile load tests are carried out, the tests measure the mean strength of the soil over the relevant volume of ground affecting the failure of the pile. Hence, when assessing the characteristic value from pile load tests, the

Table 8.4 Correlation factors to derive characteristic values from static pile load tests.

ξ for $n^* =$	1	2	3	4	≥ 5
ξ_1	1.4	1.3	1.2	1.1	1.0
ξ_2	1.4	1.2	1.05	1.0	1.0

$^*n$ = number of tested piles.

ξ factors take account of the variability of the soil and the uncertainty in the soil strength across the test site.

8.11 Design Approaches and partial factors

For ULSs involving failure in the ground, referred to as GEO ultimate limit states, Eurocode 7 provides three Design Approaches, termed Design Approaches 1, 2 and 3 (DA1, DA2 and DA3), and recommended values of the partial factors, $\gamma_F = \gamma_E$, γ_M and γ_R for each Design Approach that are applied to the representative or characteristic values of the actions, soil parameters and resistances, as appropriate, in accordance with Equations (8.3)–(8.6) to obtain the design values of these parameters. The sets of recommended partial factor values are summarized in Table 8.5. This table shows that there are two sets of partial factors for Design Approach 1: Combination 1 (DA1.C1), when the γ_F values are greater than unity and

Table 8.5 Eurocode 7 partial factors on actions, soil parameters and resistances for GEO/STR ultimate limit states in persistent and transient design situations.

Parameter	Factor	Design Approaches				
		DA1		DA2	DA3	
		DA1.C1	DA1.C2			
Partial factors on actions (γ_F) or the effects of actions (γ_E)	Set	A1	A2	A1	A1 Structural actions	A2 Geotechnical actions
Permanent unfavorable action	γ_G	1.35	1.0	1.35	1.35	1.0
Permanent favorable action	γ_G	1.0	1.0	1.0	1.0	1.0
Variable unfavorable action	γ_Q	1.5	1.3	1.5	1.5	1.3
Variable favorable action	γ_Q	0	0	0	0	0
Accidental action	γ_A	1.0	1.0	1.0	1.0	1.0
Partial factors for soil parameters (γ_M)	Set	M1	M2	M1	M2	
Angle of shearing resistance (this factor is applied to tanϕ')	$\gamma_{\tan\phi'}$	1.0	1.25	1.0	1.25	

Continued

Table 8.5 Cont'd

Parameter	Factor	Design Approaches			
		DA1		DA2	DA3
		DA1.C1	DA1.C2		
Effective cohesion c'	$\gamma_{c'}$	1.0	1.25	1.0	1.25
Undrained shear strength c_u	γ_{cu}	1.0	1.4	1.0	1.4
Unconfined strength q_u	γ_{qu}	1.0	1.4	1.0	1.4
Weight density of ground γ	γ_{γ}	1.0	1.0	1.0	1.0
Partial resistance factors (γ_R)					
Spread foundations, retaining structures and slopes	Set	R1	R1	R2	R3
Bearing resistance	$\gamma_{R;v}$	1.0	1.0	1.4	1.0
Sliding resistance, incl. slopes	$\gamma_{R;h}$	1.0	1.0	1.1	1.0
Earth resistanc	$\gamma_{R;h}$	1.0	1.0	1.4	1.0
Pile foundations – driven piles	Set	R1	R4	R2	R3
Base resistance	γ_b	1.0	1.3	1.1	1.0
Shaft resistance (compression)	γ_s	1.0	1.3	1.1	1.0
Total combined (compression)	$\gamma_{s;t}$	1.0	1.3	1.1	1.0
Shaft in tension	γ_t	1.25	1.6	1.15	1.1
Pile foundations – bored piles	Set	R1	R4	R2	R3
Base resistance	γ_{Re}	1.25	1.6	1.15	1.0
Shaft resistance (compression)	γ_b	1.0	1.3	1.1	1.0
Total combined (compression)	γ_s	1.15	1.5	1.1	1.0
Shaft in tension	γ_t	1.25	1.6	1.15	1.1

Continued

Table 8.5 Cont'd

Parameter	Factor	Design Approaches			
		DA1		DA2	DA3
		DA1.C1	DA1.C2		
Pile foundations – CFA piles	Set	R1	R4	R2	R3
Base resistance	γ_{Re}	1.1	1.45	1.1	1.0
Shaft resistance (compression)	γ_b	1.0	1.3	1.1	1.0
Total combined (compression)	γ_s	1.1	1.4	1.1	1.0
Shaft in tension	γ_t	1.25	1.6	1.15	1.1
Prestressed anchorages	Set	R1	R4	R2	R3
Temporary resistance	$\gamma_{a;t}$	1.1	1.1	1.1	1.0
Permanent resistance	$\gamma_{a:p}$	1.1	1.1	1.1	1.0

*In the design of piles, set M1 is used for calculating the resistance of piles or anchors.

the γ_M and γ_R values are equal to unity, except for the design of piles and anchors, and Combination 2 (DA1.C2), when the γ_F value on permanent actions and the γ_R values are equal to unity and the γ_M values are greater than unity, again except for the design of piles and anchorages when the γ_M values are equal to unity and the γ_R values are greater than unity. In DA2, the γ_F and γ_R values are greater than unity, while the γ_M values are equal to unity. DA3 is similar to DA1.C2, except that a separate set of γ_F factors is provided for structural loads and the γ_R values for the design of a compression pile are all equal to unity so that this Design Approach should not be used to design a compression pile using the compressive resistances obtained from pile load tests.

When, on the resistance side, partial material factors γ_M greater than unity are applied to soil parameter values, the design is referred to as a materials factor approach (MFA), whereas if partial resistance factors γ_R greater than unity are applied to resistances, the design is referred to as a resistance factor approach (RFA). Thus DA1.C2 and DA3 are both materials factor approaches while DA2 is a resistance factor approach.

The recommended values of the partial factors given in Eurocode 7 for the three Design Approaches are based on experience from a number of countries in Europe and have been chosen to give designs similar to those obtained using existing design standards. The fact that Eurocode 7 has three Design Approaches with different sets of partial factors means that use of the different Design Approaches produces designs with different overall factors of safety, as shown by the results of the spread foundation example in Section 8.13 presented in Table 8.7, and hence different reliability levels.

In the case of SLS calculations, the design values of E_d and C_d in Equation (8.2) are equal to the characteristic values, since Eurocode 7 states that the partial factors for SLS design are normally taken as being equal to 1.0.

8.11.1 Model uncertainty

It should be noted that, as well as accounting for uncertainty in the loads, material properties and the resistances, the partial factors $\gamma_F = \gamma_E$, γ_M and γ_R also account for uncertainty in the calculation models for the actions and the resistances. The influence of model uncertainty on the reliability of designs is examined in Section 8.18.

8.12 Calibration of partial factors and levels of reliability

Figure 8.2, which is an adaptation of a figure in EN 1990, presents an overview of the various methods available for calibrating the numerical values of the γ factors for use with the limit state design equations in the Eurocodes. According to this figure, the factors may be calibrated using either of two procedures.

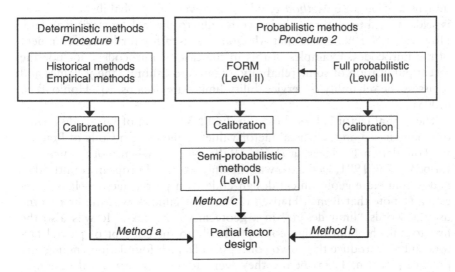

Figure 8.2 Overview of the use of reliability methods in the Eurocodes (from EN, 1990).

- *Procedure 1*: On the basis of calibration with long experience of traditional design. This is the deterministic method, referred to as *Method a* in Figure 8.2, which includes historical and empirical methods.
- *Procedure 2*: On the basis of a statistical evaluation of experimental data and field observations, which should be carried out within the framework of a probabilistic reliability theory. Two probabilistic methods are shown in Figure 8.2: *Method b*, which is a full probabilistic or Level III method, involving for example Monte Carlo simulations, and *Method c*, the first-order reliability method (FORM) or Level II method. In the Level II method, rather than considering failure as a continuous function, as in the Level III method, the reliability is checked by determining the safety index β or probability of failure P_f at only a certain point on the failure surface.

As pointed out by Gulvanessian *et al.* (2002), the term "semi-probabilistic methods (Level I)" used in Figure 8.2, which is taken from EN 1990, is not now generally accepted and may be confusing as it is not defined in EN 1990. It relates to the partial factor design method used in the Eurocodes, where the required target reliability is achieved by applying specified partial factor values to the characteristic values of the basic variables.

Commenting on the different calibration design methods, EN 1990 states that calibration of the Eurocodes and selection of the γ values has been based primarily on *Method a*, i.e. on calibration to a long experience of building tradition. The reason for this is to ensure that the safety levels of structures designed to the Eurocodes are acceptable and similar to existing designs and because *Method c*, which involves full probabilistic analyses, is seldom used to calibrate design codes due to the lack of statistical data. However, EN 1990 states that *Method c* is used for further development of the Eurocodes. Examples of this in the case of Eurocode 7 include the recent publication of some reliability analyses to calibrate the γ values and to assess the reliability of serviceability limit state designs, e.g. Honjo *et al.* (2005).

The partial factor values chosen in the ENV version of Eurocode 7 were originally based on the partial factor values in the Danish Code of Practice for Foundation Engineering (DGI, 1978), because when work started on Eurocode 7 in 1981, Denmark was the only western European country that had a limit state geotechnical design code with partial factors. It is interesting to note that Brinch Hansen (1956) of Denmark was the first to the use the words "limit design" in a geotechnical context. He was also the first to link the limit design concept closely to the concept of partial factors and to introduce these two concepts in Danish foundation engineering practice (Ovesen, 1995) before they were adopted in structural design in Europe.

Table 8.6 Target reliability index, β, and target probability of failure, p_f, values.

Limit state	Target Reliability Index, β		Target Probability of Failure, P_f	
	1 year	50 years	1 year	50 years
ULS	4.7	3.8	1×10^{-6}	7.2×10^{-5}
SLS	2.9	1.5	2×10^{-3}	6.7×10^{-2}

EN 1990 states that if Procedure 2, or a combination of Procedures 1 and 2, is used to calibrate the partial factors, then, for ULS designs, the γ values should be chosen such that the reliability levels for representative structures are as close as possible to the target reliability index, β, value given in Table 8.6, which is 3.8 for 50 years, corresponding to the low failure rate or low probability of failure, $P_f = 7.2 \times 10^{-5}$. For SLS designs the target β value is 1.5, corresponding to a higher probability of failure, $P_f = 6.7 \times 10^{-2}$, than for ULS designs. The ULS and SLS β values are 4.7 and 2.9 for 1 year, corresponding to P_f values of 1×10^{-6} and 2×10^{-3}, respectively. It should be noted that the β values of 3.8 and 4.7 provide the same reliability, the only difference being the different lengths of the reference periods. Structures with higher reliability levels may be uneconomical while those with lower reliability levels may be unsafe. Examples of the use of reliability analyses in geotechnical designs are provided in Sections 8.14 and 8.16.

As mentioned by Gulvanessian et al. (2002), the estimated reliability indices depend on many factors, including the type of geotechnical structure, the loading and the ground conditions and their variability, and consequently they have a wide scatter, as shown in Section 8.17. Also, the results of reliability analyses depend on the assumed theoretical models used to describe the variables and the limit state, as shown by the example in Section 8.18. Furthermore, as noted in EN 1990, the actual frequency of failure is significantly dependent on human error, which is not considered in partial factor design, nor are many unforeseen conditions, such as an unexpected nearby excavation, and hence the estimated β value does not necessarily provide an indication of the actual frequency of failure.

EN 1990 provides for reliability differentiation through the introduction of three consequences classes, high, medium and low, with different β values to take account of the consequences of failure or malfunction of the structure. The partial factors in Eurocode 7 correspond to the medium consequences class. In Eurocode 7, rather than modifying the partial factor values, reliability differentiation is achieved through the use of the Geotechnical Categories described in Section 8.6.

8.13 Design of a spread foundation using the partial factor method

The square pad foundation shown in Figure 8.3, based on Orr (2005), is chosen as an example of a spread foundation designed to Eurocode 7 using the partial factor method. For the ULS design, the consequence of using the different Design Approaches is investigated. The foundation supports a characteristic permanent vertical central load, G_k of 900 kN, which includes the self weight of the foundation, plus a characteristic variable vertical central load, Q_k of 600 kN, and is founded at a depth of 0.8 m in soil with $c_{u;k} = 200$ kPa, $c'_k = 0$, $\phi'_k = 35°$ and $m_{v;k} = 0.015m^2/MN$. The width of the foundation is required to support the given loads without the settlement of the foundation exceeding 25 mm, assuming the groundwater cannot rise above the foundation level.

The ULS design is carried out by checking that the requirement $E_d \leq R_d$ is satisfied. The design action effect, E_d, is equal to the design load, $F_d = \gamma_G G_k + \gamma_Q Q_k$ and obtained using the partial factors given in Table 8.5 for Design Approaches 1, 2 and 3, as appropriate. For drained conditions, the design resistance, R_d is calculated using the following equations for the bearing resistance and bearing factors for drained conditions obtained from Annex D of Eurocode 7:

$$r = R/A = (1/2)B\gamma'N_\gamma s_\gamma + q'N_q s_q \tag{8.13}$$

where:

$$N_q = e^{\pi \tan\phi'} tan(45 + \phi'/2)$$
$$N_\gamma = 2(N_q - 1)\tan\phi'$$
$$s_q = 1 + (B/L)\sin\phi'$$
$$s_\gamma = 1 - 0.3(B/L)$$

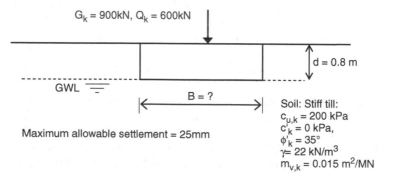

Figure 8.3 Square pad foundation design example.

Table 8.7 Design foundation widths, actions, bearing resistances and overall factors of safety for pad foundation design example for drained conditions.

	B (m)	F_k (kN)	R_k (kN)	$F_d = R_d$	OFS = R_k/F_k
DAI.CI	(1.34)*	1500	2115.0	2115.0	(1.41)*
DAI.C2	1.71	1500	3474.9	1680.0	2.32
DA2	1.56	1500	2961.0	2115.0	1.97
DA3	1.90	1500	4372.6	2115.0	2.92

* Since B for DAI.CI is less than B for DAI.C2, the B for DAI.CI is not the DAI design width.

and applying the partial factors for Design Approaches 1, 2 and 3 given in Table 8.5, either to $\tan\phi'_k$ or to the characteristic resistance, R_k, calculated using the ϕ'_k value, as appropriate. The calculated ULS design widths, B, for the different Design Approaches for drained conditions, are presented in Table 8.7, together with the design resistances, R_d, which are equal to the design loads, F_d. For Design Approach 1, DA1.C2 is the relevant combination as the design width of 1.71 m is greater than the DA1.C1 design width of 1.34 m. For undrained conditions, the DA1.C2 design width is 1.37 m compared to 1.71 m for drained conditions, hence the design is controlled by the drained conditions.

The overall factors of safety, OFS, using each Design Approach, obtained by dividing R_k by F_k, are also given in Table 8.7. The OFS values in Table 8.7 show that, for the ϕ' value and design conditions in this example, Design Approach 3 is the most conservative, giving the highest OFS value of 2.92; Design Approach 2 is the least conservative, giving the lowest OFS value of 1.97, and Design Approach 1 gives an OFS value of 2.32, which is between these values. The relationships between the characteristic and the design values of the resistances and the total loads for each Design Approach are plotted in Figure 8.4. The lines joining the characteristic and design values show how, for each Design Approach starting from the same total characteristic load, $F_k = G_k + Q_k = 900 + 600 = 1500$ kN, the partial factors cause the design resistances to decrease and/or the design loads to increase to reach the design values where $R_d = F_d$. Since this problem is linear with respect to the total load, when the ratio between the permanent and variable loads is constant and when ϕ' is constant, lines of constant OFS, corresponding to lines through the characteristic values for each Design Approach, plot as straight lines on Figure 8.4. These lines show the different safety levels, and hence the different reliability levels, obtained using the different Design Approaches for this square pad foundation design example. It should be noted that the design values given in Table 8.7 and plotted in Figure 8.4 are for ULS design and do not take account of the settlement of the foundation.

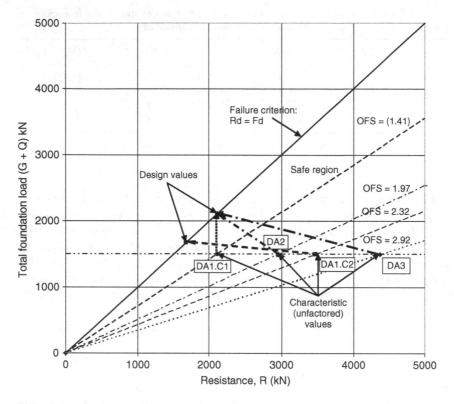

Figure 8.4 OFS values using the different Design Approaches for spread foundation example and relationship between characteristic and design values.

The settlement needs be checked by carrying out an SLS design calculation to ensure that the settlement of the foundation due to the characteristic loading, E_d, does not exceed the allowable settlement, C_d, of 25 mm, i.e. $E_d \leq C_d$. The SLS loads are calculated using the characteristic loads since the SLS partial factors are equal to unity. Two settlement components are considered: the immediate settlement, s_i, and the consolidation settlement, s_c. The immediate settlement is evaluated using elasticity theory and the following equation which has the form of that given in Annex F.2 of Eurocode 7:

$$s_i = p(1 - v_u^2)Bf/E_u \tag{8.14}$$

where p is the SLS net bearing pressure induced at the base level of the foundation, B is the foundation width, E_u is the undrained Young's modulus, v_u is undrained Poisson's ratio (equal to 0.5), and f is a settlement coefficient whose value depends on the nature of the foundation and its stiffness (for the settlement at the center of a flexible square foundation in this example,

$f = 1.12$ is used). For the stiff till, it is assumed that $E_u = 750c_u = 750 \times 200 = 150$ MPa. The consolidation settlement is calculated by dividing the ground below the foundation into layers and summing the settlement of each layer using the following equation:

$$s_c = \mu \Sigma m_v h \Delta \sigma_i' = \mu m_v h \Sigma \alpha_i p \tag{8.15}$$

where m_v is the coefficient of volume compressibility $= 0.015 m^2/MN$ (assumed constant in each layer for the relevant stress level), h is the thickness of the layer, $\Delta \sigma_i'$ is the increment in vertical effective stress in the *ith* layer due to p and is given by $\alpha_i p$ where α_i is the coefficient for the increase in vertical effective stress in the *ith* layer obtained from the Fadum chart, and μ is the settlement reduction coefficient to account for the fact that the consolidation settlement is not one-dimensional (Craig, 2004) and its value is chosen as 0.55 for the square foundation on stiff soil.

The SLS is checked for the DA1.C2 design foundation width $B = 1.71$ m. The SLS bearing pressure is:

$$p = (G_k + Q_k)/A = (900 + 600)/1.71^2 = 513.0 \text{ kPa}$$

The immediate settlement:

$$s_i = p(1 - v_u^2)Bf/E_u = 513.0 \times (1 - 0.5^2)1.71 \times 1.12/150 = 4.9 \text{ mm}$$

To calculate the consolidation settlement, the soil below foundation is divided into four layers, 1.0 m thick. The increases in vertical stress, $\Delta \sigma_1' = \alpha_i p$, at the centers of the layers due to the foundation are estimated from the α_i factors which are 0.90, 0.40, 0.18, and 0.1 giving stress increases of 463, 206, 93, and 51 kPa. Hence the consolidation settlement:

$$s_c = \mu m_v h \Sigma \alpha_i p = 0.55 \times 0.015 \times 1.0 \times \Sigma (0.90 + 0.40 + 0.18 + 0.10)514.6$$

$$= 0.55 \times 0.015 \times 1.0 \times 1.58 \times 514.6 = 6.1 \text{ mm}$$

Total settlement: $s = s_i + s_c = 4.9 + 6.1 = 11.0$ mm < 25 mm.

Therefore the SLS requirement is satisfied for the DA1.C2 foundation width of 1.71 m.

8.14 First-order reliability method (FORM) analyses

In EN 1990, a performance function, g, which is referred to as the safety margin, SM in this section, is defined as:

$$g = SM = R - E \tag{8.16}$$

where R is the resistance and E is the action effect. SM is used in Level II FORM analyses (see Figure 8.2) to investigate the reliability of a design.

Assuming E, R and SM are all random variables with Gaussian distributions, the reliability index is given by:

$$\beta = \mu(SM)/\sigma(SM) \qquad (8.17)$$

where $\mu(SM)$ is the mean value and $\sigma(SM)$ is the standard deviation of SM given by:

$$\mu(SM) = \mu(R) - \mu(E) \qquad (8.18)$$

$$\sigma(SM) = \sqrt{\sigma(R)^2 + \sigma(E)^2} \qquad (8.19)$$

The probability of failure is the probability that $SM < 0$ and is obtained from:

$$P_f = P[SM < 0] = P[R < E] = \Phi(-\beta) = 1 - \Phi(\beta) \qquad (8.20)$$

where Φ denotes the standardized normal cumulative distribution function.

The example chosen to illustrate the use of the FORM analysis is the design, for undrained conditions, of the square foundation shown in Figure 8.3. Performing a FORM analysis for drained conditions is more complex due to the nonlinear expression for the drained bearing resistance factors, N_q and N_γ (see Section 8.13). The action effect and the bearing resistance for undrained conditions are:

$$E = F_u = G + Q \qquad (8.21)$$

$$R = R_u = B^2(N_c c_u s_c + \gamma D) = B^2((\pi + 2)1.2 c_u + \gamma D) \qquad (8.22)$$

If these variables are assumed to be independent, then the means and standard deviations of F_u and R_u are:

$$\mu(F_u) = \mu(G) + \mu(Q) \qquad (8.23)$$

$$\sigma(F_u) = \sqrt{\sigma(G)^2 + \sigma(Q)^2} \qquad (8.24)$$

$$\mu(R_u) = B^2((\pi + 2)1.2\mu(c_u) + \gamma D) \qquad (8.25)$$

$$\sigma(R_u) = B^2((\pi + 2)1.2)\sigma(c_u) \qquad (8.26)$$

In order to ensure the relevance of the comparisons between the partial factor design method and the FORM analysis, the means of all the three random variables are selected so that the characteristic values are equal to the given characteristic values of $G_k = 900$ kN, $Q_k = 600$ kN, and $c_{u:k} = 200$ kPa. The mean values of these parameters, needed in the FORM analysis to determine

Table 8.8 Parameter values for reliability analyses of the spread foundation example.

	Coefficient of variation, V	Mean μ	Standard deviation σ
Actions			
G	0	900 kN	0 kN
Q	0.15	458.7 kN	68.8 kN
Soil parameters			
$\tan\phi'$	0.10	0.737 ($\equiv 36.4°$)	0.074 ($\equiv +/-3°$)
c_u	0.30	235.3 kPa	70.6 kPa
m_v	0.40	0.01875 m^2/MN	0.0075 m^2/MN

the reliability, are calculated from these characteristic values as follows using the appropriate V values given in Table 8.8:

- for the permanent load in this example, it is assumed that $V(G) = 0$, in accordance with EN 1990 as noted in Section 8.8, so that the mean permanent load is equal to the characteristic value since reorganizing Equation (8.7) gives:

$$\mu(G) = G_k/(1 + 1.645\,V(G)) = G_k = 900 \text{ kN} \qquad (8.27)$$

- for the variable load, it is assumed to be normally distributed, as noted in Section 8.8, so that, for an occurrence probability of 0.02 or 2%, the multiplication factor in Equation (8.9) is 2.054 rather than 1.654 for 5%, and hence, for $V(Q) = 0.15$, the mean value of the variable load is given by:

$$\mu(Q_m) = Q_k/(1 + 2.054\,V(Q)) \qquad (8.28)$$
$$\mu(Q) = 600/(1 + 2.054 \times 0.15) = 458.7 \text{ kN}$$

- for material parameters, using Equation (8.9) by Schneider (1997):

$$\mu(X) = X_k/(1 - 0.5\,V(X)) \qquad (8.29)$$
$$\mu(c_u) = c_{u,k}/(1 - 0.5 \times V(X)) = 200/(1 - 0.5 \times 0.30) = 235.3 \text{ kPa}$$

Although not required for the undrained FORM analysis of the square foundation, the mean values of $\tan\phi'$ and m_v, which are used in the Monte Carlo simulations in Section 8.16, are also calculated using Equation (8.29) and are given in Table 8.8, together with the standard deviation values, calculated by multiplying the mean and V values. Hence, assuming Gaussian distributions and substituting the mean G, Q and c_u values in Equations (8.23), (8.25) and (8.18) for $\mu(F_u)$ and $\mu(R_u)$ and $\mu(SM)$ and using the DA1.C1 foundation width of $B = 1.37$ m reported in Section 8.13 for undrained conditions, gives:

$$\mu(F_u) = \mu(G) + \mu(Q) = 900.0 + 458.7 = 1358.7 \text{ kN}$$

$$\mu(R_u) = B^2((\pi + 2)1.2\mu(c_u) + \gamma D)$$

$$= 1.37^2((\pi + 2)1.2 \times 235.3 + 22 \times 0.8) = 2757.8 \text{ kN}$$

$$\mu(SM) = \mu(R_u) - \mu(F_u) = 2757.8 - 1358.7 = 1399.1 \text{ kN}$$

Substituting the standard deviation values for G, Q and c_u in Equations (8.24), (8.26) and (8.19) for $\sigma(F_u)$, $\sigma(R_u)$ and $\sigma(SM)$ gives:

$$\sigma(F_u) = \sqrt{\sigma(G)^2 + \sigma(Q)^2} = \sqrt{0^2 + 68.8^2} = 68.8 \text{ kN}$$

$$\sigma(R_u) = B^2(\pi + 2)1.2\sigma(c_u) = 1.37^2(\pi + 2)1.2 \times 70.6 = 762.9 \text{ kN}$$

$$\sigma(SM) = \sqrt{\sigma(R_u)^2 + \sigma(F_u)^2} = \sqrt{762.9^2 + 68.8^2} = 766.0 \text{ kN}$$

Hence, substituting the values for $\mu(SM)$ and $\sigma(SM)$ in Equation (8.17), the reliability index is:

$$\beta = \mu(SM)/\sigma(SM) = 1399.1/766.0 = 1.83$$

and, from Equation (8.20), the probability of failure is:

$$P_{f;ULS} = \Phi(-\beta) = \Phi(-1.83) = 0.034 = 3.4 \times 10^{-2}.$$

The calculated β value of 1.83 for undrained conditions for a foundation width of 1.37 m is much less than the target ULS β value of 3.8, and hence the calculated probability of failure of 3.4×10^{-2} is much greater than the target value of 7.2×10^{-5}.

It has been seen in Table 8.7 that the design width for drained conditions is $B = 1.71$ m for DA1.C2, and hence this width controls the design. It is therefore appropriate to check the reliability of this design foundation width for drained conditions. Substituting $B = 1.71$ m in Equations (8.25) and (8.26) gives $\beta = 2.30$ and $P_{f;ULS} = 1.06 \times 10^{-2}$. Although this foundation width has a greater reliability, both values are still far from and do not satisfy the target values. The reliability indices calculated in this example may indirectly demonstrate that, in general, geotechnical design equations are usually conservative; i.e. because of model errors they underpredict the reliability index and overpredict the probability of failure. The low calculated reliability indices shown above can arise because the conservative model errors are not included. The reliability index is also affected by uncertainties in the calculation model, as discussed in Section 8.18. However, Phoon and

Kulhawy (2005) found that the mean model factor for undrained bearing capacity at the pile tip is close to unity. If this model factor is applicable to spread foundations, it will not account for the low β values reported above.

The overall factor of safety, OFS, as defined in Section 8.13, is obtained by dividing the characteristic resistance by the characteristic load. For $B = 1.71$ m, $R_k = 1.71^2((\pi + 2) \times 1.2 \times 200 + 22 \times 0.8) = 3659.8$ kN and $F_k = 900 + 600 = 1500$ kN so that OFS $= R_k/F_k = 2418.2/1500 = 2.44$. Thus, using the combination of the recommended partial factors of 1.4 on c_u and 1.3 on Q for undrained conditions with $B = 1.71$ m gives the OFS value of 2.44, which is similar in magnitude to the safety factor often used in traditional foundation design, whereas the calculated probability of failure of 1.06×10^{-2} for this design is far from negligible.

8.15 Effect of variability in c_u on design safety and reliability of a foundation in undrained conditions

To investigate how the variability of the ground properties affects the design safety and reliability of a foundation in undrained conditions, the influence of the variability of c_u on MR, where MR is ratio of the mean resistance to the mean load, $\mu(R_u)/\mu(F_u)$, is first examined. It should be noted that the MR value is not the traditional factor of safety used in deterministic designs, which is normally a cautious estimate of the resistance divided by the load, nor is it the overall factor of safety for Eurocode 7 designs defined in Section 8.13 as OFS $= R_k/F_k$. The influence of the variability of c_u on MR is investigated by varying $V(c_u)$ in the range 0 to 0.40. As in the previous calculations, the characteristic values of the variables G, Q and c_u are kept equal to the given characteristic values so that the following analyses are consistent with the previous partial factor design and the FORM analysis. An increase in $V(c_u)$ causes $\mu(R_u)$ and hence MR to increase, as shown by the graphs of MR against $V(c_u)$ in Figure 8.5 for the two foundation sizes, $B = 1.37$ m and 1.71 m. The increase in MR with increase in $V(c_u)$ is due to the greater difference between the characteristic and mean c_u values as the $V(c_u)$ value increases. For undrained conditions with $B = 1.37$ m and $V(c_u) = 0.3$, the ratio MR $= \mu(R_u)/\mu(F_u) = 2757.8/1358.7 = 2.03$, whereas for $V(c_u) = 0$, when $\mu(R_u) = R_k$, then MR $= R_k/\mu(F_u) = 2349.1/1358.7 = 1.73$.

The fact that MR increases as $V(c_u)$ increases demonstrates the problem with using MR, which is based on the mean R value, as the safety factor and not taking account of the variability of c_u, because it is not logical that the safety factor should increase as $V(c_u)$ increases. This is confirmed by the graphs of β and $P_{f;ULS}$ plotted against $V(c_u)$ in Figures 8.6 and 8.7, which show, as expected, that for a given foundation size, the reliability index, β, decreases and the ULS probability of failure of the foundation, $P_{f;ULS}$, increases as $V(c_u)$ increases. This problem with MR is overcome if the safety factor is calculated using a cautious estimate of the mean R value

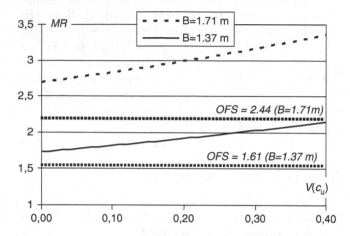

Figure 8.5 Change in mean resistance, mean load ratio, MR with change in $V(c_u)$.

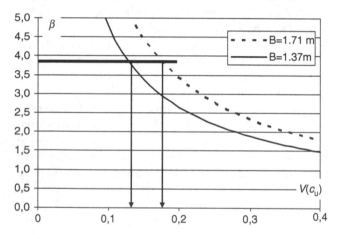

Figure 8.6 Change in reliability index, β, with change in $V(c_u)$.

that takes account of the variability of c_u, for example, using the OFS and the Eurocode 7 characteristic value as defined in Eurocode 7, rather than the mean R value. For undrained conditions with $B = 1.71$ m, OFS $= 2.44$, as calculated above, while for $B = 1.39$ m, OFS $= 1.61$. However, these OFS values are constant as $V(c_u)$ increases, as shown in Figure 8.5, and thus the OFS value provides no information on how the reliability of a design changes as $V(c_u)$ changes. The fact that the OFS value does not provide any information on how the reliability of a design changes as $V(c_u)$ changes demonstrates the importance of, and hence the interest in, reliability modeling.

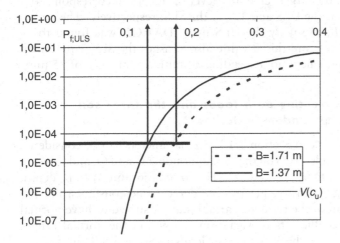

Figure 8.7 Change in probability of failure, P_f, with change in $V(c_u)$.

The graphs in Figures 8.6 and 8.7 show that, for a given foundation size, the reliability index, β decreases and the ULS probability of failure of the foundation for undrained conditions, $P_{f;UL}$, increases as $V(c_u)$ increases. The target reliability index $\beta = 3.8$ and the corresponding target probability of failure, $P_{f;ULS} = 7.2 \times 10^{-5}$, are only achieved if $V(c_u)$ does not exceed 0.17 for $B = 1.71$ m, the design width required for drained conditions, which exceeds that for undrained conditions. Since this $V(c_u)$ value is less than the lowest $V(c_u)$ value of 0.20 typically encountered in practice, as shown by the range of $V(c_u)$ in Table 8.3, this demonstrates that undrained designs to Eurocode 7, such as this spread foundation, have a calculated ULS probability of failure greater than the target value. Choosing the mean $V(c_u)$ value of 0.30 from Table 8.3 for this example with $B = 1.71$ m gives $\beta = 2.30$ with $P_{f;ULS} = 1.06 \times 10^{-2}$.

8.16 Monte Carlo analysis of the spread foundation example

Since a FORM analysis is an approximate analytical method for calculating the reliability of a structure, an alternative and better method is to use a Monte Carlo analysis, which involves carrying out a large number of simulations using the full range of combinations of the variable parameters to assess the reliability of a structure. In this section, the spread foundation example in Section 8.13 is analyzed using Monte Carlo simulations to estimate the reliability level of the design solution obtained by the Eurocode 7 partial factor design method. As noted in Section 8.12 and shown in Table 8.6, the target reliability levels in EN 1990 correspond to β values equal to 3.8 at the ULS and 1.5 at the SLS, respectively. The design foundation width, B, to satisfy the ULS using DA1.C2, was found to be 1.71 m (see Table 8.7), and this B value also satisfies the SLS requirement of the foundation settlement not exceeding a limiting value, C_d, of 25 mm.

8.16.1 Simulations treating both loads and the three soil parameters as random variables

Considering the example in Section 8.13, five parameters can be considered as being random variables, two on the "loading side": G and Q, and three on the "resistance side": $\tan\phi'$, c_u and m_v. The foundation breadth, B, depth, d, and the soil weight density, γ, are considered to be constants. For the sake of simplicity, all of the random variables are assumed to have normal distributions, with coefficients of variation, V, within the normal ranges for these parameters (see the V values for loads in Section 8.8 and for soil parameters in Table 8.3). Since the parameter values given in the example are the characteristic values, the mean values for the reference case need to be calculated, as in the case of the FORM analyses in Section 8.14, using the appropriate V values and Equations (8.27)–(8.29). The mean values and standard deviations were calculated in Section 8.14, using these equations for the five random parameters, and are given in Table 8.8.

When using Monte Carlo simulations, a fixed number of simulations is chosen (typically $N = 10^4$ or 10^5, and then, for each simulation, a set of random variables is generated and the ULS and SLS conditions are checked. It can happen that, during the generation process with normally distributed variables, when V is high, some negative parameter values can be generated. However, these values have no physical meaning. When this occurs, they are taken to be zero and the process is continued. It has been checked that these values have no significant influence on the estimated failure probability. An example of a distribution generated by this process is the cumulative distribution of ϕ' values in Figure 8.8 for the random variable $\tan\phi'$ plotted around the mean value of $\phi' = 36.4°$. Since in Eurocode 7, the ULS criterion to be satisfied is $F_d \le R_d$, it is checked in the ULS simulations that the calculated action $F[G, Q]$ does not exceed the calculated resistance $R[B, \tan\phi']$. The number of occurrences for which $F > R$ is counted (N_{FULS}) and the estimated ULS failure probability, P_{fULS} is the ratio of N_{FULS} to N:

$$P_{fULS} = N_{FULS}/N \tag{8.30}$$

Similarly, since the SLS criterion to be satisfied in Eurocode 7 is $E_d \le C_d$, in the SLS simulations it is checked that the action effect, $E[B, G, Q, c_u, m_v]$ is less than the limiting value of the deformation of the structure at the SLS, C_d. The number of occurrences for which $E > C_d$ is counted (N_{FSLS}) and the

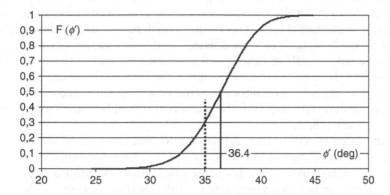

Figure 8.8 Cumulative distribution of generated ϕ' values.

Table 8.9 Estimated and target P_f and β values for spread foundation example.

	Probability of failure, P_f		Reliability Index, β	
	Estimated	Target	Estimated	Target
ULS	1.6×10^{-3}	7.2×10^{-5}	2.96	3.8
SLS	4.9×10^{-3}	6.7×10^{-2}	2.58	1.5

estimated SLS failure probability, P_{fSLS} is the ratio of N_{FSLS} to N:

$$P_{f;SLS} = N_{f;SLS}/N \tag{8.31}$$

The estimated P_f values from the ULS and SLS simulations are 1.6×10^{-3} and 4.9×10^{-3}, respectively, corresponding to estimated β values of 2.96 and 2.58. These values are presented in Table 8.9 together with the target values. These values show that the target reliability is not satisfied at the ultimate limit state, but is satisfied at the serviceability limit state, since $\beta_{ULS} = 2.96$ is less than the ULS target value of 3.8 and $\beta_{SLS} = 2.58$ is greater than the SLS target value of 1.5. This is similar to the result obtained in Section 8.14 when using the FORM analyses to calculate the P_f value for undrained conditions where for $B = 1.71$ m and $V(c_u) = 0.3$, $P_{f;ULS} = 9.69 \times 10^{-3}$ corresponding to $\beta = 2.34$. The influence of variations in the V values for the soil parameter on the P_f value obtained from the Monte Carlo analyses is examined in Section 8.16.

A limitation of the Monte Carlo simulations is that the calculated failure probabilities obtained at the end of the simulations depend not only on the variability of the soil properties, which are focused on here, but also

on the random distribution of the loading. This prevents too general conclusions being drawn about the *exact* safety level or β value corresponding
to a Eurocode design, since these values change if different assumptions are
made regarding the loading. It is more efficient, and more realistic, to analyze the possible ranges of these values, and to investigate the sensitivity of
the design to changes in the assumptions about the loading. For example,
a value of $V = 0.05$ has been chosen for the permanent load when, according
to EN 1990, a constant value equal to the mean value could be chosen as an
alternative (see Section 8.8).

Another limitation concerning the Monte Carlo simulations arises from
the assumption of the statistical independence of the three soil properties.
This assumption is not normally valid in practice, since soil with a high
compressibility is more likely to have a low modulus. It is of course possible
to perform simulations assuming some dependency, but such simulations
need additional assumptions concerning the degree of correlation between
the variables, hence more data, and the generality of the conclusions will
be lost. It should, however, be pointed out that assuming the variables are
independent usually underestimates the probabilities of unfavorable events
and hence is not a conservative assumption.

8.17 Effect of soil parameter variability on the reliability of a foundation in drained conditions

To investigate how the variability of the soil parameters affects the reliability of a drained foundation design and a settlement analysis, simulations
are carried out based on reference coefficients of variation, V_{ref}, for the five
random variables equal to the V values given in Table 8.8 that were used for
the FORM analyses. The sensitivity of the calculated probability of failure
is investigated for variations in the V values of the three soil parameters
around V_{ref} in the range V_{min} to V_{max} shown in Table 8.10. All the probability estimates are obtained with at least $N = 40,000$ simulations, this
number being doubled in the case of small probabilities. All computations
are carried out for the square foundation with $B = 1.71$ m, resulting from the
partial factor design using DA1.C2. For each random variable, the mean and
standard deviations of the Gaussian distribution are chosen so that, using
Equation 8.29, they are consistent with the given characteristic value, as

Table 8.10 Range of variation of V values for the three soil parameters.

	V_{min}	V_{ref}	V_{max}
$\tan\phi'$	0.05	0.10	0.21
c_u (kPa)	0.15	0.30	0.45
m_v (m²/kN)	0.20	0.40	0.60

explained in Section 8.14. The consequence of this is that, as V increases, the mean and standard deviation have both to increase to keep the characteristic value constant and equal to the given value.

The influence of the V value for each soil parameter on the reliability of the spread foundation is investigated by varying the V value for one parameter while keeping the V values for the other parameters constant. The results presented are:

- the influence of $V(\tan\phi')$ on the estimated $P_{f;ULS}$ and the estimated β_{ULS} in Figures 8.9 and 8.10;
- the influence of $V(m_v)$ on the estimated $P_{f;SLS}$ and the estimated β_{SLS} in Figures 8.11 and 8.12.

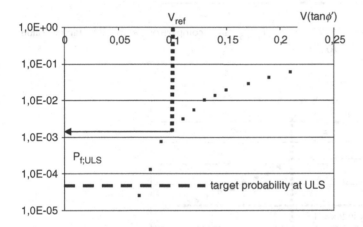

Figure 8.9 Influence of $V(\tan\phi')$ on $P_{f;ULS}$ and comparison with ULS target value ($P_f = 7.2 \times 10^{-5}$).

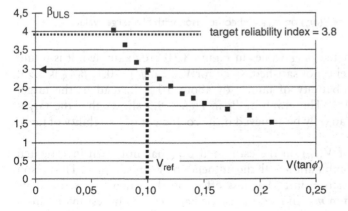

Figure 8.10 Influence of $V(\tan\phi')$ on β_{ULS} and comparison with ULS target value.

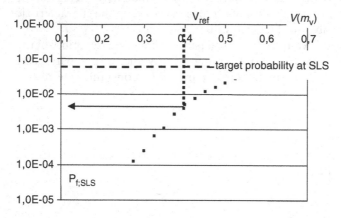

Figure 8.11 Influence of $V(m_v)$ on $P_{f;SLS}$ and comparison with SLS target value ($P_f = 6.7 \times 10^{-2}$).

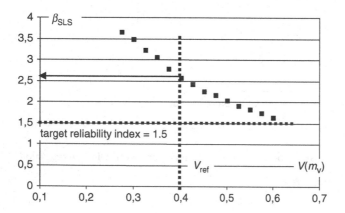

Figure 8.12 Influence of $V(m_v)$ on β_{SLS} and comparison with SLS target value.

When the estimated β_{ULS} values in Figure 8.10 are examined, it is found that the target level is not satisfied, since for $V_{ref}(\tan\phi') = 0.1$, β_{ULS} is 2.95, which gives a probability of failure of about 10^{-3} instead of the target value of 7.2×10^{-5}. The graph of P_f in Figure 8.9 shows that the target probability level can only be achieved if the coefficient of variability of $\tan\phi'$ is less than 0.08.

The influence of $V(c_u)$ on the estimated β_{SLS} has not been investigated directly, only in combination with the influence of $V(m_v)$ on β_{SLS}. The reason is because the variation in c_u has less effect on the calculated settlements than the variation in m_v. In fact, the immediate settlement, estimated from

$E_u = 750c_u$, is only $4.9/11.0 = 45\%$ of the total deterministic settlement calculated in Section 8.13, and c_u is less variable than m_v. Hence the variability of m_v would normally be the main source of SLS uncertainty. However, the situation is complicated by the fact that the immediate settlement is proportional to $1/c_u$, which is very large if c_u is near zero (or even negative if a negative value is generated for c_u) for simulations when $V(c_u)$ is very large. This results from the assumption of a normal distribution for c_u and a practical solution has been obtained by rejecting any simulated value lower than 10 kPa. Another way to calculate the settlement would have been to consider a log-normal distribution of the parameters. In all simulations, the coefficients of variation of both c_u and m_v were varied simultaneously, with $V(c_u) = 0.75V(m_v)$. This should be borne in mind when interpreting the graphs of the estimated $P_{f;SLS}$ and β_{SLS} values plotted against $V(m_v)$ in Figures 8.11 and 8.12. These show that the target $P_{f;SLS}$ and β_{SLS} values are achieved in all cases, even with $V(m_v) = 0.6$, and that when the variabilities are at their maximum, i.e. when $V(m_v) = 0.6$ and $V(c_u) = 0.45$, then $P_{f;SLS} = 5.15 \times 10^{-2}$ and $\beta_{SLS} = 1.63$.

8.18 Influence of model errors on the reliability levels

In this section, other components of safety and the homogeneity of safety and code calibration are considered. The simulations considered in this section correspond to those using the reference set of coefficients of variation in Table 8.10, which give $\beta_{SLS} = 2.58$ and $\beta_{ULS} = 2.96$ presented in Table 8.9, and discussed in Section 8.16.

As noted in Section 8.11, the partial factors in Eurocode 7 account for uncertainties in the models for the loads and resistances as well as for uncertainties in the loads, material properties and resistances themselves. Consider, for example, the model (Equation (8.15)) used to calculate the consolidation settlement of the foundation example in Section 8.13. Until now, the way the settlement was calculated has not been examined, and the calculation model has been assumed to be correct. In fact, the consolidation settlement has been calculated as the sum of the settlements of a limited number of layers below the foundation, the stress in each layer being deduced from the Fadum chart. If there are i layers, summing the coefficients of influence, α_i for each layer gives a global coefficient $\alpha = 1.58$ for the foundation settlement. Also the reduction coefficient, μ, has been taken equal to 0.55, thus the product $\alpha\mu$ is equal to $1.58 \times 0.55 = 0.87$.

Thus the consolidation settlement calculation model has some uncertainties, for instance because of the influence of the layers at greater depths and because μ is an empirical coefficient whose value is chosen more on the basis of "experience" than by relying on a reliable theoretical model. For instance, if the chosen μ value is changed from 0.55 to 0.75 in a first step and then to

1.0 in a second step, the β_{SLS} value obtained from the simulations reduces from 2.58 to 2.16 in the first step and to 1.27 in the second step, this last value being well below the target β_{SLS} value.

On the other hand, if the maximum allowable settlement is only 20 mm, instead of 25 mm, and if α and μ are unchanged, then β_{SLS} decreases from 2.58 to 2.06. These two examples illustrate the very high sensitivity of the reliability index to the values of the coefficients used in the model to calculate the settlement and to the value given or chosen for the limiting settlement. For the foundation example considered above, this sensitivity is larger than that due to a reasonable change in the coefficients of variation of the soil properties.

8.19 Uniformity and consistency of reliability

As noted in Section 8.4, the aim of Eurocode 7 is to provide geotechnical designs with the appropriate degrees of reliability. This means that the chosen target probability of failure should take account of the mode of failure, the consequences of failure and the public aversion to failure. It thus can vary from one type of structure to another, but it should also be as uniform as possible for all similar components within a given geotechnical category; for example, for all spread foundations, for all piles and for all slopes. It should also be uniform (consistent) for different sizes of a given type of structure. As pointed out in Section 8.12, this is achieved for ULS designs by choosing γ values such that the reliability levels for representative structures are as close as possible to the specified target reliability index β_{ULS} value of 3.8.

To achieve the required reliability level, the partial factors need to be calibrated. As part of this process, the influence of the size of the foundation on the probability of failure is investigated. The simplest way to carry out this is to change the magnitude of the loading while keeping all the other parameters constant. Thus, a scalar multiplying factor, λ, is introduced and the safety level of foundations designed to support the loading $\lambda(G_k + Q_k)$ is analyzed for drained conditions as the load factor λ varies. For nine λ values in the range 0.4–2.0, the design foundation width, B, is determined for DA1.C2, and constant $(G_k + Q_k)$ and soil parameter values. The resulting graphs of foundation size, B, and characteristic bearing pressure, p_k, increase monotonically as functions of λ, as shown by the curves on Figure 8.13.

The characteristic bearing pressure, p_k, increases with load factor, λ, since the characteristic load $\lambda(G_k + Q_k)$ increases at a rate greater than B^2. This occurs because B increases at a smaller rate than the load factor, as can be seen from the graph of B against λ in Figure 8.13. Examination of the graphs in Figure 8.13 shows that variation of λ in the range 0.4–2.0 results in B varying in the range 1.13–2.33 m and p_k varying in the range 472–553 kPa.

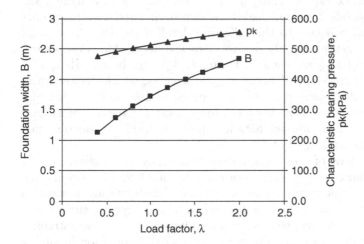

Figure 8.13 Variation in foundation width, B (m) and characteristic bearing pressure, p, with load factor λ.

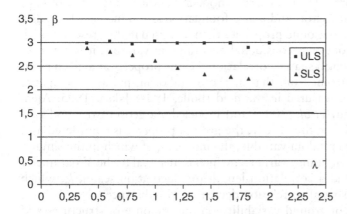

Figure 8.14 Graphs showing variation in reliability index, β, with load factor, λ, for ULS and SLS.

The variation in p_k corresponds to a variation between -8% and $+8\%$ around the 514 kPa value obtained when $\lambda = 1$.

Monte Carlo simulations, performed for the same nine λ values as above and using the V values in Table 8.8 for the five load and soil parameters treated as random variables, result in estimates of the reliability indices for the ULS and SLS for each λ value. Graphs of the estimated β_{ULS} and β_{SLS} values obtained from the simulations are plotted in Figure 8.14. These graphs show that β_{ULS} remains constant as λ varies and remains slightly below 3.0 for all λ values. Thus, for this example and this loading situation, the Eurocode 7 semi-probabilistic approach provides designs that are consistent with regard to the reliability level at the ULS (any small variations in the graph are only due to statistical noise). However, examining the graph of β_{SLS} against λ in Figure 8.14, it can be seen that, as B increases, β_{SLS} decreases, and hence $P_{f;SLS}$ increases, which demonstrates that the Eurocode 7 semi-probabilistic approach does not provide SLS designs with a consistent reliability level. In this example, the reliability level decreases as λ increases, although for λ less than 2.5, the β_{SLS} values obtained remain above the target level of 1.5.

The reliability level of geotechnical designs depends on the characteristic soil property value chosen, which, as noted in Section 8.9, is a cautious estimate of the mean value governing the occurrence of the limit state. It was also noted that the failure of a large foundation involving a large slip surface or a large volume of ground will be governed by a higher characteristic strength value than the value governing the failure of a small foundation involving a small volume of ground. Consequently, depending on the uniformity of the ground with depth, the ULS reliability level may actually increase as the foundation size increases, rather than remaining constant, as indicated by the β_{ULS} graph in Figure 8.14. The correct modeling of the change in both β_{ULS} and β_{SLS} as the load factor, and hence foundation size is increased, needs a more refined model of ground properties than that used in the present model.

If the ground property is considered as a random variable, it has also to be described as having a spatial correlation, i.e. the property at a given point is only slightly different from that at the neighboring points. The spatial correlation can be measured in the field (Bolle, 1994; Jaksa, 1995; Alén, 1998; Moussouteguy et al., 2002) and modeled via geostatistics and random field theory. Using this theory, the ground property is considered as a spatially distributed random variable, the modeling of which in the simpler cases requires the long-range variance, which is identical to the V parameter discussed above, and a correlation length (or fluctuation scale), l_c, which describes how the spatial correlation decreases with distance.

Thus the effects of ground variability on the response of structures and buildings resting on the ground can be predicted. However, it is not possible to summarize how ground variability or uncertainty in the geotechnical data affects the response, such as the settlement, of structures and buildings, since the spatial variation can induce soil–structure interaction patterns that are specific to particular ground conditions and structures. A theoretical investigation by Breysse et al., (2005) has shown that the response of a structure is governed by a dimensional ratio obtained by dividing a characteristic dimension of the structure, for instance the distance between the foundation supports or a pipe length, by the soil correlation length. Eurocode 7 requires the designer to consider the effects of soil variability. However, soil–structure interaction analyses to take account of the effects of soil variability are not a part of routine geotechnical designs, i.e. Geotechnical Category 2. Situations where the soil variability needs to be included in the analyses are Geotechnical Category 3 and hence no specific guidance is provided in Eurocode 7 for such situations, since they often involve nonlinear constitutive laws for the materials. An example of such a situation is where the effects of soil variability on the reliability of a piled raft have been analyzed and confirmed by Bauduin (2003) and Niandou and Breysse (2005). However, providing details of these analyses is beyond the scope of this chapter.

8.20 Conclusions

This chapter explains the reliability basis for Eurocode 7, the new European standard for geotechnical design. Geotechnical designs to Eurocode 7 aim to achieve the required reliability for a particular geotechnical structure by taking into account all the relevant factors at all stages of the design, by considering all the relevant limit states and by the use of appropriate design calculations. For ultimate limit state design situations, the required level of reliability is achieved by applying appropriate partial factors to characteristic values of the loads and the soil parameters or resistances. For serviceability limit state design situations, the required level of reliability is achieved by using characteristic loads and soil parameter values, with all the partial factor values equal to unity. The ultimate limit state partial factors used in design calculations to Eurocode 7 are selected on the basis of long experience and may be calibrated using reliability analyses. Whether partial factors are applied to the soil parameters or the resistances has given rise to three Design Approaches with different sets of partial factors, which result in different designs and hence different reliabilities, depending on the Design Approach adopted. The reliability of geotechnical designs has been shown to be significantly dependent on the variability of the soil parameters and on the assumptions in the calculation model. Traditional factors of safety do not take account of the variability of the soil and hence cannot provide a reliable assessment of the probability of failure or actual safety.

The advantage of reliability analyses is that they take account of the variability of the parameters involved in a geotechnical design and so present a consistent analytical framework linking the variability to the target probability of failure and hence providing a unifying framework between geotechnical and structural designs (Phoon et al., 2003). It has been shown that, due to uncertainty in the calculation models and model errors, reliability analyses tend to overpredict the probability of geotechnical designs. Reliability analyses are particularly appropriate for investigating and evaluating the probability of failure in the case of complex design situations or highly variable ground conditions and hence offer a more rational basis for geotechnical design decisions in such situations.

References

Alén, C. (1998). On probability in geotechnics. Random calculation models exemplified on slope stability analysis and ground–superstructure interaction. PhD thesis, Chalmers University, Sweden.

Bauduin, C. (2003). Assessment of model factors and reliability index for ULS design of pile foundations. In *Proceedings Conference BAP IV, Deep Foundation on Bored & Auger Piles, Ghent 2003*, Ed. W. F. Van Impe Millpress, Rotterdam, pp. 119–35.

Bolle, A. (1994). How to manage the spatial variability of natural soils. In *Probabilities and Materials*, Ed. D. Breysse. Kluwer, Dordrecht, pp. 505–16.

Breysse, D., Kouassi, P. and Poulain, D. (1999). Influence of the variability of compacted soils on slope stability of embankments. In *Proceedings International Conference on the Application of Statistics and Probability, ICASP 8*, Sydney, Eds. R. E. Melchers and M. G. Stewart. Balkema, Rotterdam, 1, pp. 367–73.

Breysse, D., Niandou, H., Elachachi, S. M. and Houy, L. (2005). Generic approach of soil–structure interaction considering the effects of soil heterogeneity. *Geotechnique*, LV(2), pp. 143–50.

Brinch Hansen, J. (1956). Limit design and safety factors in soil mechanics (in Danish with an English summary). In *Bulletin No. 1*, Danish Geotechnical Institute, Copenhagen.

Cherubini, C. and Orr, T. L. L. (1999). Considerations on the applicability of semi-probabilistic Bayesian methods to geotechnical design. In *Proceedings XX Convegno Nazionale di Geotechnica*, Parma, Associatione Geotecnica Italiana, pp. 421–6.

Craig, R. F. (2004). *Craig's Soil Mechanics*. Taylor & Francis, London.

DGI (1978). *Danish Code of Practise for Foundation Engineering*. Danish Geotechnical Institute, Copenhagen.

Duncan, J. M. (2000). Factors of safety and reliability in geotechnical engineering. *Journal of Geotechnical and Geoenvironmental Engineering, ASCE*, 126(4), 307–16.

EC (2002). *Guidance Paper L – Application and Use of Eurocodes*. European Commission, Brussels.

Gulvanessian, H., Calgaro, J.-A., and Holický, M. (2002). *Designers' Guide to EN 1990 – Eurocode 1990: Basis of Structural Design*. Thomas Telford, London.

Honjo, Y., Yoshida, I., Hyodo, J. and Paikowski, S. (2005). Reliability analyses for serviceability and the problem of code calibration. In *Proceedings International Workshop on the Evaluation of Eurocode 7*. Trinity College, Dublin, pp. 241–9.

Jaksa, M. B. (1995). The influence of spatial variability on the geotechnical design properties of a stiff, overconsolidated clay. PhD thesis, University of Adelaide, Australia.

Moussouteguy, N., Breysse, D. and Chassagne, P. (2002). Decrease of geotechnical uncertainties via a better knowledge of the soil's heterogeneity. *Revue Française de Génie Civil*, 3, 343–54.

Niandou, H. and Breysse, D. (2005). Consequences of soil variability and soil–structure interaction on the reliability of a piled raft. In *Proceedings of International Conference on Statistics, Safety and Reliability ICOSSAR'05*, Rome, Eds. G. Augusti, G. I. Schuëller and M. Ciampoli. Millpress, Rotterdam, pp. 917–24.

Orr, T. L. L. (1994). Probabilistic characterization of Irish till properties. In *Risk and Reliability in Ground Engineering*. Thomas Telford, London, pp. 126–33.

Orr, T. L. L. (2005). Design examples for the Eurocode 7 Workshop. In *Proceedings of the International Workshop on the Evaluation of Eurocode 7*. Trinity College, Dublin, pp. 67–74.

Orr, T. L. L. and Farrell, E. R. (1999). *Geotechnical Design to Eurocode 7*. Springer, London.

Ovesen, N. K. (1995). Eurocode 7: A European code of practise for geotechnical design. In *Proceedings International Symposium on Limit State Design in Geotechnical Engineering, ISLSD 93*, Copenhagen, Danish Geotechnical Institute, 3, pp. 691–710.

Ovesen, N. K. and Denver, H. (1994). Assessment of characteristic values of soil parameters for design. In *Proceedings XIII International Conference on Soil Mechanics and Foundation Engineering*, New Delhi. Balkema, 1, 437–60.

Phoon, K. K. and Kulhawy, F. H. (2005). Characterization of model uncertainties for drilled shafts under undrained axial loading. In *Contemporary Issues in Foundation Engineering (GSP 131)*, Eds J. B. Anderson, K. K. Phoon, E. Smith, J. E. Loehr. ASCE, Reston.

Phoon, K. K., Becker, D. E., Kulhawy, F. H., Honjo, Y., Ovesen, N. K. and Lo, S. R. (2003). Why consider reliability analysis for geotechnical limit state design. In *Proceedings International Workshop on Limit State Design in Geotechnical Engineering Practise*, Eds K. K. Phoon, Y. Honjo, and D. Gilbert. World Scientific Publishing Company, Singapore, pp. 1–17.

Schneider, H. R. (1997). Definition and determination of characteristic soil properties. In *Proceedings XII International Conference on Soil Mechanics and Geotechnical Engineering*, Hamburg. Balkema, Rotterdam, pp. 4, 2271–4.

Student (1908). The probable error of a mean. *Biometrica*, 6, 1–25.

Chapter 9

Serviceability limit state reliability-based design

Kok-Kwang Phoon and Fred H. Kulhawy

9.1 Introduction

In foundation design, the serviceability limit state often is the governing criterion, particularly for large-diameter piles and shallow foundations. In addition, it is widely accepted that foundation movements are difficult to predict accurately. Most analytical attempts have met with only limited success, because they did not incorporate all of the important factors, such as the in-situ stress state, soil behavior, soil–foundation interface characteristics, and construction effects (Kulhawy, 1994). A survey of some of these analytical models showed that the uncertainties involved in applying each model and evaluating the input parameters were substantial (Callanan and Kulhawy, 1985). Ideally, the ultimate limit state (ULS) and the serviceability limit state (SLS) should be checked using the same reliability-based design (RBD) principle. The magnitude of uncertainties and the target reliability level for SLS are different from those of ULS, but these differences will be addressed, consistently and rigorously, using reliability-calibrated deformation factors (analog of resistance factors).

One of the first systematic studies reporting the development of RBD charts for settlement of single piles and pile groups was conducted by Phoon and co-workers (Phoon *et al.*, 1990; Quek *et al.*, 1991, 1992). A more comprehensive implementation that considered both ULS and SLS within the same RBD framework was presented in EPRI Report TR-105000 (Phoon *et al.*, 1995). SLS deformation factors (analog of resistance factors) were derived by considering the uncertainties underlying nonlinear load–displacement relationships explicitly in reliability calibrations. Nevertheless, the ULS still received most of the attention in later developments, although interest in applying RBD for the SLS appears to be growing in the recent literature (e.g. Fenton *et al.*, 2005; Misra and Roberts, 2005; Zhang and Ng, 2005; Zhang and Xu, 2005; Paikowsky and

Lu, 2006). A wider survey is given elsewhere (Zhang and Phoon, 2006). A special session on "Serviceability Issues in Reliability-Based Geotechnical Design" was organized under the Risk Assessment and Management Track at Geo-Denver 2007, the ASCE Geo-Institute annual conference held in February 2007, in Denver, Colorado. Four papers were presented on excavation-induced ground movements (Boone, 2007; Finno, 2007; Hsiao et al., 2007; Schuster et al., 2007), and two papers were presented on foundation settlements (Roberts et al., 2007; Zhang and Ng, 2007).

The reliability of a foundation at the ULS is given by the probability of the capacity being less than the applied load. If a consistent load test interpretation procedure is used, then each measured load–displacement curve will produce a single "capacity." The ratio of this measured capacity to a predicted capacity is called a bias factor or a model factor. A lognormal distribution usually provides an adequate fit to the range of model factors found in a load test database (Phoon and Kulhawy, 2005a, b). It is natural to follow the same approach for the SLS. The capacity is replaced by an allowable capacity that depends on the allowable displacement (Phoon et al., 1995; Paikowsky and Lu, 2006). The distribution of the SLS bias or model factor can be established from a load test database in the same way. The chief drawback is that the distribution of this SLS model factor has to be re-evaluated when a different allowable displacement is prescribed. It is tempting to argue that the load–displacement behavior is linear at the SLS, and subsequently the distribution of the model factor for a given allowable displacement can be extrapolated to other allowable displacements by simple scaling. However, for reliability analysis, the applied load follows a probability distribution, and it is possible for the load–displacement behavior to be nonlinear at the upper tail of the distribution (corresponding to high loads). Finally, it may be more realistic to model the allowable displacement as a probability distribution, given that it is affected by many interacting factors, such as the type and size of the structure, the properties of the structural materials and the underlying soils, and the rate and uniformity of the movement. The establishment of allowable displacements is given elsewhere (Skempton and MacDonald, 1956; Bjerrum, 1963; Grant et al., 1974; Burland and Wroth, 1975; Burland et al., 1977; Wahls, 1981; Moulton, 1985; Boone, 2001).

This chapter employs a probabilistic hyperbolic model to perform reliability-based design checks at the SLS. The nonlinear features of the load–displacement curve are captured by a two-parameter, hyperbolic curve-fitting Equation. The uncertainty in the entire load–displacement curve is represented by a relatively simple bivariate random vector containing the hyperbolic parameters as its components. It is not necessary to evaluate a new model factor and its accompanying distribution for each allowable displacement. It is straightforward to introduce the allowable displacement as a random variable for reliability analysis. To accommodate this additional source of uncertainty using the standard model factor approach, it is necessary to determine a conditional probability distribution of the model factor as a continuous function of the allowable displacement. However, the approach proposed in this chapter is significantly simpler. Detailed calculation steps for simulation and first-order reliability analysis are illustrated using EXCEL™ to demonstrate this simplicity. It is shown that a random vector containing two curve-fitting parameters is the simplest probabilistic model for characterizing the uncertainty in a nonlinear load–displacement curve and should be the recommended first-order approach for SLS reliability-based design.

9.2 Probabilistic hyperbolic model

The basic idea behind the probabilistic hyperbolic model is to: (a) reduce the measured load–displacement curves into two parameters using a hyperbolic fit, (b) normalize the resulting hyperbolic curves using an interpreted failure load to reduce the data scatter, and (c) model the remaining scatter using an appropriate random vector (possibly correlated and/or non-normal) for the curve-fitting hyperbolic parameters. There are three advantages in this empirical approach. First, this approach can be applied easily in practice, because it is simple and does not require input parameters that are difficult to obtain. However, the uncertainties in the stiffness parameters are known to be much larger than the uncertainties in the strength parameters (Phoon and Kulhawy, 1999). Second, the predictions are realistic, because the approach is based directly on load test results. Third, the statistics of the model can be obtained easily, as described below. The first criterion is desirable but not necessary for RBD. The next two criteria, however, are crucial for RBD. If the average model bias is not addressed, different models will produce different reliability for the same design scenario as a result of varying built-in conservatism, which makes reliability calculations fairly meaningless. The need for robust model statistics should be evident, because no proper reliability analysis can be performed without these data.

9.2.1 Example: Augered Cast-in-Place (ACIP) pile under axial compression

Phoon et al. (2006) reported an application of this empirical approach to an ACIP pile load test database. The ACIP pile is formed by drilling a continuous flight auger into the ground and, upon reaching the required depth, pumping sand–cement grout or concrete down the hollow stem as the auger is withdrawn steadily. The side of the augered hole is supported by the soil-filled auger, and therefore no temporary casing or slurry is needed. After reaching the ground surface, a reinforcing steel cage, if required, is inserted by vibrating it into the fresh grout. The database compiled by Chen (1998) and Kulhawy and Chen (2005) included case histories from 31 sites with 56 field load tests conducted mostly in sandy cohesionless soils. The diameter (B), depth (D), and D/B of the piles ranged from 0.30 to 0.61 m, 1.8 to 26.1 m, and 5.1 to 68.2, respectively. The associated geotechnical data typically are restricted to the soil profile, depth of ground water table, and standard penetration test (SPT) results. The range of effective stress friction angle is 32–47 degrees and the range of horizontal soil stress coefficient is 0.6–2.4 (established by correlations). The load–displacement curves are summarized in Figure 9.1a.

The normalized hyperbolic curve considered in their study is expressed as:

$$\frac{Q}{Q_{STC}} = \frac{y}{a + by} \tag{9.1}$$

in which Q = applied load, Q_{STC} = failure load or capacity interpreted using the slope tangent method, a and b = curve-fitting parameters, and y = pile butt displacement. Note that the curve-fitting parameters are physically meaningful, with the reciprocals of a and b equal to the initial slope and asymptotic value, respectively (Figure 9.2). The slope tangent method defines the failure load at the intersection of the load–displacement curve with the initial slope (not elastic slope) line offset by $(0.15 + B/120)$ inches, in which B = shaft diameter in inches. Phoon et al. (2006) observed that

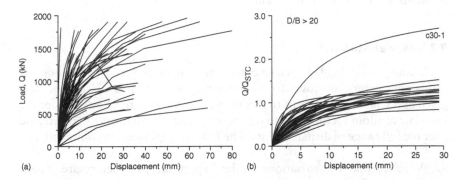

Figure 9.1 (a) Load–displacement curves for ACIP piles in compression and (b) fitted hyperbolic curves normalized by slope tangent failure load (Q_{STC}) for $D/B > 20$. From Phoon et al., 2006. Reproduced with the permission of the American Society of Civil Engineers.

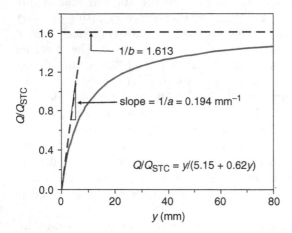

Figure 9.2 Definition of hyperbolic curve-fitting parameters.

the normalized load–displacement curves for long piles ($D/B > 20$) cluster together quite nicely (Figure 9.1b). Case 30-1 is a clear outlier within this group and is deleted from subsequent analyses. There are 40 load tests in this group of "long" piles, with diameter (B), depth (D), and D/B ranging from 0.30 to 0.61 m, 7.5 to 26.1 m, and 21.0 to 68.2, respectively. The range of effective stress friction angle is 32–43 degrees and the range of horizontal soil stress coefficient is 0.6–2.1.

The curve-fitting Equation is empirical and other functional forms can be considered, as shown in Table 9.1. However, the important criterion is to apply a curve-fitting Equation that produces the least scatter in the normalized load–displacement curves.

9.2.2 Model statistics

Each continuous load–displacement curve can be reduced to two curve-fitting parameters, as illustrated in Table 9.2 for ACIP piles. Using the data given in the last two columns of this table, one can construct an appropriate bivariate random vector that can reproduce the scatter in the normalized load over the full range of displacements. The following steps are recommended.

1. Validate that the variations in the hyperbolic parameters are indeed "random."
2. Fit the marginal distributions of a and b using a parametric form (e.g. lognormal distribution) or a non-parametric form (e.g. Hermite polynomial).

Table 9.1 Curve-fitting Equations for load–displacement curves from various foundations.

Foundation type	Loading mode	Curve-fitting Equation*	Reference source
Spread foundation	Uplift	$Q/Q_m = y/(a+by)$	Phoon et al., 1995, 2003, 2007; Phoon and Kulhawy, 2002b
Drilled shaft	Uplift	$Q/Q_m = y/(a+by)$	Phoon et al., 1995; Phoon and Kulhawy, 2002a
	Compression (undrained)	$Q/Q_m = (y/B)/[a+b(y/B)]$	Phoon et al., 1995
	Compression (drained)	$Q/Q_m = a(y/B)^b$	Phoon et al., 1995
	Lateral-moment	$Q/Q_m = (y/D)/[a+(y/D)]$	Phoon et al., 1995, Kulhawy and Phoon, 2004
Augered cast-in-place (ACIP) pile	Compression (drained)	$Q/Q_m = y/(a+by)$	Phoon et al., 2006
Pressure-injected footing	Uplift (drained)	$Q/Q_m = y/(a+by)$	Phoon et al., 2007

*Interpreted failure capacity (Q_m): tangent intersection method for spread foundations, L_1–L_2 method for drilled shafts under uplift and compression; hyperbolic capacity for drilled shafts under lateral-moment loading; slope tangent method for ACIP piles and pressure-injected footings.

3. Calculate the product–moment correlation between a and b using:

$$\rho_{a,b} = \frac{\sum_{i=1}^{n} (a_i - \bar{a})(b_i - \bar{b})}{\sqrt{\sum_{i=1}^{n} (a_i - \bar{a})^2 \sum_{i=1}^{n} (b_i - \bar{b})^2}} \tag{9.2}$$

in which (a_i, b_i) denotes a pair of a and b values along one row in Table 9.2, n = sample size (e.g. $n = 40$ in Table 9.2), and $\bar{a}$ and $\bar{b}$ = the following sample means:

$$\bar{a} = \frac{\sum_{i=1}^{n} a_i}{n}$$

$$\bar{b} = \frac{\sum_{i=1}^{n} b_i}{n} \tag{9.3}$$

Table 9.2 Hyperbolic curve-fitting parameters for ACIP piles under axial compression [modified from load test data reported by Chen (1998)].

Case no.	B (m)	D (m)	D/B	SPT-N*	Q_{STC} (kN)	a (mm)	b
c1	0.41	19.2	47.3	18.4	1957.1	8.382	0.499
c2	0.41	14.9	36.8	23.4	1757.0	10.033	0.382
c3	0.41	10.1	24.8	39.1	1761.4	3.571	0.679
c4	0.30	10.7	35.0	16.6	367.0	4.296	0.615
c5-1	0.41	17.7	43.5	21.4	2081.7	7.311	0.445
c5-2	0.41	19.8	48.8	14.6	1903.7	5.707	0.591
c6	0.41	13.7	33.8	14.6	729.5	3.687	0.679
c7	0.30	11.8	38.7	11.6	685.0	2.738	0.701
c8-1	0.36	12.0	33.9	23.7	1072.0	3.428	0.723
c9	0.36	10.4	29.2	18.4	1654.7	7.559	0.461
c10	0.41	9.1	22.5	29.0	1467.8	5.029	0.582
c11	0.36	11.9	33.4	31.6	1654.7	2.528	0.744
c12	0.41	13.1	32.3	15.3	1103.1	5.260	0.533
c13	0.36	11.6	32.6	35.8	1605.7	4.585	0.632
c14	0.41	11.8	29.0	4.3	889.6	2.413	0.770
c15	0.36	12.2	34.3	21.6	2490.9	15.646	0.328
c16-1	0.41	13.4	33.0	27.0	1067.5	4.572	0.607
c16-2	0.41	13.4	33.0	26.7	1076.4	7.376	0.613
c17	0.41	9.0	22.1	14.2	551.6	2.047	0.682
c18-1	0.36	13.9	39.0	34.4	3380.5	12.548	0.380
c19-1	0.36	19.2	54.0	27.1	1334.4	5.829	0.570
c19-2	0.36	24.2	68.2	21.0	1183.2	6.419	0.559
c19-3	0.41	26.1	64.1	20.0	1227.6	4.907	0.690
c19-4	0.41	18.6	45.8	22.4	2152.8	4.426	0.692
c20	0.52	16.6	31.9	30.4	3736.3	1.312	0.840
c21	0.36	7.8	21.9	16.5	889.6	8.128	0.510
c22-1	0.61	20.0	32.8	11.4	1779.2	10.297	0.640
c22-2	0.41	14.9	36.8	7.7	951.9	4.865	0.487
c22-3	0.41	19.9	49.0	10.0	1174.3	5.599	0.517
c23-1	0.36	7.5	21.0	22.7	1316.6	4.511	0.622
c24-1	0.41	9.5	23.4	11.7	1183.2	3.142	0.501
c24-2	0.41	9.5	23.4	11.7	1174.3	3.353	0.513
c24-3	0.41	9.5	23.4	21.5	1005.2	4.104	0.367
c24-4	0.41	9.0	22.1	8.0	862.9	4.928	0.430
c25-1	0.61	17.5	28.7	18.8	1067.5	1.006	0.780
c25-2	0.61	17.5	28.7	18.8	1263.2	1.237	0.923
c25-3	0.61	16.0	26.3	18.2	1076.4	1.352	0.932
c25-4	0.61	20.0	32.8	18.4	1112.0	4.128	1.050
c25-5	0.61	17.5	28.7	18.8	1547.9	4.862	0.748
c27-3	0.46	12.1	26.4	14.4	640.5	3.054	0.792

*Depth-averaged SPT-N value along shaft.

For the hyperbolic parameters in Table 9.2, the product–moment correlation is −0.673.

4. Construct a bivariate probability distribution for a and b using the translation model.

Note that the construction of a bivariate probability distribution for a and b is not an academic exercise in probability theory. Each pair of a and b values within one row of Table 9.2 is associated with a single load–displacement curve [see Equation (9.1)]. Therefore, one should expect a and b to be correlated as a result of this curve-fitting exercise. This is similar to the well-known negative correlation between cohesion and friction angle. A linear fit to a nonlinear Mohr–Coulomb failure envelope will create a small intercept (cohesion) when the slope (friction angle) is large, or vice versa. If one adopts the expedient assumption that a and b are statistically independent random variables, then the column of a values can be shuffled independently of the column of b values. In physical terms, this implies that the initial slopes of the load–displacement curves can be varied independently of their asymptotic values. This observation is not supported by measured load test data, as shown in Section 9.3.3. In addition, there is no necessity to make this simplifying assumption from a calculation viewpoint. Correlated a and b values can be calculated with relative ease, as illustrated below using the data from Table 9.2.

9.3 Statistical analyses

9.3.1 Randomness of the hyperbolic parameters

In the literature, it is common to conclude that a parameter is a random variable by presenting a histogram. However, it is rarely emphasized that parameters exhibiting variations are not necessarily "random." A random variable is an appropriate probabilistic model only if the variations are not explainable by deterministic variations in the database, for example, foundation geometry or soil property. Figure 9.3 shows that the hyperbolic parameters do not exhibit obvious trends with D/B or the average SPT-N value.

9.3.2 Marginal distributions of a and b

The lognormal distribution provides an adequate fit to both hyperbolic parameters, as shown in Figure 9.4 (p-values from the Anderson–Darling goodness-of-fit test, p_{AD}, are larger than 0.05). Note that S.D. is the standard deviation, COV is the coefficient of variation, and n is the sample size. Because of lognormality, a and b can be expressed in terms of *standard*

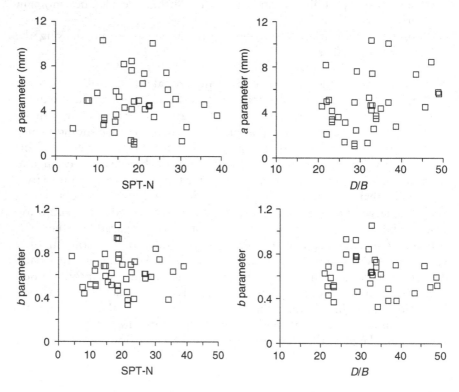

Figure 9.3 Hyperbolic parameters versus D/B and SPT-N for ACIP piles.

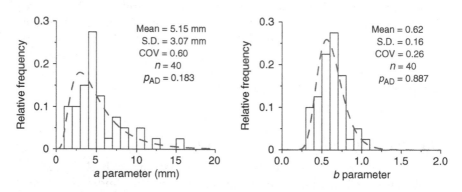

Figure 9.4 Marginal distributions of hyperbolic parameters for ACIP piles. From Phoon et al., 2006. Reproduced with the permission of the American Society of Civil Engineers.

normal variables X_1 and X_2 as follows:

$$a = \exp(\xi_1 X_1 + \lambda_1)$$
$$b = \exp(\xi_2 X_2 + \lambda_2) \tag{9.4}$$

The parameters λ_1 and ξ_1 are the equivalent normal mean and equivalent normal standard deviation of a, respectively, as given by:

$$\xi_1 = \sqrt{\ln\left(1 + s_a^2/m_a^2\right)} = 0.551$$
$$\lambda_1 = \ln(m_a) - 0.5\xi_1^2 = 1.488 \tag{9.5}$$

in which the mean and standard deviation of a are $m_a = 5.15$ mm and $s_a = 3.07$ mm, respectively. The parameters λ_2 and ξ_2 can be calculated from the mean and standard deviation of b in the same way and are given by $\lambda_2 = -0.511$ and $\xi_2 = 0.257$.

There is no reason to assume that a lognormal probability distribution will always provide a good fit for the hyperbolic parameters. A general but simple approach based on Hermite polynomials can be used to fit *any* empirical distributions of a and b:

$$a = a_{10}H_0(X_1) + a_{11}H_1(X_1) + a_{12}H_2(X_1) + a_{13}H_3(X_1)$$
$$+ a_{14}H_4(X_1) + a_{15}H_5(X_1) + \cdots \tag{9.6}$$
$$b = a_{20}H_0(X_2) + a_{21}H_1(X_2) + a_{22}H_2(X_2) + a_{23}H_3(X_2)$$
$$+ a_{24}H_4(X_2) + a_{25}H_5(X_2) + \cdots$$

in which the Hermite polynomials $H_j(.)$ are given by:

$$H_0(X) = 1$$
$$H_1(X) = X \tag{9.7}$$
$$H_{k+1}(X) = X\,H_k(X) - k\,H_{k-1}(X)$$

The advantages of using a Hermite expansion are: (a) Equation (9.6) is directly applicable for reliability analysis (Phoon and Honjo, 2005; Phoon et al., 2005); (b) a short expansion, typically six terms, is sufficient; and (c) Hermite polynomials can be calculated efficiently using the recursion formula shown in the last row of Equation (9.7). The main effort is to calculate the Hermite coefficients in Equation (9.6). Figure 9.5 shows that the entire calculation can be done with relative ease using EXCEL. Figure 9.6 compares a and b values simulated using the lognormal distribution and the Hermite expansion with the measured values in Table 9.2.

	A	B	C	D	E	F	G	H	I	J	K	L	M
1	Sorted a	Rank a	F(a)	X		H0	H1	H2	H3	H4	H5		a1
2													
3						0	1	2	3	4	5		
4	1.006	1	0.024	-1.971		1	-1.971	2.883	-1.740	-5.221	17.246		5.400
5	1.237	2	0.049	-1.657		1	-1.657	1.745	0.423	-5.935	8.143		2.967
6	1.312	3	0.073	-1.453		1	-1.453	1.110	1.293	-5.208	2.394		1.290
7	1.352	4	0.098	-1.296		1	-1.296	0.679	1.712	-4.254	-1.337		-0.138
8	2.047	5	0.122	-1.165		1	-1.165	0.358	1.914	-3.303	-3.805		0.151
9	2.413	6	0.146	-1.052		1	-1.052	0.107	1.992	-2.417	-5.423		-0.096
10	2.528	7	0.171	-0.951		1	-0.951	-0.095	1.993	-1.611	-6.440		
11	2.738	8	0.195	-0.859		1	-0.859	-0.262	1.943	-0.864	-7.014		
12	3.054	9	0.220	-0.774		1	-0.774	-0.401	1.858	-0.234	-7.251		
13	3.142	10	0.244	-0.694		1	-0.694	-0.519	1.747	0.344	-7.228		
14	3.353	11	0.268	-0.618		1	-0.618	-0.618	1.618	0.854	-7.000		
15	3.428	12	0.293	-0.546		1	-0.546	-0.702	1.474	1.303	-6.608		
16	3.571	13	0.317	-0.476		1	-0.476	-0.774	1.320	1.692	-6.085		
17	3.687	14	0.341	-0.408		1	-0.408	-0.833	1.157	2.027	-5.457		
18	4.104	15	0.366	-0.343		1	-0.343	-0.882	0.988	2.309	-4.745		
19	4.128	16	0.390	-0.279		1	-0.279	-0.922	0.814	2.540	-3.965		
20	4.296	17	0.415	-0.216		1	-0.216	-0.953	0.637	2.723	-3.135		

Column	EXCEL functions
A	Sort values of "a" using "Data > Sort > Ascending". Let f be a 40×1 vector containing values in column A.
B	Rank the values in column A in ascending order, e.g., B4 = RANK(A4,$A$4:$A$43,1)
C	Calculate empirical cumulative distribution F(a) using (rank)/(sample size +1), e.g., C4 = B4/41
D	Calculate standard normal variate using Φ^{-1}[F(a)], e.g., D4 = NORMSINV(C4)
F..K	Calculate Hermite polynomials using recursion, e.g., F4 = 1, G4 = D4, H4 = G4*$D4-(H$3-1)*F4
M	Let H be a 40×6 matrix containing values in column F to K. Calculate the Hermite coefficients in column M (denoted by a 6×1 vector $\underline{a}_1$) by: $H^T H \underline{a}_1 = H^T \underline{f}$ The EXCEL array formula is: {=MMULT(MINVERSE(MMULT(TRANSPOSE(F4:K43),F4:K43)), MMULT(TRANSPOSE(F4:K43),A4:A43))}

Figure 9.5 Calculation of Hermite coefficients for *a* using EXCEL.

Hyperbolic parameters for uplift load–displacement curves associated with spread foundations, drilled shafts, and pressure-injected footings are shown in Figure 9.7. Although lognormal distributions can be fitted to the histograms (dashed line), they are not quite appropriate for the *b* parameters describing the asymptotic uplift load–displacement behavior of spread foundations and drilled shafts (*p*-values from the Anderson–Darling goodness-of-fit test, p_{AD}, are smaller than 0.05).

9.3.3 Correlation between a and b

The random vector (*a* , *b*) is described completely by a bivariate probability distribution. The assumption of statistical independence commonly is adopted in the geotechnical engineering literature for expediency. The practical advantage is that the random components can be described completely

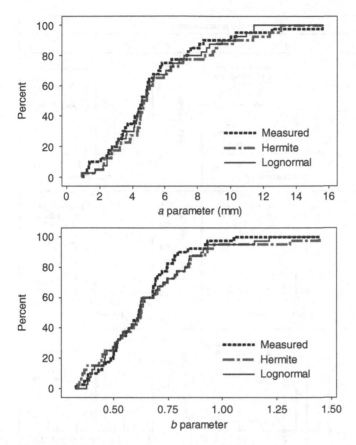

Figure 9.6 Simulation of hyperbolic parameters using lognormal distribution and Hermite expansion.

by marginal or univariate probability distributions such as those shown in Figures 9.4 and 9.7. Another advantage is that numerous distributions are given in standard texts, and the data on hand usually can be fitted to one of these classical distributions. The disadvantage is that this assumption can produce unrealistic *scatter plots* between *a* and *b* if the data are strongly correlated. For hyperbolic parameters, it appears that strong correlations are the norm, rather than the exception (Figure 9.8).

It is rarely emphasized that the multivariate probability distributions (even bivariate ones) underlying random vectors are very difficult to construct theoretically, to estimate empirically, and to simulate numerically. A cursory review of basic probability texts reveals that non-normal distributions usually are presented in the univariate form. It is not possible to generalize

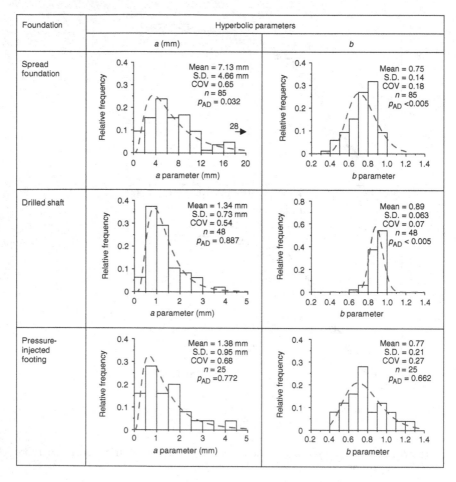

Figure 9.7 Marginal distributions of hyperbolic parameters for uplift load–displacement curves associated with spread foundations, drilled shafts, and pressure-injected footings. From Phoon *et al.*, 2007. Reproduced with the permission of the American Society of Civil Engineers.

these univariate formulae to higher dimensions unless the random variables are independent. This assumption is overly restrictive in reliability analysis where input parameters can be correlated by physics (e.g. sliding resistance increases with normal load, undrained shear strength increases with loading rate, etc.) or by curve-fit (e.g. cohesion and friction angle in a linear Mohr–Coulomb failure envelope, hyperbolic parameters from load–displacement curves). A full discussion of these points is given elsewhere (Phoon, 2004a, 2006).

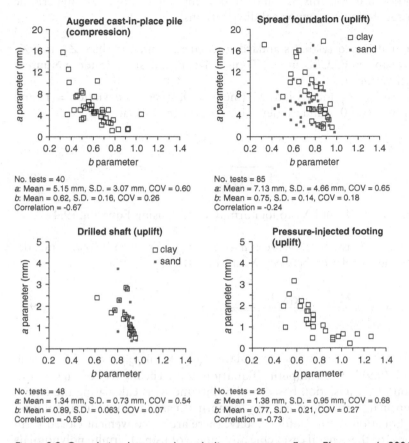

No. tests = 40
a: Mean = 5.15 mm, S.D. = 3.07 mm, COV = 0.60
b: Mean = 0.62, S.D. = 0.16, COV = 0.26
Correlation = -0.67

No. tests = 85
a: Mean = 7.13 mm, S.D. = 4.66 mm, COV = 0.65
b: Mean = 0.75, S.D. = 0.14, COV = 0.18
Correlation = -0.24

No. tests = 48
a: Mean = 1.34 mm, S.D. = 0.73 mm, COV = 0.54
b: Mean = 0.89, S.D. = 0.063, COV = 0.07
Correlation = -0.59

No. tests = 25
a: Mean = 1.38 mm, S.D. = 0.95 mm, COV = 0.68
b: Mean = 0.77, S.D. = 0.21, COV = 0.27
Correlation = -0.73

Figure 9.8 Correlation between hyperbolic parameters. From Phoon *et al.*, 2006, 2007. Reproduced with the permission of the American Society of Civil Engineers.

9.4 Bivariate probability distribution functions

9.4.1 Translation model

For bivariate or higher-dimensional probability distributions, there are few practical choices available other than the multivariate normal distribution. In the bivariate normal case, the entire dependency relationship between both random components can be described completely by a product–moment correlation coefficient [Equation (9.2)]. The practical importance of the normal distribution is reinforced further because usually there are insufficient data to compute reliable dependency information beyond the correlation coefficient.

It is expected that a bivariate probability model for the hyperbolic parameters will be constructed using a suitably correlated normal model.

The procedure for this is illustrated using an example of hyperbolic parameters following lognormal distributions.

1. Simulate uncorrelated standard normal random variables Z_1 and Z_2 (available in EXCEL under "Tools > Data Analysis > Random Number Generation").
2. Transform Z_1 and Z_2 into X_1 (mean = 0, standard deviation = 1) and X_2 (mean = 0, standard deviation = 1) with correlation $\rho_{X_1 X_2}$ by:

$$X_1 = Z_1$$

$$X_2 = Z_1 \rho_{X_1 X_2} + Z_2 \sqrt{1 - \rho_{X_1 X_2}^2} \tag{9.8}$$

3. Transform X_1 and X_2 to lognormal a and b using Equation (9.4).

Note that the correlation between a and b $(\rho_{a,b})$, as given below, is not the same as the correlation between X_1 and X_2 $(\rho_{X_1 X_2})$.

$$\rho_{a,b} = \frac{\exp(\xi_1 \xi_2 \rho_{X_1 X_2}) - 1}{\sqrt{[\exp(\xi_1^2) - 1][\exp(\xi_2^2) - 1]}} \tag{9.9}$$

Therefore, one needs to back-calculate $\rho_{X_1 X_2}$ from $\rho_{a,b}$ ($= -0.67$) using Equation (9.9) before computing Equation (9.8). The main effort in the construction of a translation-based probability model is calculating the equivalent normal correlation. For the lognormal case, a closed-form solution exists [Equation (9.9)]. Unfortunately, there are no convenient closed-form solutions for the general case. One advantage of using Hermite polynomials to fit the empirical marginal distributions of a and b [Equation (9.6)] is that a simple power series solution can be used to address this problem:

$$\rho_{a,b} = \frac{\sum_{k=1}^{\infty} k! a_{1k} a_{2k} \rho_{X_1 X_2}^k}{\sqrt{\left(\sum_{k=1}^{\infty} k! a_{1k}^2\right)\left(\sum_{k=1}^{\infty} k! a_{2k}^2\right)}} \tag{9.10}$$

The bivariate lognormal model based on $\rho_{X_1 X_2} = -0.8$ [calculated from Equation (9.9) with $\rho_{a,b} = -0.7$] produces a scatter of simulated a and b values that is in reasonable agreement with that exhibited by the measured values (Figure 9.9a). In contrast, adopting the assumption of statistical independence between a and b, corresponding to $\rho_{X_1 X_2} = 0$, produces the unrealistic scatter plot shown in Figure 9.9b, in which there are too many

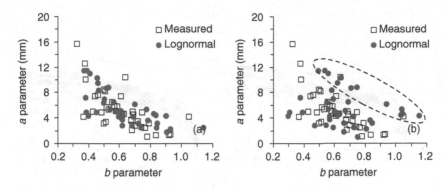

Figure 9.9 Simulated hyperbolic parameters for ACIP piles: (a) $\rho_{X_1 X_2} = -0.8$, and (b) $\rho_{X_1 X_2} = 0$.

simulated points above the observed scatter, as highlighted by the dashed oval. Note that the marginal distributions for a and b in Figures 9.9a and b are almost identical (not shown). The physical difference produced by correlated and uncorrelated hyperbolic parameters can be seen in the simulated load–displacement curves as well (Figures 9.10b–d). These simulated curves are produced easily by substituting simulated pairs of a and b into Equation (9.1).

9.4.2 Rank model

By definition, the product–moment correlation $\rho_{a,b}$ is bounded between –1 and 1. One of the key limitations of a translation model is that it is not possible to simulate the full range of possible product–moment correlations between a and b (Phoon, 2004a; 2006). Based on the equivalent normal standard deviations of a and b for ACIP piles ($\xi_1 = 0.551$ and $\xi_2 = 0.257$), it is easy to verify that Equation (9.9) with $\rho_{X_1 X_2} = -1$ will result in a minimum value of $\rho_{a,b} = -0.85$. The full relationship between translation non-normal correlation and equivalent normal correlation is shown in Figure 9.11. More examples are given in Phoon (2006), as reproduced in Figure 9.12 (note: $\rho_{a,b} = \rho_{Y_1 Y_2}$).

If a translation-based probability model is inadequate, a simple alternate approach is to match the rank correlation as follows:

1 Convert a and b to their respective ranks. Note that these ranks are calculated based on ascending values of a and b, i.e. the smallest value is assigned a rank value equal to 1.
2 Calculate the rank correlation using the same Equation as the product–moment correlation [Equation (9.2)], except the rank values are used in

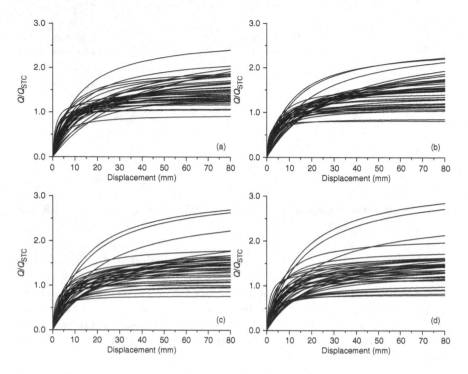

Figure 9.10 Normalized hyperbolic curves for ACIP piles: (a) measured, (b) $\rho_{X_1 X_2} = -0.8$, (c) $\rho_{X_1 X_2} = -0.4$, and (d) $\rho_{X_1 X_2} = 0$.

place of the actual a and b values. The rank correlation for Table 9.2 is –0.71.

3   Simulate correlated standard normal random variables using Equation (9.8) with $\rho_{X_1 X_2}$ equal to the rank correlation computed above.

4   Simulate independent non-normal random variables following any desired marginal probability distributions.

5   Re-order the realizations of the non-normal random variables such that they follow the rank structure underlying the correlated standard normal random variables.

The simulation procedure for hyperbolic parameters of ACIP piles is illustrated in Figure 9.13. Note that the simulated a and b values in columns M and N will follow the rank structure (columns J and K) of the correlated standard normal random variables (columns H and I) because the exponential transform (converting normal to lognormal) is a monotonic increasing function. The theoretical basis underlying the rank model is different from the translation model and is explained by Phoon *et al.* (2004)

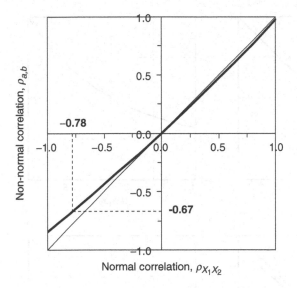

Figure 9.11 Relationship between translation non-normal correlation and equivalent normal correlation.

and Phoon (2004b). It suffices to note that there is no unique method of constructing correlated non-normal random variables, because the correlation coefficient (product–moment or rank) does not provide a full description of the association between two non-normal variables.

9.5 Serviceability limit state reliability-based design

This section presents the practical application of the probabilistic hyperbolic model for serviceability limit state (SLS) reliability-based design. The SLS occurs when the foundation displacement (y) is equal to the allowable displacement (y_a). The foundation has exceeded serviceability if $y > y_a$. Conversely, the foundation is satisfactory if $y < y_a$. These three situations can be described concisely by the following performance function:

$$P = y - y_a = y(Q) - y_a \tag{9.11}$$

An alternate performance function is:

$$P = Q_a - Q = Q_a(y_a) - Q \tag{9.12}$$

Figure 9.14 illustrates the uncertainties associated with these performance functions. In Figure 9.14a, the applied load Q is assumed to be deterministic

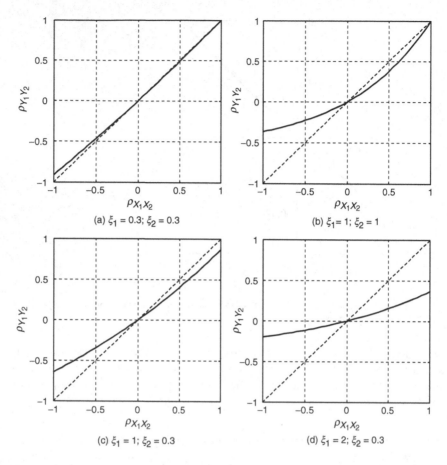

Figure 9.12 Some relationships between observed correlation ($\rho_{Y_1 Y_2}$) and correlation of underlying equivalent Gaussian random variables ($\rho_{X_1 X_2}$). From Phoon, 2006. Reproduced with the permission of the American Society of Civil Engineers.

to simplify the visualization. It is clear that the displacement follows a distribution even if Q is deterministic, because the load–displacement curve $y(Q)$ is uncertain. The allowable displacement may follow a distribution as well. In Figure 9.14b, the allowable displacement is assumed to be deterministic. In this alternate version of the performance function, the allowable load Q_a follows a distribution even if y_a is deterministic because of the uncertainty in the load–displacement curve. The effect of a random load Q and the possibility of upper tail values falling on the nonlinear portion of the load–displacement curve are illustrated in this figure.

File Edit View Insert Format Tools Data Window Help

	A	B	C	D	E	F	G	H	I	J	K	L	M	N
1	Measured parameters												Simulated parameters	
2	a	b	Rank a	Rank b		Z1	Z2	X1	X2	Rank X1	Rank X2		a	b
3	8.382	0.499	36	9		-0.206	0.347	-0.206	0.390	13	29		3.953	0.663
4	10.033	0.382	37	4		0.003	0.938	0.003	0.654	19	34		4.435	0.710
5	3.571	0.679	13	26		-0.966	-0.069	-0.966	0.628	7	31		2.600	0.705
6	4.296	0.615	17	21		-0.499	1.853	-0.499	1.652	10	39		3.364	0.917
7	7.311	0.445	32	6		0.530	-2.985	0.530	-2.474	26	1		5.930	0.318
8	5.707	0.591	29	18		-0.854	-1.763	-0.854	-0.623	8	16		2.787	0.511
9	3.697	0.679	14	25		0.259	-2.103	0.259	-1.657	24	4		5.109	0.392
10	2.738	0.701	8	30		0.936	0.331	0.936	-0.438	31	18		7.416	0.536
11	3.429	0.723	12	31		1.133	0.456	1.133	-0.491	34	17		8.289	0.529
12	7.559	0.461	34	7		0.499	0.851	0.499	0.239	25	26		5.830	0.638
13	5.029	0.582	26	17		-0.966	1.813	-0.966	1.980	8	40		2.598	0.993
14	2.528	0.744	7	32		-0.147	-0.616	-0.147						
15	5.280	0.533	27	14		1.270	0.077	1.270						
16	4.595	0.632	21	23		-2.190	-1.310	-2.190						
17	2.413	0.770	6	34		0.805	-0.519	0.805						
18	15.646	0.328	40	1		-0.308	0.258	-0.308						
19	4.572	0.607	20	19		-0.002	0.926	-0.002						
20	7.376	0.613	33	20		-0.971	0.274	-0.971						
21	2.047	0.682	5	27		-1.104	0.522	-1.104						
22	12.548	0.380	39	3		1.711	0.456	1.711						
23	5.829	0.570	30	16		0.048	-1.582	0.048						
24	8.419	0.559	31	15		0.066	0.040	0.066						
25	4.907	0.690	24	28		1.368	-0.177	1.368						
26	4.426	0.692	18	29		0.885	-0.824	0.885						
27	1.312	0.640	3	37		1.477	1.939	1.477						
28	8.128	0.510	35	11		1.234	-0.323	1.234						
29	10.297	0.640	38	24		0.205	-0.204	0.205						
30	4.885	0.487	23	8		0.681	-1.971	0.681						
31	5.599	0.517	28	13		-0.090	0.425	-0.090						

Dialog box (Random Number Generation):
- Number of Variables: 2
- Number of Random Numbers: 40
- Distribution: Normal
- Parameters
- Mean = 0
- Standard deviation = 1
- Random Seed:
- Output options
- Output Range: $F$3
- New Worksheet Ply:
- New Workbook
- OK / Cancel / Help

Column	EXCEL functions
A	Measured "a" values from Table 9.2
B	Measured "b" values from Table 9.2
C	Rank the values in column A in ascending order, e.g., C3 = RANK(A3,$A$3:$A$42,1)
D	Rank the values in column B in ascending order, e.g., D3 = RANK(B3,$B$3:$B$42,1)
F, G	Simulate two columns of independent standard normal random variables using Tools > Data Analysis > Random Number Generation > Number of Variables = 2; Number of Random Numbers = 40; Distribution = Normal; Mean = 0; Standard deviation = 1; Output Range = $F$3
H	Set $X_1 = Z_1$, e.g., H3 = F3
I	Set $X_2 = Z_1 \rho_{X_1,X_2} + Z_2(1 - \rho_{X_1,X_2}^2)^{0.5}$ with ρ_{X_1,X_2} = rank correlation between columns C and D, e.g., I3 = F3*(−0.71)+G3*SQRT(1−0.71^2)
M	Simulated "a" values, e.g., M3 = EXP(H3*0.551+1.488)
N	Simulated "b" values, e.g., N3 = EXP(I3*0.257−0.511)

Figure 9.13 Simulation of rank correlated a and b values for ACIP piles using EXCEL.

The basic objective of RBD is to ensure that the probability of failure of a component does not exceed an acceptable threshold level. For ULS or SLS, this objective can be stated using the performance function as follows:

$$p_f = \mathrm{Prob}(P < 0) \leq p_T \tag{9.13}$$

in which $\mathrm{Prob}(\cdot)$ = probability of an event, p_f = probability of failure, and p_T = acceptable target probability of failure. A more convenient alternative to the probability of failure is the reliability index (β), which is defined as:

$$\beta = -\Phi^{-1}(p_f) \tag{9.14}$$

in which $\Phi^{-1}(\cdot)$ = inverse standard normal cumulative function.

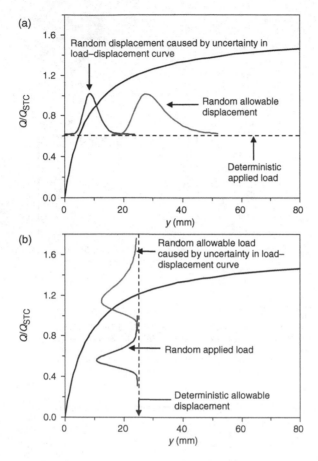

Figure 9.14 Serviceability limit state reliability-based design: (a) deterministic applied load, and (b) deterministic allowable displacement.

9.5.1 First-order reliability method (FORM)

The probability of failure given by Equation (9.13) can be calculated easily using the first-order reliability method (FORM) once the probabilistic hyperbolic model is established:

$$p_f = \text{Prob}\left(Q_a < Q\right) = \text{Prob}\left(\frac{y_a}{a + by_a}Q_m < Q\right) \tag{9.15}$$

The interpreted capacity Q_m, allowable displacement y_a, and applied load Q are assumed to be lognormally distributed with, for example, coefficients of variation (COV) = 0.5, 0.6, and 0.2, respectively. The COV of Q_m is related to the capacity prediction method and the quality of the geotechnical data. Phoon *et al.* (2006) reported a COV of 0.5 associated with the prediction of compression capacity of ACIP piles using the Meyerhof method. The COV of y_a is based on a value of 0.583 reported by Zhang and Ng (2005) for buildings on deep foundations. The COV of $Q = 0.2$ is appropriate for live load (Paikowsky, 2002). The hyperbolic parameters are negatively correlated with an equivalent normal correlation of –0.8 and are lognormally distributed with statistics given in Figure 9.4. The mean of y_a is assumed to be 25 mm. Equation (9.15) can be re-written to highlight the mean factor of safety (FS) as follows:

$$p_f = \text{Prob}\left(\frac{y_a}{a + by_a} < \frac{m_Q Q^*}{m_{Q_m} Q_m^*}\right) = \text{Prob}\left(\frac{y_a}{a + by_a} < \frac{1}{FS}\frac{Q^*}{Q_m^*}\right) \qquad (9.16)$$

in which m_{Q_m} and $m_Q =$ mean of Q_m and Q, FS $= m_{Q_m}/m_Q$, and Q_m^* and $Q^* =$ unit-mean lognormal random variables with COV $= 0.5$ and 0.2. Note that FS refers to the *ultimate limit state*. It is equivalent to the global factor of safety adopted in current practice.

Figure 9.15 illustrates that the entire FORM calculation can be performed with relative ease using EXCEL. The most tedious step in FORM is the search for the most probable failure point, which requires a nonlinear optimizer. Low and Tang (1997, 2004) recommended using the SOLVER function in EXCEL to perform this tedious step. However, their proposal to perform optimization in the physical random variable space is not recommended, because the means of these random variables can differ by orders of magnitudes. For example, the mean $b = 0.62$ for the ACIP pile database, while the mean of Q_m can be on the order of 1000 kN. In Figure 9.15, the SOLVER is invoked by changing cells $B$14:$B$18, which are random variables in standard normal space (i.e. uncorrelated, zero-mean, unit variance components). An extensive discussion of reliability analysis using EXCEL is given elsewhere (Phoon, 2004a).

For a typical $FS = 3$, the SLS reliability index for this problem is 2.21, corresponding to a probability of failure of 0.0134.

9.5.2 Serviceability limit state model factor approach

The allowable load can be evaluated from the interpreted capacity using a SLS model factor (M_S) as follows:

$$Q_a = \frac{y_a}{a + by_a}Q_m = M_S Q_m \qquad (9.17)$$

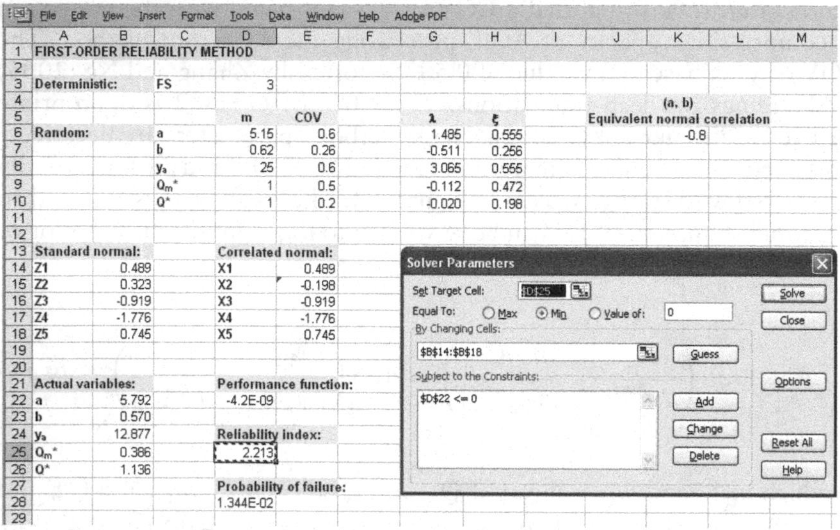

	A	B	C	D	E	F	G	H	I	J	K	L	M
1	FIRST-ORDER RELIABILITY METHOD												
2													
3	Deterministic:		FS	3									
4												(a, b)	
5				m	COV		λ	ξ			Equivalent normal correlation		
6	Random:		a	5.15	0.6		1.485	0.555				-0.8	
7			b	0.62	0.26		-0.511	0.256					
8			yₐ	25	0.6		3.065	0.555					
9			Qₘ*	1	0.5		-0.112	0.472					
10			Q*	1	0.2		-0.020	0.198					
11													
12													
13	Standard normal:			Correlated normal:									
14	Z1	0.489		X1	0.489								
15	Z2	0.323		X2	-0.198								
16	Z3	-0.919		X3	-0.919								
17	Z4	-1.776		X4	-1.776								
18	Z5	0.745		X5	0.745								
19													
20													
21	Actual variables:			Performance function:									
22	a	5.792		-4.2E-09									
23	b	0.570											
24	yₐ	12.877		Reliability index:									
25	Qₘ*	0.386		2.213									
26	Q*	1.136											
27				Probability of failure:									
28				1.344E-02									
29													

Solver Parameters

Set Target Cell: [D$25]

Equal To: ○ Max ● Min ○ Value of: 0

By Changing Cells:

$B$14:$B$18

Subject to the Constraints:

$D$22 <= 0

[Solve] [Close] [Guess] [Options] [Add] [Change] [Reset All] [Delete] [Help]

Cell	EXCEL functions
E14:E18	E14=B14; E15=B14*$K$6+B15*SQRT(1-$K$6^2); E16=B16; E17=B17; E18=B18
B22:B26	B22=EXP(E14*H6+G6); B23=EXP(E15*H7+G7); etc.
D22	B24/(B24*B23+B22)-B26/B25/D3
D25	{= SQRT(MMULT(TRANSPOSE(B14:B18),B14:B18))}
D28	= NORMSDIST(-D25)

Solver

Set Target Cell: $D$25 Equal To: Min

By Changing Cells: $B$14:$B$18

Subject to the Constraints: $D$22 <= 0

Figure 9.15 EXCEL implementation of first-order reliability method for serviceability limit state.

If M_S follows a lognormal distribution, then a closed-form solution exists for Equation (9.15):

$$\beta = \frac{\ln\left[\left(\frac{m_{M_s} m_{Qm}}{m_Q}\right)\sqrt{\frac{1+COV_Q^2}{\left(1+COV_{M_s}^2\right)\left(1+COV_{Qm}^2\right)}}\right]}{\sqrt{\ln\left[\left(1+COV_{M_s}^2\right)\left(1+COV_{Qm}^2\right)\left(1+COV_Q^2\right)\right]}} \qquad (9.18)$$

This procedure is similar to the bias or model factor approach used in ULS RBD. However, the distribution of M_S has to be evaluated for each allowable displacement. Figure 9.16 shows two distributions of M_S determined by

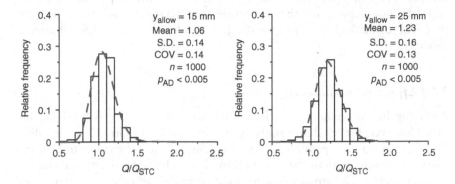

Figure 9.16 Simulated distributions of SLS model factors (Q/Q_{STC}) for ACIP piles (allowable displacement y_a assumed to be deterministic).

substituting simulated values of a and b into $y_a/(a + by_a)$, with $y_a = 15$ mm and 25 mm. In contrast, the foundation capacity is a single number (not a curve), and only one model factor distribution is needed.

Based on first-order, second-moment analysis, the mean (m_{M_s}) and COV (COV_{M_s}) of M_S can be estimated as:

$$m_{M_s} \approx \frac{y_a}{m_a + m_b y_a} \tag{9.19}$$

$$\text{COV}_{M_s} \approx \frac{\sqrt{s_a^2 + y_a^2 s_b^2 + 2 y_a \rho_{a,b} s_a s_b}}{m_a + m_b y_a} \tag{9.20}$$

For $y_a = 25$ mm, the mean and COV of M_S are estimated to be 1.21 and 0.145, which compare favorably with the simulated statistics shown in Figure 9.16 (mean = 1.23 and COV = 0.13).

For the example discussed in Section 9.5.1, the reliability index calculated using Equation (9.18) is 2.29, and the corresponding probability of failure is 0.0112. The correct FORM solution is $\beta = 2.38$ ($p_f = 0.0086$). For a deterministic $y_a = 15$ mm, Equation (9.18) gives $\beta = 2.00$ ($p_f = 0.0227$), while FORM gives $\beta = 2.13$ ($p_f = 0.0166$). These small differences are to be expected, given that the histograms shown in Figure 9.16 are not lognormally distributed despite visual agreement. The null hypothesis of lognormality is rejected because the p-values from the Anderson-Darling goodness-of-fit test (p_{AD}) are significantly smaller than 0.05. In our opinion, one practical disadvantage of the SLS model factor approach is that M_S can not be evaluated easily from a load test database if y_a follows a distribution. Given the simplicity of the FORM calculation shown in Figure 9.15, it is preferable to use the probabilistic hyperbolic model

for SLS reliability analysis. The distributions of the hyperbolic parameters do not depend on the allowable displacement, and there are no
practical difficulties in incorporating a random y_a for SLS reliability
analysis.

9.5.3 Effect of nonlinearity

It is tempting to argue that the load–displacement behavior is linear at the
SLS. However, for reliability analysis, the applied load follows a probability
distribution, and it is possible for the load–displacement behavior to be nonlinear at the upper tail of the distribution. This behavior is illustrated clearly
in Figure 9.14b. The probabilistic hyperbolic model can be simplified to a linear model by setting $b = 0$ (physically, this is equivalent to an infinitely large
asymptotic limit). For the example discussed in Section 9.5.1, the FORM
solution under this linear assumption is $\beta = 2.76$ ($p_f = 0.00288$). Note that
the solution is very unconservative, because the probability of failure is about
4.7 times smaller than that calculated for the correct nonlinear hyperbolic
model. If the mean allowable displacement is reduced to 15 mm, and the
mean factor of safety is increased to 10, then the effect of nonlinearity is less
pronounced. For the linear versus nonlinear case, $\beta = 3.50$ ($p_f = 0.000231$)
versus $\beta = 3.36$ ($p_f = 0.000385$), respectively.

9.5.4 Effect of correlation

The effect of correlated hyperbolic parameters on the load–displacement
curves is shown in Figure 9.10. For the example discussed in Section 9.5.1,
the FORM solution for uncorrelated hyperbolic parameters is $\beta = 2.05$
($p_f = 0.0201$). The probability of failure is 1.5 times larger than that calculated for the correlated case. For a more extreme case of $FS = 10$ and mean
$y_a = 50$ mm, the FORM solution for uncorrelated hyperbolic parameters
is $\beta = 4.42$ ($p_f = 4.99 \times 10^{-6}$). Now the probability of failure is significantly more conservative. It is 3.3 times larger than that calculated for the
correlated case given by $\beta = 4.67$ ($p_f = 1.49 \times 10^{-6}$). Note that assuming
zero correlation between the hyperbolic parameters does not always produce
the most conservative (or largest) probability of failure. If the hyperbolic
parameters were found to be positively correlated, perhaps with a weak
equivalent normal correlation of 0.5, the FORM solution would be $\beta = 4.25$
($p_f = 1.06 \times 10^{-5}$). The probability of failure for the uncorrelated case is now
2.1 times smaller than that calculated for the correlated case.

The effect of a negative correlation is to reduce the COV of the SLS model
factor as shown by Equation (9.20). A smaller COV_{Ms} leads to a higher reliability index as shown by Equation (9.18). The reverse is true for positively
correlated hyperbolic parameters.

9.5.5 Uncertainty in allowable displacement

The allowable displacement is assumed to be deterministic in the majority of the previous studies. For the example discussed in Section 9.5.1, the FORM solution for a deterministic allowable displacement is $\beta = 2.38$ ($p_f = 0.00864$). The probability of failure is 1.6 times smaller than that calculated for a random y_a with COV = 0.6. For a more extreme case of $FS = 10$ and mean $y_a = 50$ mm, the FORM solutions for deterministic and random y_a are $\beta = 4.69(p_f = 1.36 \times 10^{-6})$ and $\beta = 4.67(p_f = 1.49 \times 10^{-6})$, respectively. It appears that the effect of uncertainty in the allowable displacement is comparable to, or smaller than, the effect of correlation in the hyperbolic parameters.

9.5.6 Target reliability index for serviceability limit state

The Eurocode design basis, BS EN1990:2002 (British Standards Institute, 2002), specifies the safety, serviceability and durability requirements for design and describes the basis for design verification. It is intended to be used in conjunction with nine other Eurocodes that govern both structural and geotechnical design, and it specifies a one-year target reliability index of 2.9 for SLS. Phoon et al. (1995) recommended a target reliability index of 2.6 for transmission line (and similar) structure foundations based on extensive reliability calibrations with existing designs for different foundation types, loading modes, and drainage conditions. Zhang and Xu (2005) analyzed the settlement of 149 driven steel H-piles in Hong Kong and calculated a reliability index of 2.46 based on a distribution of measured settlements at the service load (mean = 19.9 mm and COV = 0.23) and a distribution of allowable settlement for buildings (mean = 96 mm and COV = 0.583).

Reliability indices corresponding to different mean factors of safety and different allowable displacements can be calculated based on the EXCEL method described in Figure 9.15. The results for ACIP piles are calculated using the statistics given in Figure 9.4 and are presented in Figure 9.17. It can be seen that a target reliability index = 2.6 is not achievable at a mean $FS = 3$ for a mean allowable displacement of 25 mm. However, it is achievable for a mean FS larger than about 4. Note that the 50-year return period load $Q_{50} \approx 1.5\ m_Q$ for a lognormal distribution with COV = 0.2. Therefore, a mean $FS = 4$ is equivalent to a nominal $FS = m_{Q_m}/Q_{50} = 2.7$. A nominal FS of 3 is probably closer to prevailing practice than a mean $FS = 3$.

It is also unclear if the allowable displacement prescribed in practice really is a mean value or some lower bound value. For a lognormal distribution with a COV = 0.6, a mean allowable displacement of 50 mm produces a mean less one standard deviation value of 25 mm. Using this interpretation, a

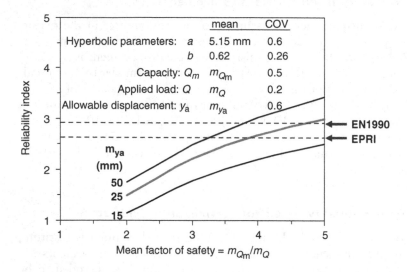

Figure 9.17 Relationship between reliability index and mean factor of safety for ACIP piles.

target reliability index of 2.6 is achievable at a mean *FS* about 3 or a nominal *FS* about 2.

Overall, the EPRI target reliability index for SLS is consistent with foundation designs resulting from a traditional global *FS* between 2 and 3.

9.6 Simplified reliability-based design

Existing implementations of RBD are based on a simplified approach that involves the use of multi-factor formats for checking designs. Explicit reliability calculations such as those presented in Section 9.5 are not done commonly in current routine design. Simplified RBD checks can be calibrated using the following general procedure (Phoon *et al.*, 1995).

1   Select realistic performance functions for the ULS and/or the SLS.
2   Assign probability models for each basic design parameter that are commensurate with available statistical support.
3   Conduct a parametric study on the variation of the reliability level with respect to each parameter in the design problem using the FORM to identify appropriate calibration domains.
4   Determine the range of reliability levels implicit in existing designs.
5   Adjust the resistance factors (for ULS) or deformation factors (for SLS) in the RBD Equations until a consistent target reliability level is achieved within each calibration domain.

The calibration of simplified RBD checks was discussed by Phoon and Kulhawy (2002a) for drilled shafts subjected to drained uplift, by Phoon *et al.* (2003) for spread foundations subjected to uplift, and by Kulhawy and Phoon (2004) for drilled shafts subjected to undrained lateral-moment loading. This section presents an example of SLS design of drilled shafts subjected to undrained uplift (Phoon *et al.* 1995). The corresponding ULS design was given by Phoon *et al.* (1995, 2000).

9.6.1 Performance function for serviceability limit state

The uplift capacity of drilled shafts is governed by the following vertical equilibrium Equation:

$$Q_u = Q_{su} + Q_{tu} + W \tag{9.21}$$

in which Q_u = uplift capacity, Q_{su} = side resistance, Q_{tu} = tip resistance, and W = weight of foundation. For undrained loading, the side resistance can be calculated as follows:

$$Q_{su} = \pi B \alpha \int_0^D s_u(z) dz \tag{9.22}$$

in which B = shaft diameter, D = shaft depth, α = adhesion factor, s_u = undrained shear strength, and z = depth. The α factor is calibrated for a specific s_u test type. For s_u determined from CIUC (consolidated-isotropically undrained compression) tests, the corresponding adhesion factor was determined by the following regression Equation:

$$\alpha = 0.31 + \frac{0.17}{(s_u/p_a)} + \varepsilon \tag{9.23}$$

in which p_a = atmospheric stress ≈ 100 kN/m^2 and ε = normal random variable with mean = 0 and standard deviation = 0.1 resulting from linear regression analysis. The undrained shear strength (s_u) was modeled as a lognormal random variable, with mean $(m_{su}) = 25 - 200$ kN/m^2 and COV $(\mathrm{COV}_{s_u}) = 10 - 70\%$. It has been well-established for a long time (e.g. Lumb, 1966) that there is no such thing as a "typical" COV for soil parameters because the degree of uncertainty depends on the site conditions, degree of equipment and procedural control, quality of the correlation model, and other factors. If geotechnical RBD codes do not consider this issue explicitly, then practicing engineers can not adapt the code design to their specific set of local circumstances, such as site conditions, measurement techniques, correlation models, standards of practice, etc. If code provisions do not allow these adaptations, then local "ad-hoc" procedures are likely to develop that basically defeat the purpose of the code in the first place. This situation is not desirable. A good code with proper guidelines is preferable.

The tip resistance under undrained loading conditions develops from suction forces and was estimated by:

$$Q_{tu} = (-\Delta u - u_i)A_{tip} \tag{9.24}$$

in which u_i = initial pore water stress at the foundation tip or base, Δu = change in pore water stress caused by undrained loading, and A_{tip} = tip or base area. Note that no suction force develops unless $-\Delta u$ exceeds u_i. Tip suction is difficult to predict accurately because it is sensitive to the drainage conditions in the vicinity of the tip. In the absence of an accurate predictive model and statistical data, it was assumed that the tip suction stress is uniformly distributed between zero and one atmosphere. [Note that many designers conservatively choose to disregard the undrained tip resistance in routine design.]

For the load–displacement curve, it was found that the simple hyperbolic model resulted in the least scatter in the normalized curves (Table 9.1). The values of a and b corresponding to each load–displacement curve were determined by using the displacement at 50% of the failure load (y_{50}) and the displacement at failure (y_f) as follows:

$$a = \frac{y_{50}y_f}{y_f - y_{50}} \tag{9.25a}$$

$$b = \frac{y_f - 2y_{50}}{y_f - y_{50}} \tag{9.25b}$$

Forty-two load–displacement curves were analyzed using Equations 9.25 (a and b). The database is given in Table 9.3, and the statistics of a and b are summarized in Table 9.4. It can be seen that the effect of soil type is small, and therefore the statistics calculated using all of the data were used. In this earlier study, a and b were assumed to be independent lognormal random variables for convenience. This assumption is conservative as noted in Section 9.5.4. The mean and one standard deviation load–displacement curves are plotted in Figure 9.18, along with the data envelopes.

Following Equation (9.12), the performance function at SLS is:

$$P = Q_{ua} - F = \frac{y_a}{a + by_a}Q_u - F \tag{9.26}$$

in which F = weather-related load effect for transmission line structure foundations = kV^2 (Task Committee on Structural Loadings of ASCE, 1991), V = Gumbel random variable with COV = 30%, and k = deterministic constant. The probability distributions assigned to each design parameter are summarized in Table 9.5.

Table 9.3 Hyperbolic curve-fitting parameters for drilled shafts under uplift (Phoon et al., 1995).

Case[a]	Soil type[b]	Q_u^c (kN)	D^d (m)	B^e (m)	y_{50}^f (mm)	y_f^g (mm)	a^h (mm)	b^h
5/1	C	320	3.17	0.63	0.48	12.70	0.50	0.96
8/1	C	311	4.57	0.61	0.90	12.70	0.96	0.92
12/5	C	311	4.57	0.61	1.02	12.70	1.10	0.91
15/1	C	285	2.44	0.61	1.34	6.71	1.68	0.75
16/1	C	338	2.44	0.91	0.61	7.32	0.66	0.91
25/NE	C	80	1.98	0.52	0.91	12.70	0.98	0.92
38/A2	C	222	3.05	0.61	0.81	12.70	0.87	0.93
38/A3	C	498	6.10	0.61	0.91	12.70	0.98	0.92
40/1	C	285	3.66	0.61	0.84	12.70	0.90	0.93
40/2	C	338	3.66	0.61	0.74	12.70	0.79	0.94
56/1	C	934	10.06	0.35	2.50	25.00	2.78	0.89
56/2	C	552	6.80	0.35	1.25	12.70	1.39	0.89
56/17	C	667	7.50	0.30	0.62	12.70	0.66	0.95
56/23	C	667	6.49	0.45	0.62	12.70	0.66	0.95
63/1	C	2225	11.58	1.52	1.40	12.70	1.58	0.87
64/4	C	1050	3.66	1.52	0.80	12.70	0.85	0.93
69/1	C	1388	20.33	0.91	1.73	6.35	2.38	0.62
73/1	C	338	2.80	0.61	0.81	11.58	0.87	0.92
73/6	C	507	3.69	0.76	1.22	10.16	1.38	0.86
14/1	S	231	2.44	0.91	0.37	12.70	0.38	0.97
18/1	S	445	3.05	0.91	1.35	16.15	1.48	0.91
20/1	S	409	3.05	0.91	1.34	24.38	1.42	0.94
41/1	S	365	3.05	0.91	0.33	5.08	0.35	0.93
44/1	S	534	8.53	0.46	1.22	12.70	1.35	0.89
46/10	S	25	1.37	0.34	1.15	12.70	1.27	0.90
47/11	S	71	2.44	0.37	0.77	12.70	0.82	0.93
48/12	S	163	3.66	0.37	1.54	12.70	1.75	0.86
49/E	S	296	4.27	0.41	1.15	12.70	1.27	0.90
51/1	S	525	3.66	0.91	0.61	12.70	0.64	0.95
62/1	S	890	6.40	1.07	0.49	6.35	0.53	0.92
62/2	S	890	6.40	1.07	0.98	14.00	1.05	0.92
70/3	S	1557	6.10	0.61	1.02	12.70	1.10	0.91
95/1	S	338	12.19	0.53	2.00	26.24	2.16	0.92
95/2	S	347	12.19	0.38	3.12	19.99	3.70	0.81
99/2	S	1086	3.05	1.22	0.66	4.19	0.78	0.81
99/3	S	1255	3.66	1.22	0.56	7.62	0.61	0.92
21/1	S + C	454	15.24	0.41	0.91	12.70	0.98	0.92
36/1	S + C	712	10.36	0.41	1.91	12.70	2.24	0.82
42/1	S + C	1629	21.34	0.46	1.34	12.70	1.49	0.88
50/1	S + C	100	2.90	0.30	1.47	8.20	1.78	0.78
50/2	S + C	400	3.81	0.30	2.54	23.11	2.85	0.88
96/1	S + C	667	12.19	0.53	1.33	12.70	1.49	0.88

[a]Case number in Kulhawy et al. (1983).
[b]S = sand; C = clay; S + C = sand and clay.
[c]Failure load determined by L_1–L_2 method (Hirany and Kulhawy, 1988).
[d]Foundation depth.
[e]Foundation diameter.
[f]Displacement at 50% of failure load.
[g]Displacement at failure load.
[h]Curve-fitted parameters for normalized hyperbolic model based on Equation (9.25).

Table 9.4 Statistics of a and b for uplift load–displacement curve. From Phoon and Kulhawy, 2002a. Reproduced with the permission of the American Society of Civil Engineers.

Soil type	No. of tests	a		b	
		Mean (mm)	COV	Mean	COV
Clay	19	1.16	0.52	0.89	0.09
Sand	17	1.22	0.67	0.91	0.05
Mixed	6	1.80	0.36	0.86	0.06
All	42	1.27	0.56	0.89	0.07

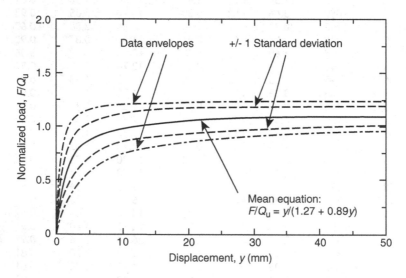

Figure 9.18 Normalized uplift load–displacement curve for drilled shafts. From Phoon and Kulhawy, 2002a. Reproduced with the permission of the American Society of Civil Engineers.

9.6.2 Parametric study

To conduct a realistic parametric study, typical values of the design parameters should be used. Herein, the emphasis was on those used in the electric utility industry, which sponsored the basic research study. The drilled shaft diameter (B) was taken as 1–3 m and the depth to diameter ratio (D/B) was taken as 3–10. The typical range of the mean wind speed (m_V) was back-calculated from the 50-year return period wind speed (v_{50}) given in Table 9.5

Table 9.5 Summary of probability distributions for design parameters.

Parameter	Description	Distribution	Mean	COV
V	Wind speed	Gumbel	30–50[a] m/s	0.30
a	Curve-fitting parameter [Equation (9.24a)]	Lognormal	1.27 mm	0.56
b	Curve-fitting parameter [Equation (9.24b)]	Lognormal	0.89	0.07
ε	Regression error associated with adhesion factor model [(Equation (9.23)]	Normal	0	0.1
s_u	Undrained shear strength	Lognormal	25–200 kN/m^2	0.1–0.7
Q_{tu}	Undrained tip suction stress	Uniform	0–1 atmos.[b]	–

[a]50-year return period wind speed; data from Task Committee on Structural Loadings (1991).
[b]Range of uniform distribution.

and the COV of V ($\text{COV}_V = 0.3$), as follows:

$$v_{50} = m_V(1 + 2.59\,\text{COV}_V) \tag{9.27}$$

The constant of proportionality, k that was used in calculating the foundation load, was determined by the following serviceability limit state design check:

$$F_{50} = kv_{50}^2 = Q_{uan} \tag{9.28}$$

in which Q_{uan} = nominal allowable uplift capacity calculated using:

$$Q_{uan} = \frac{y_a}{m_a + m_b y_a} Q_{un} \tag{9.29}$$

in which m_a = mean of $a = 1.27$ mm, m_b = mean of $b = 0.89$, and y_a = allowable displacement = 25 mm. The nominal uplift capacity (Q_{un}) is given by:

$$Q_{un} = Q_{sun} + Q_{tun} + W \tag{9.30}$$

in which Q_{sun} = nominal uplift side resistance and Q_{tun} = nominal uplift tip resistance. The nominal uplift side resistance was calculated as follows:

$$Q_{sun} = \pi BD\alpha_n m_{su} \tag{9.31}$$

$$\alpha_n = 0.31 + \frac{0.17}{\left(m_{su}/p_a\right)} \tag{9.32}$$

in which α_n = nominal adhesion factor. The nominal tip resistance (Q_{tun}) was evaluated using:

$$Q_{tun} = (\frac{W}{A_{tip}} - u_i)A_{tip} \qquad (9.33)$$

The variations of the probability of failure (p_f) with respect to the 50-year return period wind speed, allowable displacement limit, foundation geometry, and the statistics of the undrained shear strength, were given by Phoon *et al.* (1995). It was found that p_f is affected significantly by the mean and COV of the undrained shear strength (m_{su} and COV_{su}). Therefore, the following domains were judged to be appropriate for the calibration of the simplified RBD checks:

1   m_{su} = 25–50 kN/m^2 (medium clay), 50–100 kN/m^2 (stiff clay), and 100–200 kN/m^2 (very stiff clay);
2   COV_{su} = 10–30%, 30–50%, and 50–70%.

For the other, less-influential design parameters, calibration can be performed over the entire range of typical values. Note that the COVs of most manufactured structural materials fall between 10 and 30%. For example, Réthati (1988), citing the 1965 specification of the American Concrete Institute, noted that the quality of concrete can be evaluated as follows:

    COV < 10%   excellent

    COV = 10–15%   good

    COV = 15–20%   satisfactory

    COV > 20%   bad

The COVs of natural geomaterials usually are much larger and do not fall within a narrow range (Phoon and Kulhawy, 1999).

9.6.3 Load and Resistance Factor Design (LRFD)

The following LRFD format was chosen for reliability calibration:

$$F_{50} = \Psi_u Q_{uan} \qquad (9.34)$$

in which Ψ_u = deformation factor. The SLS deformation factors were calibrated using the following general procedure (Phoon *et al.*, 1995).

1   Partition the parameter space into several smaller domains following the approach discussed in Section 9.6.2. For example, the combination of

$m_{su} = 25$–50 kN/m^2 and COV$_{su} = 30$–50% is one possible calibration domain. The reason for partitioning is to achieve greater uniformity in reliability over the full range of design parameters.

2   Select a set of representative points from each domain. For example, $m_{su} = 30$ kN/m^2 and COV$_{su} = 40\%$ constitutes one calibration point. Ideally, the set of representative points should capture the full range of variation in the reliability level over the whole domain.

3   Determine an acceptable foundation design for each calibration point and evaluate the reliability index for each design. Foundation design is performed using a simplified RBD check [e.g. Equation (9.34)] and a set of trial resistance or deformation factors.

4   Adjust the resistance or deformation factors until the reliability indices for the representative designs are as close to the target level (e.g. 3.2 for ULS and 2.6 for SLS) as possible in a statistical sense.

The results of this reliability calibration exercise are shown in Table 9.6. A comparison between Figure 9.19a [existing design check based on Equation (9.28)] and Figure 9.19b [LRFD design check based on Equation (9.34)] illustrates the improvement in the uniformity of the reliability levels. It is interesting to note that deformation factors can be calibrated for broad ranges of soil property variability (e.g. COV of $s_u = 10$–30%, 30–50%, 50–70%) without compromising on the uniformity of reliability achieved. Experienced engineers should be able to choose the appropriate data category, even with limited statistical data supplemented with first-order guidelines (Phoon and Kulhawy, 1999) and reasoned engineering judgment.

Table 9.6 Undrained uplift deformation factors for drilled shafts designed using $F_{50} = \Psi_u Q_{uan}$ (Phoon et al., 1995).

Clay	Undrained shear strength		Ψ_u
	Mean (kN/m^2)	COV (%)	
Medium	25–50	10–30	0.65
		30–50	0.63
		50–70	0.62
Stiff	50–100	10–30	0.64
		30–50	0.61
		50–70	0.58
Very stiff	100–200	10–30	0.61
		30–50	0.57
		50–70	0.52

Note: Target reliability index = 2.6.

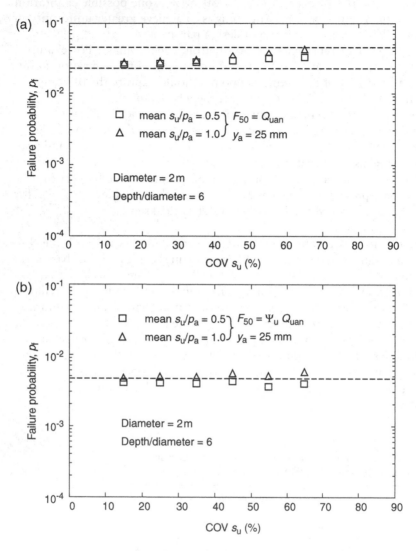

Figure 9.19 Performance of SLS design checks for drilled shafts subjected to undrained uplift: (a) existing and (b) reliability-calibrated LRFD (Phoon *et al.*, 1995).

Two key conclusions based on wider calibration studies (Phoon *et al.*, 1995) are noteworthy.

1   One resistance or deformation factor is insufficient to achieve a uniform target reliability across the full range of geotechnical COVs.

Table 9.7 Ranges of soil property variability for reliability calibration in geotechnical engineering.

Geotechnical parameter	Property variability	COV (%)
Undrained shear strength	Low[a]	10–30
	Medium[b]	30–50
	High[c]	50–70
Effective stress friction angle	Low[a]	5–10
	Medium[b]	10–15
	High[c]	15–20
Horizontal stress coefficient	Low[a]	30–50
	Medium[b]	50–70
	High[c]	70–90

[a]typical of good quality direct lab or field measurements.
[b]typical of indirect correlations with good field data, except for the standard penetration test (SPT).
[c]typical of indirect correlations with SPT field data and with strictly empirical correlations.

2   In terms of COV, three ranges of soil property variability (low, medium, and high) are sufficient to achieve reasonably uniform reliability levels for simplified RBD checks. The appropriate ranges depend on the geotechnical parameter, as shown in Table 9.7. Note that the categories for effective stress friction angle are similar to those specified for concrete.

9.7 Conclusions

This chapter discussed the application of a probabilistic hyperbolic model for performing reliability-based design checks at the serviceability limit state. The uncertainty in the entire load–displacement curve is represented by a relatively simple bivariate random vector containing the hyperbolic parameters as its components. The steps required to construct a realistic random vector are illustrated using forty measured pairs of hyperbolic parameters for ACIP piles. It is important to highlight that these hyperbolic parameters typically exhibit a strong negative correlation. A translation-based lognormal model or a more general translation-based Hermite model is needed to capture this correlation aspect correctly. The common assumption of statistical independence can circumvent the additional complexity associated with a translation model, because the bivariate probability distribution reduces to two significantly simpler univariate distributions in this special case. However, the scatter in the measured load–displacement curves can not be reproduced properly by simulation under this assumption.

The nonlinear feature of the load–displacement curve is captured by the hyperbolic curve-fitting Equation. This nonlinear feature is crucial in reliability analysis because the applied load follows a probability distribution, and it is possible for the load–displacement behavior to be nonlinear at the upper tail of the distribution. Reliability analysis for the more general case of a random allowable displacement can be evaluated easily using an implementation of the first-order reliability method in EXCEL. The proposed approach is more general and convenient to use than the bias or model factor approach because the distributions of the hyperbolic parameters do not depend on the allowable displacement. The random vector containing two curve-fitting parameters is the simplest probabilistic model for characterizing the uncertainty in a nonlinear load–displacement curve and should be the recommended first-order approach for SLS reliability-based design.

Simplified RBD checks for SLS and ULS can be calibrated within a single consistent framework using the probabilistic hyperbolic model. Because of the broad range of geotechnical COVs, one resistance or deformation factor is insufficient to achieve a uniform target reliability level (the key objective of RBD). In terms of COV, three ranges of soil property variability (low, medium, and high) are more realistic for reliability calibration in geotechnical engineering. Numerical values of these ranges vary as a function of the influential geotechnical parameter(s). Use of these ranges allows for achieving a more uniform target reliability level.

References

Bjerrum, L. (1963). Allowable settlement of structures. In *Proceedings of the 3rd European Conference on Soil Mechanics and Foundation Engineering* (3), Wiesbaden, pp. 135–7.

Boone, S. J. (2001). Assessing construction and settlement-induced building damage: a return to fundamental principles. In *Proceedings Underground Construction*, Institution of Mining and Metallurgy, London, pp. 559–70.

Boone, S. J. (2007). Assessing risks of construction-induced building damage for large underground projects. In *Probabilistic Applications in Geotechnical Engineering* (GSP 170), Eds. K. K. Phoon, G. A. Fenton, E. F. Glynn, C. H. Juang, D. V. Griffiths, T. F. Wolff and L. M. Zhang. ASCE, Reston (CDROM).

British Standards Institute (2002). *Eurocode: Basis of structural design*. BS EN 1990:2002, London.

Burland, J. B. and Wroth, C. P. (1975). *Settlement of buildings and associated damage*. Building Research Establishment Current Paper, Building Research Establishment, Watford.

Burland, J. B., Broms, B. B. and DeMello, V. F. B. (1977). Behavior of foundations and structures: state of the art report. In *Proceedings of the 9th International Conference on Soil Mechanics and Foundation Engineering* (2), Tokyo, pp. 495–546.

Callanan, J. F. and Kulhawy, F. H. (1985). *Evaluation of procedures for predicting foundation uplift movements*. Report EL–4107, Electric Power Research Institute, Palo Alto.

Chen, J.-R. (1998). *Case history evaluation of axial behavior of augered cast-in-place piles and pressure-injected footings*. MS thesis, Cornell University, Ithaca, NY.

Fenton, G. A., Griffiths, D. V. and Cavers, W. (2005). Resistance factors for settlement design. *Canadian Geotechnical Journal*, 42(5), 1422–36.

Finno, R. J. (2007). Predicting damage to buildings from excavation induced ground movements. In *Probabilistic Applications in Geotechnical Engineering* (GSP 170), Eds. K. K. Phoon, G. A. Fenton, E. F. Glynn, C. H. Juang, D. V. Griffiths, T. F. Wolff and L. M. Zhang. ASCE, Reston (CDROM).

Grant, R., Christian, J. T. and VanMarcke, E. H. (1974). Differential settlement of buildings. *Journal of Geotechnical Engineering, ASCE*, 100(9), 973–91.

Hirany, A. and Kulhawy, F. H. (1988). *Conduct and interpretation of load tests on drilled shaft foundations: detailed guidelines*. Report EL-5915(1), Electric Power Research Institute, Palo Alto.

Hsiao, E. C. L., Juang, C. H., Kung, G. T. C. and Schuster, M. (2007). Reliability analysis and updating of excavation-induced ground settlement for building serviceability evaluation. In *Probabilistic Applications in Geotechnical Engineering* (GSP 170), Eds. K. K. Phoon, G. A. Fenton, E. F. Glynn, C. H. Juang, D. V. Griffiths, T. F. Wolff and L. M. Zhang. ASCE, Reston (CDROM).

Kulhawy, F. H. (1994). Some observations on modeling in foundation engineering. In *Proceeding of the 8th International Conference on Computer Methods and Advances in Geomechanics* (1), Eds. H. J. Siriwardane and M. M. Zaman. A. A. Balkema, Rotterdam, pp. 209–14.

Kulhawy, F. H. and Chen, J.-R. (2005). Axial compression behavior of augered cast-in-place (ACIP) piles in cohesionless soils. In *Advances in Designing and Testing Deep Foundations* (GSP 129), Eds. C. Vipulanandan and F. C. Townsend. ASCE, Reston, pp. 275–289. [Also in *Advances in Deep Foundations* (GSP 132), Eds. C. Vipulanandan and F. C. Townsend. ASCE, Reston, 13 p.]

Kulhawy, F. H. and Phoon, K. K. (2004). Reliability-based design of drilled shafts under undrained lateral-moment loading. In *Geotechnical Engineering for Transportation Projects* (GSP 126), Eds. M. K. Yegian and E. Kavazanjian. ASCE, Reston, pp. 665–76.

Kulhawy, F. H., O'Rourke, T. D., Stewart, J. P. and Beech, J. F. (1983). *Transmission line structure foundations for uplift-compression loading: load test summaries*. Report EL-3160, Electric Power Research Institute, Palo Alto.

Low, B. K. and Tang, W. H. (1997). Efficient reliability evaluation using spreadsheet. *Journal of Engineering Mechanics, ASCE*, 123(7), 749–52.

Low, B. K. and Tang, W. H. (2004). Reliability analysis using object-oriented constrained optimization. *Structural Safety*, 26(1), 69–89.

Lumb, P. (1966). Variability of natural soils. *Canadian Geotechnical Journal*, 3(2), 74–97.

Misra, A. and Roberts, L. A. (2005). Probabilistic axial load displacement relationships for drilled shafts. In *Contemporary Issues in Foundation Engineering* (GSP 131), Eds. J. B. Anderson, K. K. Phoon, E. Smith and J. E. Loehr. ASCE, Reston (CDROM).

Moulton, L. K. (1985). *Tolerable movement criteria for highway bridges.* Report FHWA/RD-85/107, Federal Highway Administration, Washington, DC.

Paikowsky, S. G. (2002). Load and resistance factor design (LRFD) for deep foundations. In *Proceedings of the International Workshop on Foundation Design Codes and Soil Investigation in View of International Harmonization and Performance Based Design,* Tokyo. A. A. Balkema, Lisse, pp. 59–94.

Paikowsky, S. G. and Lu, Y. (2006). Establishing serviceability limit state in design of bridge foundations. In *Foundation Analysis and Design: Innovative Methods* (GSP 153), Eds. R. L. Parsons, L. M. Zhang, W. D. Guo, K. K. Phoon and M. Yang. ASCE, Reston, pp. 49–58.

Phoon, K. K. (1995). *Reliability-based design of foundations for transmission line structures,* Ph.D. thesis, Cornell University, Ithaca, NY.

Phoon, K. K. (2004a). *General non-Gaussian probability models for first-order reliability method (FORM): a state-of-the-art report.* ICG Report 2004-2-4 (NGI Report 20031091-4), International Centre for Geohazards, Oslo.

Phoon, K. K. (2004b). Application of fractile correlations and copulas to non-Gaussian random vectors. In *Proceedings of the 2nd International ASRANet (Network for Integrating Structural Safety, Risk, and Reliability) Colloquium,* Barcelona (CDROM).

Phoon, K. K. (2006). Modeling and simulation of stochastic data. In *Geo-Congress 2006: Geotechnical Engineering in the Information Technology Age,* Eds. D. J. DeGroot, J. T. DeJong, J. D. Frost and L. G. Braise. ASCE, Reston (CDROM).

Phoon, K. K. and Honjo, Y. (2005). Geotechnical reliability analyses: towards development of some user-friendly tools. In *Proceedings of the 16th International Conference on Soil Mechanics and Geotechnical Engineering,* Osaka (CDROM).

Phoon, K. K. and Kulhawy, F. H. (1999). Characterization of geotechnical variability. *Canadian Geotechnical Journal,* 36(4), 612–24.

Phoon, K. K. and Kulhawy, F. H. (2002a). Drilled shaft design for transmission line structures using LRFD and MRFD. In *Deep Foundations 2002 (GSP116),* Eds. M. W. O'Neill and F. C. Townsend. ASCE, Reston, pp. 1006–17.

Phoon, K. K. and Kulhawy, F. H. (2002b). EPRI study on LRFD and MRFD for transmission line structure foundations. In *Proceedings International Workshop on Foundation Design Codes and Soil Investigation in View of International Harmonization and Performance Based Design,* Tokyo. Balkema, Lisse, pp. 253–61.

Phoon K. K. and Kulhawy, F. H. (2005a). Characterization of model uncertainties for laterally loaded rigid drilled shafts. *Geotechnique,* 55(1), 45–54.

Phoon, K. K. and Kulhawy, F. H. (2005b). Characterization of model uncertainties for drilled shafts under undrained axial loading. In *Contemporary Issues in Foundation Engineering* (GSP 131), Eds. J. B. Anderson, K. K. Phoon, E. Smith and J. E. Loehr, ASCE, Reston, 13 p.

Phoon, K. K., Chen, J.-R. and Kulhawy, F. H. (2006). Characterization of model uncertainties for augered cast-in-place (ACIP) piles under axial compression. In *Foundation Analysis and Design: Innovative Methods* (GSP 153), Eds. R. L. Parsons, L. M. Zhang, W. D. Guo, K. K. Phoon and M. Yang. ASCE, Reston, pp. 82–9.

Phoon, K. K., Chen, J.-R. and Kulhawy, F. H. (2007). Probabilistic hyperbolic models for foundation uplift movements. In *Probabilistic Applications in Geotechnical Engineering* (GSP 170), Eds. K. K. Phoon, G. A. Fenton, E. F. Glynn, C. H. Juang, D. V. Griffiths, T. F. Wolff and L. M. Zhang. ASCE, Reston (CDROM).

Phoon, K. K., Kulhawy, F. H. and Grigoriu, M. D. (1995). *Reliability-based design of foundations for transmission line structures*. Report TR-105000, EPRI, Palo Alto.

Phoon, K. K., Kulhawy, F. H. and Grigoriu, M. D. (2000). Reliability-based design for transmission line structure foundations. *Computers and Geotechnics*, 26(3–4), 169–85.

Phoon, K. K., Kulhawy, F. H. and Grigoriu, M. D. (2003). Multiple resistance factor design (MRFD) for spread foundations. *Journal of Geotechnical and Geoenvironmental Engineering, ASCE*, 129(9), 807–18.

Phoon, K. K., Nadim, F. and Lacasse, S. (2005). First-order reliability method using Hermite polynomials. In *Proceedings of the 9th International Conference on Structural Safety and Reliability*, Rome (CDROM).

Phoon, K. K., Quek, S. T., Chow, Y. K. and Lee, S. L. (1990). Reliability analysis of pile settlement. *Journal of Geotechnical Engineering, ASCE*, 116(11), 1717–35.

Phoon, K. K., Quek, S. T. and Huang, H. W. (2004). Simulation of non-Gaussian processes using fractile correlation. *Probabilistic Engineering Mechanics*, 19(4), 287–92.

Quek, S. T., Chow, Y. K. and Phoon, K. K. (1992). Further contributions to reliability-based pile settlement analysis. *Journal of Geotechnical Engineering, ASCE*, 118(5), 726–42.

Quek, S. T., Phoon, K. K. and Chow, Y. K. (1991). Pile group settlement: a probabilistic approach. *International Journal of Numerical and Analytical Methods in Geomechanics*, 15(11), 817–32.

Réthati, L. (1988). *Probabilistic Solutions in Geotechnics*. Elsevier, New York.

Roberts, L. A., Misra, A. and Levorson, S. M. (2007). Probabilistic design methodology for differential settlement of deep foundations. In *Probabilistic Applications in Geotechnical Engineering* (GSP 170), Eds. K. K. Phoon, G. A. Fenton, E. F. Glynn, C. H. Juang, D. V. Griffiths, T. F. Wolff & L. M. Zhang. ASCE, Reston (CDROM).

Schuster, M. J., Juang, C. H., Roth, M. J. S., Hsiao, E. C. L. and Kung, G. T. C. (2007). Serviceability limit state for probabilistic characterization of excavation-induced building damage. In *Probabilistic Applications in Geotechnical Engineering* (GSP 170), Eds. K. K. Phoon, G. A. Fenton, E. F. Glynn, C. H. Juang, D. V. Griffiths, T. F. Wolff and L. M. Zhang. ASCE, Reston (CDROM).

Skempton, A. W. and MacDonald, D. H. (1956). The allowable settlement of buildings. *Proceedings of the Institution of Civil Engineers*, Part 3(5), 727–68.

Task Committee on Structural Loadings (1991). *Guidelines for Electrical Transmission Line Structural Loading. Manual and Report on Engineering Practice 74*. ASCE, New York.

Wahls, H. E. (1981). Tolerable settlement of buildings. *Journal of Geotechnical Engineering Division, ASCE*, 107(GT11), 1489–504.

Zhang, L. M. and Ng, A. M. Y. (2005). Probabilistic limiting tolerable displacements for serviceability limit state design of foundations. *Geotechnique*, 55(2), 151–61.

Zhang, L. M. and Ng, A. M. Y. (2007). Limiting tolerable settlement and angular distortion for building foundations. In *Probabilistic Applications in Geotechnical Engineering* (GSP 170), Eds. K. K. Phoon, G. A. Fenton, E. F. Glynn, C. H. Juang, D. V. Griffiths, T. F. Wolff and L. M. Zhang. ASCE, Reston (CDROM).

Zhang, L. M. and Phoon K. K. (2006). Serviceability considerations in reliability-based foundation design. In *Foundation Analysis and Design: Innovative Methods* (GSP 153), Eds. R. L. Parsons, L. M. Zhang, W. D. Guo, K. K. Phoon and M. Yang. ASCE, Reston, pp. 127–36.

Zhang, L. M. and Xu, Y. (2005). Settlement of building foundations based on field pile load tests. In *Proceedings of the 16th International Conference on Soil Mechanics and Geotechnical Engineering*, Osaka (CDROM).

Chapter 10

Reliability verification using pile load tests

Limin Zhang

10.1 Introduction

Proof pile load tests are an important means to cope with uncertainties in the design and construction of pile foundations. In this chapter, the term load test refers to a static load test following specifications such as ASTM D1143, BS8004 (BSI, 1986), ISSMFE (1985), and PNAP 66 (Buildings Department, 2002). Load tests serve several purposes. Some of their functions include verification of design parameters (this is especially important if the geotechnical data are uncertain), establishment of the effects of construction methods on foundation capacities, and provision of data for the improvement of design methodologies in use and for research purposes (Passe *et al.*, 1997; Zhang and Tang, 2002). In addition to these functions, savings may be derived from load tests (Hannigan *et al.*, 1997; Passe *et al.*, 1997). With load tests, lower factors of safety (FOS) or higher-strength parameters may be used. Hence, there are cost savings even when no changes in the design are made after the load tests. For example, the US Army Corps of Engineers (USACE) (1993), the American Association of State Highway and Transportation Officials (AASHTO) (1997), and Geotechnical Engineering Office (2006) recommend the use of different FOSs, depending on whether load tests are carried out or not to verify the design. The FOSs recommended by the USACE are shown in Table 10.1. The FOS under the usual load combination may be reduced from 3.0 for designs based on theoretical or empirical predictions to 2.0 for designs based on the same predictions that are verified by proof load tests.

Engineers have used proof pile tests in the allowable stress design (ASD) as follows. The piles are sized using a design method suitable for the soil conditions and the construction procedure for the project considered. In designing the piles, a FOS of 2.0 is used when proof load tests are utilized. Preliminary piles or early test piles are then constructed, necessary design modifications made based on construction control, and proof tests on the test piles or some working piles conducted. If the test piles do not reach a prescribed failure criterion at the maximum test load (i.e. twice the design load), then the

Table 10.1 Factor of safety for pile capacity (after USACE, 1993).

Method of determining capacity	Loading condition	Minimum factor of safety	
		Compression	Tension
Theoretical or empirical prediction to be verified by pile load test	Usual	2.0	2.0
	Unusual	1.5	1.5
	Extreme	1.15	1.15
Theoretical or empirical prediction to be verified by pile driving analyzer	Usual	2.5	3.0
	Unusual	1.9	2.25
	Extreme	1.4	1.7
Theoretical or empirical prediction not verified by load test	Usual	3.0	3.0
	Unusual	2.25	2.25
	Extreme	1.7	1.7

design is validated; otherwise, the design load for the piles must be reduced or additional piles or pile lengths must be installed.

In a reliability-based design (RBD), the value of proof load tests can be further maximized. In a RBD incorporating the Bayesian approach (e.g. Ang and Tang, 2007), the same load test results reveal more information. Predicted pile capacity using theoretical or empirical methods can be very uncertain. Results from load tests not only suggest a more realistic pile capacity value, but also greatly reduce the uncertainty of the pile capacity, since the error associated with load-test measurements is much smaller than that associated with predictions. In other words, the reliability of a design for a test site can be updated by synthesizing existing knowledge of pile design and site-specific information from load tests.

For verification of the reliability of designs and calibration of new theories, it is preferable that load tests be conducted to failure. Load tests are often used as proof tests; however, the maximum test load is usually specified as twice the anticipated design load (ASTM D 1143; Buildings Department, 2002) unless failure occurs first. In many cases, the pile is not brought to failure at the maximum test load. Ng et al. (2001a) analyzed more than 30 cases of static load tests on large-diameter bored piles in Hong Kong. They found that the Davisson failure criterion was reached in less than one half of the cases and in no case was the 10%-diameter settlement criterion suggested by ISSMFE (1985) and BSI (1986) reached. In the Hong Kong Driven Pile Database developed recently (Zhang et al., 2006b), only a small portion of the 312 static load tests on steel H-piles were conducted to failure as specified by the Davisson criterion.

Kay (1976), Baecher and Rackwitz (1982), Lacasse and Goulois (1989), and Zhang and Tang (2002) investigated updating the reliability of piles using load tests conducted to failure. Zhang (2004) studied how the majority of proof tests not conducted to failure increases the reliability of piles, and how the conventional proof test method can be modified to allow for a greater value in the RBD framework.

This chapter consists of three parts. A systematic method to incorporate the results of proof load tests into foundation design is first described. Second, illustrative acceptance criteria for piles based on proof load tests are proposed for use in the RBD. Finally, modifications to the conventional proof test procedures are suggested so that the value derived from the proof tests can be maximized. Reliability and factor of safety are used together in this chapter, since they provide complementary measures of an acceptable design (Duncan, 2000), as well as better understanding of the effect of proof tests on the costs and reliability of pile foundations. This chapter will focus on ultimate limits design. Discussion on serviceability limit states is beyond the scope of the chapter.

10.2 Within-site variability of pile capacity

In addition to uncertainties with site investigation, laboratory testing, and prediction models, the values of capacity of supposedly identical piles within one site also vary. Suppose several "identical" test piles are constructed at a seemingly uniform site and are load-tested following an "identical" procedure. The measured values of the ultimate capacity of the piles would usually be different due to the so-called "within-site" variability following Baecher and Rackwitz (1982). For example, Evangelista et al. (1977) tested 22 "identical" bored piles in a sand–gravel site. The piles were "all" 0.8 m in diameter and 20 m in length, and construction of the piles was assisted with bentonite slurry. The load tests revealed that the coefficient of variation (COV) of the settlement of these "identical" piles at the intended working load was 0.21, and the COV of the applied loads at the mean settlement corresponding to the intended load was 0.13.

Evangelista et al. (1977) and Zhang and Tang (2002) described several sources of the within-site variability of the pile capacity: inherent variability of properties of the soil in the influence zone of each pile, construction effects, variability of pile geometry (length and diameter), variability of properties of the pile concrete, and soil disturbance caused by pile driving and afterward set-up. The effects of both construction and set-up are worth a particular mention, because they can introduce many mechanisms that cause large scatter of data. One example of construction effects is the influence of drilling fluids on the capacity of drilled shafts. According to O'Neill and Reese (1999), improper handling of bentonite slurry alone could reduce the beta factor for pile shaft resistance from a common range of 0.4–1.2 to below 0.1. Set-up effects refer to the phenomenon that the pile capacity increases with time following pile installation. Chow et al. (1998), Shek et al. (2006) and many others revealed that the capacity of driven piles in sand can increase from a few percent to over 200% after the end of initial driving. Such effects make the pile capacity from a load test a "nominal" value.

The variability of properties of the soil and the pile concrete is affected by the space over which the properties are estimated (Fenton, 1999). The issue of spatial variability of soils is the subject of interest in several chapters of this book. Note that, after including various effects, the variability of the soils at a site may not be the same as the variability of the pile capacity at the same site. Dasaka and Zhang (2006) investigated the spatial variability of a weathered ground using random field theory and geostatistics. The scale of fluctuation of the depth of competent rock (moderately decomposed rock) beneath a building block is approximately 30 m, whereas the scale of fluctuation of the as-build depth of the piles that provide the same nominal capacity is only approximately 10 m.

Table 10.2 lists the values of the coefficient of variation (COV) of the capacity of driven piles from load tests in eight sites. These values range from 0.12 to 0.28. Note that the variability reported in the table is among test piles in one site; the variability among production piles and among different sites may be larger (Baecher and Rackwitz, 1982). In this chapter, a mean value of COV = 0.20 is adopted for analysis. This assumes that the standard deviation value is proportional to the mean pile capacity.

Kay (1976) noted that the possible values of the pile capacity within a site favor a log-normal distribution. This may be tested with the data from 16 load tests in a site in southern Italy reported by Evangelista et al. (1977).

Table 10.2 Within-site viability of the capacity of driven piles in eight sites.

Site	Number of piles	Diameter (m)	Length (m)	Soil	COV	References
Ashdod, Israel	12	–	–	Sand	0.22	Kay 1976
Bremerhaven, Germany	9	–	–	Sand	0.28	Kay 1976
San Francisco, USA	5	–	–	Sand and clay	0.27	Kay 1976
Southern Italy	12	0.40	8.0	Sand and gravel	0.25	Evangelista et al., 1977
Southern Italy	4	0.40	12.0	Sand and gravel	0.12	Evangelista et al., 1977
Southern Italy	17	0.52	18.0	Sand and gravel	0.19	Evangelista et al., 1977
Southern Italy	3	0.36	7.3	Sand and gravel	0.12	Evangelista et al., 1977
Southern Italy	4	0.46	7.0	Sand and gravel	0.14	Evangelista et al., 1977
Southern Italy	16	0.50	15.0	Clay, sand and gravel	0.20	Evangelista et al., 1977

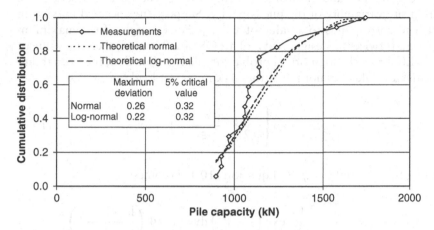

Figure 10.1 Observed and theoretical distributions of within-site pile capacity.

Figure 10.1 shows the measured and assumed normal and log-normal distributions of the pile capacity in the site. Using the Kolmogorov–Smirnov test, neither the normal nor the log-normal theoretical curves appear to fit the observed cumulative curve very well. However, it is acceptable to adopt a log-normal distribution for mathematical convenience, since the maximum deviation in Figure 10.1 is still smaller than the 5% critical value. The tail part of the cumulative distribution of the pile capacity is of more interest to designers, since the pile capacity in that region is smaller. In Figure 10.1, it can be seen that the assumed log-normal distribution underestimates the pile capacity at a particular cumulative percentage near the tail. Therefore, the assumed distribution will lead to results on the conservative side.

From the definition, the within-site variability is inherent in a particular geological setting and a geotechnical construction procedure at a specific site. The within-site variability represents the minimum variability for a construction procedure at a site, which cannot be reduced using load tests.

10.3 Updating pile capacity with proof load tests

10.3.1 Proof load tests that pass

For proof load tests that are not carried out to failure, the pile capacity values are not known, although they are greater than the maximum test loads. Define the variate, x, as the ratio of the measured pile capacity to the predicted pile capacity (called the "bearing capacity ratio" hereafter). At a particular site, x can be assumed to follow a log-normal distribution (Whitman, 1984; Barker *et al.*, 1991) with the mean and standard deviation of x being μ and σ and those of $\ln(x)$ being η and ξ where σ or ξ describes the within-site variability of the pile capacity. Suppose the specified maximum test load corresponds to a value of $x = x_T$. For example, if the maximum test load is twice the design load and a FOS of 2.0 is used, then the value of x_T is 1.0. From the log-normal probability density function, the probability that the test pile does not fail at the maximum test load (i.e., $x > x_T$) is

$$p(x \geq x_T) = \int_{x_T}^{\infty} \frac{1}{\sqrt{2\pi}\xi x} \exp\left\{ -\frac{1}{2}\left(\frac{\ln(x) - \eta}{\xi} \right)^2 \right\} dx \tag{10.1}$$

Noting that $y = (\ln(x) - \eta)/\xi$, Equation (10.1) becomes:

$$p(x \geq x_T) = \int_{\frac{\ln(x_T) - \eta}{\xi}}^{\infty} \frac{1}{\sqrt{2\pi}} \exp\left(-\frac{1}{2}y^2 \right) dy = 1 - \Phi\left(\frac{\ln(x_T) - \eta}{\xi} \right)$$

$$= \Phi\left(-\frac{\ln(x_T) - \eta}{\xi} \right) \tag{10.2}$$

where Φ is the cumulative distribution function of the standard normal distribution. Suppose that n proof tests are conducted and none of the test piles fails at x_T. If the standard deviation, ξ of $\ln(x)$, is known but its mean, μ or η, is a variable, then the probability that all of the n test piles do not fail at x_T is

$$L(\mu) = \prod_{i=1}^{n} p_X(x \geq x_T)\Big|_{\mu} = \Phi^n\left(-\frac{\ln(x_T) - \eta}{\xi} \right) \tag{10.3}$$

$L(\mu)$ is also called the "likelihood function" of μ. Given $L(\mu)$, the updated distribution of the mean of the bearing capacity ratio is (e.g. Ang and Tang, 2007)

$$f''(\mu) = k\Phi^n\left(-\frac{\ln(x_T) - \eta}{\xi} \right) f'(\mu) \tag{10.4}$$

where $f'(\mu)$ is the prior distribution of μ, which can be constructed based on the empirical log-normal distribution of x and the within-site variability information (Zhang and Tang, 2002), and k is a normalizing constant:

$$k = \left[\int_{-\infty}^{\infty} \Phi^n\left(-\frac{\ln(x_T) - \eta}{\xi} \right) f'(\mu) d\mu \right]^{-1} \tag{10.5}$$

Given an empirical distribution $N(\mu_X, \sigma_X)$, the prior distribution can also be assumed as a normal distribution with the following parameters:

$$\mu' = \mu_X \tag{10.6}$$

$$\sigma' = \sqrt{\sigma_X^2 - \sigma^2} \tag{10.7}$$

The updated distribution of the bearing capacity ratio, x, is thus (e.g. Ang and Tang, 2007)

$$f_X(x) = \int_{-\infty}^{\infty} f_X(x|\mu) f''(\mu) d\mu \tag{10.8}$$

where $f_X(x|\mu)$ is the distribution of x given the distribution of its mean. This distribution is assumed to be log-normal, as mentioned earlier.

10.3.2 Proof load tests that do not pass

More generally, suppose only r out of n test piles do not fail at $x = x_T$. The probability that this event occurs now becomes

$$L(\mu) = \binom{n}{r} \left[p(x \geq x_T) \big|_\mu \right]^r \left[1 - p(x \geq x_T) \big|_\mu \right]^{(n-r)} \tag{10.9}$$

and the normalizing constant, k, is

$$k = \left[\int_{-\infty}^{\infty} \binom{n}{r} \left[p(x \geq x_T) \big|_\mu \right]^r \left[1 - p(x \geq x_T) \big|_\mu \right]^{(n-r)} f'(\mu) d\mu \right]^{-1} \tag{10.10}$$

Accordingly, the updated distribution of the mean of the bearing capacity ratio is

$$f''(\mu) = k \binom{n}{r} \left[p(x \geq x_T) \big|_\mu \right]^r \left[1 - p(x \geq x_T) \big|_\mu \right]^{(n-r)} f'(\mu) \tag{10.11}$$

Equation (10.11) has two special cases. When all tests pass (i.e. $r = n$), Equation (10.11) reduces to Equation (10.4). If none of the tests passes (i.e. $r = 0$), a dispersive posterior distribution will result. The updated distribution of the bearing capacity ratio can be obtained using Equation (10.8).

10.3.3 Proof load tests conducted to failure

Let us now consider the proof load tests carried out to failure, i.e. cases where the pile capacity values are known. To start with, the log-normally distributed pile capacity assumed previously is now transformed into a normal variate. If the test outcome is a set of n observed values representing a random sample following a normal distribution with a known standard deviation σ and a mean of $\overline{x}$, then the distribution of the pile capacity can be updated directly using Bayesian sampling theory. Given a known prior normal distribution, $N_\mu(\mu', \sigma')$, the posterior density function of the bearing capacity ratio, $f_X''(x)$, is also normal (Kay, 1976; Zhang and Tang, 2002; Ang and Tang, 2007),

$$f_{X''}(x) = N_X(\mu_{X''}, \sigma_{X''}) \tag{10.12}$$

where

$$\mu_X'' = \frac{\overline{x}(\sigma')^2 + \mu'\left(\frac{\sigma^2}{n}\right)}{(\sigma')^2 + \left(\frac{\sigma^2}{n}\right)} \tag{10.13}$$

$$\sigma_X'' = \sigma\sqrt{1 + \frac{\left(\frac{\sigma'^2}{n}\right)}{(\sigma')^2 + \left(\frac{\sigma^2}{n}\right)}} \tag{10.14}$$

10.3.4 Multiple types of tests

In many cases, multiple test methods are used for construction quality assurance at a single site. The construction control of driven H-piles in Hong Kong (Chu et al., 2001) and the seven-step construction control process of Bell et al. (2002) are two examples. In the Hong Kong practice, both dynamic formula and static analysis are adopted for determining the pile length. Early test piles or preliminary test piles are required at the early stage of construction to verify the construction workmanship and design parameters. In addition, up to 10% of working piles may be tested using high-strain dynamic tests, e.g. using a Pile Driving Analyzer (PDA), and one-half of these PDA tests should be analyzed by a wave equation analysis such as the CASe Pile Wave Analysis Program (CAPWAP). Finally, at the end of construction, 1% of working piles should be proof-tested using static loading tests. With these construction control measures in mind, a low factor of safety is commonly used. Zhang et al. (2006a) described the piling practice in the language of Bayesian updating. A preliminary pile length is first set based primarily on a dynamic formula. The pile performance is then verified by PDA tests during the final setting tests. The combination of the dynamic formula (prior) and the PDA tests will result in a posterior distribution. Taking the posterior distribution after the PDA tests as a prior, the pile performance is further updated based on the outcome of the CAPWAP analyses, and finally updated based on the outcome of the proof load tests. This exercise can be repeated if more indirect or direct verification tests are involved.

10.4 Reliability of piles verified by proof tests

10.4.1 Calculation of reliability index

In a reliability-based design (RBD), the "safety" of a pile can be described by a reliability index, β. To be consistent with current efforts in code development, such as AASHTO's LRFD Bridge Design Specifications (Barker et al., 1991; AASHTO, 1997; Witham et al., 2001; Paikowsky, 2002) or the National Building Code of Canada (NRC, 1995; Becker, 1996a, b), the first-order reliability method (FORM) is used for calculating the reliability index. If both resistance and load effects are log-normal variates, then the reliability index for a linear performance function can be written as (Whitman, 1984)

$$\beta = \frac{\ln\left(\frac{\overline{R}}{\overline{Q}}\sqrt{\frac{1+\text{COV}_Q^2}{1+\text{COV}_R^2}}\right)}{\sqrt{\ln[(1+\text{COV}_R^2)(1+\text{COV}_Q^2)]}} \tag{10.15}$$

where $\overline{Q}$ and $\overline{R}$ = the mean values of load effect and resistance, respectively; COV_Q and COV_R = the coefficients of variation (COV) for the load effect and resistance, respectively. If the only load effects to be considered are dead and live loads, Barker et al. (1991), Becker (1996b), and Witham et al. (2001) have shown that

$$\frac{\overline{R}}{\overline{Q}} = \frac{\lambda_R \text{FOS}\left(\frac{Q_D}{Q_L}+1\right)}{\lambda_{QD}\frac{Q_D}{Q_L}+\lambda_{QL}} \tag{10.16}$$

and

$$\text{COV}_Q^2 = \text{COV}_{QD}^2 + \text{COV}_{QL}^2 \tag{10.17}$$

where Q_D and Q_L = the nominal values of dead and live loads, respectively; λ_R, λ_{QD}, and λ_{QL} = the bias factors for the resistance, dead load, and live load, respectively, with the bias factor referring to the ratio of the mean value to the nominal value; COV_{QD}, and COV_{QL} = the coefficients of variation for the dead load and live load, respectively; and FOS = the factor of safety in the traditional allowable stress design.

If the load statistics are prescribed, Equations (10.15) and (10.16) indicate that the reliability of a pile foundation designed with a FOS is a function of λ_R and COV_R. As will be shown later, if a few load tests are conducted at a site, the values of λ_R and COV_R associated with the design analysis can be updated using the load test results. If the results are favorable, then the updated COV_R will decrease but the updated λ_R will increase. The reliability level or the β value will therefore increase. Conversely, if a target reliability, β_T is specified, the FOS required to achieve the β_T can be calculated using Equations (10.15)–(10.17) and the costs of construction can be reduced if the load test results are favorable.

10.4.2 Example: design based on an SPT method and verified by proof tests

For illustration purposes, a standard penetration test (SPT) method proposed by Meyerhof (1976) for pile design is considered. This method uses the average corrected SPT blow count near the pile toe to estimate the toe resistance and the average uncorrected blow count along the pile shaft to estimate the shaft resistance. According to statistical studies conducted by Orchant *et al.* (1988), the bias factor and COV of the pile capacity from the SPT method are $\lambda_R = 1.30$ and $COV_R = 0.50$, respectively. The within-site variability of the pile capacity is assumed to be $COV = 0.20$ (Zhang and Tang, 2002), which is smaller than the COV_R of the design analysis. This is because the COV_R of the prediction includes more sources of errors such as model errors and differences in construction effects between the site where the model is applied and the sites from which the information was extracted to formulate the model (Zhang *et al.*, 2004).

Calculations for updating the probability distribution can be conducted using an Excel spreadsheet. Figure 10.2 shows the empirical distribution of the bearing capacity ratio, x, based on the given λ_R and COV_R values and the updated distributions after verification by proof tests. The translated mean η and standard deviation ξ of $\ln(x)$, calculated based on the given λ_R and COV_R, are used to define the empirical log-normal distribution. In Figure 10.2(a), all proof tests are positive (i.e. the test piles do not fail at twice the design load); the mean value of the updated pile capacity after verification by the proof tests increases with the number of tests while the COV value decreases. Specifically, the updated mean increases from 1.30 for the empirical distribution to 1.74 after the design has been verified by three positive tests. In Figure 10.2(b), the cases in which no test, one test, two tests, and all three tests are positive are considered. As expected, the updated mean decreases significantly when the number of positive tests decreases. The updated mean and COV values with different outcomes from the proof tests are summarized in Table 10.3. The updated distributions in Figure 10.2 may be approximated by the log-normal distribution for the convenience of reliability calculations using Equation (10.15). The log-normal distribution fits the cases with some non-positive tests very well. For the cases in which all tests are either positive or negative, the log-normal distribution slightly exaggerates the scatter of the distributions, as shown in Figure 10.2(b).

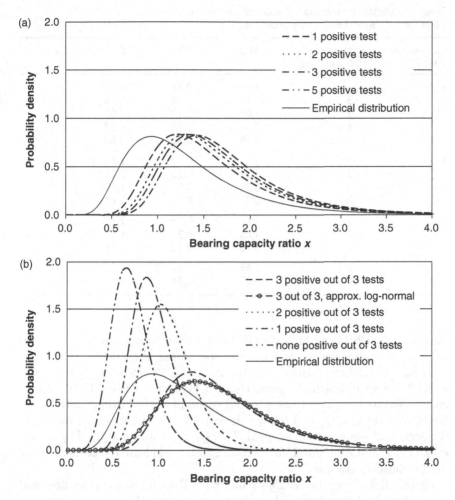

Figure 10.2 Distributions of the pile capacity ratio after verification by (a) proof tests that pass at twice the design load, and (b) proof tests in which some piles fail.

The use of the log-normal distribution is therefore slightly on the conservative side. Because of the differences in the distributions updated by proof tests of different outcomes, the COV values in Table 10.3 do not change in a descending or ascending manner as the number of positive tests decreases.

The following typical load statistics in the LRFD Bridge Design Specifications (AASHTO, 1997) are adopted for illustrative reliability analyses: $\lambda_{QD} = 1.08$, $\lambda_{QL} = 1.15$, $COV_{QD} = 0.13$, and $COV_{QL} = 0.18$. The dead-to-live load ratio, Q_D/Q_L, is structure-specific. Investigations by

Table 10.3 Updated mean and COV of the bearing capacity ratio based on SPT
(Meyerhof, 1976) and verified by proof tests.

Total number of tests	Number of positive tests	Mean	COV	β at FOS = 2.0
4	4	1.77	0.38	2.67
	3	1.19	0.24	2.38
	2	1.04	0.24	2.00
	1	0.91	0.24	1.57
	0	0.69	0.30	0.60
3	3	1.74	0.38	2.58
	2	1.14	0.25	2.21
	1	0.96	0.25	1.70
	0	0.72	0.31	0.67
2	2	1.68	0.39	2.45
	1	1.06	0.26	1.91
	0	0.75	0.32	0.77
1	1	1.59	0.41	2.23
	0	0.82	0.34	0.95

Barker et al. (1991) and McVay et al. (2000) show that β is relatively insensitive to this ratio. For the SPT method considered, the calculated β values are 1.92 and 1.90 for distinct Q_D/Q_L values of 3.69 (75 m span length) and 1.58 (27 m span length), respectively, if a FOS of 2.5 is used. The difference between the two β values is indeed small. In the following analyses, the larger value of $Q_D/Q_L = 3.69$, which corresponds to a structure on which the dead loads dominate, is adopted. This Q_D/Q_L value was used by Barker et al. (1991) and Zhang and Tang (2002), and would lead to factors of safety on the conservative side.

Figure 10.3 shows the reliability index β values for single piles designed with a FOS of 2.0 and verified by several proof tests. Each of the curves in this figure represents the reliability of the piles when they have been verified by n proof tests out of which r tests are positive [see the likelihood function, Equation, (10.9)]. The β value corresponding to the empirical distribution (see Figure 10.2) and a FOS of 2.0 is 1.49. If one conventional proof test is conducted to verify the design and the test is positive, then the β value will be updated to 2.23. The updated β value will continue to increase if more proof tests are conducted and if all the tests are positive. In the cases when not all tests are positive, the reliability of the piles will decrease with the number of tests that are not positive. For instance, the β value of the piles verified by three positive tests is 2.58. If one, two, or all of the three test piles fail, the β values decrease to 2.21, 1.70, and 0.67, respectively. If multiple proof tests are conducted, the target reliability marked by the shaded zone

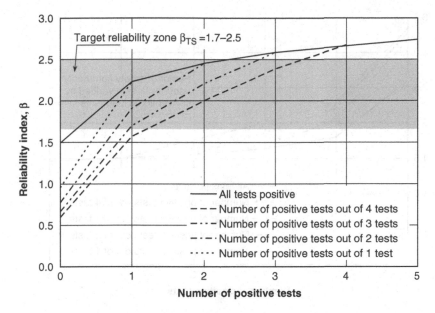

Figure 10.3 Reliability of single driven piles designed with a FOS of 2.0 and verified by conventional proof tests of different outcomes.

can still be satisfied, even if some of the tests are not positive. Issues on target reliability will be discussed later in this chapter.

The FOS that will result in a sufficient level of reliability is of interest to engineers. Figure 10.4 shows the calculated FOS values required to achieve a β value of 2.0 for the pile foundation verified by proof tests with different outcomes. It can be seen that a FOS of 2.0 is sufficient for piles that are verified by one or more consecutive positive tests, or by three or four tests in which no more than one test is not positive. However, larger FOS values should be used if the only proof test or one out of two tests is not positive.

The updating effects of the majority of the proof tests conducted to twice the design load and the tests in which the piles have a mean measured capacity of twice the design load are different. Table 10.4 compares the updated statistics of the two types of tests. For the proof tests that are not conducted to failure, the updated mean bearing capacity ratio and β values significantly increase with the number of tests. For the tests that are conducted to failure at $\bar{x} = 1.0$, the updated mean value will approach 1.0 and the updated COV value will approach the assumed within-site COV value of 0.20, as expected when the number of tests is sufficiently large. This is because test outcomes carry a larger weight than that of the prior information as described in Equations (10.13) and (10.14).

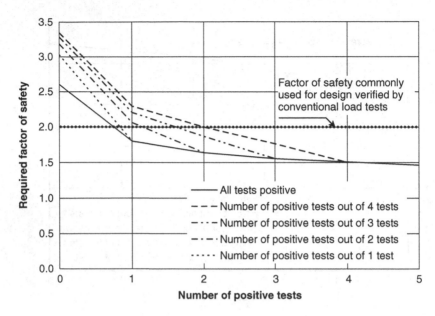

Figure 10.4 Factor of safety for single driven piles that are required to achieve a target reliability index of 2.0 after verification by proof load tests.

Table 10.4 Comparison of the updating effects of proof tests that pass at $x_T = 1.0$ and tests in which the test piles fail at a mean bearing capacity ratio of $x_T = 1.0$.

Number of tests	Tests that pass at $x_T = 1.0$			Tests that fail at $x_T = 1.0$		
	Mean	COV	β	Mean	COV	β
0	1.30	0.50	1.49	1.30	0.50	1.49
1	1.59	0.41	2.23	1.06	0.27	1.89
2	1.68	0.39	2.45	1.04	0.24	1.98
3	1.74	0.38	2.58	1.04	0.23	2.02
4	1.77	0.38	2.67	1.03	0.22	2.04
5	1.80	0.37	2.74	1.03	0.22	2.05
10	1.89	0.36	2.94	1.02	0.21	2.09

Note
Mean and COV of the bearing capacity ratio x.

10.4.3 Effect of accuracy of design methods

The example in the previous section focuses on a single design method. One may ask if the observations from the example would apply to other design methods. Table 10.5 presents statistics of a number of commonly used methods for design and construction of driven piles (Zhang *et al.*, 2001).

Table 10.5 Reliability statistics of single driven piles (after Zhang et al., 2001).

Method category	Prediction method	Number of cases	Bias factor λ_R	Coefficient of variation COV_R	ASD factor of safety FOS	Reliability index β	References
Dynamic methods with measurements (Construction stage)	CAPWAP (EOD, Florida database)	44	1.60	0.35	2.5	3.11	McVay et al. (2000)
	CAPWAP (BOR, Florida database)	79	1.26	0.35	2.5	2.55	McVay et al. (2000)
	CAPWAP (EOD, national database)	125	1.63	0.49	2.5	2.39	Paikowsky and Stenersen (2000)
	CAPWAP (BOR, national database)	162	1.16	0.34	2.5	2.38	Paikowsky and Stenersen (2000)
	Energy approach (EOD, Florida database)	27	1.11	0.34	2.5	2.29	McVay et al. (2000)
	Energy approach (BOR, Florida database)	72	0.84	0.36	3.25	2.11	McVay et al. (2000)
	Energy approach (EOD, national database)	128	1.08	0.40	2.5	1.94	Paikowsky and Stenersen (2000)
	Energy approach (BOR, national database)	153	0.79	0.37	3.25	1.92	Paikowsky and Stenersen (2000)
Static methods (Design stage)	Alpha method, clay type I	–	1.10	0.21	2.5	3.08	Sidi (1985), Barker et al. (1991)
	Alpha method, clay type II	–	2.34	0.57	2.5	2.72	Sidi (1985), Barker et al. (1991)
	Beta method	–	1.03	0.21	2.5	2.82	Sidi (1985), Barker et al. (1991)
	CPT method	–	1.03	0.36	2.5	1.98	Orchant et al. (1988), Barker et al. (1991), Schmertmann (1978)
	Lambda method, clay type I	–	1.02	0.41	2.5	1.74	Sidi (1985), Barker et al. (1991)
	Meyerhof's SPT method	–	1.30	0.50	2.5	1.92	Orchant et al. (1988), Barker et al. (1991)

Note
Type I refers to soils with undrained shear strength $S_u < 50$ kPa; Type II refers to soils with $S_u > 50$ kPa.

In this table, failure of piles is defined by the Davisson criterion. For each design method in the table, the driven pile cases are put together regardless of ground conditions or types of pile response (i.e. end-bearing or floating). Ideally, the cases should be organized into several subsets according to their ground conditions and types of pile response. The statistics in the table are therefore only intended to be used to illustrate the proposed methodology. The bias factors of the methods in the table vary from 0.79 to 2.34 and the COV values vary from 0.21 to 0.57. Indeed, the ASD approach results in designs with levels of safety that are rather uneven from one method to another (i.e. $\beta = 1.74$–3.11). If a FOS of 2.0 is used for all these methods and each design analysis is verified by two positive proof tests conducted to twice the design load (i.e. $x_T > 1.0$), the statistics of these methods can be updated as shown in Table 10.6. The updated bias factors are greater but the updated COV_R values are smaller than those in Table 10.5. The updated β values of all these methods fall into a narrow range (i.e. $\beta = 2.22$–2.89) with a mean of approximately 2.5.

Now consider the case when a FOS of 2.0 is applied to all these methods in Table 10.5 and the design is verified by two proof tests conducted to failure at an average load of twice the design load (i.e. $x_T = 1.0$). The corresponding updated statistics using the Bayesian sampling theory [Equations (10.12)–(10.14)] are also shown in Table 10.6. Both the updated bias factors ($\lambda_R = 0.97$–1.12) and the updated COV_R values ($COV_R = 0.21$–0.24) fall into narrow bands. Accordingly, the updated β values also fall into a narrow band (i.e. $\beta \doteq 1.78$–2.31) with a mean of approximately 2.0.

The results in Table 10.6 indicate that the safety level of a design verified by proof tests is less influenced by the accuracy of the design method. This is logical in the context of Bayesian statistical theory, in which the information of the empirical distribution will play a smaller role when more measured data at the site become available. This is consistent with foundation engineering practice. In the past, reliable designs had been achieved by subjecting design analyses of varying accuracies to proof tests and other quality control measures (Hannigan et al., 1997; O'Neill and Reese, 1999). This also shows the effectiveness of the observational method (Peck, 1969), with which uncertainties can be managed and acceptable safety levels can be maintained by acquiring additional information during construction. However, the importance of the accuracy of a design method in sizing the pile should be emphasized. A more accurate design method has a smaller COV_R and utilizes a larger percentage of the actual pile capacity (McVay et al., 2000). Hence, the required safety level can be achieved more economically. It should also be noted that the within-site COV is an important parameter. If the within-site COV at the site is larger than 0.2, as assumed in this chapter, the updated reliability will be lower than the results presented in Table 10.6.

Table 10.6 Reliability statistics of single driven piles designed using a FOS of 2.0 and verified by two load tests.

Prediction method	Verified by two positive tests, $x_T > 1.0$			Verified by two tests with a mean $x_T = 1.0$		
	Bias factor λ_R	Coefficient of variation COV_R	Reliability index β	Bias factor λ_R	Coefficient of variation COV_R	Reliability index β
CAPWAP (EOD, Florida database)	1.71	0.32	2.89	1.12	0.24	2.21
CAPWAP (BOR, Florida database)	1.46	0.30	2.62	1.07	0.24	2.07
CAPWAP (EOD, national database)	1.89	0.42	2.58	1.07	0.24	2.05
CAPWAP (BOR, national database)	1.38	0.29	2.54	1.05	0.24	2.02
Energy approach (EOD, Florida database)	1.34	0.28	2.50	1.04	0.24	1.99
Energy approach (BOR, Florida database)	1.22	0.28	2.26	0.98	0.24	1.81
Energy approach (EOD, national database)	1.42	0.31	2.46	1.03	0.24	1.95
Energy approach (BOR, national database)	1.20	0.28	2.22	0.97	0.24	1.78
Alpha method, clay type I	1.12	0.21	2.40	1.09	0.21	2.31
Alpha method, clay type II	2.54	0.52	2.69	1.09	0.24	2.09
Beta method	1.07	0.21	2.22	1.03	0.21	2.10
CPT method	1.33	0.29	2.43	1.02	0.24	1.93
Lambda method, clay type I	1.40	0.32	2.42	1.02	0.24	1.92
Meyerhof's SPT method	1.68	0.39	2.45	1.04	0.24	1.98

Note
(1) Type I refers to soils with undrained shear strength $S_u < 50$ kPa; Type II refers to soils with $S_u > 50$ kPa.
(2) x_T =Ratio of the maximum test load to the predicted pile capacity.

10.5 Acceptance criterion based on proof load tests

10.5.1 General principle

In the conventional allowable stress design approach, acceptance of a design using proof tests is based on whether the test piles fail before they are loaded to twice the design load. Presumably, the level of safety of the designs that are verified by "satisfactory" proof tests is not uniform. In a reliability-based design, a general criterion for accepting piles should be that the pile foundation as a whole, after verification by the proof tests, meets the prescribed target reliability levels for both ultimate limit states and serviceability limit states. Discussion on serviceability limit states is beyond the scope of this chapter.

In order to define acceptance conditions, "failure" of piles has to be defined first. Many failure criteria exist in the literature. Hirany and Kulhawy (1989), Ng et al. (2001b), and Zhang and Tang (2001) conducted extensive review of failure criteria for piles and drilled shafts and discussed the bias in the axial capacity arising from failure criteria. A particular failure criterion may be applicable only to certain deep foundation types and ground conditions. In discussing a particular design method, a failure criterion should always be specified explicitly. In the examples presented in this chapter, only driven piles are involved and only the Davisson failure criterion is used. These examples are for illustrative purposes. Suitable failure criteria should be selected in establishing acceptance conditions for a particular job.

The next step in defining acceptance conditions is to select the target reliability of the pile foundation. The target reliability of the pile foundation should match that of the superstructure and of other types of foundations. Meyerhof (1970), Barker et al. (1991), Becker (1996a), O'Neill and Reese (1999), Phoon et al. (2000), Zhang et al. (2001), Paikowsky (2002) and others have studied the target reliability of foundations based on calibration of the ASD practice. A target reliability index, β_T, between 3.0 and 3.5 appears to be suitable for pile foundations.

The reliability of a pile group can be significantly higher than that of single piles (Tang and Gilbert, 1993; Bea et al., 1999; Zhang et al., 2001). Since proof tests are mostly conducted on single piles, it is necessary to find a target reliability for single piles that matches the target reliability of the foundation system. Based on calibration of the ASD practice, Wu et al. (1989), Tang (1989), and Barker et al. (1991) found a range of reliability index values between 1.4 and 3.1 and Barker et al. (1991), ASSHTO (1997) and Withiam et al. (2001) recommended target reliability index values of $\beta_{TS} = 2.0$–2.5 for single driven piles and $\beta_{TS} = 2.5$–3.0 for single drilled shafts. Zhang et al. (2001) attributed the increased reliability of pile foundation systems to group effects and system effects. They calculated β_{TS} values for single driven piles to achieve a β_T value of 3.0. For pile groups larger than four piles, the calibrated β_{TS} values are mostly in the range of 2.0–2.8 if no system effects are considered. The β_{TS} values decrease to 1.7–2.5 when a system factor of 1.25 is considered, and further to 1.5–2.0 when a larger system factor of 1.5 is considered. For a structure supported by four or fewer far-apart piles, the pile group effect and system effect may not be dependable (Zhang et al., 2001; Paikowsky, 2002). The target reliability of single piles should therefore be the same as that of the foundation system, say $\beta_{TS} = 3.0$.

10.5.2 Reliability-based acceptance criteria

In the ASD approach, it is logical to use a maximum test load of twice the design load since a FOS of 2.0 is commonly used for designs verified by proof load tests. Table 10.6 shows that the ASD approach using two proof tests conducted to twice the design load can indeed lead to a uniform target reliability level around $\beta = 2.5$ regardless of the accuracy of the design method. Even if the test piles fail at an average load of twice the design load, a uniform reliability level around $\beta = 2.0$ can still be obtained. Compared with recommended target reliability indices, the reliability of the current ASD practice for large pile groups may be considered sufficient. In the RBD approach, the target reliability is a more direct acceptance indicator. From Equations (10.15) and (10.16), a target reliability can be achieved by many combinations of the proof test parameters such as the number of tests, the test load, the FOS, and the test outcomes. Thus, the use of other FOS values and test loads is equally feasible.

Analyses with the SPT method (Meyerhof, 1976) are again conducted to illustrate the effects of the load test parameters. Figure 10.5 shows the updated reliability index of piles designed using a FOS of 2.0 and verified by various numbers of tests conducted to 1.0–3.0 times the design load. If the proof tests were carried out to the design load ($k = 1.0$), the tests would have a negligible effect on the reliability of the piles; hence, the effectiveness of the tests is minimal. This is because the probability that the piles do not fail at the design load is 96% based on the empirical distribution, and the proof tests only prove an event of little uncertainty. If the test load were set at 1.5 times ($k = 1.5$) the design load, the effectiveness of the proof tests is still limited since a β_T value of 2.5 cannot be verified with a reasonable number of proof tests. The test load of twice ($k = 2.0$) the design load provides a better solution: when the design is verified by one or two positive proof tests, the updated β values will be in the range of 1.7–2.5, which may be considered acceptable for large pile systems. If the test load is set at three times ($k = 3.0$) the design load, then a high target reliability of $\beta_T = 3.0$ can be verified by one or two successful proof tests. Figure 10.6 further defines suitable maximum test loads that are needed to achieve specific target

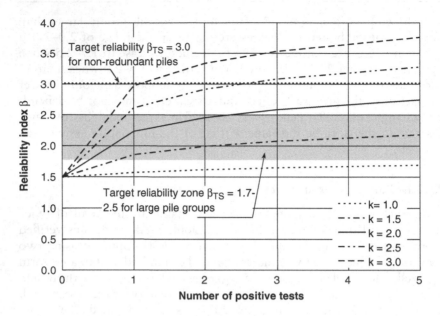

Figure 10.5 Reliability index of single driven piles designed with a FOS of 2.0 and verified by proof tests conducted to various maximum test loads (in *k* times the design load).

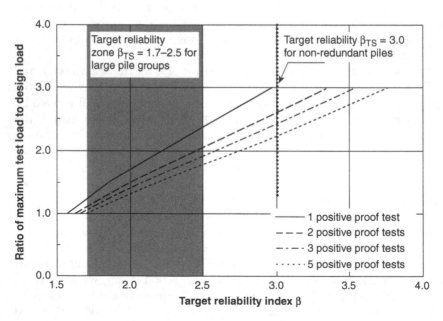

Figure 10.6 Maximum test load that is required to verify the SPT design method using a FOS of 2.0.

reliability values using a limited number of proof tests. A maximum test load of 2.5 times or three times the design load is necessary for non-redundant pile groups where group effects and system effects are not present, otherwise the required target reliability of $\beta_T = 3.0$ cannot be verified by a reasonable number of conventional proof tests conducted only to twice the design load.

Figure 10.7 shows the FOS values needed to achieve a β value of 2.5. As the verification test load increases, the required FOS decreases. In particular, if two proof tests are conducted to 1.5 times ($x_T = 1.5$) the predicted pile capacity, a β_T of 2.5 can be achieved with a FOS of approximately 1.5, compared with a FOS of 2.0 if a test load equal to the predicted pile capacity ($x_T = 1.0$) is adopted.

For codification purposes, the effects of proof test loads on the reliability of various design methods should be studied and carefully collected databases are needed to characterize the variability of the design methods. The analysis of the design methods in Table 10.5 is for illustrating the methodology only. Limitations with the reliability statistics in the table have been pointed out earlier. Table 10.7 presents the FOS values required to achieve β_T values of 2.0, 2.5, and 3.0 when the design is verified by two proof tests conducted to different test loads. Given a proof test load, the FOS values required for

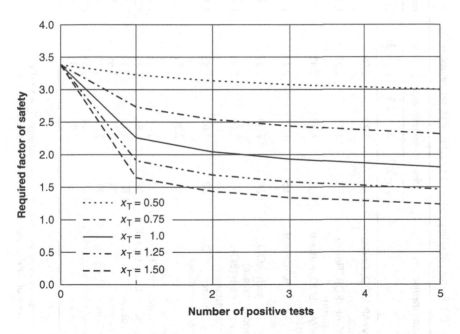

Figure 10.7 Factor of safety that is required to achieve a target reliability level of $\beta_T = 2.5$ after verification by proof tests conducted to various maximum test loads.

Table 10.7 Factors of safety required to achieve target reliability index values of $\beta_T = 2.0$, $\beta_T = 2.5$, and $\beta_T = 3.0$.

Prediction method	Verified by two tests with $x_T > 1.0$			Verified by two tests with $x_T > 1.25$			Verified by two tests with $x_T > 1.5$		
	$\beta_T = 2.0$	$\beta_T = 2.5$	$\beta_T = 3.0$	$\beta_T = 2.0$	$\beta_T = 2.5$	$\beta_T = 3.0$	$\beta_T = 2.0$	$\beta_T = 2.5$	$\beta_T = 1.5$ ... $\beta_T = 3.0$
CAPWAP (EOD, Florida database)	1.42	1.72	2.08	1.27	1.52	1.83	1.13	1.35	1.62
CAPWAP (BOR, Florida database)	1.59	1.91	2.29	1.39	1.65	1.97	1.23	1.46	1.73
CAPWAP (EOD, national database)	1.53	1.93	2.42	1.31	1.62	2.01	1.13	1.39	1.71
CAPWAP (BOR, national database)	1.65	1.97	2.36	1.43	1.70	2.02	1.26	1.50	1.78
Energy approach (EOD, Florida database)	1.68	2.00	2.39	1.45	1.73	2.05	1.29	1.52	1.80
Energy approach (BOR, Florida database)	1.83	2.18	2.59	1.55	1.84	2.19	1.36	1.61	1.90
Energy approach (EOD, national database)	1.68	2.03	2.45	1.43	1.71	2.05	1.24	1.48	1.77
Energy approach (BOR, national database)	1.85	2.20	2.62	1.57	1.86	2.21	1.37	1.62	1.91
Alpha method, clay type I	1.77	2.06	2.39	1.74	2.02	2.34	1.70	1.97	2.29
Alpha method, clay type II	1.38	1.81	2.36	1.22	1.57	2.03	1.07	1.36	1.74
Beta method	1.87	2.18	2.54	1.81	2.10	2.45	1.75	2.03	2.37
CPT method	1.71	2.05	2.45	1.47	1.75	2.08	1.29	1.53	1.81
Lambda method, clay type I	1.71	2.06	2.49	1.44	1.73	2.07	1.25	1.49	1.78
Meyerhof's SPT method	1.64	2.04	2.54	1.37	1.69	2.07	1.18	1.43	1.75
Mean value	1.67	2.01	2.43	1.46	1.75	2.10	1.30	1.55	1.85

Note
(1) Type I refers to soils with undrained shear strength $S_u < 50$ kPa; Type II refers to soils with $S_u > 50$ kPa.
(2) x_T = Ratio of the maximum test load to the predicted pile capacity.

Table 10.8 Illustrative values of recommended factor of safety and maximum test load.

$\beta_T = 2.0$			$\beta_T = 2.5$			$\beta_T = 3.0$		
x_T	FOS	k	x_T	FOS	k	x_T	FOS	k
1.00	1.70	1.7	1.00	2.00	2.0	1.00	2.45	2.5
1.25	1.50	1.9	1.25	1.75	2.2	1.25	2.10	2.7
1.50	1.30	2.0	1.50	1.55	2.4	1.50	1.85	2.8

Note
(1) Based on the reliability statistics in Table 10.5 for driven piles.
(2) Pile failure is defined by the Davisson criterion.
(3) Two proof tests are assumed and the outcomes are assumed to be positive.
(4) x_T = ratio of the maximum test load to the predicted pile capacity.
(5) FOS = rounded factor of safety.
(6) k = recommended ratio of the maximum test load to the design load (rounded values).

the design methods to achieve a prescribed target reliability level are rather uniform. Thus, the mean of the FOS values for all the methods offers a suitable FOS for design. Table 10.8 summarizes the rounded average FOS values that could be used with proof tests, based on the limited reliability statistics in Table 10.5. Rounded values of the ratio of the maximum test load to the design load, k, which is the multiplication of FOS by x_T, are also shown in Table 10.8.

Table 10.8 provides illustrative criteria for the design and acceptance of driven pile foundations using proof tests. The criteria may be implemented in the following steps:

1 calculate the design load and choose a target reliability, say $\beta_T = 2.5$;
2 size the piles using a FOS (e.g. 1.75), bearing in mind that proof load tests will be conducted to verify the design;
3 construct a few trial piles;
4 proof test the trial piles to the required test load (e.g. 2.2 times the design load);
5 accept the design if the proof tests are positive.

It can be argued that proof load tests should be conducted to a large load (e.g. $x_T = 1.50$ or 2.8 times the design load in the case of FOS = 1.85 and $\beta_T = 3.0$). First, this is required for verifying the reliability of non-redundant pile foundations in which the group effects and system effects are not present. Second, the large test load warrants the use of a lower FOS (e.g. smaller than 2.0) and considerable savings since the extra costs for testing a pile to a larger load are relatively small. Third, more test piles will fail at a larger test load (i.e. 71% of test piles may fail at $x_T = 1.5$ based on the empirical distribution in Figure 10.2), and hence the tests will have greater research value for improvement of practice. Malone (1992) has made a similar suggestion in studying the performance of piles in weathered granite.

Because of the differences in the acceptance criteria adopted in the ASD and RBD approaches, some "acceptable proof tests" in the ASD may become unacceptable in the RBD. For example, the reliability of a column supported by a single large diameter bored pile may not be sufficient because the β value of the pile can still be smaller than $\beta_T = 3.0$, even if the pile has been proof-tested to twice the design load. On the other hand, some "unsatisfactory" piles in the ASD may turn out to be satisfactory in the RBD provided that the design analysis and the outcome of the proof tests together result in sufficient system reliability. For example, the requirement of $\beta_T = 2.0$ can be satisfied when the test load is somewhat smaller than twice the design load (see Table 10.8) or when one out of four tests fails (see Figure 10.3).

10.6 Summary

It has been shown that pile load tests are an integral part of the design and construction process of pile foundations. Pile load tests can be used as a means for verifying reliability and reducing costs. Whether a load test is carried out to failure or not, the test outcome can be used to ensure that the acceptance reliability is met using the methods described in this chapter. Thus, contributions of load tests can be included in a design in a systematic manner. Although various analysis methods could arrive at considerably different designs, the reliability of the designs associated with these analysis methods is rather uniform if the designs adopt the same FOS of 2.0 and are verified by consecutive positive proof tests.

In the reliability-based design, the acceptance criterion is proposed to be that the pile foundation as a whole, after verification by the proof tests, meets the specified target reliability levels for both ultimate limit states and serviceability limit states. This is different from the conventional allowable stress design in which a test pile will not be accepted if it fails before it has been loaded to twice the design load. Since the reliability of a single pile is not the same as the reliability of a pile foundation system, the target reliability for acceptance purposes should be structure-specific.

If the maximum test load were only around the design load, little information from the test could be derived. The use of a maximum test load of one times the predicted pile capacity is generally acceptable for large pile group foundations. However, a larger test load of 1.5 times the predicted pile capacity is recommended. First, it is needed for verifying the reliability of the pile foundation in which group effects and system effects are not present. Second, it warrants the use of a smaller FOS and hence savings in pile foundation costs. Third, the percentage of test piles that may fail at 1.5 times the predicted pile capacity is larger. The tests will thus have greater research value for further improvement of practice.

Acknowledgment

The research described in this chapter was substantially supported by the Research Grants Council of the Hong Kong Special Administrative Region (Project Nos. HKUST6035/02E and HKUST6126/03E).

References

AASHTO (1997). *LRFD Bridge Design Specifications*. American Association of State Highway and Transportation Officials, Washington, DC.

Ang, A. H-S. and Tang, W. H. (2007). *Probability Concepts in Engineering-Emphasis on Applications to Civil and Environmental Engineering*, 2nd edition. John Wiley & Sons, New York.

Baecher, G. R. and Rackwitz, R. (1982). Factors of safety and pile load tests. *International Journal for Numerical and Analytical Methods in Geomechanics*, 6, 409–24.

Barker, R. M., Duncan, J. M., Rojiani, K. B., Ooi, P. S. K., Tan, C. K. and Kim, S. G. (1991). *Manuals for the Design of Bridge Foundations, NCHRP Report 343*. Transportation Research Board, National Research Council, Washington, DC.

Bea, R. G., Jin, Z., Valle, C. and Ramos, R. (1999). Evaluation of reliability of platform pile foundations. *Journal of Geotechnical and Geoenvironmental Engineering, ASCE*, 125, 696–704.

Becker, D. E. (1996a). Eighteenth Canadian geotechnical colloquium: limit states design for foundations. Part I. An overview of the foundation design process. *Canadian Geotechnical Journal*, 33, 956–83.

Becker, D. E. (1996b). Eighteenth Canadian geotechnical colloquium: limit states design for foundations. Part II. Development for the national building code of Canada. *Canadian Geotechnical Journal*, 33, 984–1007.

Bell, K. R., Davie, J. R., Clemente, J. L. and Likins, G. (2002). Proven success for driven pile foundations. In *Proceedings of the International Deep Foundations Congress, Geotechnical Special Publication No. 116*, Eds. M. W. O'Neill, and F. C. Townsend. ASCE, Reston, VA, pp. 1029–37.

BSI (1986). *BS 8004 British Standard Code of Practice for Foundations*. British Standards Institution, London.

Buildings Department (2002). *Pile foundations-practice notes for authorized persons and registered structural engineers* PNAP66. Buildings Department, Hong Kong.

Chow, F. C., Jardine, R. J., Brucy, F. and Nauroy, J. F. (1998). Effect of time on the capacity of pipe piles in dense marine sand. *Journal of Geotechnical and Geoenvironmental Engineering, ASCE*, 124(3), 254–64.

Chu, R. P. K., Hing, W. W., Tang, A. and Woo, A. (2001). Review and way forward of foundation construction requirements from major client organizations. In *New Development in Foundation Construction*. The Hong Kong Institution of Engineers, Hong Kong.

Dasaka, S. M. and Zhang, L. M. (2006). Evaluation of spatial variability of weathered rock for pile design. In *Proceedings of International Symposium on New Generation Design Codes for Geotechnical Engineering Practice*, 2–3 Nov. 2006, Taipei. World Scientific, Singapore (CDROM).

Duncan, J. M. (2000). Factors of safety and reliability in geotechnical engineering. *Journal of Geotechnical and Geoenvironmental Engineering*, ASCE, 126, 307–16.

Evangelista, A., Pellegrino, A. and Viggiani, C. (1977). Variability among piles of the same foundation. In *Proceedings of the 9th International Conference on Soil Mechanics and Foundation Engineering*, Tokyo, pp. 493–500.

Fenton, G. A. (1999). Estimation for stochastic soil models. *Journal of Geotechnical Engineering*, ASCE, 125, 470–85.

Geotechnical Engineering Office (2006). *Foundation Design and Construction*, GEO Publication 1/2006. Geotechnical Engineering Office, Hong Kong.

Hannigan, P. J., Goble, G. G., Thendean, G., Likins, G. E. and Rausche, F. (1997). *Design and Construction of Driven Pile Foundations, Workshop Manual, Vol. 1*. Publication No. FHWA-HI-97-014. Federal Highway Administration, Washington, DC.

Hirany, A. and Kulhawy, F. H. (1989). Interpretation of load tests on drilled shafts – Part 1: Axial compression. In *Foundation Engineering: Current Principles and Practices*, Vol. 2, Ed. F. H. Kulhawy. ASCE, New York, pp. 1132–49.

ISSMFE Subcommittee on Field and Laboratory Testing (1985). Axial pile loading test – Part 1: Static loading. *Geotechnical Testing Journal*, 8, 79–89.

Kay, J. N. (1976). Safety factor evaluation for single piles in sand. *Journal of Geotechnical Engineering*, ASCE, 102, 1093–108.

Lacasse, S. and Goulois, A. (1989). Uncertainty in API parameters for predictions of axial capacity of driven piles in sand. In *Proceedings of the 21st Offshore Technology Conference*, Houston, TX, pp. 353–8.

Malone, A. W. (1992). Piling in tropic weathered granite. In *Proceedings of International Conference on Geotechnical Engineering*. Malaysia Technological University, Johor Bahru, pp. 411–57.

McVay, M. C., Birgisson, B., Zhang, L. M., Perez, A. and Putcha, S. (2000). Load and resistance factor design (LRFD) for driven piles using dynamic methods – a Florida perspective. *Geotechnical Testing Journal ASTM*, 23, 55–66.

Meyerhof, G. G. (1970). Safety factors in soil mechanics. *Canadian Geotechnical Journal*, 7, 349–55.

Meyerhof, G. G. (1976). Bearing capacity and settlement of pile foundations. *Journal of Geotechnical Engineering*, ASCE, 102, 195–228.

Ng, C. W. W., Li, J. H. M. and Yau, T. L. Y. (2001a). Behaviour of large diameter floating bored piles in saprolitic soils. *Soils and Foundations*, 41, 37–52.

Ng, C. W. W., Yau, T. L. Y., Li, J. H. M. and Tang, W. H. (2001b). New failure load criterion for large diameter bored piles in weathered geomaterials. *Journal of Geotechnical and Geoenvironmental Engineering*, ASCE, 127, 488–98.

NRC (1995). *National Building Code of Canada*. National Research Council Canada, Ottawa.

O'Neill, M. W. and Reese, L. C. (1999). *Drilled Shafts, Construction Procedures and Design Methods*. Publication No. FHWA-IF-99-025. Federal Highway Administration, Washington, DC.

Orchant, C. J., Kulhawy, F. H. and Trautmann, C. H. (1988). *Reliability-based foundation design for transmission line structures*, Vol. 2, Critical evaluation of in-situ test methods. EL-5507 Final Report, Electrical Power Institute, Palo Alto.

Paikowsky, S. G. (2002). Load and resistance factor design (LRFD) for deep foundations. In *Foundation Design Codes – Proceedings, IWS Kamakura 2002*, Eds. Y. Honjo, O. Kusakabe, K. Matsui, M. Kouda, and G. Pokharel. A. A. Balkema, Rotterdam, pp. 59–94.

Paikowsky, S. G. and Stenersen, K. L. (2000). The performance of the dynamic methods, their controlling parameters and deep foundation specifications. In *Proceedings of the Sixth International Conference on the Application of Stress-Wave Theory to Piles*, Eds. S. Niyama and J. Beim. A. A. Balkema, Rotterdam, pp. 281-304.

Passe, P., Knight, B. and Lai, P. (1997). Load tests. In *FDOT Geotechnical News*, Vol. 7, Ed. P. Passe. Florida Department of Transportation, Tallahassee.

Peck, R. B. (1969). Advantages and limitations of the observational method in applied soil mechanics. *Geotechnique*, 19, 171–87.

Phoon, K. K., Kulhawy, F. H. and Grigoriu, M. D. (2000). Reliability-based design for transmission line structure foundations. *Computers and Geotechnics*, 26, 169–85.

Schmertmann, J. H. (1978). *Guidelines for cone penetration test, performance, and design*. Report No. FHWA-TS-78-209. Federal Highway Administration, Washington, DC.

Shek, M. P., Zhang, L. M. and Pang, H. W. (2006). Setup effect in long piles in weathered soils. *Geotechnical Engineering, Proceedings of the Institution of Civil Engineers*, 159(GE3), 145–52.

Sidi, I. D. (1985). *Probabilistic prediction of friction pile capacities*. PhD thesis, University of Illinois, Urbana-Champaign.

Tang, W. H. (1989). Uncertainties in offshore axial pile capacity. In *Geotechnical Special Publication No. 27, Vol. 2*, Ed. F. H. Kulhawy: ASCE, New York, pp. 833–47.

Tang, W. H. and Gilbert, R. B. (1993). Case study of offshore pile system reliability. In *Proceedings 25th Offshore Technology Conference*, Houston, TX, pp. 677–86.

US Army Corps of Engineers (1993). *Design of Pile Foundations*. ASCE Press, New York.

Whitman, R. V. (1984). Evaluating calculated risk in geotechnical engineering. *Journal of Geotechnical Engineering, ASCE*, 110, 145–88.

Withiam, J. L., Voytko, E. P., Barker, R. M., Duncan, J. M., Kelly, B. C., Musser, S. C. and Elias, V. (2001). *Load and resistance factor design (LRFD) for highway bridge substructures*, Report No. FHWA HI-98-032. Federal Highway Administration, Washington, DC.

Wu, T. H., Tang, W. H., Sangrey, D. A. and Baecher, G. B. (1989). Reliability of offshore foundations – state of the art. *Journal of Geotechnical Engineering, ASCE*, 115, 157–78.

Zhang, L. M. (2004). Reliability verification using proof pile load tests. *Journal of Geotechnical and Geoenvironmental Engineering, ASCE*, 130, 1203–13.

Zhang, L. M. and Tang, W. H. (2001). Bias in axial capacity of single bored piles arising from failure criteria. In *Proceedings ICOSSAR'2001, International Association for Structural Safety and Reliability*, Eds. R. Crotis, G. Schuller and M. Shinozuka. A. A. Balkema, Rotterdam (CDROM).

Zhang, L. M. and Tang, W. H. (2002). Use of load tests for reducing pile length. In *Geotechnical Special Publication No. 116*, Eds. M. W. O'Neill and F. C. Townsend. ASCE, Reston, pp. 993–1005.

Zhang, L. M., Li, D. Q. and Tang, W. H. (2006a). Level of construction control and safety of driven piles. *Soils and Foundations*, 46(4), 415–25.

Zhang, L. M., Shek, M. P., Pang, W. H. and Pang, C. F. (2006b). Knowledge-based pile design using a comprehensive database. *Geotechnical Engineering, Proceedings of the Institution of Civil Engineers*, 159(GE3), 177–85.

Zhang, L. M., Tang, W. H. and Ng, C. W. W. (2001). Reliability of axially loaded driven pile groups. *Journal of Geotechnical and Geoenvironmental Engineering, ASCE*, 127, 1051–60.

Zhang, L. M., Tang, W. H., Zhang, L. L. and Zheng, J. G. (2004). Reducing uncertainty of prediction from empirical correlations. *Journal of Geotechnical and Geoenvironmental Engineering, ASCE*, 130, 526–34.

Chapter 11

Reliability analysis of slopes

Tien H. Wu

11.1 Introduction

Slope stability is an old and familiar problem in geotechnical engineering. Since the early beginnings, impressive advances have been made and a wealth of experience has been collected by the profession. The overall performance of the general methodology has been unquestionably successful. Nevertheless, the design of slopes remains a challenge, particularly with new and emerging problems. As with all geotechnical projects, a design is based on inputs, primarily loads and site characteristics. The latter includes the subsoil profile, soil moisture conditions and soil properties. Since all of the above are estimated from incomplete data and cannot be determined precisely for the lifetime of the structure, an uncertainty is associated with each component. To account for this, the traditional approach is to apply a safety factor, derived largely from experience. This approach has worked well, but difficulty is encountered where experience is inadequate or nonexistent. Then reliability-based design (RBD) offers the option of evaluating the uncertainties and estimating their effects on safety. The origin of RBD can be traced to partial safety factors. Taylor (1948) explained the need to account for different uncertainties about the cohesional and frictional components of soil strength, and Lumb (1970) used the standard deviation to represent the uncertainty and expressed the partial safety factors in terms of the standard deviations. Partial safety factors were incorporated into design practice by Danish engineers in the 1960s (Hansen, 1965).

This chapter reviews reliability methods as applied to slope stability and uncertainties in loads, pore pressure, strength and analysis. Emphasis is directed toward the simple methods, the relevant input data, and their applications. The use of the first-order, second-moment method is illustrated by several detailed examples since the relations between the variables are explicit and readily understood. Advance methods are reviewed and the reader is referred to the relevant articles for details.

11.2 Methods of stability and reliability analysis

Three types of analysis may be used for deterministic prediction of failure: limit equilibrium analysis, displacement analysis for seismic loads, and finite element analysis. Reliability analysis can be applied to all three types.

11.2.1 Probability of failure

Reliability is commonly expressed as a probability of failure, P_f, or a reliability index, β, defined as

$$P_f = 1 - \Phi[\beta] \tag{11.1a}$$

If F_s has normal distribution,

$$P_f = 1 - \Phi\left[\frac{\overline{F}_s - 1}{\sigma(F_s)}\right], \quad \beta = \frac{\overline{F}_s - 1}{\sigma(F_s)} \tag{11.1b}$$

and if F_s has log-normal distribution,

$$P_f = 1 - \Phi\left[\frac{\ln \overline{F}_s - 0.5\Delta(F_s)^2}{\Delta(F_s)}\right], \quad \beta = \frac{\ln \overline{F}_s - 0.5\Delta(F_s)^2}{\Delta(F_s)} \tag{11.1c}$$

$$\Phi(\beta) = \frac{1}{\sqrt{2\pi}} \int_{-\infty}^{\beta} e^{\left(-\frac{1}{2}z^2\right)} dz \tag{11.1d}$$

F_s = safety factor and the overbar denotes the mean value, $\sigma()$ = standard deviation, $\Delta = ()$ coefficient of variation (COV), and $\Delta(F_s) = \sigma(F_s)/\overline{F}_s$. Equation (11.1c) is approximate and should be limited to $\Delta < 0.5$.

The term $\overline{F}_s - 1$, Equation (11.1b), or $\ln \overline{F}_s - 0.5\,\Delta(F_s)^2$, Equation (11.1c), denotes the distance between the mean safety factor and $F_s = 1$, or failure. The term $\sigma(F_s)$, or $\Delta(F_s)$, which represents the uncertainty about F_s, is reflected by the spread of the probability density function (pdf) $f(F_s)$. Figure 11.1 shows the pdf for normal distribution. Curves 1 and 2 represent two designs, with the same mean safety factor $\overline{F}_s$, but with $\sigma(F_s) = 0.4$ and 1.2, respectively. The shaded area under each curve is P_f. It can be seen that P_f for design 2 is several times greater than for design 1, because of the greater uncertainty. The reliability index β in Equations (11.1b) and (11.1c) is a convenient expression that normalizes the safety factor with respect to its uncertainty. Two designs with the same β will have the same P_f. Both normal and ln-normal distributions have been used to represent soil properties and safety factors. The ln-normal has the advantage that it does not have negative values. However, except for very large values of β, the difference between the two distributions is not large.

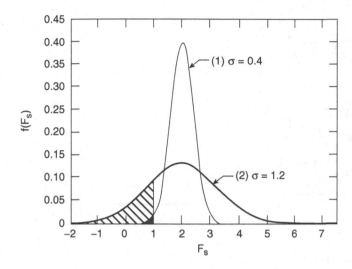

Figure 11.1 Probability distribution of F_s.

While the mathematical rules for computation of probability are well established, the meaning of probability is not well defined. It can be interpreted as the relative frequency of occurrence of an event or a degree of belief. The general notion is that it denotes the "chance" that an event will occur. The reader should refer to Baecher and Christian (2003) for a thorough treatment of this issue. What constitutes an acceptable P_f depends on the consequence or damage that would result from a failure. This is a decision-making problem and is described by Gilbert in Chapter 5. The current approach in Load Resistance Factor Design (LRFD) is to choose a "target" β by calibration against current practice (Rojiani *et al.*, 1991).

11.2.2 Limit equilibrium method

The method of slices (Fellenius, 1936), with various modifications, is widely used for stability analysis. Figure 11.2 shows the forces on a slice, for a very long slope failure or plane strain. The general equations for the safety factor are (Fredlund and Krahn, 1977)

$$F_{s,m} = \frac{M_r}{M_o} = \frac{\sum \{c'lr + (P - ul)r\tan\phi'\}}{\sum Wx - \sum Pf + \sum kWe}, \tag{11.2a}$$

$$F_{s,f} = \frac{F_r}{F_o} = \frac{\sum c'l\cos\alpha + \sum (P - ul)\tan\phi'\cos\alpha}{\sum P\sin\alpha + \sum kW} \tag{11.2b}$$

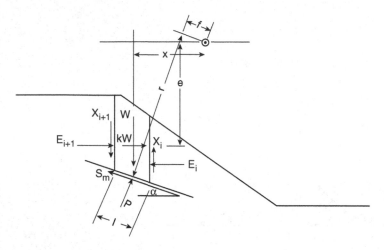

Figure 11.2 Forces on failure surface.

where subscripts m and f denote equilibrium with respect to moment and force; M_o, M_r = driving and resisting moments; F_o, F_r = horizontal driving force and resistance; W = weight of slice; P = normal force on base of slice, whose length is l; r, f = moment arms of the shear and normal forces on base of slice; k = seismic coefficient; e = moment arm of inertia force; α = angle of inclination of base. The expression for P depends on the assumption used to render the problem statically determinate. The summation in Equations (11.2) is taken over all the slices. For unsaturated soils, the pore pressure consists of the poreair pressure u_a and the porewater pressure u_w. The equations for $F_{s,m}$ and $F_{s,f}$ are given in Fredlund and Rahardjo (1993).

The first-order, second moment method (FOSM) is the simplest method to calculate the mean and variance of F_s. For a function $Y = g(X_1, \ldots, X_n)$, in which $X_1, \ldots, X_n$ = random variables, the mean and variance are

$$\overline{Y} = g[\overline{X}_1, \ldots, \overline{X}_n)] \tag{11.3a}$$

$$V(Y) = \sum_i \left(\frac{\partial g}{\partial X_i} \right)^2 V(X_i) + \ldots \sum_i \sum_j \left(\frac{\partial g}{\partial X_i} \right) \left(\frac{\partial g}{\partial X_j} \right) \mathrm{Cov}(X_i X_j)$$

$$= \rho_{ij} \sigma(X_i) \sigma(X_j) \tag{11.3b}$$

where $\partial g / \partial X_i$ = sensitivity of g to X_i, Cov $(X_i X_j)$ = covariance of $X_i X_j$, ρ_{ij} = correlation coefficient. Given the mean and variance of the input variables X_i

to Equations (11.2), Equations (11.3) are used to calculate $\overline{F}_s$ and $V(F_s)$. The distribution of Y is usually unknown except for very simple forms of the function g. If the normal or ln-normal distribution is assumed, β and P_f can be calculated via Equations (11.1).

In many geotechnical engineering problems, the driving moment is well defined so that its uncertainty is small and may be ignored. Then only $V(M_r)$ and $V(F_r)$ need to be evaluated. Let the random variables be c', $\tan\phi'$ and u. Application of Equation (11.3b) to the numerator in Equation (11.2a) yields

$$V(M_r) = \Sigma r^2 l^2 V(c') + [\Sigma r^2 (P - \overline{u}l)^2] V(\tan\varphi')$$

$$+ \Sigma r^2 l(P - \overline{u}l)\rho_{c,\phi}\,\sigma(c')\sigma(\tan\varphi') + [-rl\overline{\tan\varphi'}]^2 V(u) \qquad (11.4a)$$

If a total stress analysis is performed for undrained loading, or $\phi = 0$,

$$V[M_r] = \Sigma r^2 l^2 V(c) \qquad (11.4b)$$

Tang et al. (1976) and Yucemen and Tang (1975) provide examples of application of FOSM to total stress analysis and effective stress analysis of slopes. Some of the simplifications that have been used include the assumption of independence between X_i and X_j and use of the F_s for the critical circle determined by limit equilibrium analysis to calculate P_f.

Example 1

The slope failure in Chicago, described by Tang et al. (1976) and Ireland (1954), Figure 11.3, is used to illustrate the relationship between the mean safety factor, $\overline{F}_s$, the uncertainty about F_s and the failure probability P_f.

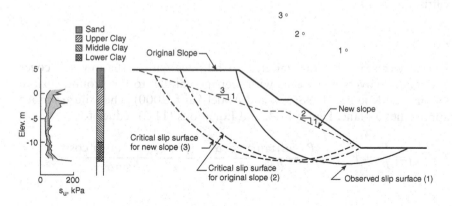

Figure 11.3 Slope failure in Chicago (adapted from Ireland (1954) and Tang et al. (1976)). s_u = undrained shear strength.

Table 11.1 Reliability index and failure probabilities, Chicago slope.

Slope	Slip surface	$\overline{F}_s$	$\Delta(M_r)$	Ln-normal		Normal	
				β	P_f	β	P_f
Existing	Observed	1.18	0.18	0.82	0.20	0.84	0.21
	Critical	0.86	0.18	−0.92	0.82	−0.10	0.82
Redesigned	Critical	1.29	0.16	1.62	0.05	1.49	0.07

The mean safety factor, calculated with the means of the undrained shear strengths, is 1.18 for the observed slip surface. The uncertainty about F_s comes from the uncertainty about the resisting moment, expressed as $\Delta(M_R)$, which, in turn, comes from the uncertainty about the shear strength of the soil and inaccuracies in the stability analysis. The total uncertainty is $\Delta(M_R) = \Delta(F_s) = 0.18$. Equation (11.2c), which assumes the ln-normal distribution for F_s, gives $\beta = 0.82$, $P_f = 0.2$ (Table 11.1). For the critical slip surface determined by limit equilibrium analysis, similar calculations yield $\beta = -0.92$ and $P_f = 0.82$. A redesigned slope with $\overline{F}_s = 1.29$ has $\beta = 1.62$ and $P_f = 0.05$. For comparison, the values of β and P_f calculated with the normal distribution are also shown in Table 11.1. The evaluation of the uncertainties and $\Delta(M_R)$ is described in Example 3.

With respect to the difference between the critical failure surface (2) and the observed failure surface (1), it should be noted that all of the soil samples were taken outside the slide area (Ireland, 1954). Hence, the strengths along the failure surfaces remain questionable. In probabilistic terms, the pertinent question now becomes: given that failure has, or has not, occurred, what is the shear strength? This issue is addressed in Section 11.3.7 (see also Christian, 2004).

Example 2

During wet seasons, slope failures on hillsides with a shallow soil cover and landfill covers may occur with seepage parallel to the ground surface (Soong and Koerner, 1996; Wu and Abdel-Latif, 2000). The failure surface approaches a plane, Figure 11.4, and Equations (11.2) reduce to

$$F_{s,f} = \frac{F_r}{F_o} = F_{s,m} = \frac{(P - ul)\tan\varphi'\cos\alpha}{P\sin\alpha} = \frac{(\gamma h - \gamma_w h_w)\tan\varphi'\cos\alpha}{\gamma h \sin\alpha}$$

(11.5)

with $P = \gamma h \cos\alpha$, $ul = \gamma_w h_w \cos\alpha$, γ = unit weight of soil, γ_w = unit weight of water, and h and h_w are as shown in Figure 11.4. It is assumed that γ is

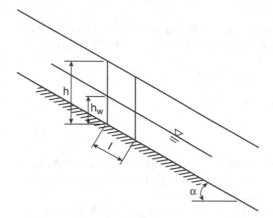

Figure 11.4 Infinite slope.

the same above and below the water table, and $k = 0$. The major hazard comes from the rise in the water level, h_w. Procedures for estimating h_w are described in Section 11.2.4.

For the present example, consider a slope with $h = 0.6$ m, $\alpha = 22°$, $c' = 0$, $\varphi' = 30°$. Assume that the annual maximum of h_w has a mean and standard deviation equal to $\overline{h}_w = 0.12$ m and $\sigma(h_w) = 0.075$ m. Substituting the mean values into Equation (11.5) gives $\overline{F}_R = 0.5$ kN and $\overline{F}_s = 1.28$. Following Equation (11.3b), the variance of F_R is

$$V(F_R) = [-\gamma_w \cos \alpha \,\overline{\tan \varphi'}]^2 V(h_w) + [(\gamma h - \gamma_w \overline{h}_w) \cos \alpha]^2 V(\tan \varphi') \tag{11.6a}$$

If $V(\tan \varphi) = 0$,

$$V(F_R) = [\gamma_w \cos 22° \tan 30°]^2 [0.075]^2 = 0.0014 \, kN^2 \tag{11.6b}$$

$$\Delta(F_R) = \sqrt{0.0014}/0.5 = 0.075 \tag{11.6c}$$

Then $\beta = 3.25$, $P_f = 0.001$, according to Equations (11.1c) and (11.1d). This is the annual failure probability.

For a design life of 20 years, the failure probability may be found with the binomial distribution. The probability of x occurrences in n independent trials, with a probability of occurrence p per trial, is

$$P(X = x) = \frac{n!}{x!(n-x)!} p^x (1-p)^{n-x} \tag{11.7}$$

With $p = 0.001$ and $n = 20$, $x = 0$, $P(X = 0) = 0.99$, which is the probability of no failure. The probability of failure is $1 - 0.99 = 0.01$. To illustrate the importance of h_w, let $\sigma(h_w) = 0.03$ m. Similar calculations lead to $\beta = 8.5$ and $P_f \approx 0$.

Consider next the additional uncertainty due to uncertainty about $\tan \varphi'$. Using $\sigma(\tan \varphi') = 0.02$ in Equation (11.6a) with $\sigma(h_w) = 0.075$ m gives

$$V(F_R) = [(0.6\gamma - 0.12\gamma_w)\cos 22°]^2 (0.02)^2$$

$$+ [\gamma_w \cos 22° \tan 30°]^2 [0.075]^2$$

$$= 0.0017 \, kN^2 \tag{11.8}$$

The annual failure probability becomes 0.0013. Note that this cannot be used in Equation (11.7) to compute the failure probability in 20 years because $\tan \varphi'$ in successive years are not independent trials.

While FOSM is simple to use, it has shortcomings, which include the assumption of linearity in Equations (11.3). It is also not invariant with respect to the format used for safety. Baecher and Christian (2003) give some examples. The Hasofer–Lind method, also called the first-order reliability method (FORM), is an improvement over FOSM. It uses dimensionless variables

$$X_i^* = \{X_i - \overline{X}_i\}/\sigma[X_i] \tag{11.9a}$$

and reliability is measured by the distance of the design point from the failure state,

$$g(X_1^*, \ldots, X_i^*) = 0 \tag{11.9b}$$

which gives the combination of $X_1, \ldots, X_n$ that will result in failure. For limit equilibrium analysis, $g = F_s - 1 = 0$. Figure 11.5 shows g as a function of X_1^* and X_2^*. The design point is shown as a, and ab is the minimum distance to $g = 0$. Except for very simple problems, the search for the minimum ab is done via numerical solution. Low $et\ al.$ (1997) show the use of spread sheet

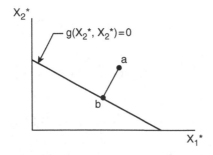

$Figure\ 11.5$ First-order reliability method.

to search for the critical circle. A detailed treatment on the use of spread sheet and an example are given by Low in Chapter 3.

The distribution of F_s and P_f can be obtained by simulation. Values of the input parameters (e.g. c', φ', etc.) that represent the mean values along the potential slip surface are random variables generated from the distribution functions and are used to calculate the distribution of F_s. From this, the failure probability $P(F_s \leq 1)$ can be determined. Examples include Chong *et al.* (2000) and El-Ramly *et al.* (2003). Several commercial software packages also have an option to calculate β and P_f by simulation. Most do simulations on the critical slip surface determined from limit equilibrium analysis.

11.2.3 Displacement analysis

The Newmark (1965) method for computing displacement of a slope during an earthquake considers a block sliding on a plane slip surface (Figure 11.6a). Figure 11.6b shows the acceleration calculated for a given ground acceleration record. Slip starts when a_i exceeds the yield acceleration a_y, at $F_s \leq 1$. The displacement during each slip is

$$y_i = (a_i - a_y)t_i^2/2 \tag{11.10a}$$

where a_i = acceleration during time interval t_i. For n cycles with $F_s \leq 1$, the displacement is

$$Y = \sum_i^n y_i \tag{11.10b}$$

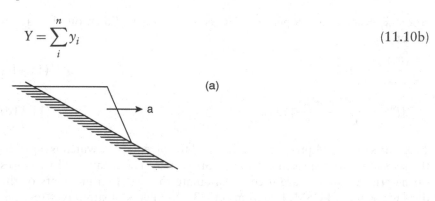

(a)

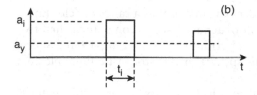

(b)

Figure 11.6 Displacement by Newmark's method.

The displacement analysis assumes that the soil strength is constant. Liquefaction is not considered. Displacement analysis has been used to evaluate the stability of dams and landfills. Statistics of ground motion records were used to compute distribution of displacements by Lin and Whitman (1986) and Kim and Sitar (2003). Kramer (1996) gives a comprehensive review of slopes subjected to seismic hazards.

11.2.4 Finite element method

The finite element method (FEM) is widely used to calculate stress, strain, and displacement in geotechnical structures. In probabilistic FEM, material properties are represented as random variables or random fields.

A simplified case is where only the uncertainties about the average material properties need be considered. Auvinet et al. (1996) show a procedure which treats the properties x_i as random variables. Consider the finite element equations for an elastic material

$$\{P\} = [K]\{Y\} \tag{11.11a}$$

where $\{P\}$ = applied nodal forces, $[K]$ = stiffness matrix, $\{Y\}$ = nodal displacements. Differentiating and transposing terms yields

$$[K]\frac{\partial\{Y\}}{\partial x_i} = \frac{\partial\{P\}}{\partial x_i} - \frac{\partial\{K\}}{\partial x_i}\{Y\} \tag{11.11b}$$

where x_i = material properties. This is analogous to Equation (11.11a) if we let

$$\frac{\partial\{Y\}}{\partial x_i} = \{Y^*\} \tag{11.11c}$$

$$\{P^*\} = \frac{\partial\{P\}}{\partial x_i} - \frac{\partial\{K\}}{\partial x_i}\{Y\} \tag{11.11d}$$

Then the same FEM program can be used to solve for Y^*, which is equal to the sensitivity of displacement to the properties. The means and variances of material properties are used to calculate the first two moments of the displacement via FOSM, Equations (11.3). Moments of strain or stress can be calculated in the same manner.

The perturbation method considers the material as a random field. Equations (11.11) are extended to second-order and used to compute the first two moments of displacement, stress and strain (e.g. Liu et al., 1986). Sudret and Berveiller describe an efficient spectral-based stochastic finite element method in Chapter 7.

Another approach is simulation. A FEM analysis is done with material properties generated from the probability distribution of the properties.

Griffith and Fenton (2004) studied the influence of spatial variation in soil strength on failure probability for a saturated soil slope subjected to undrained loading. The strength is represented as a random field and the generated element properties account for spatial variation and correlation.

For dynamic loads, deterministic analyses with FEM and various material models have been made (e.g. Seed *et al.*, 1975; Lacy and Prevost, 1987; Zerfa and Loret, 2003) to yield pore pressure generation and embankment deformation for given earthquake motion records. See Popescu *et al.* (Chapter 6) for stochastic FEM analysis of liquefaction.

11.2.5 Pore pressure, seepage and infiltration

For deep-seated failures, where the pore pressure is controlled primarily by groundwater, it can be estimated from measurements or from seepage calculations. It is often assumed that the pore pressure in the zone above the water table, the vadose zone, is zero. This may be satisfactory where the vadose zone is thin. Estimation of pore pressure or suction (ψ) in the vadose zone is complex. Infiltration from the surface due to rainfall and evaporation is needed to calculate the suction. Where the soil layer is thin, one-dimensional solution may be used to calculate infiltration. A commonly used solution is Richards (1931) equation, which in one-dimension is,

$$\frac{\partial \theta}{\partial \psi}\frac{\partial \psi}{\partial t} = \frac{\partial}{\partial z}\left[K\frac{\partial \psi}{\partial z}\right] + Q \tag{11.12}$$

where θ = volumetric moisture content, ψ = soil suction, K = permeability, Q = source or sink term, $\partial \theta / \partial \psi$ = soil moisture versus suction relationship. Numerical solutions are usually used. This equation has been used for moisture flow in landfill covers (Khire *et al.*, 1999) and slopes (Anderson *et al.*, 1988). Two-dimensional flow may be solved via FEM. Fredlund and Rahardjo (1993) provide the details on the numerical solutions.

The soil moisture versus suction relation in Equation (11.12), also called the soil water characteristic curve, is a material property. A widely used relationship is the van Genuchten (1980) equation,

$$\theta = \theta_r + \frac{\theta_s - \theta_r}{[1 + (a\psi)^n]^m} \tag{11.13}$$

where θ_r, θ_s = residual and saturated volumetric water contents, a, m and n = shape parameters. The Fredlund and Xing (1994) Equation, which has a rational basis, is

$$\theta = \theta_s\left[\frac{1}{\ln[e + (\psi/a)^n}\right]^m \tag{11.14}$$

where a and m = shape parameters.

The calculated pore pressures are used in a limit equilibrium or finite element analysis. Examples of application to landslides on unsaturated slopes subjected to rainfall infiltration include Fourie *et al.* (1999) and Wilkinson *et al.* (2000).

Perturbation methods have been applied to continuity equations, such as Equation (11.12), to study flow through soils with spatial variations in permeability (e.g. Gelhar, 1993). Simulation was used by Griffith and Fenton (1997) for saturated flow and by Chong *et al.* (2000) and Zhang *et al.* (2005) to study the effect of uncertainties about the hydraulic properties of unsaturated soils on soil moisture and slope stability. A more complicated issue is the presence of macro-pores and fractures (e.g. Beven and Germann, 1982). This is difficult to incorporate into seepage analysis, largely because their geometry is not well known. Field measurements (e.g. Pierson, 1980; Johnson and Sitar, 1990) have shown seepage patterns that differ substantially from those predicted by simple models. Although the causes were not determined, flow through macropores is one likely cause.

11.3 Uncertainties in stability analysis

Uncertainties encountered in predictions of stability are considered under three general categories: material properties, boundary conditions, and analytical methods. The uncertainties involved in estimating the relevant soil properties are familiar to geotechnical engineers. Boundary conditions include geometry, loads, and groundwater levels, infiltration and evaporation at the boundaries, and ground motions during earthquakes. Errors and associated uncertainties may be random or systematic. Uncertainties due to data scatter are random, while model errors are systematic. Random errors decrease when averaged over a number of samples, but systematic errors apply to the average value.

11.3.1 Soil properties

The uncertainties about a soil property s, can be expressed as

$$N_s s_m = s + \zeta \tag{11.15}$$

where s_m = value measured in a test, N_s = bias or inaccuracy of the test method, ζ = testing error, which is the precision of the test procedure. The random variable, s_m, has mean = $\bar{s}_m$, variance = $V(s_m)$, and COV = $\Delta(s_m)$ = $\sigma(s_m) / \bar{s}_m$. It includes the natural variability of the material, with COV $\Delta(s)$ and variance $V(s)$, and testing error with zero mean and variance $V(\zeta)$. The model error, N_s, is the combination of N_i's ($i = 1, 2, \ldots$) that represent bias from the ith source with mean $\bar{N}_i$ and COV Δ_i. The true bias is, of course, unknown, since the perfect test is not available.

For n independent samples tested, the uncertainty about $\bar{s}_m$, due to insufficient samples, has a COV

$$\Delta_0 = \Delta(s_m)/\sqrt{n} \tag{11.16}$$

For high-quality laboratory and in-situ tests, the COV of ζ is around 0.1 (Phoon and Kulhawy, 1999). Hence, unless n is very small, the testing error is not likely to an important source of uncertainty.

Input into analytical models requires the average s over a given distance (e.g. limiting equilibrium analysis) or a given area or volume (e.g. FEM). Consider the one-dimensional case. The average of s_m over a distance L, $\bar{s}_m(L)$, has a mean and variance $\bar{s}_m$ and $V[\bar{s}_m(L)]$. To evaluate $V[\bar{s}_m(L)]$, the common procedure is to fit a trend to the data to obtain the mean as a function of distance. The departures from the trend are considered to be random with variance $V(s_m)$. To evaluate the uncertainty about $\bar{s}_m(L)$ due to the random component, it is necessary to consider its spatial nature. The departures from the mean at points i and j, separated by distance r, are likely to be correlated. To account for this correlation, VanMarcke (1977) introduced the variance reduction factor

$$\Gamma^2 = V[\bar{s}_m(L)]/V[s_m], \tag{11.17a}$$

A simple expression is

$$\Gamma^2 = 1, \text{ if } L \leq \delta \tag{11.17b}$$

$$\Gamma^2 = \delta/L, \text{ if } L > \delta \tag{11.17c}$$

where $\delta =$ correlation distance, which measures the decay in the correlation coefficient

$$\rho = e^{-2r/\delta} \tag{11.18}$$

L/δ can also be considered as the equivalent number of independent samples, n_e, contained within L. Then the variance and COV are,

$$V[\bar{s}_m(L)] = \Gamma^2 V[s_m] = V[s_m]/n_e \tag{11.19a}$$

$$\Delta_e = \Gamma \Delta(s_m) = \Delta(s_m)/\sqrt{n_e} \tag{11.19b}$$

Extensions to two and three dimensions are given in VanMarcke (1977). It should be noted that Δ_0 applies to the average strength and is systematic, while Δ_e changes with L and n_e. A more detailed treatment of this subject is given by Baecher and Christian in Chapter 2. Data on $V(s_m)$, $\Delta(s_m)$, and δ for various soils have been summarized by Lumb (1974),

DeGroot and Baecher (1993), Phoon and Kulhawy (1999) and Baecher and Christian (2003). Data on some unsaturated soils have been reviewed by Chong *et al.* (2000) and Zhang *et al.* (2005).

In limit equilibrium analysis, the average in-situ shear strength to be used in Equations (11.2) contains all of the uncertainties described above. It can be expressed as (Tang *et al.*, 1976)

$$s = N_s \hat{S} \tag{11.20a}$$

where $\hat{s}$ = estimated shear strength and is usually taken as the average of the measured values , $\bar{s}_m$. The combined uncertainty is represented by N

$$N = N_0 N_e N_s = N_0 N_e \prod_i N_i \tag{11.20b}$$

where N_0 = uncertainty about the population mean, N_e = uncertainty about the mean value along the slip surface, and N_i = uncertainty due to ith source of model error. Equations (11.3) are used to obtain the mean and COV

$$\bar{s} = \prod_i \bar{N}_i \bar{s}_m = \bar{N}_s \bar{s}_m \tag{11.21a}$$

$$\Delta(s) = [\Delta_0^2 + \Delta_e^2 + \Sigma \Delta_i^2]^{1/2} \tag{11.21b}$$

Note that $\bar{N}_0 = \bar{N}_e = 1$. It is assumed that the model errors are independent.

Two examples of uncertainties about soil properties due to data scatter $\Delta(s_m)$ are shown in Table 11.2. All of the data scatter is attributed to spatial variation. The values of Δ_0 and Δ_e, representing the systematic and random components, are also given. For the Chicago clay, the value of δ was obtained from borehole samples and represents δ_y. For the James Bay site, both δ_x and δ_y were obtained. Because a large segment of the slip surface is nearly horizontal, δ is large. Consequently Γ is small. This illustrates the significance of δ.

Table 11.2 Uncertainties due to data scatter.

Site	Soil	$\bar{s}_m$ (kPa)	$\Delta(s_m)$	Δ_0	δ(m)	Δ_e	Ref.
Chicago	upper clay	51.0	0.51	0.083	0.2	0.096	Tang et al.
	middle clay	30.0	0.26	0.035	0.2	0.035	(1976)
	lower clay	37.0	0.32	0.056	0.2	0.029	
James Bay	lacustrine clay	31.2	0.27	0.045	40+		deGroot and Baecher (1993)
	slip surface*		0.14		24	0.063	Christian et al. (1994)

$^+\delta_x$; *for circular arc slip surface and $H = 12$ m.

Examples of estimated bias in the strength model, made between 1970 and 1993, are shown in Table 11.3. Since the true model error is unknown, the estimates are based on data from laboratory or field studies and are subjective in nature because judgment is involved in the interpretation of published data. Note that different individuals considered different possible sources of error. For the James Bay site, the model error of the field vane shear tests consists of the data scatter in Bjerrum's (1972) correction factor. The correction factors $\overline{N}_i$ for the various sources of errors are mostly between 0.9 and 1.1, with an average close to 1.0. The Δ_i's for the uncertainties from the different sources range between 0.03 and 0.14 and the total uncertainties $(\Sigma \Delta_i^2)^{1/2}$ are about the same order of magnitude as those from data scatter (Table 11.2).

When two or more parameters, describing a soil property, are derived by fitting a curve to test data, the correlation between the parameters may be important. Data on correlations are limited. Examples are the correlation between c' and ϕ' (Lumb, 1970) and between parameters a, n and k_s in the Fredlund and Xing (1994) equation (Zhang et al., 2003).

To illustrate the influence of correlation between properties, consider the slope in a c', φ' material (Figure 11.7), analyzed by Fredlund and Krahn (1977) by limit equilibrium and by Low et al. (1997) by FORM. The random variables X_1, X_2 in Equation (11.9) are c', φ'. The correlation coefficient $\rho_{c,\varphi}$ is −0.5. The critical circles from the limit equilibrium analysis (LEA) and that from FORM are also shown. Note that FORM searches for the most unfavorable combination of c' and φ', or point b in Figure 11.5, which are $c' = 0.13$ kPa and $\varphi' = 23.5°$. The shallow circle is the result of the low c' and high φ' (Taylor, 1948). The difference reflects the uncertainty about the material model and may be expected to decrease with decreasing values of $\rho_{c,\varphi}$ and the COVs of c' and φ'. Lumb's (1970) study on three residual soils show that $\rho_{c,\varphi} \leq -0.2$, $\Delta(c) \approx 0.15$, and $\Delta(\varphi') \approx 0.05$, which are considerably less than the values used by Low et al. It should be added that software packages that use simulation on the critical slip surface from limit equilibrium analysis cannot locate point b in Figure 11.5, and will not properly account for correlation.

11.3.2 Loads and geometry

The geometry and live and dead loads for most geotechnical structures are usually well defined except for special cases. When there is more than one load on the slope, a simple procedure is to use the sum of the loads. However, this may be overly conservative if the live loads have limited durations. Then failure can be modeled as a zero-crossing process (Wen, 1990) with the loads represented as Poisson processes. An application to landslide hazard is given by Wu (2003).

Table 11.3 Uncertainties due to errors in strength model.

Material	Test	$\bar{N}_i, \Delta_i$						$\bar{N}_s, (\Sigma \Delta_i^2)^{1/2}$	Ref.
		Sample disturbance	Stress state	Anisotropy	Sample size	Strain rate	Progress failure	Total	
Detroit clay	unconf. comp.	1.15,0.08	1.00,0.03	0.86,0.09	X	X	X	0.99, 0.12	Wu and Kraft (1970)
Chicago Upper clay	unconf. comp.	(a)1.38,0.024 (b)1.05,0.02	1.05,0.03	1.0, 0.03	0.75, 0.09	0.80,0.14	0.93,0.03		Tang et al. (1976)
Middle clay		(a)1.38,0.024 (b)1.05,0.02	1.05,0.03	1.0, 0.03	0.93, 0.05	0.80,0.14	0.97,0.03		
Lower clay		(a)1.38,0.024 (b)1.05,0.02	1.05,0.03	1.0, 0.03	0.93, 0.05	0.80,0.14	0.97,0.03	1.05, 0.17	
Labrador clay	field vane	X	X	X	X	X	X	1.0, 0.15	Christian et al. (1994)

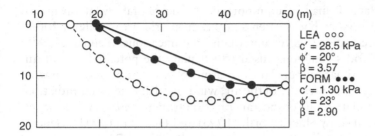

Figure 11.7 FORM analysis of a slope (adapted from Low *et al.* (1997)).

11.3.3 Soil boundaries

Subsoil profiles are usually constructed from borehole logs. It is common to find the thickness of individual soil layers to vary spatially over a site. Where there is no sharp distinction between different soil types, judgment is required to choose the boundaries. Uncertainties about boundary and thickness result in uncertainty about the mean soil property. Both systematic and random errors may be present. Their representation depends on the problem at hand. An example is given in Christian *et al.* (1994). Errors in the subsoil model, which result from failure to detect important geologic details could have important consequences (e.g. Terzaghi, 1929), but are not included here because available methods cannot adequately account for errors in judgment.

11.3.4 Flow at boundaries

Calculation of seepage requires water levels or flow rates at the boundaries. For levees and dams, flood levels and durations at the upstream slope are derived from hydrological studies (e.g. Maidman, 1992). Impermeable boundaries, often used to represent bedrock or clay layers, may be pervious because of fractures or artesian pressures. Modeling these uncertainties is difficult.

For calculation of infiltration, rainfall depth and duration for return period, T_r, are available (Hershfield, 1961). However, rainfall and duration are correlated and simulation using rainfall statistics is necessary (Richardson, 1981). Under some conditions, the precipitation per storm controls (Wu and Abdel-Latif, 2000). Then this can be used in a seepage model to derive the mean and variance of pore pressure. In unsaturated soils, failure frequently occurs due to reduction in suction from infiltration (e.g. Krahn *et al.*, 1989; Fourie *et al.*, 1999; Aubeny and Lytton, 2004, and others). Usually both rainfall and duration are needed to estimate the suction or critical depth (e.g. Fourie *et al.*, 1999). Then a probabilistic analysis requires simulation.

Evapotranspiration is a term widely used for water evaporation from a vegetated surface. It may be unimportant during rainfall events in humid and temperate regions but is often significant in arid areas (Blight, 2003). The Penman equation, with subsequent modifications (Penman, 1948; Monteith, 1973), is commonly used to calculate the potential evapotranspiration. The primary input data are atmospheric conditions: saturation vapor pressure, vapor pressure of the air, wind speed and net solar radiation. Methods to estimate actual evapotranspiration from potential evapotranspiration are described by Shuttleworth (1993) and Dingman (2002). Generation of evapotranspiration from climate data (Ritchie, 1972) has been used to estimate moisture in landfill covers (Schroeder *et al.*, 1994). Model error and uncertainty about input parameters are not well known.

11.3.5 Earthquake motion

For pseudostatic analysis, empirical rules for seismic coefficients are available but the accuracy is poor. Seismic hazard curves that give the probability that an earthquake of a given magnitude or given maximum acceleration will be exceeded during given time exposures (T) are available for the US (Frankel *et al.*, 2002). Frankel and Safak (1998) reviewed the uncertainties in the hazard estimates. The generation of ground motion parameters and records for design is a complicated process and subject of active current research. It is beyond the scope of this chapter. A comprehensive review of this topic is given in Sommerville (1998). The errors associated with the various procedures are largely unknown.

It should be noted that, in some problems, the magnitude and acceleration of future earthquakes may be the predominant uncertainty and estimates of this uncertainty with current methods are often sufficient for making design decisions. An example is the investigation of the seismic stability of the Mormon Island Auxiliary Dam in California (Sykora *et al.*, 1991). Limit equilibrium analysis indicated that failure is likely if liquefaction occurs in a gravel layer. Dynamic FEM analysis indicated that an acceleration $k = 0.37$ from a magnitude 5.25 earthquake, or $k = 0.25$ from a magnitude 6.5 earthquake, would cause liquefaction. However, using the method of Algermissan *et al.* (1976), the annual probability of occurrence of an earthquake with a magnitude of 5.25 or larger is less than 0.0009, which corresponds to a return period of 1100 years. Therefore, the decision was made not to strengthen the dam.

11.3.6 Analytical models

In limit equilibrium analysis, different methods usually give different safety factors. For plane strain, the safety factors obtained by different methods, with the exception of the Fellenius (1936) method, are generally within ±7%

Table 11.4 Errors in analytical model.

Analysis		$\overline{N}_i, \Delta_i$			$\overline{N}_a, \Delta_a$	Ref.
		3D	Slip surf.	Numerical	Total	
Cut	$\varphi = 0$	1.05,0.03	0.95,0.06	X	1.0, 0.067	Wu and Kraft (1970)
Cut	$\varphi = 0$	X	X	X	0.98, 0.087	Tang et al. (1976)
Embankment	$\varphi = 0$	1.1,0.05	1.0,0.05	1.0,0.02	1.0,0.07	Christian et al. (1994)
Landfill	c', φ'	1.1,0.16	X	X	1.1,0.16	Gilbert et al. (1998)

(Whitman and Bailey, 1967; Fredlund and Krahn, 1977). A more important error is the use of the plane-strain analysis for failure surfaces which are actually three-dimensional. Azzouz *et al.* (1983), Leshchinsky *et al.* (1985), and Michalowski (1989) show that the error introduced by using a two-dimensional analysis has a bias factor of $N_a = 1.1 - 1.3$ for $\varphi = 0 - 25°$. Table 11.4 shows estimates of various errors in the limit equilibrium model. The largest model error, $\Delta_a = 0.16$, is for three-dimensional (3D) effect in a landfill (Gilbert *et al.*, 1998).

Ideally, a 3D FEM analysis using the correct material model should give the correct result. However, even fairly advanced material models are not perfect. The model error is unknown and has been estimated by comparison with observed performance. The review by Duncan (1996) shows various degrees of success of FEM predictions of deformations under static loading. In general, Type A predictions (Lambe, 1973) for static loading cannot be expected to come within ±10% of observed performances. This is where the profession has plenty of experience. The model error and the associated uncertainty may be expected to be much larger in problems where experience is limited. One example is the prediction of seepage and suction in unsaturated soils.

11.3.7 Bayesian updating

Bayesian updating allows one to combine different sources of information to obtain an "updated" estimate. Examples include combining direct and indirect measurements of soil strength (Tang, 1971), and properties of landfill materials measured in different tests (Gilbert *et al.*, 1998). It can also be used to calibrate predictions with observation. In geotechnical engineering, back-analysis or solving the inverse problem is widely used to verify design assumptions when measured performance is available. An example is a slope failure. This observation ($F_s = 1$) may be used to provide an updated estimate of the soil properties or other input parameters.

In Bayes theorem, the posterior probability is

$$P = P[A_i|B] = \frac{P[B|A_i]P[A_i]}{P[B]} \tag{11.22a}$$

where

$$P[B] = \sum_{i=1}^{n} P[B|A_i]P[A_i] \tag{11.22b}$$

$P[A_i|B]$ = probability of A_i (shear strength is s_i), given B (failure) has been observed, $P[B|A_i]$ = probability that event B occurs (slope fails) given event A_i occurs (the correct strength is used), $P[A_i]$ = prior probability that A_i occurs. In a simple case, $P[A_i]$ represents the uncertainty about strength due to data scatter, etc., $P[B|A_i]$ represents the uncertainty about the safety factor due to uncertainties about the strength model, stability analysis, boundary conditions, etc. Example 5 illustrates a simple application. Gilbert et al. (1998) describe an application of this method to the Kettleman Hills landfill, where there are several sources of uncertainty.

Regional records of slope performance, when related to some significant parameter, such as rainfall intensity, are valuable as an initial estimate of failure probability. Cheung and Tang (2005) collected data on failure probability of slopes in Hong Kong as a function of age. This was used as prior probability in Bayesian updating to estimate the failure probability for a specific slope. The failure probability based on age is $P[A_i]$ and $P[B|A_i]$ is the pdf of β as determined from investigations for failed slopes, and β is the reliability index for the specific slope under investigation. To apply Equation (11.22a),

$$P[B] = P[B|A_i]P[A_i] + P[B|A_i']P[A_i'] \tag{11.22c}$$

$$P[A_i'] = 1 - P[A_i]$$

where $P[B|A_i']$ = pdf of β for stable slopes. $\tag{11.22d}$

Example 3

The analysis of the slope failure in Chicago by Tang et al. (1976) is reviewed here to illustrate the combination of uncertainties using Equations (11.20) and (11.21). The undrained shear strength, s, was measured by unconfined compression tests on 5.08 cm Shelby tube samples. The subsoil consists mainly of "gritty blue clay," whose average strength varies with depth, Figure 11.3. The clay was divided into three layers with values of $\bar{s}_m$ and

$\Delta(s_m)$ as given in Table 11.2. The value of δ is the same for the three clay layers. The values of Δ_0 calculated according to Equation (11.16) are given in Table 11.2. The correlation coefficients ρ_{ij} between the strength of the upper and middle, middle and lower, and upper and lower layers are: 0.66, 0.59, and 0.40, respectively.

The sources of uncertainties about the shear strength model are listed in Table 11.3. The errors were evaluated by review of published data, mostly laboratory studies, on the influence of the various factors on strength. These were used to establish a range of upper and lower limits for N_i. To calculate $\overline{N}_i$ and Δ_i, a distribution of N_i is necessary. The rectangular and triangular distributions between the upper and lower limits were used, and the mean and variance of N_i were calculated with the relations given in Ang and Tang (1975). Clearly, judgment was exercised in choosing the sources of error and the ranges and distributions of N_i. The estimated values of $\overline{N}_i$ and Δ_i are shown in Table 11.3. Two sets of values are given for sample disturbance; (a) denotes the strength change due to mechanical disturbance and (b) denotes that due to stress changes from sampling.

The resisting moment in Equations (11.2) can be written as

$$M_R = r \sum_i \bar{s}_i l_i \tag{11.23a}$$

where i denotes the soil layer. According to Equation (11.3a), the mean is

$$\overline{M}_R = \overline{N} r \sum_i \bar{s}_i l_i \tag{11.23b}$$

According to Equation (11.21a),

$$\overline{N} = 1.38 \times 1.05 \times 1.05 \times 1.0 \times 0.75 \times 0.80 \times 0.93 = 0.85 \tag{11.23c}$$

for the upper clay and the values of $\overline{N}_i$ are as given in Table 11.3. The calculated $\overline{M}_R$ for the different segments are given in Table 11.5 and the sum is $\overline{M}_R = 27.17\,MN - m/m$.

To evaluate the variance or COV of M_r, the first step is to evaluate $\Delta(\bar{s})$ for the segments of the slip surface passing through the different soil layers. For the upper clay layer, substituting Δ_0 in Table 11.2 and the Δ_i's in Table 11.3 into Equation (11.21b) gives

$$\Delta(\bar{s}_1) = [(0.083)^2 + (0.024)^2 + (0.02)^2 + (0.03)^2 + (0.03)^2 + (0.09)^2$$
$$+ (0.14)^2 + (0.03)^2]^{1/2} = 0.20 \tag{11.24a}$$

for failure surface 1. Since n_e is large, the contribution of Δ_e is ignored. The values of $\Delta(s) = \Delta(M_R)$ and $\sigma(M_R)$ for the clay layers are given in

Table 11.5 Mean and standard deviation of resisting moment, Chicago slope.

Layer	$\overline{M}_R$, MN $-$ m/m	$\overline{N}$	$\Delta(M_R)$	$\sigma(M_R)$, MN $-$ m/m
Upper clay	4.01	0.85	0.20	0.80
Middle clay	5.70	1.10	0.16	0.91
Lower clay	16.90	1.0	0.17	2.88
Sand	0.56			
Σ	27.17		0.15	4.12
			0.18*	

*With model error Δ_a.

Table 11.5. Substitution into Equation (11.3b) gives,

$$V(M_R) = (0.80)^2 + (0.91)^2 + (2.88)^2 + 2[(0.80 \times 0.91 \times 0.66)$$

$$+ (0.80 \times 2.88 \times 0.59) + (0.91 \times 2.88 \times 0.4)]$$

$$= 17.06(MN - m/m)^2 \tag{11.24b}$$

$$\Delta(M_R) = [17.06]^{1/2}/27.17 = 0.152 \tag{11.24c}$$

To include the error, N_a, in the stability analysis model, $\overline{N}_a$ and Δ_a given in Table 11.3. are applied to M_R. The combined uncertainty is

$$\Delta(M_R) = [\Delta_a^2 + \Delta_s^2]^{1/2} = [(0.087)^2 + (0.152)^2]^{1/2} = 0.18 \tag{11.24d}$$

The calculation of β and P_f are described in Example 1.

Example 4

The failure of a slope in Cleveland (Wu et al., 1975), Figure 11.8, is used to illustrate the choice of soil properties and its influence on predicted failure probability. The subsoil consists of layers of silty clay and varved clay. Failure occurred shortly after construction and the shape of the failure surface (1), shown in Figure 11.8, suggested the presence of a weak layer near elev. 184 m. An extensive investigation following failure revealed the presence of one or more thin layers of gray clay. Most of these were encountered near elev. 184 m. Many samples of this material contained slickensides and had very low strengths. It was assumed that the horizontal portion of the slip surface passed through this weak material. Because no samples of the weak material were recovered during the initial site exploration, its presence was not recognized. A simplified analysis of the problem is used for this example. The layers of silty clay and varved clay are considered as one unit and the weak gray clay is treated as thin layer at elev. 184 m (Figure 11.8).

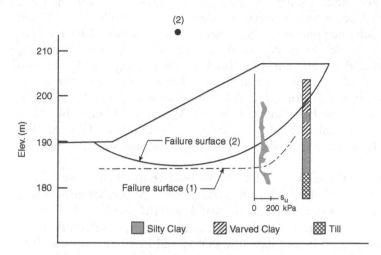

Figure 11.8 Slope failure in Cleveland (adapted from Wu *et al.*, 1975). s_u = undrained shear strength.

Table 11.6 Strength properties, Cleveland slope.

Analysis	One material	Two materials	
	Combined data	Varved clay and silty clay	Weak gray clay layer
$\bar{s}_m$, kPa	67.1	83.1	39.5
σ_m (s_m), kPa	38.3	38.4	16.8
$\Delta(s_m)$	0.57	0.46	0.42
n	19	12	7
δ_y, m	3.3	3.3	–
δ_x, m	–	–	33
Δ_0	0.13	0.13	0.16
n_e	56	41	2.4
Δ_e	0.076	0.072	0.27
$[(\Delta_0)^2 + (\Delta_e)^2]^{1/2}$	0.15	0.15	0.31

The undrained shear strengths, s_m, of the two materials, as measured by the unconfined compression test, are summarized in Table 11.6. The correlation distance of the silty clay and varved clay, δ_y, is determined by VanMarcke's (1977) method. The number of samples of the weak gray clay is too small to allow calculation of δ_x. Hence, it is arbitrarily taken as $10\delta_y$.

Consider first the case where the silty clay and varved clay and the weak material are treated as one material. All the measured strengths are treated as samples from one population ("combined data" in Table 11.6). For simplification, the model errors, N_i and Δ_i, are ignored. Then only Δ_0 and Δ_e contribute to $\Delta(s)$ in Equation (11.21b). The values of Δ_0 and Δ_e are evaluated as described in Example 2 and given in Table 11.6. Limit equilibrium analysis gives $\overline{F}_s = 1.36$ for the circular arc shown as (2) in Figure 11.8. The values of β and P_f, calculated by FOSM, are given in Table 11.7. Values of β and P_f calculated by simulation with SLOPE/W (GEO-SLOPE International), where s is modeled as a normal variate, are also given in Table 11.7.

Next, consider the subsoil to be composed of two materials, with $\bar{s}_m$ and $\Delta(s_m)$ as given in Table 11.6. The values of Δ_0 and Δ_e are evaluated as before and given in Table 11.6. The larger values of Δ_e and $[(\Delta_0)^2 + (\Delta_e)^2]^{1/2}$ are the result of the large δ_x of the weak material. Values of, $\overline{F}_s$, β, and P_f for the composite failure surface (1), calculated by FOSM and by simulation are given in Table 11.7. The P_f is much larger than that for the circular arc (2). This illustrates the importance of choosing the correct subsoil model. Figure 11.9 shows the distribution of F_s from simulation. Note that SLOPE/W uses the normal distribution to represent the simulated F_s. Hence the mean of the normal distribution may differ from the F_s obtained by limit equilibrium analysis for the critical slip surface.

In the preceding analysis of the two-material model, it is assumed that the strengths of the two materials are uncorrelated. The available data are insufficient to evaluate the correlation coefficient and geologic deductions could be made in favor of either correlation or no correlation. To illustrate the influence of weak correlation, it is assumed that $\rho_{ij} = 0.2$. The β calculated by FOSM is reduced from 1.64 to 1.49.

Table 11.7 Calculated reliabilities, Cleveland slope.

Analysis		One material	Two materials
Limit equil.	$\overline{F}_s$	1.36	1.26
FOSM,	$\Delta(F_s)$	0.15	0.14
no model	β	1.97	1.64
error	P_f	0.025	0.08
Simulation,	$\overline{F}_s$	1.36	1.31
no model	β	2.48	1.64
error	P_f	0.006	0.05
FOSM,	$[\Sigma \Delta_j^2]^{1/2}$	0.14	0.14
with model	Δ_a	0.035	0.035
error	$\Delta(F_s)$	0.21	0.20
	β	1.37	1.05
	P_f	0.09	0.15

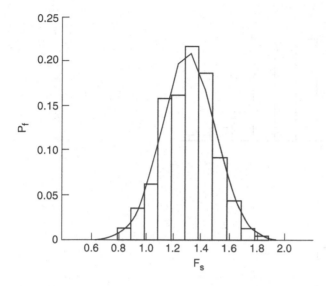

Figure 11.9 Distribution of F_s, Cleveland slope.

Finally, consider model errors. The total uncertainty about strength model based on the unconfined compression test is estimated to be $\overline{N}_s = 1.0$ and $(\Sigma \Delta_i^2)^{1/2} = 0.14$. Assume that the analytical model error has $N_a = 1.0$, $\Delta_a = 0.035$. Calculation with FOSM gives $\beta = 1.37$ and 1.05 for the one- and two-material models, respectively, (Table 11.7).

Example 5

If the slope in Example 4 is to be redesigned, it is helpful to revise the shear strength with the observation that failure has occurred. Bayes' Theorem, Equations (11.22), is used to update the shear strength of the weak layer in Example 4. Let $P[B|A_i]$ = probability of failure given the shear strength s_i, $P[A_i]$ = prior probability that the shear strength is s_i. In this simplified example, the distribution of s_i is discretized as shown in Figure 11.10a, where $P[A_i]$ = probabilities that the shear strength falls within the respective ranges. Then, $P[B|A_i]$ is found from the distribution of P_f calculated with SLOPE/W. Because of errors in the stability analysis, it is assumed that failure will occur when the calculated F_s falls between 1.00 ± 0.05. Figure 11.9 shows that for $s_i = 39.5$ kPa, $P[B|A_i] = P[0.95 < F_s < 1.05] \approx 0.05$. Other values of $P[B|A_i]$ are found in the same way and plotted in Figure 11.10b. Substitution of the values in Figures 11.10a and 11.10b into Equations (11.22) gives the posterior distribution P'', which is shown in Figure 11.10c. The mean of the updated s is 30.3 kPa, with a standard deviation of 8.3 kPa and COV = 0.27.

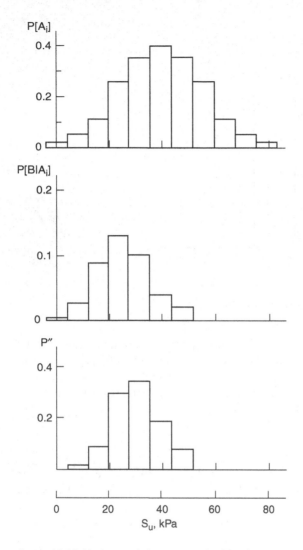

Figure 11.10 Updating of shear strength, Cleveland slope.

Note that $\Delta(s)$ after updating is still large. This reflects the large COV of $P[B|A_i]$ in Figure 11.10b, due to the large $\Delta(s_m)$ of the gray silty clay and varved clay, and expresses the dependence of the back-calculated strength on the accuracy of the stability analysis and other input parameters (Leroueil and Tavanas, 1981; Leonards, 1982). Inclusion of model errors would further increase the uncertainty. Gilbert *et al.* (1998) provide a detailed example of the evaluation of $P[B|A_i]$ for the strength of a geomembrane–clay interface.

11.4 Case histories

Results from earlier studies are used to illustrate the relative importance of uncertainties from different sources and their influence on the failure probability. Consider the uncertainties in a limit equilibrium analysis for undrained loading of slopes in intact clays. This is a common problem for which the profession has plenty of experience. Where the uncertainties about live and dead loads are negligible, then there remain only the uncertainties about soil properties due to data scatter and the strength model, and the analytical model. Tables 11.2–11.4 show estimated uncertainties for several sites. All three sources, data scatter, model errors for shear strength and stability analysis contribute to P_f.

Comparison of estimated failure probability with actual performance provides a calibration of the reliability method. Table 11.8 shows some relatively simple cases with estimated uncertainties about input parameters. The failure probabilities, calculated for mean safety factors larger than 1.3, with values of $N \approx 1$ and $\Delta(F_s) \approx 0.1$, are well below 10^{-2}. The FEM analysis of Griffith and Fenton (2004), in which the undrained shear strength is a random field with $\Delta(s) = 0.25$ and $n_e = 15$, shows that for $\overline{F}_s = 1.47$, the failure probability is less than 10^{-2}. This serves to confirm the results of the simple analyses cited above. In addition, their analysis searches for the failure surface in a spatially variable field, which is not necessarily a circle. The departure from the critical circle in a limit equilibrium analysis depends on the correlation distance $(0 < \delta < L)$. An important implication is that when $\delta_x \gg \delta_y$ and δ_x is a very large, the problem approaches that of a weak zone (Leonards, 1982), as illustrated in Example 4.

Table 11.8 Estimated failure probability and slope performance.

Site	Material	Analysis	Failure surface	F_s	$\Delta(F_s)$	P_f	Stability	Ref.
Cleveland	Clay	Total stress	Composite	1.18	0.26	0.3	Failure	Wu et al., 1975
Chicago	Clay	Total stress	Circular	1.18	0.18	0.2*	Failure	Tang et al., 1976
James Bay	Clay	Total stress	Circular	1.53	0.13	0.003	Stable	Christian et al., 1994
Alberta	Clay shale	Effective stress	Composite	1.31	0.12	0.0016	Stable	El-Ramly et al., 2003

*Observed failure surface.

The success of current practice, using a safety factor of 1.5 or larger, is enough to confirm that for simple cuts and embankments on intact clay, the uncertainties (Tables 11.2–11.4) are not a serious problem. Hence, reliability analysis and evaluation of uncertainties should be reserved for analysis of more complex site conditions or special design problems. An example of the latter is the design of the dams at James Bay (Christian *et al.*, 1994). A reliability analysis was made to provide the basis for a rational decision on the safety factors for the low dams (6 or 12 m), built in one stage, and the high dam (23 m), built by staged construction.

The analysis using FOSM showed that, for $\overline{F}_s \approx 1.5$, β for the high dam is much larger than that for the low dams. However, the uncertainty about F_s for the low dams comes largely from spatial variation, which is random, while that for the high dam comes largely from systematic errors. Failures due to spatial variation (e.g. low strength) are likely to be limited to a small zone, but failures due to systematic error would involve the entire structure and may extend over a long segment. Clearly, the consequences of failure of a large section in a high dam are much more serious than those of failure of small sections in a low dam. Consideration of this difference led to the choice of target β's of 0.01, 0.001, and 0.0001 for the 6 m, 12 m, and 23 m dams, respectively. The corresponding F_s's are 1.63, 1.53, and 1.43 when model error is not included. These numbers are close enough to 1.50 that $F_s = 1.50$ was recommended for all three heights in the feasibility study and preliminary cost estimates.

Where experience is limited, the estimation of uncertainties becomes difficult because model errors are not well known. The significance of uncertainties about properties of unsaturated soils has only recently been investigated. Chong *et al.* (2000) studied the influence of uncertainties about parameters in the van Genuchten equation, Equation (11.13). The parameters α and n have the largest influence on $\overline{F}_s$ and $\Delta(F_s)$; α alone can lead to a $\Delta(F_s) \approx 0.1$ and a reduction of $\overline{F}_s$ from 1.4 to 1.2. Zhang *et al.* (2005) used data on decomposed granite to evaluate the uncertainties about parameters a and n in the Fredlund–Xing equation, Equation (11.14), K_s = saturated permeability, θ_s = saturated moisture content, and four parameters of the critical-state model. The correlations between a, n, K_s, and θ_s were determined from test data. The values of $\Delta(\ln a)$, $\Delta(\ln n)$ and $\Delta(\ln K_s)$ are larger than 1.0. Under constant precipitation, the calculated $\Delta(F_s)$ is 0.22 for $\overline{F}_s = 1.42$. These are model errors and are systematic. It can be appreciated that $\Delta(F_s)$ and the corresponding failure probability are significantly larger than those for the slopes in saturated soils given in Table 11.8. Uncertainties about precipitation and evapotranspiration and spatial variation would further increase the overall uncertainty about pore pressure and stability. These two examples should serve notice that for slopes on unsaturated soils, the practice of $F_s = 1.5$ should be viewed with caution and reliability analysis deserves serious attention.

11.5 Summary and conclusions

The limitations of reliability analysis are:

1 Published values on $\Delta(s_m)$ and δ of soils (e.g. Barker *et al.*, 1991; Phoon and Kulhawy, 1999) are based on data from a limited number of sites (e.g. Appendix A, Barker *et al.* 1991) and show a wide range, even for soils with similar classification. Therefore, the choice of design parameters involves unknown errors. Even estimation of these values from site exploration data involves considerable errors, unless careful sampling schemes are used (DeGroot and Baecher, 1993; Baecher and Christian, 2003). The calculated failure probabilities may be expected to depend strongly on these properties.

2 Errors in strength and analysis models, which result from simplifications, are often unknown. Estimates are usually made from case histories, which are limited in number, particularly where there is little experience. Errors in climate and earthquake models are difficult to evaluate. Hence, estimation of uncertainties about input variables contain a large subjective element.

3 The reliability analyses reviewed in this chapter consider only a limited number of failure modes. For a complex system, where there are many sources of uncertainties, the system reliability can be evaluated by expanding the scope via event trees. This is beyond the scope of this chapter. See Baecher and Christian (2003) for a detailed treatment.

4 Available methods of reliability analysis cannot account for omissions or errors of judgment. A particularly serious problem in geotechnical engineering is misjudgment of the site conditions.

The limitations cited above should not discourage the use of reliability analysis. Difficulties due to inadequate information about site conditions and model errors are familiar to geotechnical engineers and apply to both deterministic and probabilistic design. Some examples of use of reliability analysis in design are given below.

1 Reliability can be used to compare different design options and help decision-making (e.g. Christian *et al.*, 1994).

2 Even when a comprehensive reliability analysis cannot be made, sensitivity analysis can be used to identify the random variables that contribute most to failure probability. The design process can then concentrate on these issues (e.g. Sykora *et al.*, 1991).

3 Bayes's theorem provides a logical basis for combining different sources of information (e.g. Cheung and Tang, 2005).

4 The methodology of reliability is well established, and it should be very useful in evaluating the safety of new and emerging geotechnical

problems, as has been illustrated in the case of unsaturated soils (Chong *et al.*, 2000; Zhang *et al.*, 2005).

Acknowledgments

The writer thanks K. K. Phoon for valuable help and advice during the writing of this chapter, and W. H. Tang for sharing his experience and insight through many discussions and comments.

References

Algermissen, S. T., Perkins, D. M., Thenhaus, P. C., Hanson, S. L. and Bender, B. L. (1982). *Probabilistic estimates of maximum acceleration and velocity in rock in the contiguous United States*. Open/file Report 82-10033, Geological Survey US Department of the Interior, Denver.

Anderson, M. G., Kemp, M. J. and Lloyd, D. M. (1988). Application of soil water difference models to slope stability problems. In *5th International Symposium on Landslides*. A. A. Balkema, Rotterdam, Lausanne, 1, pp. 525–30.

Ang, A. H-S. and Tang, W. H. (1975). *Probability Concepts in Engineering Planning and Design*, Vol. 1. John Wiley and Sons, New York.

Aubeny, C. P. and Lytton, R. L. (2004). Shallow slides in compacted high plasticity clay slopes. *Journal of Geotechnical and Geoenvironmental Engineering*, 130, 717–27.

Auvinet, G., Bouyad, A., Orlandi, S. and Lopez, A. (1996). Stochastic finite element method in geomechanics. In *Uncertainty in the Geologic Environment*, Eds C. D. Shackelford, P. P. Nelson and M. J. S. Roth, Geotechnical Special Publications 58, ASCE, Reston VA, 2, pp. 1239–53.

Azzouz, A. S., Baligh, M.M. and Ladd, C. C. (1983). Corrected field vane strength for embankment design. *Journal of Geotechnical Engineering*, 120, 730–4.

Baecher, G. B. and Christian, J. T. (2003). *Reliability and Statistics in Geotechnical Engineering*. John Wiley and Sons, Chichester, UK.

Barker, R. M., Duncan, J. M., Rojiani, K. B., Ooi, P. S. K., Tan, C. K. and Kim, S. G. (1991). *Manuals for the design of bridge foundations*. NCHRP Report 343, Transportation Research Board, Washington, DC.

Beven, K. and Germann, P. (1982). Macropores and water flow through soils. *Water Resources Research*, 18, 1311–25.

Bjerrum, L. (1972). Embankments on soft ground: state of the art report. In *Proceedings, Specialty Conference on Performance of Earth and Earth-supported Structures*, ASCE, New York, 2, pp. 1–54.

Blight, G. B. (2003). The vadose zone soil-water balance and transpiration rates of vegetation. *Geotechnique*, 53, 55–64.

Cheung, R. W. M. and Tang, W. H. (2005). Realistic assessment of slope stability for effective landslide hazard management. *Geotechnique*, 55, 85–94.

Chong, P. C., Phoon, K. K. and Tan, T. S. (2000). Probabilistic analysis of unsaturated residual soil slopes. In *Applications of Statistics and Probability*, Eds R. E. Melchers and M. G. Stewart. Balkema, Rotterdam, pp. 375–82.

Christian, J. T. (2004). Geotechnical engineering reliability: how well do we know what we are doing?, *Journal of Geotechnical and Geoenvironmental Engineering*, 130, 985–1003.

Christian, J. T., Ladd, C. C. and Baecher G. B. (1994). Reliability applied to slope stability analysis. *Journal of Geotechnical Engineering*, 120, 2180–207.

DeGroot, D. J. and Baecher, G. B. (1993). Estimating autocovariance of in-situ soil properties. *Journal of Geotechnical Engineering*, 119, 147–67.

Dingman, S. L. (2002). *Physical Hydrology*. Prentice Hall, Upper Saddle River, NJ.

Duncan, J. M. (1996). State of the art: limit equilibrium and finite element analysis of slopes. *Journal Geotechnical and Geoenvironmental Engineering*, 122, 577–97.

El-Ramly, H., Morgenstern, N. R. and Cruden, D. M. (2003). Probabilistic stability analysis of a tailings dyke on pre-sheared clay-shale. *Canadian Geotechnical Journal*, 40, 192–208.

Fellenius, W. (1936). Calculation of the stability of earth dams. *Proceedings Second Congress on Large Dams*, 4, 445–63.

Fourie, A. B., Rowe, D. and Blight, G. E. (1999). The effect of infiltration on the stability of the slopes of a dry ash dump. *Geotechnique*, 49, 1–13.

Frankel, A. D. and Safak, E. (1998). Recent trends and future prospects in seismic hazard analysis. In *Geotechnical Earthquake Engineering and Soil Dynamics III*, Geotechnical Special Pub. 75, Eds P. Dakoulas, M. Yegian and R. D. Holtz. ASCE, Reston, VA.

Frankel, A. D., Petersen, M. D., Mueller, C. S., Haller, K. M., Wheeler, R. L., Leyendecker, E. V., Wesson, R. L., Harmsen, S. C., Cramer, C. H., Perkins, D. M. and Rukstales, K. S. (2002). *Documentation for the 2002 update of the national seismic hazard maps*. U. S. Geological Survey Open-file report 02-420.

Fredlund, D. G. and Krahn, J. (1977). Comparison of slope stability methods of analysis. *Canadian Geotechnical Journal*, 14, 429–39.

Fredlund, D. G. and Rahardjo, H. (1993). *Soil Mechanics for Unsaturated Soils*. John Wiley and Sons, New York.

Fredlund, D. G. and Xing, A. (1994). Equations for the soil-water suction curve. *Canadian Geotechnical Journal*, 31, 521–32.

Gelhar, L. W. (1993). *Stochastic Subsurface Hydrology*. Prentice Hall, Englewood Cliffs, NJ.

Gilbert, R. B., Wright, S. G. and Liedke, E. (1998). Uncertainty in back analysis of slopes: Kettleman Hills case history. *Journal of Geotechnical and Geoenvironmental Engineering*, 124, 1167–76.

Griffith, V. D. and Fenton, G. A. (1997). Three-dimensional seepage through spatially random soil. *Journal of Geotechnical and Geoenvironmental Engineering*, 123, 153–60.

Griffith, V. D. and Fenton, G. A. (2004). Probabilistic slope stability analysis by finite elements. *Journal of Geotechnical and Geoenvironmental Engineering*, 130, 507–18.

Hansen, J. Brinch (1965). The philosophy of foundation design: design criteria, safety factors, and settlement limits. In *Bearing Capacity and Settlement of Foundations*, Ed. A. S. Vesic. Duke University, Durham, pp. 9–14.

Hershfield, D. M. (1961). *Rainfall frequency atlas of the United States*. Tech. Paper 40, US Department of Commerce, Washington, DC.

Ireland, H. O. (1954). Stability analysis of the Congress Street open cut in Chicago. *Geotechnique*, 40, 161–80.

Johnson, K. A. and Sitar, N. (1990). Hydrologic conditions leading to debris-flow initiation. *Canadian Geotechnical Journal*, 27, 789–801.

Khire, M. V., Benson, C. H. and Bosscher, P. J. (1999). Field data from a capillary barrier and model predictions with UNSAT-H. *Journal of Geotechnical and Geoenvironmental Engineering*, 125, 518–27.

Kim, J. and Sitar, N. (2003). Importance of spatial and temporal variability in the analysis of seismically induced slope deformation. In *Applications of Statistics and Probability in Civil Engineering*, Eds A. Der Kiureghian, S. Madanat and J. M. Pestana. Balkema, Rotterdam, pp. 1315–22.

Krahn, J., Fredlund, D. G. and Klassen, M. J. (1989). Effect of soil suction on slope stability at Notch Hill. *Canadian Geotechnical Journal*, 26, 2690–78.

Kramer, S. L. (1996). *Geotechnical Earthquake Engineering*. Prentice Hall, Upper Saddle, NJ.

Lacy, S. J. and Prevost, J. H. (1987). Nonlinear seismic analysis of earth dams. *Soil Dynamics and Earthquake Engineering*, 6, 48–63.

Lambe, T. W. (1973). Predictions in soil engineering. *Geotechnique*, 23, 149–202.

Leonards, G. A. (1982). Investigation of failures. *Journal of Geotechnical Engineering*, 108, 185–246.

Leroueil, S. and Tavanas, F. (1981). Pitfalls of back-analysis. In *10th International Conference on Soil Mechanics and Foundation Engineering*. A. A. Balkema, Rotterdam, 1, pp. 185–90.

Leshchinsky, D., Baker, R. and Silver, M.L. (1985). Three dimensional analysis of slope stability. *International Journal Numerical and Analytical Methods in Geomechanics*, 9, 199–223.

Lin, J.-S. and Whitman, R. V. (1986). Earthquake induced displacements of sliding blocks. *Journal of Geotechnical Engineering*, 112, 44–59.

Liu, W. K., Belytschko, T. and Mani, A. (1986). Random field finite elements. *International Journal of Numerical Methods in Engineering*, 23, 1831–45.

Low, B. F., Gilbert, R. B. and Wright, S. G. (1998). Slope reliability analysis using generalized method of slices. *Journal of Geotechnical and Geoenvironmental Engineering*, 124, 350–62.

Lumb, P. (1970). Safety factors and the probability distribution of strength. *Canadian Geotechnical Journal*, 7, 225–42.

Lumb, P. (1974). Application of statistics in soil mechanics. In *Soil Mechanics – New Horizons*, Ed. I.K. Lee. Elsevier, New York, pp. 44–111.

Maidman, D. R. (1992). *Handbook of Hydrology*. McGraw-Hill, New York.

Michalowski, R. L. (1989). Three-dimensional analysis of locally loaded slopes. *Geotechnique*, 39(1), 27–38.

Monteith, J. L. (1973). *Principles of Environmental Physics*. Elsevier, New York.

Newmark, N. M. (1965). Effects of earthquakes on dams and embankments. *Geotechnique*, 15, 139–60.

Penman, H. L. (1948). Natural evapotranspiration from open water, bare soil and grass. *Proceedings of the Royal Society A*, 193, 120–46.

Phoon, K. K. and Kulhawy, F. H. (1999). Characterization of geotechnical variability. *Canadian Geotechnical Journal*, 36, 612–24.

Pierson, T. C. (1980). Piezometric response to rainstorms in forested hillslope drainage depression. *Journal of Hydrology (New Zealand)*, 19, 1–10.

Richards, L. A. (1931). Capillary conduction of liquids through porous medium. *Journal of Physics*, 72, 318–33.

Richardson, C. W. (1981). Stochastic simulation of daily precipitation, temperature, and solar radiation. *Water Resources Research*, 17, 182–90.

Ritchie, J. T. (1972). A model for predicting evaporation from a row crop with incomplete plant cover. *Water Resources Research*, 8, 1204–13.

Rojiani, K. B., Ooi, P. S. K. and Tan, C. K. (1991). Calibration of load factor design code for highway bridge foundations. In *Geotechnical Engineering Congress*, Geotechnical Special Publication No. 217. ASCE, Reston, VA, 2, pp. 1353–64.

Schroeder, P. R., Morgan, J. M., Walski, T. M. and Gibson, A. C. (1994). *Hydraulic Evaluation of Landfill Performance – HELP*. EPA/600/9-94/xxx, US EPA Risk Reduction Engineering Laboratory Cincinnati, OH.

Seed, H. B., Lee, K. L., Idriss, I. M. and Makdisi, F. I. (1975). The slides in the San Fernando dams during the earthquake of Feb. 9, 1971. *Journal Geotechnical Engineering Division, ASCE*, 10, 651–88.

Shuttleworth, W. J. (1993). Evaporation. In *Handbook of Hydrology*, Ed. D. R. Maidman. McGraw-Hill, New York.

Sommerville, P. (1998). Emerging art: earthquake ground motion. In *Geotechnical Earthquake Engineering and Soil Dynamics III*, Eds P. Dakoulas, M. Yegian and R. D. Holtz. Geotechnical Special Pub. 75, ASCE, Reston, VA, pp. 1–38.

Soong, T-Y. and Koerner, R. M. (1996). Seepage induced slope instability. *Journal of Geotextiles and Geomembranes*, 14, 425–45.

Sykora, D. W., Koester, J. P. and Hynes, M. E. (1991). Seismic hazard assessment of liquefaction potential at Mormon Island Auxiliary Dam, California, USA. In *Proceedings 23rd US-Japan Joint panel Meeting on Wind and Seismic Effects*, NIST SP 820, Gaithersburg, National Institute of Standards and Technology, pp. 247–67.

Tang, W. H. (1971). A Bayesian evaluation of information for foundation engineering design. In *Statistics and Probability in Civil Engineering*, Ed. P. Lumb. Hong Kong University Press, Hong Kong, pp. 173–85.

Tang, W. H., Yucemen, M. S. and Ang, A. H-S. (1976). Probability-based short term design of soil slopes. *Canadian Geotechnical Journal*, 13, 201–15.

Taylor, D. W. (1948). *Fundamentals of Soil Mechanics*. John Wiley and Sons, New York.

Terzaghi, K. (1929). *Effect of minor geologic details on the safety of dams*. Technical Pub. 215. American Institute of Mining and Metallurgical Engineering, New York, NY, pp. 31–44.

van Genuchten, M. Th. (1980). A closed-form equation for predicting hydraulic conductivity of unsaturated soils. *Soil Science Society of America Journal*, 44, 892–8.

VanMarcke, E. H. (1977). Probabilistic modeling of soil profiles. *Journal of the Geotechnical Engineering Division, ASCE*, 103, 1227–46.

Wilkinson, P. L., Brooks, S. M. and Anderson, M. G. (2000). Design and application of an automated non-circular slip surface search within a combined hydrology and stability model (CHASM). *Hydrological Processes*, 14, 2003–17.

Wen, Y. K. (1990). *Structural Load Modeling and Combinations for Performance and Safety Evaluation*. Elsevier, Amsterdam.

Whitman, R. V. and Bailey, W. A. (1967). Use of computers for slope stability analysis. *Journal of Soil Mechanics and Foundations Division, ASCE*, 93, 475–98.

Wu, T. H. (2003). Assessment of landslide hazard under combined loading. *Canadian Geotechnical Journal*, 40, 821–9.

Wu, T. H. and Abdel-Latif, M. A. (2000). Prediction and mapping of landslide hazard. *Canadian Geotechnical Journal*, 37, 579–90.

Wu, T. H. and Kraft, L. M. (1970). Safety analysis of slopes. *Journal of Soil Mechanics and Foundations Division, ASCE*, 96, 609–30.

Wu, T. H., Thayer, W. B. and Lin, S. S. (1975). Stability of embankment on clay. *Journal of Geotechnical Engineering Division, ASCE*, 101, 913–32.

Yucemen, M. S. and Tang, W. H. (1975). Long-term stability of soil slopes – a reliability approach. In *Applications of Statistics and Probability in Soil and Structural Engineering*. Deutsche Gesellschaft fur Erd-und Grundbau e.V., Essen, 2, pp. 215–30.

Zerfa, F. Z. and Loret, B. (2003). Coupled dynamic elastic–plastic analysis of earth structures. *Soil Dynamics and Earthquake Engineering*, 23, 4358–454.

Zhang, L. L., Zhang, L. M. and Tang, W. H. (2005). Rainfall-induced slope failure considering variability of soil properties. *Geotechnique*, 55, 183–8.

Notations

c = cohesion
F = force
F_s = safety factor
h = depth of soil layer
k = seismic coefficient
K = permeability coefficient
l = length
M = moment
N = model bias
P = normal force
P_f = failure probability
Q = source or sink term
r = distance
s = soil property
s_m = measured soil property
u = pore pressure
T = return period
$V()$ = variance
W = weight
X, Y = random variable
α = angle of inclination of failure surface
β = reliability index

γ = unit weight
δ = correlation distance
ζ = testing error
θ = volumetric water content
ρ = corelation coefficient
$\sigma()$ = standard deviation
φ = angle of internal friction
ψ = suction
$\Delta()$ = coefficient of variation
Γ = variance reduction function

Chapter 12

Reliability of levee systems

Thomas F. Wolff

12.1 About levees

12.1.1 Dams, levees, dikes and floodwalls

Dams and levees are two types of flood protection structures. Both consist primarily of earth or concrete masses intended to retain water. However, there are a number of important differences between the two, and these differences drive considerations about their reliability. As shown in Figure 12.1, *dams* are constructed perpendicular to a river, and are seldom more than 1 or 2 km long. They reduce flooding of downstream floodplains by retaining runoff from storms or snowmelt behind the dam and releasing it in a controlled manner over a period of time. The peak stage of the flood is reduced, but downstream stages will be higher than natural for some time afterward. Because dam sites are fairly localized, they can be chosen very deliberately, and a high level of geotechnical exploration, testing and analysis is usually done during design. Many dams are multiple-use structures, with one or more of the following beneficial uses: flood control, water supply, power supply, and recreation. These often provide substantial economic benefits, permitting substantial costs to be expended in design and construction. Finally, they are generally used on a daily basis, and have some operating staff that can provide some level of regular observation.

In contrast, *levees* are constructed parallel to rivers and systems of levees are typically tens to thousands of kilometers long (see Figure 12.1). They prevent flooding of the adjacent landward floodplain by keeping the river confined to one side of the levee, or where levees are present on both sides of a river, between the levees. As levees restrict the overbank flow area, they may actually increase river stages, making a flood peak even higher than it would naturally be. However, they can be designed high enough to account for this effect, and to protect landside areas against some design flood event, even though that event will occur at a higher stage with the levee in place. Because levees are long and parallel to riverbanks, designers often have a much

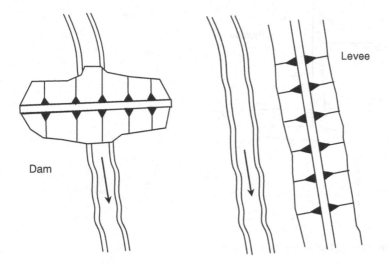

Figure 12.1 Dams and levees.

more limited choice of locations than for dams. Furthermore, riverbank soil deposits may be complex and highly variable, alternating between weak and strong, and pervious and impervious. Levees tend to be single-use (flood control) projects that may go unneeded and unused for long periods of time. Compared to dam design, an engineer designing a levee may face greater levels of uncertainties, yet proportionately smaller budgets for exploration and testing. Given all of these uncertainties, it would seem that levees are inherently more fraught with uncertainty than dams. If there are any "balancing factors" to this dilemma, it may be that most dams tend to be much higher than most levees, and that levees designed for very rare events have a very low probability of loading.

There are some structures that lie in a "gray area" between dams and levees. Like levees, they tend to be very long and parallel to watercourses or water bodies. But like dams, they may permanently retain water to some part of their height, and provide further protection against temporary high waters. Notable examples are the levees around New Orleans, Louisiana, the dikes protecting much of the Netherlands, and the Herbert Hoover Dike around Lake Okeechobee, Florida. If more precise terminology is desired, the author suggests using *dike* to refer to long structures that run parallel to a water body, provide permanent retention of water to some level, and periodic protection against higher water levels.

Finally, another related structure is a *floodwall* (see Figure 12.2). A floodwall performs the same function as a levee, but is constructed of concrete, steel sheetpile, or a combination. As these materials provide much

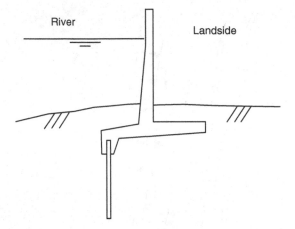

Figure 12.2 Inverted-T floodwall with sheetpile.

greater structural strength, floodwalls are much more slender and require much less width than levees. They are used primarily in urban areas, where the cost of the floodwall would be less than the cost of real estate required for a levee.

12.1.2 Levee terminology

Figure 12.3 illustrates the terminology commonly used to describe the various parts of the levee. Directions perpendicular to the levee are stated as *landside* and *riverside*. The *crown* refers to the flat area at the top of the levee, or sometimes to the edges of that area. The *toe* of a levee refers to the point where the slope meets the natural ground surface. The *foreshore* is the area between the levee and the river. Construction material for the levee may be dredged from the river, but often is taken from *borrow pits*, which are broad, shallow excavations. These are commonly located

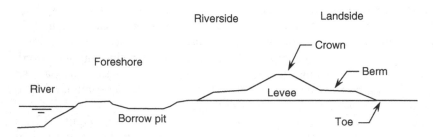

Figure 12.3 Levee terminology.

riverside of the levee, within the foreshore, but occasionally may be found on the landside.

As one proceeds along the length of a levee, the embankment geometry of the levee may vary, as may the subsurface materials and their configuration. A levee *reach* is a length of levee (often several hundred meters) for which the geometry and subsurface conditions are sufficiently similar that they can be represented for analysis and design by a single two-dimensional cross-section and foundation profile. Using the cross-section representing the reach, analyses are performed to develop a *design template* for use throughout that reach. A design template specifies slope inclination, crown and berm widths and crown and berm elevations.

The *line of protection* refers to the continuous locus of levees or other features (floodwalls, gates, or temporary works such as sandbags) that separate the flooded area from the protected area. The line of protection must either tie to high ground or connect back to itself. The protected area behind a single, continuous line of protection is referred to as a *polder*, especially in the Netherlands.

An unintended opening in the line of protection is referred to as a *breach*. When destructive flow passes through a breach, creating significant erosive damage to the levee and significant landward flooding, the breach is sometimes called a *crevasse*.

The following appurtenances may be used to provide increased protection against failure:

- *Stability berms* are extensions of the primary embankment section. Moving downslope from the crown, the slope is interrupted by a nearly flat area of some width (the berm), and then the slope continues. The function of a stability berm is to provide additional weight and shear resistance to prevent sliding failures.
- *Seepage berms* may be similar in shape to stability berms, but their function is to provide weight over the landside toe area to counteract upward seepage forces due to underseepage.
- *Cutoffs* are impervious, vertical features provided beneath a levee. They may consist of driven sheet piles, slurry walls, or compacted clay trenches. They reduce underseepage problems by physically reducing seepage under the levee, and lengthening the seepage path.
- *Relief wells* are vertical, cylindrical drains placed into the ground near the landside toe. If piezometric levels in the landside foundation materials exceed the ground surface, water will flow from the relief wells, keeping uplift pressure buildup to a safe level as flood levels continue to rise.
- Riverside *slope protection* such as stone *riprap* or paving may be provided to prevent wave erosion of the levee where there is potential for wave attack.

12.2 Failure Modes

Levees are subject to failure by a variety of modes. The following paragraphs summarize the five failure modes most commonly considered, and which will be addressed further.

12.2.1 Overtopping

The simplest and most obvious failure mode is overtopping. The levee is built to some height, and the flood water level exceeds that height, flooding the protected area. While the probability of flooding for water above the top of a levee will obviously be near unity, it may be somewhat lower as the levee might be raised with sandbags during a floodfight operation.

12.2.2 Slope instability

Embankments or berms may slide due to the soil strength being insufficient to resist the driving forces from the embankment weight and water loading. Of most concern is an earth slide during flood loading that would lower the levee crown below the water elevation, resulting in breaching and subsequent failure. Usually such a slide would occur on the landside of the levee. Figure 12.4 shows the slide at the 17th St. Canal levee in New Orleans following the storm surge from Hurricane Katrina in 2005. At that site, a portion of the levee slid landward about 15–20 m landward as a relatively intact block. "Drawdown" slope failures may occur on the riverside slope of a levee when flood waters recede more quickly than pore pressures in the foundation soils dissipate. Slope failures may occur on either slope

Figure 12.4 Slide at 17th St. Canal levee, New Orleans, 2005 (photo by T. Wolff).

during non-flood conditions; one scenario involves levees built of highly plastic clays cracking during dry periods, and the cracks becoming filled with water during rainy periods, leading to slides. Slope slides during non-flood periods can leave a deficient levee height if not repaired before the next flood event.

12.2.3 Underseepage

During floods, the higher water on the riverside creates a gradient and subsequent seepage water under the levee, as shown in Figure 12.5. Seepage may emerge near-vertically from the ground on the landside of the levee. Where foundation conditions are uniform, the greatest pressure, upward gradient and flow will be concentrated at the landside levee toe. In some instances, the water may seep out gently, leaving the landside ground surface soft and spongy, but not endangering the levee. Where critical combinations of water levels, soil types, and foundation stratigraphy (layering) are present, water pressures below the surface or the velocity of the water at the surface may be sufficient to move the soil near the levee toe. *Sand boils* (Figure 12.6) may appear; these are small volcano-like cones of sand, built from foundation sands carried to the surface by the upward-flowing water. Continued erosion of the subsoil may lead to settlement of the levee, permitting overtopping, or to catastrophic uncontrolled subsurface erosion, sometimes termed a *blowout*.

12.2.4 Throughseepage

Throughseepage refers to detrimental seepage coming through the levee above the ground surface. When seepage velocity is sufficient to move materials, the resulting internal erosion is called *piping*. Piping can occur if there are cracks or voids in the levee due to hydraulic fracturing, tensile stresses, decay of vegetation, animal activity, low density adjacent to conduits, or any similar location where there may be a preferential path of seepage. For piping to occur, the tractive shear stress exerted by the flowing water must exceed the critical tractive shear stress of the soil. In addition to internal discontinuities

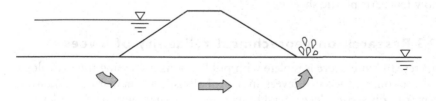

Figure 12.5 Underseepage.

Figure 12.6 Sand boil (photo by T. Wolff).

leading to piping, high exit gradients on the downstream face of the levee may cause piping and progressive backward erosion.

12.2.5 Surface erosion

Erosion of the levee section during flood events can lead to failure by reduction of the levee section, especially near the crown. As flood stages increase, the potential increases for surface erosion from two sources: (1) excessive current velocities parallel to the levee slope, and (2) erosion due to wave attack directly against the levee slope. Protection is typically provided by a thick grass cover, occasionally with stone or concrete revetment at locations expected to be susceptible to wave attack. During floods, additional protection may be provided where necessary using dumped rock, snow fence, or plastic sheeting.

12.3 Research on geotechnical reliability of levees

The research literature is replete with publications discussing the hydrologic and economic aspects of levees in a probabilistic framework. These focus primarily on setting levee heights in a manner that optimizes the level of protection against parameters such as costs, benefits, damage, and possible loss of life. In many of these studies, the geotechnical reliability is either neglected, suggesting that the relative probability of geotechnical failure is negligible, or left for others to provide suitable probabilities of failure for geotechnical failure modes.

Duncan and Houston (1983) estimated failure probabilities for California levees constructed from a heterogeneous mixture of sand, silt, and peat, and founded on peat of uncertain strength. Stability failure due to riverside water load was analyzed using a horizontal sliding block model. The factor of safety was expressed as a function of the shear strength, which is a random variable due to its uncertainty, and the water level, for which there is a defined annual exceedance probability. Values for the annual probability of failure for 18 islands in the levee system were calculated by integrating numerically over the joint events of high-water levels and insufficient shear strength. The obtained probability of failure values were adjusted based on several practical considerations; first, they were normalized with respect to length of levee reach modeled (longer reaches should be more likely to fail than shorter one) and second, they were adjusted from relative probability values to more absolute values by adjusting them with respect to the observed number of failures.

Peter (1982), in *Canal and River Levees*, provides the most complete reference-book treatise on levee design, based on work in Slovakia. Notable among Peter's work is a more up-to-date and extended treatment of mathematical and numerical modeling than in most other references. Under-seepage safety is cast as a function of particle size and size distribution, and not just gradient alone. While Peter's work does not address probabilistic methods, it is of note because it provides a very comprehensive collection of analytical models for a large number of potential failure modes which could be incorporated into probabilistic analyses.

Vrouwenvelder (1987) provides a very thorough treatise on a proposed probabilistic approach to the design of dikes and levees in the Netherlands. Notable aspects of this work include the following.

- It is recognized that exceedance frequency of the crest elevation is not taken as the frequency of failure; there is some probability of failure for lower elevations, and there is some probability of no failure or inundation above this level as an effort might be made to raise the protection.
- Considered failure modes are overflowing and overtopping, macro-instability (deep sliding), micro-instability (shallow sliding or erosion of the landside slope due to seepage), and piping (as used, equivalent to underseepage as used herein). In an example, 11 parameters are taken as random variables which are used in conjunction with relatively simple mathematical physical models. Aside from overtopping, piping (under-seepage) is found to be the governing mode for the section studied; slope

stability is of little significance to probability of failure. Surface erosion due to wave attack or parallel currents is not considered.

- The probabilistic procedure is aimed at optimizing the height and slope angle of new dikes with respect to total costs including construction and expected losses, including property and life. In the example, macro-instability (slope failure) of the inner slope was found to have a low risk, much less than 8×10^{-8} per year. Piping was found to be sensitive to seepage path length; probabilities of failure for piping varied, but were several orders of magnitude higher (10^{-2} to 10^{-3} per year). Micro-instability (landside sloughing due to seepage) was found to have very low probabilities of failure. Based on these results, it was considered that only overtopping and piping need be considered in a combined reliability evaluation.
- The "length problem" (longer dikes are less reliable than equivalent short ones) is discussed.

The U.S. Army Corps of Engineers first introduced probabilistic concepts to levee evaluation in the United States in their Policy Guidance Letter No. 26 (USACE, 1991). Prior to that time, planning studies for federally funded levee improvements to existing, non-federal levees were based on the assumption that the existing levee was essentially absent and provided no protection. Following 1991, it was assumed that the levee was present with some probability, which was a function of water elevation. The function was defined as a straight line between two points. The probable failure point (PFP) was taken as the water elevation for which the probability of levee failure was estimated to be 0.85, and the probable non-failure point (PNP) was defined as the water elevation for which the probability of failure was estimated at 0.15. The only guidance for setting the coordinates of these points was a "template method," based only on the geometry of the levee section without regard for geotechnical conditions, or the engineer's judgment based on other studies.

To better define the shape of the function relating probability of failure to water level, research by Wolff (1994) led to a report for the U.S. Army Corps of Engineers entitled *Evaluating the Reliability of Existing Levees*. It reviewed past and current practice for deterministic analysis and design of levees and illustrated how they could be incorporated into a probabilistic model. Similar to Vrouwenvelder (1987), the report considers slope stability, underseepage, and through seepage: it also considers surface erosion and a provision for incorporating other information through judgmental probabilities. The overall probability of failure considering multiple failure modes is treated by assuming the failure modes form a series system. The report was later incorporated as Appendix B in the Corps' Engineer Technical Letter ETL 1110-2-556 "Risk-Based Analysis in Geotechnical Engineering for Support of Planning Studies" (USACE, 1999). It should be noted that the guidance was intended to apply only to the problem of characterizing existing levees in probabilistic-based economic and planning studies to evaluate the feasibility of levee improvements; as of this writing, the Corps has not issued any guidance recommending that probabilistic procedures be used for the geotechnical aspects of levee design. Some of the report was also briefly summarized by Wolff *et al.* (1996). This chapter draws heavily on Wolff (1994) and USACE (1999), with some additional consideration of other failure mode models and subsequent work by others.

In 1999, the Corps' Policy Guidance Letter No. 26 (USACE, 1999) was updated and that version is posted on the Internet at the time of this writing. The current version recommends consideration of multiple failure modes and permits the PFP to be assigned probability values between 0.85 and 0.99, and the PNP values between 0.01 and 0.85.

In 2000, the National Research Council published *Risk and Uncertainty in Flood Damage Reduction Studies* (National Research Council, 2000), a critical review of the Corps of Engineers' approach to the application of probabilistic methods in flood control planning. The recommendations of the report included the following:

> the Corps' risk analysis method (should) evaluate the performance of a levee as a spatially distributed system. Geotechnical evaluation of a levee, which may be many miles long, should account for the potential of failure at any point along the levee during a flood. Such an analysis should consider multiple modes of levee failure (e.g. overtopping, embankment instability), correlation of embankment and foundation properties, hazards associated with flood stage (e.g. debris, waves, flood duration) and the potential for multiple levee section failures during a flood. The current procedure treats a levee within each damage reach as independent and distinct from one reach to the next. Further, within a reach, the analysis focuses on the portion of each levee that is most likely to fail. This does not provide a sufficient analysis of the performance of the entire levee.

Further, it notes

> The Corps' new geotechnical reliability model would benefit greatly from field validation. ... The committee recommends that the Corps undertake statistical ex post studies to compare predictions of geotechnical levee failure probabilities made by the reliability model against frequencies of actual levee failures during floods.

The report noted that the simplified concepts of the Corps' Policy Guidance Letter No. 26 had been updated via the procedures in the 1999 guidance. It noted that "the numerical difference in risk analysis results compared to the initial model (PFP and PNP) may not be large. The updated model, however, supports a more complete geotechnical analysis and should replace the initial model." It further notes that the Corps' geotechnical model does not separate natural variability and model uncertainty (the former could be uncorrelated from reach to reach, with the latter perfectly correlated), does not include duration of flooding, and remains to be calibrated against the frequency of actual levee failures. Nevertheless, no further changes have been made to the Corps' methodology as of this writing in 2006.

Apel *et al.* (2004) describe a comprehensive probabilistic approach to flood risk assessment, across the entire spectrum from rainfall way to economic damage. They incorporate the probability of failure vs. water elevation functions introduced by Wolff (1994) in USACE (1999), but extend them from 2D curves to 3D surfaces to include duration of flooding.

Buijs *et al.* (2003) describe the application of the reliability-based design tools (PC-Ring) developed in the Netherlands to levees in the UK, and provides some comparison of the issues around flood defenses in the two countries. In this study, overtopping was found to be the dominating failure mode.

A doctoral thesis by Voortman (2003) summarizes the history of dike design in the Netherlands. Voortman notes that a complete probabilistic analysis considering all variables was explored for dike design in the 1980s, but that "the legislative safety requirements are still prescribed as probabilities of exceedance of the design water level and thus the full probabilistic approach has not been officially adopted to date. Probabilistic methods are sometimes applied, but a required failure probability for flood defenses is not defined in Dutch law." The required level of safety is still couched in terms of a dike or levee being "safe" against a flood elevation of some statistical frequency.

In summary, the literature to date largely consists of probabilistic models for evaluation of the most tractable levee failure modes, such as slope stability and underseepage. Rudimentary methods of systems probability have been used to consider multiple failure modes and long systems of levee reaches. Work remains to better model throughseepage, spatial variability, systems reliability and duration of flooding, and to calibrate predicted probabilities of failure to the frequency of actual failures.

12.4 A framework for levee reliability analysis

12.4.1 Accuracy of probabilistic measures

Before proceeding, a note of caution is in order. The application of probabilistic analysis in civil engineering is still an emerging technology, more so in geotechnical engineering and even more so in application to levees. Much experience remains to be gained, and the appropriate form and shape of probability distributions for most of the relevant parameters are not known with certainty. The methods described herein should not be expected to provide "true" or "absolute" probability-of-failure values, but can provide consistent measures of *relative reliability* when reasonable assumptions are employed. Such comparative measures can be used to indicate for example which reach (or length) of levee, which typical section, or which alternative design may be more reliable than another. They also can be used to judge which of several performance modes (seepage, slope stability, etc.) governs the reliability of a particular levee.

12.4.2 Calibration of models

Any reliability-based evaluation must be *calibrated*; i.e. tested against a sufficient number of well-understood engineering problems to ensure that it produces reasonable results. Performance modes known to be problematical (such as seepage) should be found to have a lower reliability than those for which problems are seldom observed; larger and more stable sections should be found to be more reliable than smaller, less stable sections, etc. As recommended by the National Research Council (2000), as additional analyses are performed by researchers and practitioners, on a wide range of real levee cross sections using real data, refinements in the procedures may be needed to produce reasonable results.

12.4.3 The conditional probability-of-failure function

For the purposes at hand, *failure* will be defined as the unintended flooding of the protected area, and thus includes both overtopping and breaching of the levee at water elevations below the crown. For an existing levee subjected to a flood, the probability of failure p_f can be expressed as a function of the flood water elevation and other factors, including soil strength, permeability, embankment geometry, foundation stratigraphy, etc. The analysis of levee reliability will start with the development of a *conditional* probability of failure function given the flood water elevation, which will be constructed using engineering estimates of the probability functions or moments of the other relevant variables.

The conditional probability of failure can be written as:

$$p_f = \Pr(failure \mid FWE) = f(FWE, X_1, X_2 \ldots X_n) \tag{12.1}$$

In the above equation, the first expression (denoting probability of failure) will be used as a shorthand version of the second term. The symbol "$|$" is read *given* and the variable *FWE* is the flood water elevation. In the second expression, the random variables X_1 through X_n denote relevant parameters such as soil strength, conductivity, top stratum thickness, etc.

Equation (12.1) can be restated as follows: "The probability of failure, given the flood water elevation, is a function of the flood water elevation and other random variables."

Two extreme values of the function can be readily estimated by engineering judgment:

- For flood water at the same level as the landside toe (base elevation) of the levee, the levee is not loaded; hence, $p_f = 0$.
- For flood water at or near the levee crown (top elevation), $p_f \rightarrow 1.00$.

In principle, the probability of failure value may be something less than 1.0 with water at the crown elevation, as additional protection can be provided by emergency measures.

The question of primary interest, however, is the shape of the function between these extremes. Quantifying this shape is the focus of the procedures to follow.

Reliability (R) is defined as:

$$R = 1 - p_f \tag{12.2}$$

hence, for any flood water elevation, the probability of failure and reliability must sum to unity.

For the case of flood water partway up a levee, p_f could be very near zero or very near unity, depending on engineering factors such as levee geometry, soil strength, hydraulic conductivity, foundation stratigraphy, etc. In turn, these differences in the conditional probability of failure function could result in very different scenarios. Four possible shapes of the p_f and R functions are illustrated in Figure 12.7. For a well-designed and constructed "good"

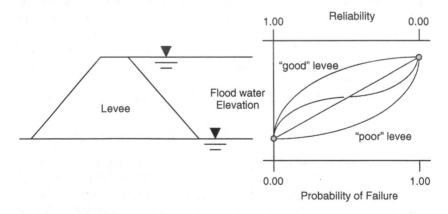

Figure 12.7 Possibble probability of failure vs. flood water elevation functions.

levee, the probability of failure may remain low and the reliability remain high until the flood water elevation is rather high. In contrast, a "poor" levee may experience greatly reduced reliability when subjected to even a small flood head. It is hypothesized that some real levees may follow the intermediate curve, which is similar in shape to the "good" case for small floods, but reverses to approach the "poor" case for floods of significant height.

12.4.4 Steps in levee system evaluation

To evaluate the annual probability of failure for an entire levee system, four steps are required.

Step 1

A set of conditional probability of failure versus flood water elevation functions is developed. For each levee reach, this requires a separate function for each of the following failure modes: slope stability, underseepage, through-seepage, and surface erosion. Overtopping and other failure modes (e.g. human error, failure to close a gate) may be added as desired. This leads to N functions, where $N = rm$, r is the number of reaches, and m is the number of failure modes considered.

Step 2

For each reach, the functions for the various failure modes developed in step 1 are systematically combined into one composite conditional probability of failure function for each reach. This leads to r composite functions.

Step 3

Given the composite reach function and a probability of exceedance function for the flood water elevation, the annual probability of levee failure for the reach can be determined by integrating the product of flood probability and conditional probability of failure over the range of water elevations from the levee toe to the crown.

Step 4

Using the results of step 3, an overall annual probability of levee system failure can be developed by combining the annual probabilities for the entire set of reaches. If only a few reaches dominate (by having comparatively high probabilities of failure), this step might be simplified by considering only those selected reaches. In any case, *length effects* must first be considered before the number of reaches is finalized.

12.5 Division into reaches

The first step in any levee analysis or design, deterministic or probabilistic, is division of the levee into reaches for geotechnical analysis. Note that those analyzing the economic benefits of a levee may divide the levee into "damage reaches" based on interior topography and value of property at risk, and those studying interior drainage may use another system of dividing a levee into reaches for their purposes. In a comprehensive risk analysis, reach identification may require careful coordination among an analysis team to ensure proper combination of probabilities of failure and associated consequences.

Analysis of slope stability and seepage conditions is usually performed using a two-dimensional perpendicular cross-section of the levee and foundation topography and stratigraphy. These are in fact an idealization of the actual conditions, which have some uncertainty at any cross-section considered, and furthermore are variable in the direction of the levee. A geotechnical *reach* is a length of levee for which the topography and soil conditions are sufficiently similar that the engineer considers they can be represented by a single cross-section for analysis. This is illustrated in Figure 12.8, where the solid lines represent the best understanding (also uncertain) of the foundation soil profile under the levee (often along the landside toe), and the dashed lines represent the idealization into reaches for analysis. For each reach, a cross-section is developed, and assumed to be representative of the conditions along the reach (See Figure 12.9). Calculated performance measures (factor of safety, exit gradients) for the section are assumed to be valid for the length of the reach. In design practice, reaches are typically several hundred meters long, but the lengths and endpoints should be carefully selected based on topographic and geotechnical information, not on a specified length. Within a reach, the design cross-section is often selected at a point considered to be the most critical, such as the greatest height.

In a probabilistic analysis, it may be required to further subdivide levee reaches into statistically independent shorter reaches for analysis of overall system reliability. This will be further discussed later under length effects.

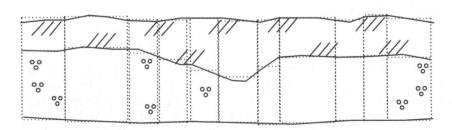

Figure 12.8 Division into reaches.

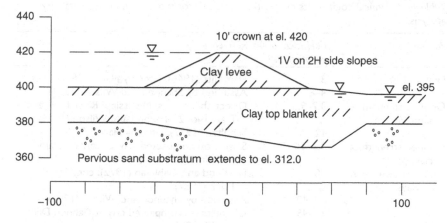

Figure 12.9 Two-dimensional cross-section representing a reach (from Wolff, 1994).

12.6 Step 1: failure mode models

12.6.1 Slope instability

It is desired to define the conditional probability of slope failure as a function of the flood water elevation, *FWE*. To do so, the relevant parameters are characterized as random variables, a performance function is defined that expresses the slope safety as a function of those variables, and a probabilistic model is selected to determine the probability of failure.

12.6.1.1 Random variables

Probabilistic slope stability analysis typically involves modeling most or all of the following parameters as random variables: unit weight, drained strength of sand and other pervious materials, and drained and undrained strength of cohesive materials such as clays. In addition, some researchers have included geometric uncertainty of the soil boundaries, and model uncertainty. Table 12.1 summarizes some typical coefficients of variation for these parameters. The coefficients of variation shown are for the *point strength*, characterizing the strength at a random point. However, the uncertainty in the *average strength* over some spatial distance may govern the stability more than the point strength. To consider the effects of spatial correlation in a slope stability analysis, it may be necessary to reduce these variances, or provide some other modification in the analysis model.

In slope stability problems, uncertainty in unit weight usually provides little contribution to the overall uncertainty, which is dominated by soil strength. For stability problems, it can usually be taken as a

Table 12.1 Typical coefficients of variation for soil properties used in slope stability analysis.

Parameter	Coefficient of variation (%)	Reference
Unit weight, γ	3	Hammitt (1966), cited by Harr (1987)
	4–8	Assumed by Shannon and Wilson (1994)
Drained strength of sand, $\phi\prime$	3.7–9.3	Direct shear tests, Mississippi River Lock and Dam No. 2, Shannon and Wilson (1994)
	12	Schultze (1972), cited by Harr (1987)
Drained strength of clay, $\phi\prime$	7.5–10.1	S tests on compacted clay at Cannon Dam (Wolff, 1985)
Undrained strength of clays, s_u	40	Fredlund and Dahlman (1972), cited by Harr (1987)
	30–40	Assumed by Shannon and Wilson (1994)
	11–45	UU tests on compacted clay at Cannon Dam, Wolff (1985)
Strength-to-effective stress ratio, s_u/σ'_{vo}	31	Clay at Mississippi River Lock and Dam No. 2, Shannon and Wilson (1994)

deterministic variable to reduce the number of random variables and simplify calculations.

Reported coefficients of variation for the drained friction angle (ϕ) of sands are in the range of 3–12%. Lower values can be used where there is some confidence that the materials considered are of consistent quality and relative density, and the higher values should be used where there is considerable uncertainty regarding material type or density. For the direct shear tests on sands from Lock and Dam No. 2, cited in Table 12.1 (Shannon and Wilson and Wolff, 1994), the lower coefficients of variation correspond to higher confining stresses, and vice versa.

Ladd *et al.* (1977) and others have shown that the undrained strength s_u (or c) of clays of common geologic origin can be normalized with respect to effective overburden stress σ'_{vo} and overconsolidation ratio OCR and defined in terms of the ratio s_u/σ'_{vo}. Analysis of test data on clay at Mississippi River Lock and Dam No. 2 (Shannon and Wilson and Wolff, 1994) showed that it was reasonable to characterize uncertainty in clay strength in terms of the probabilistic moments of the s_u/σ'_v parameter. The ratio of s_u/σ'_{vo} for 24 tested samples was found to have a mean value of 0.35, a standard deviation of 0.11, and a coefficient of variation of 31%.

12.6.1.2 Deterministic model and performance function

For slope stability, the choice of a deterministic analysis model is straightforward. A conventional limit-equilibrium slope stability model is used that yields a factor of safety defined in terms of the shear strength of the soil.

Many well documented programs are available to run the most popular slope stability analysis methods (e.g. Bishop's method, Spencer's method, and the Morgenstern–Price method). For a specific failure surface, many studies have shown that resulting factors of safety differ little for any of the methods that satisfy moment equilibrium or complete equilibrium. However, finding the geometry of the critical failure surface is another matter. As levees are typically located over stratified alluvial foundations, the selected deterministic model and computer program should be capable of analyzing failure surfaces with planar bases (e.g. "wedge" surfaces) and other general shapes, not only circular arcs. Furthermore, there remains a philosophical question as to what exactly is the "critical surface." In many probabilistic slope stability studies, the critical deterministic surface (lowest FS) has been found first, and the probability of failure for that surface has been calculated. However, it can be argued that one should locate the surface with the highest probability of failure, which can be greatly different when soils with different levels of uncertainty are present. Hassan and Wolff (1999) have investigated this problem and presented an approximate empirical method to locate the critical probabilistic surface when the available programs can search only for the surface of minimum FS.

Depending on the probabilistic model being used, it may be sufficient to be able to simply determine the factor of safety for various realizations of the random variables, and use these to determine the probability that the factor of safety is less than one. However, some methods, such as the advanced first-order second-moment method (AFOSM), may require a performance function that assumes a value of zero at the limit state. For the latter, the performance function for slope stability can be taken as

$$FS - 1 = 0 \qquad\qquad\qquad\qquad\qquad (12.3)$$

or

$$\ln FS = 0 \qquad\qquad\qquad\qquad\qquad (12.4)$$

The second expression is preferred as it works consistently with the assumption that the factor of safety is lognormally distributed, which is appealing as FS cannot assume negative values.

12.6.1.3 Probabilistic modeling

Having defined a set of random variables, a deterministic model and performance function, a probabilistic model is then needed to determine the probability of failure (or probability that the performance function assumes a negative value). Given the probability distributions, or at least the probabilistic moments of the random variables, the probabilistic model determines the distribution or at least the probabilistic moments of the factor of safety or performance function. Details are described elsewhere in this book, but common models include the first-order reliability method (FORM), the advanced first-order, second-moment method (AFOSM), both based on Taylor's series expansion, and simulation (or Monte Carlo) methods.

In the examples in this chapter, a variation called the Taylor's series – finite difference method, used by the Corps of Engineers (USACE, 1995), is used to calculate the mean and standard deviation of the factor of safety, knowing the mean and standard deviations of the random variables. The Taylor's series is expanded around the mean value of the FS function, not the "design point" found by iteration in the AFOSM. Hence, the nonlinearity of the performance function is not directly considered. This deficiency is partly mitigated by determining the required partial derivatives numerically over a range of plus to minus one standard deviation, rather than a tangent at the mean. Using this large increment captures some of the information about the functional shape.

12.6.1.4 Example problem

The example to follow is taken from Wolff (1994). The assumed levee cross-section is shown in Figure 12.10 and consists of a sand levee on a thin clay top stratum overlaying thick foundation sands. Units are in English system feet, consistent with the original reference (1 ft = 0.3048 m). Despite their high conductivity, sand levees have been used in a number of locations where clay materials are scarce; a notable example is along the middle Mississippi River in western Illinois. Three random variables were defined and their assumed probabilistic moments were assigned as shown in Table 12.2.

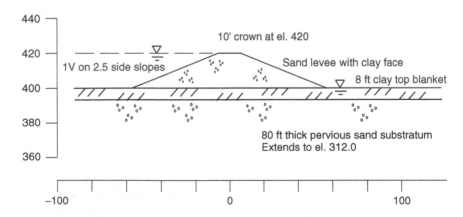

Figure 12.10 Cross-section for slope stability and underseepage example (elevations and distances in feet).

Table 12.2 Random variables for slope stability example problem.

Parameter	Expected value	Standard deviation	Coefficient of variation (%)
Friction angle of sand levee embankment, ϕ_{emb}	32 deg	2 deg	6.7
Undrained strength of clay foundation, c or s_u	800 lb/ft^2	320 lb/ft^2	40
Friction angle of sand foundation, ϕ_{found}	34 deg	2 deg	5.9

For slope stability analysis, the piezometric surface in the embankment sand was approximated as a straight line from the point where the flood water intersects the riverside slope to the landside levee toe. The piezometric surface in the foundation sands was taken as that obtained for the expected value condition in the underseepage analysis reported later in this chapter. If desired, the piezometric surface could be modeled as an additional random variable using the results of a probabilistic seepage analysis.

Using the Taylor's series–finite difference method, seven runs of a slope stability program are required for each flood water level considered; one for the expected value case, and two runs to determine the variance component of each random variable.

The seven resulting solutions and partial variance terms are summarized for $FWE = 400$ ft (base of the levee) in Table 12.3. The seven failure surfaces are nearly coincident as seen in Figure 12.11. The expected value of the factor of safety is the factor of safety calculated using the expected values of all variables:

$$E[FS] = 1.568 \tag{12.5}$$

The variance of the factor of safety is calculated per USACE (1995), using a finite-difference increment of 2σ:

$$
\begin{aligned}
VarFS &= \left(\frac{\partial FS}{\partial \phi_e}\right)^2 \sigma_{\phi_e}^2 + \left(\frac{\partial FS}{\partial c}\right)^2 \sigma_c^2 + \left(\frac{\partial FS}{\partial \phi_f}\right)^2 \sigma_{\phi_f}^2 \\
&= \left(\frac{FS_+ - FS_-}{2\sigma_{\phi_e}}\right)^2 \sigma_{\phi_e}^2 + \left(\frac{FS_+ - FS_-}{2\sigma_c}\right)^2 \sigma_c^2 + \left(\frac{FS_+ - FS_-}{2\sigma_{\phi_f}}\right)^2 \sigma_{\phi_f}^2 \\
&= \left(\frac{FS_+ - FS_-}{2}\right)^2 + \left(\frac{FS_+ - FS_-}{2}\right)^2 + \left(\frac{FS_+ - FS_-}{2}\right)^2 \\
&= \left(\frac{1.448 - 1.693}{2}\right)^2 + \left(\frac{1.365 - 1.568}{2}\right)^2 + \left(\frac{1.568 - 1.567}{2}\right)^2 \\
&= 0.025309 \tag{12.6}
\end{aligned}
$$

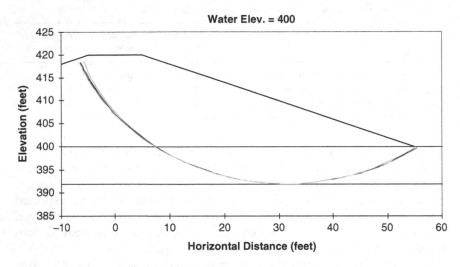

Figure 12.11 Failure surfaces for slope stability example.

Table 12.3 Example slope stability problem, undrained conditions, water at elevation 400 ($H = 0$ ft).

Run	φ (levee)	c (clay)	φ(found)	FS	Variance	Percent of total variance
1	32	800	34	1.568		
2	30	800	34	1.448		
3	34	800	34	1.693	0.015006	59.29
4	32	480	34	1.365		
5	32	1120	34	1.568	0.0.010302	40.71
6	32	800	32	1.568		
7	32	800	36	1.567	2.5×10^{-7}	0.0
			Total		.025309	100.0

In the above expression, ϕ_e is the friction angle of the embankment sand, c is the undrained strength (or cohesion) of the top blanket clay, and ϕ_f is the friction angle of the foundation sand.

The factor of safety is assumed to be a lognormally distributed random variable with $E[FS] = 1.568$ and $\sigma_{FS} = 0.025309^{1/2} = 0.159$. From the properties of the lognormal distribution, the coefficient of variation of the factor of safety is

$$V_{FS} = \frac{\sigma_{FS}}{E[FS]} = \frac{0.159}{1.568} = 0.1015 \qquad (12.7)$$

The standard deviation of ln FS is

$$\sigma_{\ln FS} = \sqrt{\ln(1 + V_{FS}^2)} = \sqrt{\ln(1 + 0.1015^2)} = 0.1012 \qquad (12.8)$$

The expected value of ln FS is

$$E[\ln FS] = \ln E[FS] - \frac{\sigma_{\ln FS}^2}{2} = \ln 1.568 - \frac{0.0102}{2} = 0.447 \qquad (12.9)$$

The reliability index is then:

$$\beta = \frac{E[\ln FS]}{\sigma_{\ln FS}} = \frac{0.447}{0.1012} = 4.394 \qquad (12.10)$$

As FS is assumed lognormally distributed, ln FS is normally distributed. From the cumulative distribution function of the standard normal distribution evaluated at $-\beta$, the conditional probability of failure for water at elevation 400 is:

$$Pr_f = 6 \times 10^{-6}$$

This is illustrated in Figure 12.12. The results for all water elevations are summarized in Table 12.4.

12.6.1.5 Interpretation

Note that the calculated probability of failure infers that the existing levee is taken to have approximately a six in one million probability of

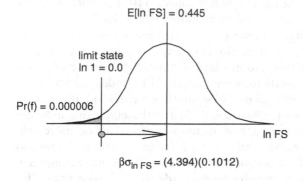

Figure 12.12 Probability of failure for slope stability example; water at elevation 400.

Table 12.4 Example slope stability problem, undrained conditions, conditional probability of failure function.

Water elevation	E[FS]	σ_{FS}	β	Pr_f
400.0	1.568	0.159	4.394	6×10^{-6}
405.0	1.568	0.161	4.351	7×10^{-6}
410.0	1.568	0.170	4.114	1.9×10^{-5}
415.0	1.502	0.107	5.699	6×10^{-9}
420.0	1.044	0.084	0.499	0.3087

instability with flood water to its base elevation of 400, even though it may in fact be existing and observed stable under such conditions. The reliability index model was developed for the analysis of yet-unconstructed structures. When applied to existing structures, it will provide probabilities of failure greater than zero. This can be interpreted as follows: given a large number of different levee reaches, each with the same geometry and with the variability in the strength of their soils distributed according to the same density functions, about six in one million of those levees might be expected to have slope stability problems. Expressing the reliability of existing structures in this manner provides a consistent probabilistic framework for use in economic evaluation of improvements to those structures.

12.6.1.6 Conditional probability of failure function

The probabilities of failure for all water elevations are summarized in Table 12.4, and plotted in Figure 12.13. Contrary to what might be expected, the reliability index reaches a maximum and the probability of failure reaches a minimum at a flood water elevation of 415.0, or three-quarters the levee height. For water above 410, the critical failure surface moves from the clay foundation materials up into the embankment sands. As this occurs, the factor of safety becomes more dependent on the shear strength of the embankment sands and less dependent on the shear strength of the foundation clays. Although the factor of safety drops as water rises from 410 to 415, the reliability index increases. As there is more certainty regarding the strength of the sand (the coefficient of variations are about 6%versus 40% for the clay), this indicates that a sand embankment with a low factor of safety can be more reliable than a clay embankment with a higher factor of safety. Finally, as the water surface reaches the top of the levee at 420, the increasing seepage forces and reduction in effective stress leads to the lowest values for FS and β.

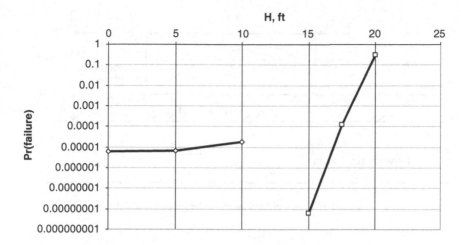

Figure 12.13 Conditional probability of failure function for slope stability.

12.6.2 Underseepage

12.6.2.1 Deterministic model

Underseepage is a common failure mode for levees and floodwalls that has been widely researched and for which analysis methods have been widely published (Peter, 1982; USACE, 2000). The example presented here uses an equation-based deterministic method developed by the Corps of Engineers (USACE, 2000) for a specific but common set of idealized foundation conditions: a relatively thin top semi-pervious top stratum of uniform thickness overlying a thick pervious substratum of uniform thickness. A typical cross-section is shown in Figure 12.14. For this case, the exit gradient i_o at the landside toe can be calculated as

$$i_o = \frac{h_o}{z} \tag{12.12}$$

where

h_o is the residual head at the levee toe, and
z is the effective thickness of the landside top blanket

The residual head can be calculated as

$$h_o = \frac{H x_3}{x_1 + x_2 + x_3} \tag{12.13}$$

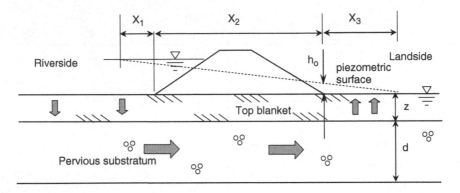

Figure 12.14 Terms used in Crops of Engineers' underseepage model.

where

H is the net height of water on levee
x_1 is the effective seepage entrance distance
x_2 is the base width of levee, and
x_3 is the effective seepage exit distance

For uniform blanket conditions of infinite landside length, x_3 can be calculated as

$$x_3 = \sqrt{\frac{k_f}{k_b}zd} \qquad (12.14)$$

where k_f is the horizontal permeability of the pervious substratum, k_b is the vertical permeability of the semi-pervious top blanket, z is the thickness of the top blanket and d is the thickness of the substratum. The coefficient of permeability k is a combined property of the soil and the permeant and is now more commonly referred to as hydraulic conductivity. The earlier nomenclature is retained for consistency with the original references.

Procedures for determining these values for a variety of conditions can be found in USACE (2000).

The exit gradient i_o is compared to the critical gradient i_c which can be calculated as follows:

$$i_c = \frac{\gamma'}{\gamma_w} \qquad (12.15)$$

where

γ' is the effective of submerged unit weight of the top stratum and γ_w is unit weight of water.

An exit gradient in excess of the critical gradient implies failure. Exit gradients can also be calculated for much more complex foundation conditions using methods such as finite element analysis.

The formulation above considers the initiation of sand boils or piping to be related only to gradient conditions. Other researchers (e.g. Peter, 1982) have proposed analysis models that include other parameters, notably grain size and grain size distribution.

12.6.2.2 Random variables

Table 12.5 provides some typical coefficients of variation for the parameters used in underseepage analysis.

12.6.2.3 Example problem

The levee cross section for this example is the same as that previously shown in Figure 12.10. Four random variables are considered, k_f, k_b, z and d. The assigned probabilistic moments for these variables are given in Table 12.6.

As borings are not available at every possible cross-section, there is some uncertainty regarding the thicknesses of the soil strata at the critical location. Hence, z and d are modeled as random variables. Their standard deviations are set by engineering judgment regarding the probable range of actual values at the site. For the blanket thickness z, assigning the standard deviation at 2.0 ft models a high probability that the actual blanket thickness will be between 4.0 and 12.0 ft (±2 standard deviations) and a very high probability that the blanket thickness will be between 2.0 and 14.0 ft (±3 standard

Table 12.5 Typical coefficients of variation for soil parameters used in underseepage analysis.

Parameter	Coefficient of variation (%)	Reference
Coefficient of permeability, k	90	For saturated soils, Nielson et al. (1973), cited by Harr (1987)
Permeability of top blanket clay, k_b	20	Derived from assumed distribution, Shannon and Wilson (1994)
Permeability of foundation sands, k_f	20–27.5	For average permeability over thickness of aquifer, Shannon and Wilson (1994)
Permeability ratio, k_f/k_b	40	Derived using 30% for k_f and k_b; see Appendix A

Table 12.6 Random variables for underseepage example problem.

Parameter	Expected value	Standard deviation	Coefficient of variation(%)
Substratum permeability, k_f	1000×10^{-4} cm/s	300×10^{-4} cm/s	30
Top blanket permeability, k_b	1×10^{-4} cm/s	0.3×10^{-4} cm /s	30
Blanket thickness, z	8.0 ft	2.0 ft	25
Substratum thickness, d	80 ft	5 ft	6.25

deviations). For the substratum thickness d, the two standard deviation range is 70–90 ft and the three standard deviation range is 65–95 ft. For analysis of real levee systems, it is suggested that the engineer review the geologic history and stratigraphy of the area and assign a range of likely strata thicknesses that are considered the thickest and thinnest probable values. These can then be taken to correspond to plus and minus 2.5 or 3.0 standard deviations from the expected value.

As the exit gradient is a function of the permeability *ratio*, k_f/k_b, and not the absolute magnitude of the values, the number of analyses can be reduced by treating the permeability ratio as a single random variable. To do so, it is necessary to determine the coefficient of variation of the permeability ratio given the coefficient of variation of the two permeability values. Using several methods to determine the probabilistic moments of the ratio of two random variables, it appears reasonable to take the expected value of the permeability ratio as 1000 and its coefficient of variation as 40%. This corresponds to a standard deviation of 400 for k_f/k_b.

The performance function is taken as the exit gradient landside of the levee, and the value of the critical gradient, assumed to be 0.85, is taken as the limit state.

For each water elevation, the exit gradient is calculated using the Taylor's series–finite difference method. This requires considering seven combinations of the input parameters. Results for a 20 ft head on the levee are summarized in Table 12.7.

For Run 1, the three random variables are all taken at their expected values. First the effective exit distance x_3 is calculated as:

$$x_3 = \sqrt{\frac{k_f}{k_b} \cdot z \cdot d} = \sqrt{1000 \cdot 8 \cdot 80} = 800 \text{ft} \qquad (12.16)$$

As the problem is symmetrical, the distance from the riverside toe to the effective source of seepage entrance, x_1, is also 800 ft.

From the geometry of the given problem, the base width of the levee, x_2, is 110 ft.

Table 12.7 Underseepage example, Taylor's series method, water at elevation 420 ($H =$ 20 ft).

Run	k_f/k_b	z	d	h_o	i	Variance	Percent of total variance
1	1000	8.0	80.0	9.357	1.170		
2	600	8.0	80.0	9.185	1.148		
3	1400	8.0	80.0	9.451	1.181	0.000276	0.30
4	1000	6.0	80.0	9.265	1.544		
5	1000	10.0	80.0	9.421	0.942	0.090606	99.69
6	1000	8.0	75.0	9.337	1.167		
7	1000	8.0	85.0	9.375	1.172	0.000006	0.01
				Total		0.090888	100.0

The net residual head at the levee toe is:

$$h_o = \frac{Hx_3}{x_1 + x_2 + x_3} = \frac{20 \cdot 800}{800 + 110 + 800} = 9.357 \text{ft} \tag{12.17}$$

and the landside toe exit gradient is:

$$i = \frac{h_o}{z} = \frac{9.357}{8.0} = 1.170 \tag{12.18}$$

For the second and third analyses, the permeability ratio is adjusted to the expected value plus and minus one standard deviation, while the other two variables are held at their expected values. These are used to determine the component of the total variance related to the permeability ratio:

$$\left(\frac{\partial i}{\partial(k_f/k_b)}\right)^2 \sigma_{k_f/k_b}^2 \approx \left(\frac{i_+ - i_-}{2\sigma_{k_f/k_b}}\right)^2 \sigma_{k_f/k_b}^2 = \left(\frac{i_+ - i_-}{2}\right)^2$$

$$= \left(\frac{1.181 - 1.148}{2}\right)^2 = 0.000276 \tag{12.19}$$

A similar calculation is performed to determine the variance components contributed by the other random variables.

When the variance components are summed, the total variance of the exit gradient is obtained as 0.090888. Taking the square root of the variance gives the standard deviation of 0.301.

The exit gradient is assumed to be a lognormally distributed random variable with probabilistic moments $E[i] = 1.170$ and $\sigma_i = 0.301$. Using the properties of the lognormal distribution, the equivalent normally distributed random variable has moments $E[\ln i] = 0.124$ and $\sigma_{\ln i} = 0.254$.

The critical exit gradient is assumed to be 0.85. The probability of failure is then:

$$Pr_f = Pr\,(\ln i > \ln 0.85) \tag{12.20}$$

This probability can be found by first calculating the standard normalized variate z:

$$z = \frac{\ln i_{\mathrm{crit}} - E[\ln i]}{\sigma_{\ln i}} = \frac{-0.16252 - 0.12449}{0.253629} = -1.132 \tag{12.21}$$

For this value, the cumulative distribution function $F(z)$ is 0.129, and represents the probability that the gradient is below critical. The probability that the gradient is above critical is

$$Pr_f = 1 - F(z) = 1 - 0.129 = 0.871 \tag{12.22}$$

Note that the z value is analogous to the reliability index β, and it could be stated that $\beta = -1.13$. The probability calculation is illustrated in Figure 12.15.

Repeating this procedure for a range of flood water elevations, the conditional probability of failure function can be plotted as shown in Figure 12.16. The probability of failure is very low until the head on the levee exceeds about 8 ft, after which it curves up sharply. It reverses curvature when the head reaches about 15 ft and the probability of failure is near 50%. When the flood water elevation is near the top of the levee, the conditional probability of failure approaches 87%.

The results of one intermediate calculation should be noted. As indicated by the relative size of the variance components shown in Table 12.7, virtually all of the uncertainty is in the top blanket thickness. A similar effect was found in other underseepage analyses by the writer reported in the Upper Mississippi River report (Shannon and Wilson and Wolff, 1994); where

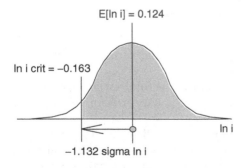

Figure 12.15 Probability of failure for underseepage example, water at el. 420.

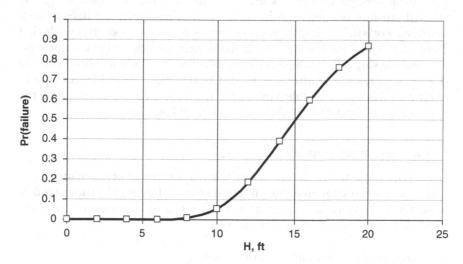

Figure 12.16 Conditional probability of failure function for underseepage example.

the top blanket thickness was treated as a random variable, its uncertainty dominated the problem. This has two implications:

- Probability of failure functions for preliminary economic analysis might be developed using a single random variable, the top blanket thickness z.
- In expending resources to design levees against underseepage failure, obtaining more data to determine the blanket thickness profile may be better justified than obtaining more data on material properties.

12.6.3 Throughseepage

Three types of internal erosion or *piping* can occur as a result of seepage through a levee:

- Cracks in a levee due to hydraulic fracturing, tensile stresses, decay of vegetation, animal activity, along the contours of hydraulic structures, etc., can all provide a preferential seepage path, along which piping may occur. For piping (movement of soil material) to occur, the tractive shear stress exerted by the flowing water must exceed the critical tractive shear stress of the soil.
- High exit gradients on the downstream face of the levee may cause piping and possible progressive backward erosion.
- Internal erosion or removal of fine grains by excessive seepage forces may occur. This type of piping occurs when the seepage gradient exceeds a critical value.

Well-constructed clay levees are generally considered resistant against internal erosion, but such erosion can occur where there are defects such as above. For sand levees, throughseepage is a common occurrence during flood. Water flowing through the embankment will cut small channels on the landside slope. If the slope is sufficiently flat, the levee will hold and the slope can be re-dressed after the flood. Construction of sand levees is relatively common on the middle Mississippi River for levees constructed by the Rock Island District of the U.S. Army Corps of Engineers.

12.6.3.1 Deterministic models

There is no single widely accepted analytical technique or performance function in common use for predicting internal erosion. Review of various models indicates that erosion susceptibility may be taken to be a function of some combination of the following parameters:

- permeability or hydraulic conductivity, k;
- hydraulic gradient, i;
- porosity, n;
- critical tractive stress, τ_c (the shear stress required for flowing water to dislodge a soil particle);
- particle size, expressed as some representative size such as D_{50} or D_{85}; and
- friction angle, ϕ, or angle of repose.

Essentially, the models use the gradient, critical tractive stress and particle size to determine whether the shear stresses induced by seepage head loss are sufficient to dislodge soil particles, and use the gradient, permeability, and porosity to determine whether the seepage flow rate is sufficient to carry away or transport the particles once they have been dislodged.

Very fine sands and silt-sized materials are among the most erosion-susceptible soils. This arises from their having a critical balance of relatively high permeability, low particle weight and low critical tractive stress. Particles larger than fine sand sizes are generally too heavy to be moved easily, as particle weight increases with the cube of the diameter. Particles smaller than silts (i.e. clay sizes), although of light weight, may have relatively large electro-chemical forces acting on them, which can substantially increase the critical tractive stress, τ_c, and also have sufficiently small permeability as to inhibit particle transport in significant quantity.

12.6.3.2 Schwartz's method

As analytical models for throughseepage are complex and not well proven, an illustrative example, based on a design procedure for the landside slope of sand levees developed by Schwartz (1976) will be used herein. Where there is experience with other erosion models, they could be substituted for the illustrated method, using the same approach of defining the probability of failure as the probability that the performance function crosses the limit state. Schwartz's method includes some elements of erosion analysis. However, the result of the method is a parameter to determine the need for providing toe berms according to semi-empirical criteria rather than to directly determine the threshold of erosion conditions or predict whether erosion will occur. Presumably, some conservatism is present in the berm criteria and thus the criteria do not represent a true limit state. Accordingly, the actual probability of failure should be somewhat lower than calculated.

The procedure involves the calculation of two parameters, the maximum erosion susceptibility, M, and the relative erosion susceptibility, R. The calculated values are compared to critical combinations for which toe berms are considered necessary. The parameters are functions of the embankment geometry and soil properties. First, the vertical distance of the seepage exit point on the downstream slope, y_c, is determined using the well-known solution for the "basic parabola" by L. Casagrande. Two parameters, λ_1 and λ_2, are then calculated as:

$$\lambda_1 = \cos\beta - \frac{\gamma_w}{\gamma_b}\sin\beta\tan(\beta-\delta) - \frac{\gamma_{sat}}{\lambda_b}\frac{\sin\beta}{\tan\phi} \tag{12.23}$$

$$\lambda_2 = \gamma_w \sin^{0.7}\beta \left(\frac{n}{1.49}\right)^{0.6} \left[k\tan(\beta-\delta)\right]^{0.6} \tag{12.24}$$

where β is the downstream slope angle

δ is taken as zero for a horizontal exit gradient
n is Manning's coefficient for sand, typically 0.02
γ_{sat} is the saturated density of the sand in lb/ft^3
γ_b is the submerged effective density of the sand in lb/ft^3
k is the sand permeability in ft/s
ϕ is the friction angle

It is important to note that the parameter λ_2 is not dimensionless, and the units stated above must be used.

The erosion susceptibility parameters are then calculated as:

$$M = \frac{\lambda_2 y_e^{0.6}}{\lambda_1 \tau_{co}} \tag{12.25}$$

$$R = \frac{y_e - \left(\frac{\lambda_1 \tau_{co}}{\lambda_2}\right)^{1.67}}{H} \tag{12.26}$$

In the above equations, τ_{co} is the critical tractive stress, which can be taken as about 0.03 lb/ft² (14.36 dynes/cm²) for medium sand, and H is the full embankment height, measured in feet. According to Schwartz's criteria, toe berms are recommended when M and R values fall above the shaded region shown in Figure 12.17. To simplify probabilistic analysis, Shannon and Wilson and Wolff (1994) suggested replacing this region with a linear approximation (also shown in Figure 12.17), and taken to be the limit state. The approximation is:

$$M + 14.4R - 13.0 = 0 \qquad\qquad (12.27)$$

Positive values of the expression to the left of the equals sign indicate the need for toe berms.

12.6.3.3 Example problem

The same embankment section previously analyzed will be used as an example of a throughseepage analysis. Random variables were characterized as shown in Table 12.8. The method, which assesses erosion at the landside seepage face, was numerically unstable (λ_1 becomes negative) for the slopes previously assumed. To make the problem stable for purposes of illustration, the slopes had to be flattened to $1V$ on $3H$ riverside and $1V$ on $5H$ landside.

The results for a 20 ft water height are summarized in Table 12.9. The most significant random variables, based on descending order of their variance components, are the unit weight, the friction angle, and the permeability.

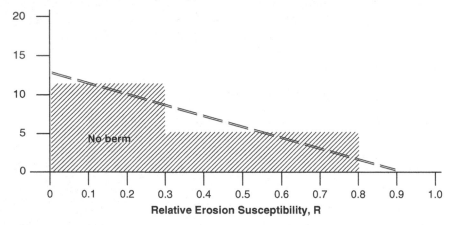

Figure 12.17 Schwartz's method berm criteria, and assumed linear limit state.

Table 12.8 Random variables for throughseepage example.

Variable	Expected value	Coefficient of variation(%)
Manning's coefficient, n	0.02	10
Unit weight, γ_{sat}	125 lb/ft^3	8
Friction angle, ϕ	30 deg	6.7
Coefficient of permeability, k	2000×10^{-4} cm/s	30
Critical tractive stress, τ	18 dynes/cm^2	10

Table 12.9 Results of internal erosion analysis, example problem (modified to flatter slopes) H = 20 ft.

n	γ_{sat}	ϕ	k	τ	Performance function	Variance component	Percent of total variance
0.02	125	30	2000	18	17.524		
0.022	125	30	2000	18	18.491		
0.018	125	30	2000	18	16.515	0.9761	2.1
0.02	135	30	2000	18	14.798		
0.02	115	30	2000	18	23.667	19.6648	42.9
0.02	125	32	2000	18	14.817		
0.02	125	28	2000	18	22.179	13.5498	29.5
0.02	125	30	2600	18	20.321		
0.02	125	30	1400	18	14.339	8.961	19.5
0.02	125	30	2000	19.8	16.046		
0.02	125	30	2000	16.2	19.369	2.7606	6.0
					Total	45.8974	100.0

The effects of Manning's coefficient and the critical tractive stress, at least for the coefficients of variation assumed, are relatively insignificant.

The reliability index was calculated as the expected value of the performance function divided by the standard deviation of the performance function. The probability of failure was calculated using the reliability index and assuming the performance function was normally distributed. (Note that the assumption of a lognormally distributed performance function was not used for this failure mode as negative values of the performance function are permitted.)

When the probabilities of failure for various flood water elevations are plotted, the conditional probability of failure function shown in Figure 12.18 is obtained. Again, it takes the expected reverse-curve shape. Below heads 10 ft, or about half the levee height, the probability of failure against through seepage failure is virtually nil. The probability of failure becomes greater than 0.5 for a head of about 16.5 feet, and approaches unity at the full head of 20 ft.

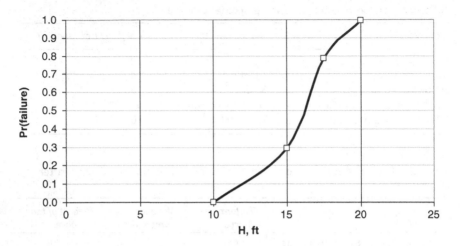

Figure 12.18 Conditional probability of failure function for throughseepage example.

12.6.3.4 Surface erosion

As flood stages increase, the potential increases for surface erosion from two sources:

- erosion due to excessive current velocities parallel to the levee slope; and
- erosion due to wave attack directly against the levee slope.

Preventing surface erosion requires providing adequate slope protection, typically a thick grass cover, and stone revetment at locations expected to be susceptible to wave attack. During flood emergencies, additional protection may be provided where necessary using dumped rock, snow fence, or plastic sheeting.

12.6.3.5 Analytical model

Although there are established criteria for determining the need for slope protection and for designing slope protection, they are not in the form of a limit state or performance function (i.e. one does not typically calculate a factor of safety against scour). To perform a reliability analysis, one needs to define the problem as a comparison between the probable velocity or wave height and the velocity or wave height that will result in damaging scour. As a first approximation for the purpose of illustration, this section will use a simple adaptation of Manning's formula for average flow velocity and assume that the critical velocity for a grassed slope can be expressed by an expected value and coefficient of variation.

For channels that are very wide relative to their depth (width $> 10\times$ depth), the flow velocity can be expressed as:

$$V = \frac{1.486y^{2/3}S^{1/2}}{n} \qquad (12.28)$$

where y is the depth of flow,
   S is the slope of the energy line, and
   n is Manning's roughness coefficient.

It will be assumed that the velocity of flow parallel to a levee slope for water heights from 0 to 20 ft can be approximated using the above formula with y taken from 0 to 20 ft. For real levees in the field, it is likely that better estimates of flow velocities at the location of the riverside slope can be obtained by more detailed hydraulic models.

The following probabilistic moments are assumed. More detailed and site-specific studies would be necessary to determine appropriate values.

$$E[S] = 0.0001 \quad V_S = 10\%$$

$$E[n] = 0.03 \quad V_n = 10\%$$

It is assumed that the critical velocity that will result in damaging scour can be expressed as:

$$E[V_{\text{crit}}] = 5.0 ft/s \quad V_{\text{vcrit}} = 20\%$$

Further research is necessary to develop guidance on appropriate values for prototype structures.

The Manning equation is of the form

$$G(x_1, x_2, x_3, \ldots) = ax_1^{g1} x_2^{g2} x_3^{g3} \ldots \qquad (12.29)$$

For equations of this form, Harr (1987) shows that the probabilistic moments can be determined using a special form of Taylor's series approximation he refers to as the *vector equation*. In such cases, the expected value of the function is evaluated as the function of the expected values. The coefficient of variation of the function can be calculated as:

$$V_G^2 = g_1^2 V^2(x_1) + g_2^2 V^2(x_2) + g_3^2 V^2(x_3) + \cdots \qquad (12.30)$$

For the case considered, the coefficient of variation of the flow velocity is then:

$$V_V = \sqrt{V_n^2 + (\tfrac{1}{4})V_S^2} \qquad (12.31)$$

Note that, although the velocity increases with flood water height y, the coefficient of variation of the velocity is constant for all heights.

Knowing the expected value and standard deviation of the velocity and the critical velocity, a performance function can be defined as the ratio of critical velocity to the actual velocity (i.e. the factor of safety) and the limit state can be taken as this ratio equaling the value 1.0. If the ratio is assumed to be lognormally distributed, the reliability index is:

$$\beta = \frac{\ln\left(\dfrac{E[C]}{E[D]}\right)}{\sqrt{V_C^2 + V_D^2}} = \frac{\ln\left(\dfrac{E[V_{\text{crit}}]}{E[V]}\right)}{\sqrt{V_{V\text{crit}}^2 + V_V^2}} \qquad (12.32)$$

and the probability of failure can be determined from the cumulative distribution function for the normal distribution.

The assumed model and probabilistic moments were used to construct the conditional probability of failure function in Figure 12.19. It is again observed that a typical levee may be highly reliable for water levels up to about one-half the height, and then the probability of failure may increase rapidly.

12.7 Step 2: Reliability of a reach – combining the failure modes

Having developed a conditional probability of failure function for each considered failure mode, the next step is to combine them to obtain a total or composite conditional probability of failure function for the reach that

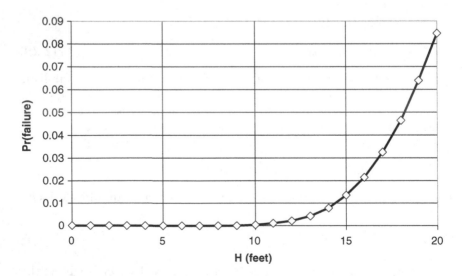

Figure 12.19 Conditional probability of failure function for surface erosion example.

combines all modes. As a first approximation, it may be assumed that the failure modes are independent and hence uncorrelated. This is not necessarily true, as some of the conditions increasing the probability of failure for one mode may likely increase the probability of failure by another. Notably, increasing pressures due to underseepage will adversely affect slope stability. However, there is insufficient research to better quantify such possible correlation between modes. Assuming independence considerably simplifies the mathematics involved and may be as good a model as can be expected at present.

12.7.1 Judgmental evaluation of other modes

Before combining failure mode probabilities, consider the probability of failure due to other circumstances not readily treated by analytical models. During a field inspection, one might observe other items and features that might compromise the levee during a flood event. These might include animal burrows, cracks, roots, and poor maintenance that might impede detection of defects or execution of flood-fighting activities. To factor in such information, a judgment-based conditional probability function could be added by answering the following question:

> *Discounting the likelihood of failure accounted for in the quantitative analyses, but considering observed conditions, what would an experienced levee engineer consider the probability of failure of this levee for a range of water elevations?*

For the example problem considered herein, the function in Table 12.10 was assumed. While this may appear to be guessing, leaving out such information has the greater danger of not considering the obvious. Formalized techniques for quantifying expert opinion exist and merit further research for application to levees.

Table 12.10 Assumed conditional probability of failure function for judgmental evaluation of observed conditions.

Flood water elevation	Probability of failure
400.0	0
405.0	0.01
410.0	0.02
415.0	0.20
417.5	0.40
420.0	0.80

12.7.2 Composite conditional probability of failure function for a reach

For N independent failure modes, the reliability or probability of no failure involving any mode is the probability of no failure due to mode 1 *and* no failure due to mode 2, *and* no failure due to mode 3, etc. As *and* implies multiplication, the overall reliability at a given flood water elevation is the product of the modal reliability values for that flood elevation, or:

$$R = R_{SS}R_{US}R_{TS}R_{SE}R_J \tag{12.33}$$

where the subscripts refer to the identified failure modes. Hence the probability of failure at any flood water elevation is:

$$
\begin{aligned}
Pr(f) &= 1 - R \\
&= 1 - (1 - p_{SS})(1 - p_{US})(1 - p_{TS})(1 - p_{SE})(1 - p_J)
\end{aligned}
\tag{12.34}
$$

The total conditional probability of failure function is shown in Figure 12.20. It is observed that probabilities of failure are generally quite low for water elevations less than one-half the levee height, then rise sharply as water levels approach the levee crest. While there is insufficient data to judge whether this shape is a general trend for all levees, it has some basis in experience and intuition.

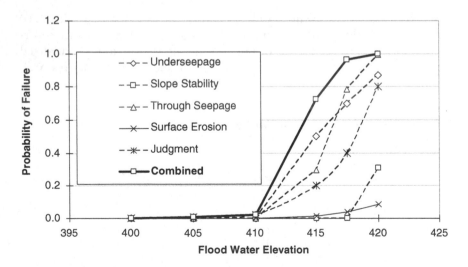

Figure 12.20 Combined conditional probability of failure function for a reach.

12.8 Step 3: Annualizing the probability of failure

Returning to Equation (12.1), the probability of levee failure is conditioned on the flood water elevation FWE, which is a random variable that varies with time. Flood water elevation is usually characterized by an *annual exceedance probability function*, which gives the probability that the highest water level in a given year will exceed a given elevation.

It is desired to find the annual probability of the joint event of flooding to any and all levels *and* a levee failure given that level. An example of the methodology is provided in Table 12.11. The first column shows the *return period* and the second shows the annual probability of exceedance for the water level shown in the third column. The return period is simply the inverse of the annual exceedance probability. These values are obtained from hydrologic studies. In the fourth column, titled *lumped increment*, all water elevations between those in the row above and the row below are assumed to be lumped as a single elevation at the increment midpoint. The annual probability of the maximum water level in a given year being within that increment can be taken as the difference of the annual exceedance probabilities for the top and bottom of the increment. For example, the annual probability of the maximum flood water elevation being between 408 and 410 is

$$Pr(408 < FWE < 410) = Pr(FWE > 408) - Pr(FWE > 410)$$
$$= 0.500 - 0.200 \qquad (12.35)$$
$$= 0.300$$

Table 12.11 Annual probability of flooding for a reach.

Return period (yr)	Annual Pr(Exceed)	Water elev.	Lumped increment	Annual Pr(FWE)	Pr(F |FWE)	Pr (F)
			<= 408	0.505	0.02	0.01010
2	0.500	408				
			409 +/− 1	0.300	0.03	0.00900
5	0.200	410				
			411 +/− 1	0.100	0.08	0.00800
10	0.100	412				
			413 +/− 1	0.050	0.45	0.02250
20	0.050	414				
			415 +/− 1	0.030	0.72	0.02160
50	0.020	416				
			417 +/− 1	0.010	0.86	0.00860
100	0.010	418				
			419 +/− 1	0.005	0.97	0.00485
200	0.005	420				
			Sum	1.000		0.08465

The flood water elevations for this increment are lumped at elevation 409.0, and hence the annual probability of the maximum flood water being elevation 409.0 is 0.300. The annual probability of each flood water elevation is multiplied by the conditional probability of failure given that flood water elevation (sixth column) to obtain the annual probability of levee failure for that FWE (seventh column). For this example, the values in the sixth column are taken from Figure 12.16. The annual probability of the joint event of maximum flood being between 408 and 410 *and* the levee failing under this load is

$$Pr(408 < FWE < 410) \times Pr(\text{failure}|FWE) = 0.300 \times 0.03 = 0.0090$$
$$(12.36)$$

Finally, the values in column seven are summed to obtain the annual probability of the joint event of flooding and failure for all possible water elevations. This can be expressed as

$$\text{Annual} Pr(f) = \sum_{\text{all } FWE} \text{Annual} Pr(FWE) \times Pr(\text{failure}|FWE) \Delta FWE$$
$$(12.37)$$

As the water level increment becomes small, the equation above approaches

$$\text{Annual} Pr(f) = \int_{\text{all } FWE} \text{Annual} Pr(FWE) \times Pf(\text{failure}|FWE) dFWE$$
$$(12.38)$$

From Table 12.11, the annual probability of failure for this reach is 0.08465, which considers all flood levels (for which annual probabilities decrease with height) and the probability of failure associated with each flood level (which increase with height). Note that this is a rather high probability; the example problem was deliberately selected to be of marginal safety to provide relative large numbers to easily view the mathematics.

Baecher and Christian (2003: 488–500) use an event-tree approach to illustrate an approach similar to that of Equations (12.37) and (12.38). The process is started with probabilities of different river discharges Q. For each discharge, river stage may be uncertain. For each river stage, the levee may or may not fail by various modes. Following all branches of the event tree leads to one of two events, failure or non-failure. Summing up all the branch probabilities leading to failure is equivalent to the equations above.

12.9 Step 4: The Levee as a system – combining the reaches

Having an annual probability of failure for each reach, the final step is to determine the overall annual probability of failure for the entire levee system, which is a set of reaches.

If a levee were "perfectly uniform" in cross-section and soil profile along its length, intuition would dictate that the probability of a failure for a very long levee should be greater than that for a short identical levee, as more length of levee is at risk.

First it will be assumed that each reach is statistically independent of all others, and then that assumption will be examined further. For each reach, an annual probability of failure has been determined. Subtracting these values from unity, the annual reach reliability R_i is obtained. If the levee system is considered to be a series system of discrete independent reaches, such as links in a chain, the system reliability is the product of the reliabilities for each link, and the same methods can be used for combining probabilities for reaches as was used for combining modes, hence:

$$R = R_1 R_2 R_3 \ldots R_N \tag{12.39}$$

where the subscripts refer to the separate reaches. Hence, for a given water level, the probability of failure for the system is:

$$
\begin{aligned}
Pr(\mathrm{f}) &= 1 - R \\
&= 1 - (1 - p_1)(1 - p_2)(1 - p_3)\ldots(1 - p_N)
\end{aligned} \tag{12.40}
$$

Where the p values are "very small," the above equation approaches the sum of the p_i values.

The problem remaining is to determine what is the distance along the levee beyond which soil properties are statistically independent. At short distances, soil properties are highly correlated, and if failure conditions are realized, they might occur over the entire length. At long distances in a very long, statistically homogeneous reach, soil properties may be very different from each other due to inherent randomness. There could be a failure due to weak strength values in one area, but values could be very strong in another. The long reach may need to be subdivided into an equivalent number of statistically independent reaches to model the appropriate total number of statistically independent reaches in Equations (12.39) and (12.40).

Much research has been done in the areas of spatial correlation, autocorrelation functions, variance reduction functions, etc., which have a direct bearing on this problem. However, there are seldom sufficient data to accurately quantify such functions. Spatial variability can be most simply expressed by using a "correlation distance," δ, discussed by a number of researchers, notably VanMarcke (1977a). For distances less than δ, soil properties can be considered highly correlated, i.e. values are uncertain, but nearly the same from point to point. For points separated by distances considerably greater than δ, soil property values are independent and uncorrelated. The degree of correlation between soil properties at separated points is assumed to follow some function which is scaled by the δ distance. For a levee, the longer a "statistically homogeneous" levee is, as a multiple of the correlation distance, the greater the probability of failure.

For a levee with non-homogeneous foundation conditions already divided into a set discrete reaches, based on the local geology, the concept of correlation distance can be included as follows.

- Where the reach length is less than the correlation distance δ, the probability of failure for the two-dimensional section could simply be taken as the probability of failure for the reach. Alternatively, if the reach length is quite small relative to the correlation distance (not common), reaches could be combined and represented by the conditional probability function for the most critical reach.
- Where the reach length is greater than δ, the probability of failure of that reach must be increased to account for length effects. This can be approximated by subdividing the reach into an additional integral number of equivalent reaches, with each reach of length δ and one additional reach for the fractional remainder.

The required correlation distance δ is a difficult parameter to estimate, as it requires a detailed statistical analysis of a relatively large set of equally spaced data, which is generally not available. The reported values for soil strength in the horizontal direction are typically a few hundred feet (e.g. VanMarcke, 1977b, uses 150 ft). On the other hand, for levees, Vrouwenvelder (1987) uses 500 m for Dutch levees. While this is a considerable difference, maximum reach lengths of 100–300 m would seem to be reasonable in levee systems analysis; there should always be a new reach assumed when the levee geometry or foundation conditions change significantly.

12.10 Remaining shortcomings

One significant shortcoming remaining is the effect of flood duration. As the duration of a flood increases, the probability of failure inevitably increases, as extended flooding increases pore pressures, and increases the likelihood and intensity of damaging erosion. The analyses herein essentially assume that the flood has been of sufficient duration that steady-state seepage conditions have developed in pervious substratum materials and pervious embankment materials, but no pore pressure adjustment has occurred in impervious clayey foundation and embankment materials. These are reasonable assumptions for economic analysis of most levees. Further research will be required to provide a rational basis for modifying these functions for flood duration.

Other shortcomings are enumerated by the National Research Council (2000) and have been discussed earlier.

12.11 Epilogue: New Orleans 2005

During the preparation of this chapter, Hurricane Katrina struck the Gulf Coast of the United States, creating storm surges that led to numerous levee breaks, flooding much of New Orleans, Louisiana, causing over 100 billion dollars of damage, and the loss of over 1000 lives. Much of New Orleans is situated below sea level, protected by a system of levees (more properly called dikes) and pumping stations.

A preliminary joint report on the performance of the levees and floodwalls was prepared by two field investigation teams, one from the American Society of Civil Engineers and one sponsored by the National Science Foundation and led by civil engineering faculty at the University of California at Berkeley (Seed *et al.*, 2005). The report noted that three levee/floodwall failures (one at the 17th St. Canal and two along the London Avenue Canal) occurred at water levels lower than the top of the protection. These were attributed to some combination of sliding instability and inadequate seepage control. Three additional major breaks along the Inner Harbor Navigation Canal were attributed to overtopping, and a large number of additional failures more remote from the urban part of the city occurred due to overtopping and/or wave erosion.

Subsequently, an extensive analysis of the New Orleans levee failures was conducted by the Interagency Project Evaluation Team (IPET, 2006), which involved a number of government agencies, academics and consultants, led by the U.S. Army Corps of Engineers. The IPET report was in turn reviewed by a second team from the American Society Society of Civil Engineers (ASCE), the External Review Panel (ERP). The IPET report found that one of the breaches on the Inner Harbor Navigation Canal also likely failed prior to overtopping.

This event, the most catastrophic levee failure in recent history, raises questions as to how the New Orleans levee system would have been viewed from a reliability perspective, and how application of reliability-based principles might have possibly reduced the level of the damage. The writer was in the area shortly after the failures, as a member of the first ASCE levee assessment team. However, the observations below should be construed only as the opinions of the writer, and not to reflect any position of the ASCE, which has published its own recommendations at its web site (www.asce.org). ·

12.11.1 Level of protection and probability of system failure

The system of levees and floodwalls forming the New Orleans hurricane protection system were reported to be constructed to heights that would correspond to return periods of 200–300 years, or an equivalent annual probability of 0.005–0.0033. While this is several times the expected lifetime of a person, the probability of such an event occurring in one's lifetime is not negligible. From the Poisson distribution, the probability of at least one event in an interval of t years is given as

$$Pr(x > 0) = 1 - e^{-\lambda t} \tag{12.41}$$

where x is the number of events in the interval t, and λ is a parameter denoting the expected number of events per unit time. For an annual probability of 1 in 200, $\lambda = p = 0.005$.

Hence, the probability of getting at least one 200-year event in a 50-year time period (a reasonable estimate of the length of time one might reside behind the levee) is

$$Pr(x > 0) = 1 - e^{(-0.005)(50)} = 0.221$$

which is between 1 in 4 and 1 in 5. It is fair to say the that the public perception of the following statement

> There is between a 1 in 4 and 1 in 5 chance that your home will be flooded sometime in your life

is much different than

> Your home is protected by a levee designed higher than the 200 year flood.

A 200-year level of protection for an urban area leaves a significant chance of flood exceedance in any 50-year period. In comparison, the primary dikes protecting the Netherlands are set to height corresponding to 10,000 year return period (Voortman, 2003) and the interior levees protecting against the Rhine are set to a return period of about 1250 years (Vrouwenvelder, 1987).

In addition to the annual probability of the flood water exceeding the levee, there is some additional probability that the levee will fail at water levels below the top of the levee. Hence, flood return periods only provide an upper bound on reliability, and perhaps a poor one. Estimating conditional failure probabilities to determine an overall probability of levee failure (as opposed to probability of height exceedance) is the purpose of the methods discussed in this chapter, as well as by Wolff (1994), Vrouwenvelder (1987), the National Research Council (2000), and others. The three or four failures in New Orleans that occurred without overtopping illustrate the importance of considering material variability, model uncertainty, and other sources of uncertainty.

Finally, the levee system in New Orleans is extremely long. Local press reports indicated that system was nearly 170 miles long. The probability of at least one failure in the system can be calculated from Equation (12.40). Using a large amount of conservatism, assume that there are 340 statistically independent reaches of one half mile each, and that conditional probability of failure for each reach with water at the top of the levee for each reach is 10^{-3}. The probability of system failure with water at the top of all levees is then

$$Pr(f) = 1 - (1 - 0.001)^{340}$$
$$= 1 - 0.7116$$
$$= 0.2884$$

or about 29%. But if the conditional probability of failure for each reach drops to 10^{-2}, the probability of system failure rises to almost 97%. It is evident from the above that very long levees protecting developed areas need to be designed to very high levels of reliability, with conditional probabilities of failure on the order of 10^{-3} or smaller, to ensure a reasonably small probability of system failure.

12.11.2 Reliability considerations in project design

In hindsight, applying several principles of reliability engineering may have prevented some damage and enhanced response in the Katrina disaster.

- *Parallel (redundant) systems are inherently much more reliable than series systems.* Long levee systems are primarily series systems; a failure at any one point is a failure of the system, leading to widespread flooding. Had the interior of New Orleans been subdivided into a set of compartments by interior levees, only a fraction of the area may have been flooded. Levees forming containment compartments are common around tanks in petroleum tank farms. Such interior levees would undoubtedly run against public perception, which would favor building a larger, stronger levee that "could not fail" over interior levees on dry land, far from the water, that would limit interior flooding.
- *However small the probability of the design event, one should consider the consequences of even lower probability events.* The levee systems in New Orleans included no provisions for passing water in a controlled,

non-destructive manner for water heights exceeding the design event (Seed *et al.* 2005). Had spillways or some other means of landside hardening been provided, the landside area would have still been flooded, but some of the areas would not have breached. This would have reduced the number of buildings flattened by the walls of water coming through the breaches, and facilitated the ability to begin pumping out the interior areas after the storm surge had passed. It can be perceived that the 200–300-year level of protection may have been perceived as such a low probability event that it would never occur.

- *Overall consequences can be reduced by designing critical facilities to a higher level of reliability.* With the exception of tall buildings, essentially all facilities in those parts of New Orleans below sea level were lower than the tops of the levees. Had critical facilities such as police and fire stations, military bases, medical care facilities, communications facilities, and pumping station operating floors been constructed on high fills or platforms, the emergency response may have been significantly improved.

References

Apel, H., Thieken, A. H., Merz, B. and Bloschl, G. (2004). Flood risk assessment and associated uncertainty. *Natural Hazards and Earth System Sciences*, 4; 295–308.

Baecher, G. B. and Christian, J. T. (2003). *Reliability and Statistics in Geotechnical Engineering.* J. Wiley and Sons, Chichester, UK.

Buijs, F. A., van Gelder, H. A. J. M. and Hall, J. W. (2003). Application of reliability-based flood defence design in the UK. *ESREL 2003 – European Safety and Reliability Conference, 2003,* Maastricht, Netherlands. Available online at: http://heron.tudelft.nl/2004_1/Art2.pdf.

Duncan, J. M. and Houston, W. N. (1983). Estimating failure probabilities for California levees. *Journal of Geotechnical Engineering, ASCE,* 109; 2.

Fredlund, D. G. and Dahlman, A. E. (1972). Statistical geotechnical properties of glacial lake edmonton sediments. In *Statistics and Probability in Civil Engineering.* Hong Kong University Press, Hong Kong.

Hammitt, G. M. (1966). Statistical analysis of data from a comparative laboratory test program sponsored by ACIL. *Miscellaneous Paper 4-785,* U.S. Army Engineer Waterways Experiment Station, Corps of Engineers.

Harr, M. E. (1987). *Reliability Based Design in Civil Engineering.* McGraw-Hill, New York.

Hassan, A. M. and Wolff, T. F. (1999). Search algorithm for minimum reliability index of slopes. *Journal of Geotechnical and Geoenvironmental Engineering, ASCE,* 124(4); 301–8.

Interagency Project Evaluation Team (IPET, 2006). *Performance Evaluation of the New Orleans and Southeast Louisiana Hurricane Protection System.* Draft Final Report of the Interagency Project Evaluation Team, 1 June 2006. Available online at https://ipet.wes.army.mil.

Ladd, C. C., Foote, R., Ishihara, K., Schlosser, F. and Poulos, H. G. (1977). Stress-deformation and strength characteristics. State-of-the-Art report. In *Proceedings of the Ninth International Conference on Soil Mechanics and Foundation Engineering*, Tokyo. *Soil Mechanics and Foundation Engineering*, 14(6); 410–14.

National Research Council (2000). *Risk Analysis and Uncertainty in Flood Damage Reduction Studies*. National Academy Press, Washington, DC. http://www.nap.edu/books/0309071364/html

Peter P. (1982). *Canal and River Levees*. Elsevier Scientific Publishing Company, Amsterdam.

Schultze, E. (1972). "Frequency Distributions and Correlations of Soil Properties," in *Statistics and Probability in Civil Engineering*. Hong Kong.

Schwartz, P. (1976). *Analysis and performance of hydraulic sand-hill levees*. PhD dissertation, Iowa State University, Anes, IA.

Seed, R. B., Nicholson, P. G., et. al. (2005). *Preliminary Report on the Performance of the New Orleans Levee Systems in Hurricane Katrina on August 29, 2005*. National Science Foundation and American Society of Civil Engineers. University of California at Berkeley Report No. UCB/CITRIS – 05/01, 17 November, 2005. Available online at: http://www.asce.org/files/pdf/katrina/teamdatareport1121.pdf.

Shannon and Wilson, Inc. and Wolff, T. F. (1994). *Probability models for geotechnical aspects of navigation structures*. Report to the St. Louis District, U.S. Army Corps of Engineers.

U.S. Army Corps of Engineers (USACE) (1991). Policy Guidance Letter No. 26, *Benefit Determination Involving Existing Levees*. Department of the Army, Office of the Chief of Engineers, Washington, D.C. Available online at: http://www.usace.army.mil/inet/functions/cw/cecwp/branches/guidance _dev/pgls/pgl26.htm.

U.S. Army Corps of Engineers (USACE) (1995). *Introduction to Probability and Reliability Methods for Use in Geotechnical Engineering*. Engineering Technical Letter 1110-2-547, 30 September 1995. Available online at: http://www.usace.army.mil/inet/usace-docs/eng-tech-ltrs/etl1110-2-547/entire.pdf.

U.S. Army Corps of Engineers (USACE) (1999). *Risk-Based Analysis in Geotechnical Engineering for Support of Planning Studies*, Engineering Technical Letter 1110-2-556, 28 May 1999. Available online at: http://www.usace.army.mil/inet/usace-docs/eng-tech-ltrs/etl1110-2-556/

U.S. Army Corps of Engineers (USACE) (2000). *Design and Construction of Levees*. Engineering Manual 1110-2-1913. April 2000. Available online at: http://www.usace.army.mil/inet/usace-docs/eng-manuals/em1110-2-1913/toc.htm

VanMarcke, E. (1977a). Probabilistic modeling of soil profiles. *Journal of the Geotechnical Engineering Division*, ASCE, 103; 1227–46.

VanMarcke, E. (1977b). Reliability of earth slopes. *Journal of the Geotechnical Engineering Division*, ASCE, 103; 1247–65.

Voortmann, H. G. (2003). *Risk-based design of flood defense systems*. Doctoral dissertation, Delft Technical University. Available online at: http://www.waterbouw.tudelft.nl/index.php?menu_items_id=49

Vrouwenvelder, A. C. W. M. (1987). *Probabilistic Design of Flood Defenses*. Report No. B-87-404. IBBC-TNO (Institute for Building Materials and Structures of the Netherlands Organization for Applied Scientific Research), The Netherlands.

Wolff, T. F. (1985). *Analysis and design of embankment dam slopes: a probabilistic approach*. PhD thesis, Purdue University, Lafayette, IN.

Wolff, T. F. (1994). *Evaluating the Reliability of Existing Levees*. prepared for U.S. Army Engineer Waterways Experiment Station, Geotechnical Laboratory, Vicksburg, MS, September 1994. This is also Appendix B to U.S. Army Corps of Engineers ETL 1110-2-556, Risk-Based Analysis in Geotechnical Engineering for Support of Planning Studies, 28 May 1999.

Wolff, T. F., Demsky, E. C., Schauer, J. and Perry, E. (1996). Reliability assessment of dike and levee embankments. In *Uncertainty in the Geologic Environment: From Theory to Practice, Proceedings of Uncertainty '96*, ASCE Geotechnical Special Publication No. 58, eds C. D. Shackelford, P. P. Nelson and M. J. S. Roth. ASCE, Reston, VA, pp. 636–50.

Chapter 13

Reliability analysis of liquefaction potential of soils using standard penetration test

Charng Hsein Juang, Sunny Ye Fang, and David Kun Li

13.1 Introduction

Earthquake-induced liquefaction of soils may cause ground failure such as surface settlement, lateral spreading, sand boils, and flow failures, which, in turn, may cause damage to buildings, bridges, and lifelines. Examples of such structural damage due to soil liquefaction have been extensively reported in the last four decades. As stated in Kramer (1996), "some of the most spectacular examples of earthquake damage have occurred when soil deposits have lost their strength and appeared to flow as fluids." During liquefaction, "the strength of the soil is reduced, often drastically, to the point where it is unable to support structures or remain stable."

Liquefaction is "the act of process of transforming any substance into a liquid. In cohesionless soils, the transformation is from a solid state to a liquefied state as a consequence of increased pore pressure and reduced effective stress" (Marcuson, 1978). The basic mechanism of the initiation of liquefaction may be elucidated from the observation of behavior of a sand sample undergoing cyclic loading in a cyclic triaxial test. In such laboratory tests, the pore water pressure builds up steadily as the cyclic deviatoric stress is applied and eventually approaches the initially applied confining pressure, producing an axial strain of about 5% in double amplitude (DA). Such a state has been referred to as "initial liquefaction" or simply "liquefaction." Thus, the onset condition of liquefaction or cyclic softening is specified in terms of the magnitude of cyclic stress ratio required to produce 5% DA axial strain in 20 cycles of uniform load application (Seed and Lee, 1966; Ishihara, 1993; Carraro *et al.*, 2003).

From an engineer's perspective, three aspects of liquefaction are of particular interest; they include (1) the likelihood of liquefaction occurrence or triggering of a soil deposit in a given earthquake, referred to herein as *liquefaction potential*; (2) the effect of liquefaction (i.e. the extent of ground failure caused by liquefaction); and (3) the response of foundations in a liquefied soil. In this chapter, the focus is on the evaluation of liquefaction potential.

The primary factors controlling the liquefaction of a saturated cohesionless soil in level ground are the intensity and duration of earthquake shaking and the density and effective confining pressure of the soil. Several approaches are available for evaluating liquefaction potential, including the cyclic stress-based approach, the cyclic strain-based approach, and the energy-based approach. In the cyclic strain-based approach (e.g. Dobry *et al.*, 1982), both "loading" and "resistance" are described in terms of cyclic shear strain. Although the cyclic strain-based approach has an advantage over the cyclic stress-based approach in that pore water pressure generation is more closely related to cyclic strains than cyclic stresses, cyclic strain amplitudes cannot be predicted as accurately as cyclic stress amplitudes, and equipment for cyclic strain-controlled testing is less readily available than equipment for cyclic stress-controlled testing (Kramer and Elgamal, 2001). Thus, the cyclic strain-based approach is less commonly used than the cyclic stress-based approach. The energy-based approach is conceptually attractive, as the dissipated energy reflects both cyclic stress and strain amplitudes. Several investigators have established relationships between the pore pressure development and the dissipated energy during ground shaking (Davis and Berrill, 1982; Berrill and Davis, 1985; Figueroa *et al.*, 1994; Ostadan *et al.*, 1996). The initiation of liquefaction can be formulated by comparing the calculated unit energy from the time series record of a design earthquake with the resistance to liquefaction in terms of energy based on in-situ soil properties (Liang *et al.*, 1995; Dief, 2000). The energy-based methods, however, are also less commonly used than the cyclic stress-based approach. Thus, the focus in this chapter is on the evaluation of liquefaction potential using the cyclic stress-based methods.

Two general types of cyclic stress based-approach are available for assessing liquefaction potential. One is by means of laboratory testing (e.g., cyclic triaxial test and cyclic simple shear test) of undisturbed samples, and the other involves use of empirical relationships that relate observed field behavior with in situ tests such as standard penetration test (SPT), cone penetration test (CPT), shear wave velocity measurement (Vs) and the Becker penetration test (BPT). Because of the difficulties and costs associated with high-quality undisturbed sampling and subsequent high-quality testing of granular soils, use of in-situ tests along with the case histories-calibrated empirical relationships (i.e. liquefaction boundary curves) has been, and is still, the dominant approach in engineering practise.

The most widely used cyclic stress-based method for liquefaction potential evaluation in North America and throughout much of the world is the simplified procedure pioneered by Seed and Idriss (1971).

The simplified procedure was developed based on field observations and field and laboratory tests with a strong theoretical basis. Case histories of liquefaction/no-liquefaction were collected from sites on level to gently sloping ground, underlain by Holocene alluvial or fluvial sediments at shallow depths (< 15 m). In a case history, the occurrence of liquefaction was primarily identified with surface manifestations such as lateral spread, ground settlement, and sand boils. Because the simplified procedure was eventually "calibrated" based on such case histories, the "occurrence of liquefaction" should be interpreted accordingly, that is, the emphasis is on the surface manifestations. This definition of liquefaction does not always correspond to the initiation of liquefaction defined based on the 5% DA axial strain in 20 cycles of uniform load typically adopted in the laboratory testing. The stress-based approach that follows the simplified procedure by Seed and Idriss (1971) is considered herein.

The state of the art for evaluating liquefaction potential was reviewed in 1985 by a committee of the National Research Council. The report of this committee became the standard reference for practicing engineers in North America (NRC, 1985). About 10 years later, another review was sponsored by the National Center for Earthquake Engineering Research (NCEER) at the State University of New York at Buffalo. This workshop focused on the stress-based simplified methods for liquefaction potential evaluation. The NCEER Committee issued a report in 1997 (Youd and Idriss, 1997), but continued to re-assess the state of the art and in 2001 published a summary paper (Youd et al., 2001), which represents the current state of the art on the subject of liquefaction evaluation. It focuses on the fundamental problem of evaluating the potential for liquefaction in level or nearly level ground, using in-situ tests to characterize the resistance to liquefaction and the Seed and Idriss (1971, 1982) simplified method to characterize the duration and intensity of the earthquake shaking. Among the methods recommended for determination of liquefaction resistance, only the SPT-based method is examined herein, as the primary purpose of this chapter is on the reliability analysis of soil liquefaction. The SPT-based method is used only as an example to illustrate the probabilistic approach.

In summary, the SPT-based method as described in Youd et al . (2001) is adopted here as the deterministic model for liquefaction potential evaluation. This method is originated by Seed and Idriss (1971) but has gone through several stages of modification (Seed and Idriss, 1982; Seed et al., 1985; Youd et al., 2001). In this chapter, a limit state of liquefaction triggering is defined based on this SPT-based method, and the issues of parameter and model uncertainties are examined in detail, followed by probabilistic analyses using reliability theory. Examples are presented to illustrate both the deterministic and the probabilistic approaches.

13.2 Deterministic approach

13.2.1 Formulation of the SPT-based method

In the current state of knowledge, the seismic loading that could cause a soil to liquefy is generally expressed in terms of cyclic stress ratio (CSR). Because the simplified stress-based methods were all developed based on calibration with field data with different earthquake magnitudes and over-burden stresses, CSR is often "normalized" to a reference state with moment magnitude $M_w = 7.5$ and effective overburden stress $\sigma'_v = 100$ kPa. At the reference state, the CSR is denoted as $CSR_{7.5,\sigma}$, which may be expressed as (after Seed and Idriss, 1971):

$$CSR_{7.5,\sigma} = 0.65 \left(\frac{\sigma_v}{\sigma'_v}\right) \left(\frac{a_{max}}{g}\right) (r_d)/MSF/K_\sigma \tag{13.1}$$

where σ_v = the total overburden stress at the depth of interest (kPa), σ'_v = the effective stress at the depth of interest (kPa), g = the unit of the acceleration of gravity, a_{max} = the peak horizontal ground surface acceleration (a_{max}/g is dimensionless), r_d = the depth-dependent stress reduction factor (dimensionless), MSF = the magnitude scaling factor (dimensionless), and K_σ = the overburden stress adjustment factor for the calculated CSR (dimensionless). For the peak horizontal ground surface acceleration, the geometric mean is preferred for use in engineering practice, although use of the larger of the two orthogonal peak accelerations is conservative and allowable (Youd et al., 2001).

For routine practice and no critical projects, the following equations may be used to estimate the values of r_d (Liao and Whitman, 1986):

$$r_d = 1.0 - 0.00765d \quad \text{for} \quad d < 9.15\text{m}, \tag{13.2a}$$

$$r_d = 1.174 - 0.0267d \quad \text{for} \quad 9.15\text{m} < d \leq 20\text{m} \tag{13.2b}$$

where d = the depth of interest (m). The variable MSF may be calculated with the following equation (Youd et al., 2001):

$$MSF = \left(M_w/7.5\right)^{-2.56} \tag{13.3}$$

It should be noted that different formulas for r_d and MSF have been proposed by many investigators (e.g. Youd et al., 2001; Idriss and Boulanger, 2006; Cetin et al., 2004). To be consistent with the SPT-based deterministic method presented herein, use of Equations (13.2) and (13.3) is recommended.

As noted previously, the variable K_σ is a stress adjustment factor used to adjust CSR to the effective overburden stress of $\sigma'_v = 100$ kPa. This is different from the overburden stress correction factor (C_N) that is applied to the SPT blow count (N_{60}), which is described later. The adjustment factor K_σ is defined as follows (Hynes and Olsen, 1999):

$$K_\sigma = (\sigma_v'/P_a)^{(f-1)} \tag{13.4}$$

where $f \approx 0.6 - 0.8$ and Pa is the atmosphere pressure (≈ 100 kPa). Data from the existing database are insufficient for precise determination of the coefficient f. For routine practice and no critical projects, $f = 0.7$ may be assumed, and thus the exponent in Equation (13.4) would be -0.3.

For the convenience of presentation hereinafter, the normalized cyclic stress ratio $CSR_{7.5,\sigma}$ is simply labeled as CSR whenever no confusion would be caused by such use. For liquefaction potential evaluation, CSR (as the seismic *loading*) is compared with liquefaction *resistance*, expressed as cyclic resistance ratio (CRR). As noted previously, the simplified stress-based methods were all developed based on calibration with field observations. Such calibration process is generally based on the concept that cyclic resistance ratio (CRR) is the *limiting* CSR beyond which the soil will liquefy. Based primarily on this concept and with engineering judgment, the following equation is recommended by Youd *et al.* (2001) for the determination of CRR using SPT data:

$$CRR = \frac{1}{34 - N_{1,60cs}} + \frac{N_{1,60cs}}{135} + \frac{50}{[10 \cdot N_{1,60cs} + 45]^2} - \frac{1}{200} \tag{13.5}$$

where $N_{1,60cs}$ (dimensionless) is the clean-sand equivalence of the overburden stress-corrected SPT blow count, defined as follows:

$$N_{1,60cs} = \alpha + \beta N_{1,60} \tag{13.6}$$

where α and β are coefficients to account for the effect of fines content, defined later, and $N_{1,60}$ is the SPT blow count normalized to the *reference* hammer energy efficiency of 60 % and effective overburden stress of 100 kPa, defined as:

$$N_{1,60} = C_N N_{60} \tag{13.7}$$

where N_{60} = the SPT blow count at 60% hammer energy efficiency and corrected for rod length, sampler configuration, and borehole diameter (Skempton, 1986; Youd *et al.*, 2001):

$$C_N = (P_a/\sigma_v')^{0.5} \leq 1.7 \tag{13.8}$$

The coefficients α and β in Equation (13.6) are related to fines content (FC) as follows:

$$\alpha = 0 \text{ for } FC \leq 5\% \tag{13.9a}$$

$$\alpha = \exp\left[1.76 - (190/FC^2)\right] \text{ for } 5\% < FC < 35\% \tag{13.9b}$$

$$\alpha = 5.0 \text{ for } FC \geq 35\% \tag{13.9c}$$

$$\beta = 1.0 \text{ for } FC \leq 5\% \tag{13.10a}$$

$$\beta = [0.99 + (FC^{1.5}/1000)] \text{ for } 5\% < FC < 35\% \tag{13.10b}$$

$$\beta = 1.2 \text{ for } FC \geq 35\% \tag{13.10c}$$

Equations (13.1) through (13.10) *collectively* represent the SPT-base deterministic model for liquefaction potential evaluation recommended by Youd *et al.* (2001). This model is recognized as the current state of the art for liquefaction evaluation using SPT. The reader is referred to Youd *et al.* (2001) for additional details on this model and its parameters. In a deterministic evaluation, factor of safety (FS), defined as $FS = CRR/CSR$, is used to "measure" liquefaction potential. In theory, liquefaction is said to occur if $FS \leq 1$, and no liquefaction if $FS > 1$. However, caution must be exercised when interpreting the calculated FS. In the back analysis of a case history or in a post-earthquake investigation analysis, use of $FS \leq 1$ to judge whether liquefaction had occurred could be misleading as the existing simplified methods tend to be conservative (in other words, there could be model bias toward the conservative side). Because of model and parameter uncertainties, $FS > 1$ does not always correspond to no-liquefaction, and $FS \leq 1$ does not always correspond to liquefaction.

The selection of a minimum required FS value for a particular project in a design situation depends on factors such as the perceived level of model and parameter uncertainties, the consequence of liquefaction in terms of ground deformation and structures damage potential, the importance of the structures, and the economic consideration. Thus, the process of selecting an appropriate FS is not a trivial exercise. In a design situation, a factor of safety of 1.2–1.5 is recommended by the Building Seismic Safety Council (1997) in conjunction with the use of the Seed *et al.* (1985) method for liquefaction evaluation. Since the Youd *et al.* (2001) method is essentially an updated version of, and is perceived as conservative as, the Seed *et al.* (1985) method, the recommended range of FS by the Building Seismic Safety Council (1997) should be applicable. In recent years, however, there is growing trend to assess liquefaction potential in terms of probability of liquefaction (Liao *et al.*, 1988; Juang *et al.*, 2000, 2002; Cetin *et al.*, 2004). To facilitate the use of probabilistic methods, calibration of the calculated probability to the previous engineering experience is needed. In a previous study by

Juang *et al.* (2002), a factor of safety of 1.2 in the Youd *et al.* (2001) method was found to correspond *approximately* to a mean probability of 0.30. In this chapter, further calibration of the calculated probability of liquefaction is presented later.

13.2.2 Example No. 1: deterministic evaluation of a non-liquefied case

This example concerns a non-liquefied case. Field observation of the site, which is designated as *San Juan B-5* (Idriss *et al.*, as cited in Cetin, 2000), indicated no occurrence of liquefaction during the 1977 Argentina earthquake. The mean values of seismic and soil parameters at the critical depth (2.9 m) are given as follows: $N_{60} = 8.0$, $FC = 3\%$, $\sigma_v' = 38.1$ kPa, $\sigma_v = 45.6$ kPa, $a_{max} = 0.2$ g, and $M_w = 7.4$ (Cetin, 2000). First, *CRR* is calculated as follows:

Using Equation (13.8),

$$C_N = (P_a/\sigma_v')^{0.5} = (100/38.1)^{0.5} = 1.62 < 1.7$$

Using Equation (13.7),

$$N_{1,60} = C_N N_{60} = (1.62)(8.0) = 13.0$$

Since $FC = 3\% < 5\%$, thus $\alpha = 0$ and $\beta = 1$ according to Equations (13.9) and (13.10). Thus, according to Equation (13.6),

$$N_{1,60cs} = \alpha + \beta N_{1,60} = 13.0.$$

Finally, using Equation (13.5), we have

$$CRR = \frac{1}{34 - N_{1,60cs}} + \frac{N_{1,60cs}}{135} + \frac{50}{[10 \cdot N_{1,60cs} + 45]^2} - \frac{1}{200}$$

$$= 0.141$$

Next, the intermediate parameters of *CSR* are calculated as follows:

$$MSF = \left(M_w/7.5\right)^{-2.56} = (7.4/7.5)^{-2.56} = 1.035$$

$$K_\sigma = (\sigma_v'/P_a)^{(f-1)} = (38.1/100)^{-0.3} = 1.335$$

$$r_d = 1.0 - 0.00765d = 1.0 - 0.00765(2.9) = 0.978$$

Finally, using Equation (13.1), we have:

$$CSR_{7.5,\sigma} = 0.65 \left(\frac{\sigma_v}{\sigma_v'}\right)\left(\frac{a_{max}}{g}\right)(r_d)/MSF/K_\sigma$$

$$= 0.65 \left(\frac{45.6}{38.1}\right)(0.2)(0.978)/[(1.035)(1.335)]$$

$$= 0.110$$

The factor of safety is calculated as follows:

$$FS = CRR/CSR = 0.141/0.110 = 1.28$$

As a back analysis of a case history, this FS value would suggest no liquefaction, which agrees with the filed observation. However, a probabilistic analysis might be necessary or desirable to complement the judgment based on the calculated FS value.

13.2.3 Example No. 2: deterministic evaluation of a liquefied case

This example concerns a liquefied case. Field observation of the site, designated as *Ishinomaki-2* (Ishihara *et al.*, as cited in Cetin, 2000), indicated occurrence of liquefaction during the 1978 Miyagiken-Oki earthquake. The mean values of seismic and soil parameters at the critical depth (3.7 m) are given as follows: $N_{1,60} = 5$, $FC = 10\%$, $\sigma_v' = 36.28$ kPa, $\sigma_v = 58.83$ kPa, $a_{max} = 0.2$ g, and $M_w = 7.4$ (Cetin, 2000).

Similar to the analysis performed in Example 1, the calculations of the CRR, CSR, and FS are carried out as follows:

$$\alpha = \exp\left[1.76 - (190/FC^2)\right] = \exp[1.76 - (190/10^2)] = 0.869$$

$$\beta = [0.99 + (FC^{1.5}/1000)] = [0.99 + (10^{1.5}/1000)] = 1.022$$

$$N_{1,60cs} = \alpha + \beta N_{1,60} = 5.98$$

$$CRR = \frac{1}{34 - N_{1,60cs}} + \frac{N_{1,60cs}}{135} + \frac{50}{[10 \cdot N_{1,60cs} + 45]^2} - \frac{1}{200}$$

$$= \frac{1}{34 - 5.98} + \frac{5.98}{135} + \frac{50}{[10 \times 5.98 + 45]^2} - \frac{1}{200}$$

$$= 0.080$$

$$MSF = \left(M_w/7.5\right)^{-2.56} = (7.4/7.5)^{-2.56} = 1.035$$

$$K_\sigma = (\sigma_v'/P_a)^{(f-1)} = (36.28/100)^{-0.3} = 1.356$$

$$r_{\rm d} = 1.0 - 0.00765d = 1.0 - 0.00765(3.7) = 0.972$$

$$CSR = 0.65 \left(\frac{\sigma_{\rm v}}{\sigma_{\rm v}'}\right) \left(\frac{a_{\max}}{g}\right)(r_{\rm d})/MSF/K_\sigma = 0.146$$

$$FS = CRR/CSR = 0.545$$

The calculated FS value is much below 1. As a back analysis of a case history, the calculated FS value would confirm the field observation with a great certainty. However, a probabilistic analysis might still be desirable to complement the judgment based on the calculated FS value.

13.3 Probabilistic approach

Various models for estimating the probability of liquefaction have been proposed (Liao et al., 1988; Juang et al., 2000, 2002; Cetin et al., 2004). These models are all data-driven, meaning that they are established based on statistical analyses of the databases of case histories. To calculate the probability using these empirical models, only the best estimates (i.e. the mean values) of the input variables are required; the uncertainty in the model, termed *model uncertainty*, and the uncertainty in the input variables, termed *parameter uncertainty*, are excluded from the analysis. Thus, the calculated probabilities might be subject to error if the effect of model and/or parameter uncertainty is significant. A more fundamental approach to this problem would be to adopt a reliability analysis that considers both model and parameter uncertainties. The formulation and procedure for conducting a rigorous reliability analysis is described in the sections that follow.

13.3.1 Limit state of liquefaction triggering

In the context of reliability analysis presented herein, the limit state of liquefaction triggering is essentially the boundary curve that separates "region" of liquefaction from the region of no-liquefaction. An example of a limit state is shown in Figure 13.1, where the SPT-based boundary curve recommended by Youd et al. (2001) is shown with 148 case histories. As reflected in the scattered data shown in Figure 13.1, uncertainty exists as to where the boundary curve should be "positioned." This uncertainty is the model uncertainty mentioned previously. The issue of model uncertainty is discussed later. At this point, the limit state may be expressed symbolically as follows:

$$h(\mathbf{x}) = CRR - CSR = 0 \tag{13.11}$$

where $\mathbf{x}$ is a vector of input variables that consist of soil and seismic parameters that are required in the calculation of CRR and CSR, and $h(\mathbf{x}) < 0$ indicates liquefaction.

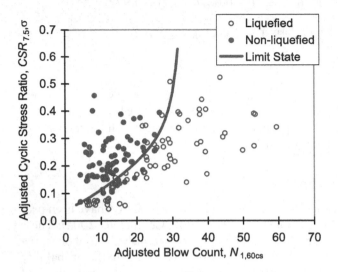

Figure 13.1 An example limit state of liquefaction triggering.

As noted previously, Equations (13.1) through (13.10) collectively represent the SPT-base deterministic model for liquefaction potential evaluation recommended by Youd *et al.* (2001). Two parameters $N_{1,60}$ and *FC* are required in the calculation of *CRR* and both are assumed to be random variables. Since *MSF* is a function of M_w, and K_σ is a function of σ_v', five parameters, including $a_{max}, M_w, \sigma_v', \sigma_v,$ and r_d, are required for calculating *CSR*. The first four parameters, $a_{max}, M_w, \sigma_v',$ and σ_v are assumed to be random variables. The parameter r_d is a function of depth (d) and is not considered as a random variable (since *CSR* is evaluated for soil at a given d). Based on the above discussions, a total of six random variables are identified in the deterministic model by Youd *et al.* (2001) described previously. Thus, the limit state of liquefaction based on this deterministic model may be expressed as follows:

$$h(\mathbf{x}) = CRR - CSR = h(N_{1,60}, FC, M_w, a_{max}, \sigma_v', \text{ and } \sigma_v) = 0 \quad (13.12)$$

It should be noted that while the parameter r_d is not considered as a random variable, the uncertainty does exist in the model for r_d (Equation (13.2)), just as the uncertainty exists in the model for *MSF* (Equation (13.3)) and in the model for K_σ (Equation (13.4)). The uncertainty in these "component" models contributes to the uncertainty of the calculated *CSR*, which, in turn, contributes to the uncertainty of the *CRR* model, since *CRR* is considered as the *limiting CSR* beyond which the soil will liquefy. Rather than dealing with the uncertainty of each component model, where there is a lack of data for calibration, the uncertainty of the entire limit state model is characterized as a whole, since the only data available for model calibration are field observations of liquefaction (indicated by $h(\mathbf{x}) \leq 0$) or no liquefaction (indicated by $h(\mathbf{x}) > 0$). Thus, the limit state model that considers model uncertainty may be rewritten as:

$$h(\mathbf{x}) = c_1 CRR - CSR = h(c_1, N_{1,60}, FC, M_w, a_{max}, \sigma_v', \text{ and } \sigma_v) = 0$$

$$(13.13)$$

where random variable c_1 represents the uncertainty of the limit state model that is yet to be characterized. Use of a single random variable to characterize the uncertainty of the entire limit state model is adequate, since CRR is defined as the *limiting CSR* beyond which the soil will liquefy, as noted previously. Thus, only one random variable c_1 applied to CRR is required. In fact, Equation (13.13) may be interpreted as: $h(\mathbf{x}) = c_1 CRR - CSR = c_1 FS - 1 = 0$. Thus, the uncertainty of the entire limit state model is seen here as the uncertainty in the calculated FS, where data (field observations) are available for calibration. For convenience of presentation hereinafter, the random variable c_1 is referred to as the model bias factor or simply, *model factor*.

13.3.2 Parameter uncertainty

For a realistic estimate of liquefaction probability, the reliability analysis must consider both parameter and model uncertainties. The issue of model uncertainty is discussed later. Thus, for reliability analysis of a future case, the uncertainties of input random variables must first be assessed. For each input variable, this process involves the estimation of the mean and standard deviation if the variable is assumed to follow normal or lognormal distribution. The engineer usually can make a pretty good estimate of the mean of a variable even with limited data. This probably has to do with the well-established statistics theory that the "sample mean" is a best estimate of the "population mean." Thus, the following discussion focuses on the estimation of standard deviation of each input random variable.

Duncan (2000) suggested that the standard deviation of a random variable may be obtained by one of the following three methods: (1) direct calculation from data, (2) estimate based on published coefficient of variation (COV); and (3) estimate based on the "three-sigma rule" (Dai and Wang, 1992). In the last method, the knowledge of the highest conceivable value (HCV) and the lowest conceivable value (LCV) of the variable is used to calculate the standard deviation σ as follows (Duncan, 2000):

$$\sigma = \frac{HCV - LCV}{6}$$

$$(13.14)$$

It should be noted that the engineer tends to under-estimate the range of a given variable (and thus, the standard deviation), particularly if the estimate was based on very limited data and judgment was required. Thus, for a small sample size, a value of less than 6 should be used for the denominator in Equation (13.14). Whenever in doubt, a sensitivity analysis should be conducted to investigate the effect of different levels of COV of a particular variable on the results of reliability analysis.

Typical ranges of COVs of the input variables according to the published data are listed in Table 13.1. It should be noted that the COVs of the earthquake parameters, a_{max} and M_w, listed in Table 13.1, are based on values reported in the published databases of case histories where recorded strong ground motions and/or locally calibrated data were available. The COV of a_{max} based on general attenuation relationships could easily be as high as 0.50 (Haldar and Tang, 1979). According to Youd et al. (2001), for a future case, the variable a_{max} may be estimated using one of the following methods:

1   Using empirical correlations of a_{max} with earthquake magnitude, distance from the seismic energy source, and local site conditions.
2   Performing local site response analysis (e.g. using SHAKE or other software) to account for local site effects.
3   Using the USGS National Seismic Hazard web pages and the NEHRP amplification factors.

Table 13.1 Typical coefficients of variation of input random variables.

Random variable	Typical range of COV^a	References
$N_{1,60}$	0.10–0.40	Harr (1987); Gutierrez et al. (2003); Phoon and Kulhawy (1999)
FC	0.05–0.35	Gutierrez et al. (2003)
σ_v'	0.05–0.20	Juang et al. (1999)
σ_v	0.05–0.20	Juang et al. (1999)
a_{max}	0.10–0.20[b]	Juang et al. (1999)
M_w	0.05–0.10	Juang et al. (1999)

Note
[a]The word "typical" here implies the range approximately bounded by the 15th percentile and the 85th percentile, estimated from case histories in the existing databases such as Cetin (2000). Published COVs are also considered in the estimate given here. The actual COV values could be higher or lower, depending on the variability of the site and the quality and quantity of data that are available.

[b]The range is based on values reported in the published databases of case histories where recorded strong ground motions and locally calibrated data were available. However, the COV of a_{max} based on general attenuation relationships or amplification factors could easily be as high as or over 0.50.

Use of the amplification factors approach is briefly summarized in the following. The USGS National Hazard Maps (Frankel *et al.*, 1997) provide rock peak ground acceleration (*PGA*) and spectral acceleration (*SA*) for a specified locality based on latitude/longitude or zip code. The USGS web page (http://earthquake.usgs.gov/research/hazmaps) provides *PGA* value and *SA* values at selected spectral periods (For example, $T = 0.2$, 0.3, and 1.0 s) each with six levels of probability of exceedance, including 1% probability of exceedance in 50 years (annual rate of 0.0002), 2% probability of exceedance in 50 years (annual rate of 0.0004), 5% probability of exceedance in 50 years (annual rate of 0.001), 10% probability of exceedance in 50 years (annual rate of 0.002), and 20% probability of exceedance in 50 years (annual rate of 0.004), and 50% probability of exceedance in 75 years (annual rate of 0.009). The six levels of probability of exceedance are often referred to as the six seismic hazard levels, with corresponding earthquake return periods of 4975, 2475, 975, 475, 224, and 108 years, respectively. For a given locality, a *PGA* can be obtained for a specified probability of exceedance in an exposure time from the USGS National Seismic Hazard Maps.

For liquefaction analysis, the rock *PGA* needs to be converted to peak ground surface acceleration at the site, a_{max}. Ideally, the conversion should be carried out based on site response analysis. Various simplified procedures are also available for an estimate of a_{max} (e.g. Green, 2001; Gutierrez *et al.*, 2003; Stewart *et al.*, 2003; Choi and Stewart, 2005). As an example, a simplified procedure for estimating a_{max}, perhaps in the simplest form, is expressed as follows:

$$a_{max} = F_a (PGA) \tag{13.15}$$

where F_a is the amplification factor, which, in a simplest form, may be expressed as a function of rock *PGA* and the NEHRP site class (NEHRP 1998). Figure 13.2 shows an example of a simplified chart for the amplification factor. The NEHRP site classes used in Figure 13.2 are based on the mean shear wave velocity of soils in the top 30 m, as listed in Table 13.2.

Choi and Stewart (2005) developed a more sophisticated model for ground motion amplification that is a function of the average shear wave velocity over the top 30 m of soils V_{S30} and "rock" reference *PGA*. The amplification factors are defined relative to "rock" reference motions from several attenuation relationships for active tectonic regions, including those of Abrahamson and Silva (1997), Sadigh *et al.* (1997), and Campbell and Bozorgnia (2003). The databases used in model development cover the parameter spaces $V_{S30} = 130 \sim 1300$ m/s and $PGA = 0.02 \sim 0.8$ g, and the model is considered valid *only* in these ranges of parameters. The Choi

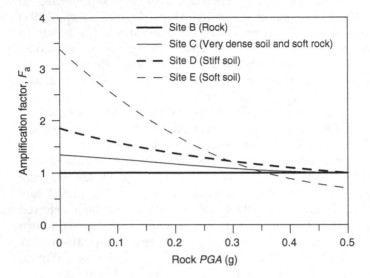

Figure 13.2 Amplification factor as a function of rock *PGA* and the NEHRP site class (reproduced from Gutierrez *et al.*, 2003).

Table 13.2 Site classes (categories) in NEHRP provisions.

NEHRP category (soil profile type)	Description
A	Hard rock with measured mean shear wave velocity in the top 30 m, $\bar{v}_s > 1500$ m/s
B	Rock with 760 m/s $< \bar{v}_s \leq 1500$ m/s
C	Dense soil and soft rock with 360 m/s $< \bar{v}_s \leq 760$ m/s
D	Stiff soil with 180 m/s $< \bar{v}_s \leq 360$ m/s
E	Soil with $\bar{v}_s \leq 180$ m/s or any profile with more than 3 m of soft clay (plasticity index PI > 20, water content $w > 40\%$ and undrained shear strength $s_u < 25$ kPa)
F	Soils requiring a site-specific study, e.g. liquefiable soils, highly sensitive clays, collapsible soils, organic soils, etc.

and Stewart (2005) model for amplification factor (F_{ij}) is expressed as follows:

$$\ln(F_{ij}) = c \ln\left(\frac{V_{S30_{ij}}}{V_{\text{ref}}}\right) + b \ln\left(\frac{PGA_{r_{ij}}}{0.1}\right) + \eta_i + \varepsilon_{ij} \qquad (13.16)$$

where PGA_r is the rock PGA expressed in units of g; b is a function of regression parameters; c and V_{ref} are regression parameters; η_i is a random effect term for earthquake event i; and ε_{ij} represents the intra-event model residual for motion j in event i.

Choi and Stewart (2005) provided many sets of empirical constants for use of Equation (13.16). As an example, for a site where the attenuation relationship by Abrahamson and Silva (1997) is applicable and a spectral period $T = 0.2$ sec is specified, Equation (13.16) becomes:

$$\ln(F_a) = -0.31 \ln\left(\frac{V_{S30}}{453}\right) + b \ln\left(\frac{PGA_r}{0.1}\right) \tag{13.17}$$

In Equation (13.17), F_a is the amplification factor; V_{S30} is obtained from site characterization; PGA_r is obtained for reference rock conditions using the attenuation relationship by Abrahamson and Silva (1997); and b is defined as follows (Choi and Stewart, 2005):

$$b = -0.52, \text{ for Site Category E} \tag{13.18a}$$

$$b = -0.19 - 0.000023\,(V_{S30} - 300)^2, \text{ for } 180 < V_{S30} < 300\,(\text{m/s}) \tag{13.18b}$$

$$b = -0.19, \text{ for } 300 < V_{S30} < 520\,(\text{m/s}) \tag{13.18c}$$

$$b = -0.19 + 0.00079\,(V_{S30} - 520), \text{ for } 520 < V_{S30} < 760\,(\text{m/s}) \tag{13.18d}$$

$$b = 0, \text{ for } V_{S30} > 760\,(\text{m/s}) \tag{13.18e}$$

The total standard deviation for the amplification factor F_a obtained from Equation (13.17) came from two sources: the inter-event standard deviation of 0.27, and the intra-event standard deviation of 0.53. Thus, for the given scenario (the specified spectral period and the chosen attenuation model), the total standard deviation is $\sqrt{(0.27)^2 + (0.53)^2} = 0.59$. The peak ground surface acceleration a_{max} can be obtained from Equation (13.15).

For subsequent reliability analysis, a_{max} obtained from Equation (13.15) may be considered as the mean value. For a specified probability of exceedance (and thus a given PGA), the variation of this mean a_{max} is primarily caused by the uncertainty in the amplification factor model. Use of simplified amplification factors for estimating a_{max} tends to result in a large variation and thus, for important projects, concerted effort to reduce this uncertainty using more accurate methods and/or better quality data should be made whenever possible. The reader is referred to Bazzurro and Cornell (2004 a,b) and Juang et al. (2008) for detailed discussions of this subject.

The magnitude of M_w can also be derived from the USGS National Seismic Hazard web pages through a de-aggregation procedure. Detailed information may be obtained from http://eqint.cr.usgs.gov/deaggint/2002.index.php. A summary of the procedure is provided in the following.

The task of seismic hazard de-aggregation involves the determination of earthquake parameters, principally magnitude and distance, for use in a seismic-resistant design. In particular, calculations are made to determine the statistical mean and modal sources for any given US site for the six hazard levels (or the corresponding probabilities of exceedance).

The seismic hazard presented in the USGS Seismic Hazard web page is de-aggregated to examine the "contribution to hazard" (in terms of frequency) as a function of magnitude and distance. These plots of "contribution to hazard" as a function of magnitude and distance are useful for specifying design earthquakes. On the available de-aggregation plots from the USGS website, the height of each bar represents the percent contribution of that magnitude and distance pair (or bin) to the specified probabilities of exceedance. The distribution of the heights of these bars (i.e. frequencies) is essentially a joint probability mass function of magnitude and distance. When this joint mass function is "integrated" along the axis of distance, the "marginal" or conditional probability mass function of the magnitude is obtained. This distribution of M_w is obtained for the same specified probability of exceedance as the one from which the PGA is derived. The distribution (or the uncertainty) of M_w here is due primarily to the uncertainty in seismic sources.

It should also be noted that for selection of a design earthquake in a deterministic approach, the de-aggregation results are often described in terms of the *mean* magnitude. However, use of the *modal* magnitude is preferred by many engineers because the mode represents the *most likely* source in the seismic-hazard model, whereas the mean might represent an unlikely or even unconsidered source, especially in the case of a strongly bimodal distribution.

In summary, a pair of PGA and M_w may be selected at a specified hazard level or probability of exceedance. The selected PGA is converted to a_{max}, and the pair of a_{max} and M_w is then used in the liquefaction evaluation. For reliability analysis, the values of a_{max} and M_w determined as described previously are taken as the mean values, and the variations of these variables are estimated and expressed in terms of the COVs. The reader is referred to Juang *et al.* (2008) for additional discussions on this subject.

13.3.3 Correlations among input random variables

The *correlations* among the input random variables should be considered in a reliability analysis. The correlation coefficients may be estimated empirically using statistical methods. Except for the pair of a_{max} and M_w, the correlation coefficient between each pair of input variables used in the limit state model is estimated based on an analysis of the actual data in the existing

Table 13.3 Coefficients of correlation among the six input random variables.

Variable \ Variable	$N_{1,60}$	FC	σ'_v	σ_v	a_{max}	M_w
$N_{1,60}$	1	0	0.3	0.3	0	0
FC	0	1	0	0	0	0
σ'_v	0.3	0	1	0.9		0
σ_v	0.3	0	0.9	1	0	0
a_{max}	0	0	0	0	1	0.9^a
M_w	0	0	0	0	0.9^a	1

Note

[a]This is estimated based on local attenuation relationships calibrated to given historic earthquakes. This correlation may be used for back-analysis of a case history. The correlation of the two parameters at a locality subject to uncertain sources, as in the analysis of a future case, could be much lower or even negligible.

databases of case histories. The correlation coefficient between a_{max} and M_w is taken to be 0.9, which is based on statistical analysis of the data generated from the attenuation relationships (Juang *et al.*, 1999). The coefficients of correlation among the six input random variables are shown in Table 13.3. The correlation between the model uncertainty factor (c_1 in Equation 13.13) and each of the six input random variables is assumed to be 0.

It should be noted that the correlation matrix as shown in Table 13.3 must be symmetric and "positive definite" (Phoon, 2004). If this condition is not satisfied, a negative variance might be obtained, which would contradict the definition of the variance. In Excel^{TM}, the condition can be easily checked using "MAT_CHOLESKY." It should be noted that MAT_CHOLESKY can be executed with a free Excel^{TM} add-in, "matrix.xla," which must be installed once by the user. The file "matrix.xla" may be downloaded from http://digilander.libero.it/foxes/index.htm. For the correlation matrix shown in Table 13.3, the diagonal entries of the matrix of Cholesky factors are all positive; thus, the condition of "positive definiteness" is satisfied.

13.3.4 Model uncertainty

The issue of model uncertainty is important but rarely emphasized in the geotechnical engineering literature, perhaps because it is difficult to address. Instead of addressing this issue directly, Zhang *et al.* (2004) suggested a procedure for reducing the uncertainty of model prediction using Bayesian updating techniques. However, since a large quantity of liquefaction case histories (Cetin, 2000) is available for calibration of the calculated reliability indexes, an estimate of the model factor (c_1 in Equation (13.13)) is possible, as documented previously by Juang *et al.* (2004). The procedure for estimating model factor involves two steps: (1) deriving a Bayesian mapping function based on the database of case histories, and (2) using the calibrated Bayesian mapping function as a reference to back-figure the model factor c_1. The detail of this procedure is not repeated herein; only a brief summary of the results obtained from the calibration of the limit state model (Equation (13.13)) is provided, and the reader is referred to Juang *et al.* (2004, 2006) for details of the procedure.

The model factor c_1 is assumed to follow lognormal distribution. With the assumption of lognormal distribution, the model factor c_1 can be characterized with a mean μ_{c1} and a standard deviation (or coefficient of variation, COV). In a previous study (Juang *et al.*, 2004), the effect of the COV of the model factor (c_1) on the final probability obtained through reliability analysis was found to be insignificant, *relative* to the effect of μ_{c1}. Thus, for the calibration (or estimation) of mean model factor μ_{c1}, an assumption of $COV = 0$ is made. It should be noted, however, that because of the assumption of $COV = 0$ and the effect of other factors such as data scatter, there will be variation on the calibrated μ_{c1}, which would be reflected in its standard deviation, $\sigma_{\mu c1}$.

The first step in the model calibration process is to develop Bayesian mapping functions based on the distributions of the values of reliability index β for the group of liquefied cases and the group of non-liquefied cases (Juang *et al.*, 1999):

$$P_{\mathrm{L}} = P(L|\beta) = \frac{P(\beta|L)P(L)}{P(\beta|L)P(L) + P(\beta|NL)P(NL)} \tag{13.19}$$

where $P(L|\beta)$ = probability of liquefaction for a given β; $P(\beta|L)$ = probability of β given that liquefaction did occur; $P(\beta|NL)$ = probability of β given that liquefaction did not occur; $P(L)$ = prior probability of liquefaction; $P(NL)$ = prior probability of no-liquefaction.

The second step in the model calibration process is to back-figure the model factor μ_{c1} using the developed Bayesian mapping functions. By means of a trial-and-error process with varying μ_{c1} values, the uncertainty of the limit state model (Equation (13.13)) can be calibrated using the probabilities interpreted from Equation (13.19) for a large number of case histories. The essence of the calibration here is to find an "optimum" model factor μ_{c1} so that the calibrated nominal probabilities match the best with the *reference* Bayesian mapping probabilities for all cases in the database. Using the database compiled by Cetin (2000), the mean model factor is calibrated to be $\mu_{c1} = 0.96$ and the variation of the calibrated mean model factor is reflected in the estimated standard deviation of $\sigma_{\mu c1} = 0.04$.

With the given limit state model (Equation (13.13) and associated equations) and the calibrated model factor, reliability analysis of a future case can be performed once the mean and standard deviation of each input random variable are obtained.

13.3.5 First-order reliability method

Because of the complexity of the limit state model (Equation (13.13)) and the fact that the basic variables of the model are non-normal and correlated, no closed-form solution for reliability index of a given case is possible. For reliability analysis of such problems, numerical methods such as the first-order reliability method (FORM) are often used. The general approach undertaken by the FORM is to transform the original random variables into independent, standard normal random variables, and the original limit state function into its counterpart in the transformed or "standard" variable space. The reliability index β is defined as the shortest distance between the limit state surface and the origin in the standard variable space. The point on the limit state surface that has the shortest distance from the origin is referred to as the design point. FORM requires an optimization algorithm to locate the design point and to determine the reliability index. Several algorithms are available and the reader is referred to the literature (e.g. Ang and Tang, 1984; Melchers, 1999; Baecher and Christian, 2003) for details of these algorithms. Once the reliability index β is obtained using FORM, the nominal probability of liquefaction, P_L, can be determined as:

$$P_L = 1 - \Phi(\beta) \tag{13.20}$$

where Φ is the standard normal cumulative distribution function. In Microsoft Excel™, numerical value of $\Phi(\beta)$ can be obtained using the function $NORMSDIST(\beta)$.

The FORM procedure can easily be programmed (e.g. Yang, 2003). Efficient implementation of the FORM procedure in Excel™ was first introduced by Low (1996), and many geotechnical applications are found in the literature (e.g. Low and Tang, 1997; Low, 2005; Juang et al., 2006). The spreadsheet solution introduced by Low (1996) is a clever solution of reliability index based on the formulation by Hasofer and Lind (1974) on the original variable space. It utilized a feature of Excel™, called "Solver," for performing the optimization process. Phoon (2004) developed a similar spreadsheet solution using "Solver;" however, the solution of reliability index was obtained in the standard variable space, which tends to produce more stable numerical results. Both spreadsheet approaches yield a solution (reliability index) that is practically identical to each other and to the solution obtained by a dedicated computer program

(Yang, 2003) that implement the well-accepted algorithms for reliability index using FORM.

The mean probability of liquefaction, P_L, may be obtained by the FORM analysis considering the mean model factor μ_{c1}. To determine the variation of the estimated mean probability as a result of the variation in μ_{c1}, in terms of standard deviation σ_{P_L}, a large number of μ_{c1} values may be "sampled" within the range of $\mu_{c1} \pm 3\sigma_{\mu c1}$. The FORM analysis with each of these μ_{c1} values will yield a large number of corresponding P_L values, and then the standard deviation σ_{P_L} can be determined. Alternatively, a simplified method may be used to estimate σ_{P_L}. Two additional FORM analyses may be performed, one with $\mu_{c1} + 1\sigma_{\mu c1}$ as the model factor, and the other with $\mu_{c1} - 1\sigma_{\mu c1}$ as the model factor. Assuming that the FORM analysis using $\mu_{c1} + 1\sigma_{\mu c1}$ as the model factor yields a probability of P_L^+, and the FORM analysis using $\mu_{c1} - 1\sigma_{\mu c1}$ as the model factor yields a probability of P_L^-, then the standard deviation σ_{P_L} may be estimated *approximately* with the following equation (after Gutierrez *et al.*, 2003):

$$\sigma_{P_L} = (P_L^+ - P_L^-)/2 \tag{13.21}$$

13.3.6 Example No. 3: probabilistic evaluation of a non-liquefied case

This example concerns a non-liquefied case that was analyzed previously using the deterministic approach (See Example No. 1 in Section 13.2.2). As described previously, field observation of the site indicated no occurrence of liquefaction during the 1977 Argentina earthquake. The mean values of seismic and soil parameters at the critical depth (2.9 m) are given as follows: $N_{1,60} = 13$, $FC = 3\%$, $\sigma_v{}' = 38.1$ kPa, $\sigma_v = 45.6$ kPa, $a_{max} = 0.2$ g, and $M_w = 7.4$, and the corresponding coefficients of variation of these parameters are assumed to be 0.23, 0.333, 0.085, 0.107, 0.075, and 0.10, respectively (Cetin, 2000).

In the reliability analysis based on the limit state model expressed in Equation (13.13), the parameter uncertainty and the model uncertainty are considered along with the correlation between each pair of input random variables. However, no correlation is assumed between the model factor c_1 and each of the six input random variables of the limit state model. This assumption is supported by the finding of a recent study by Phoon and Kulhawy (2005) that the model factor is *weakly* correlated to the input variables. To facilitate the use of this reliability analysis, a spreadsheet that implements the procedure of the FORM is developed. This spreadsheet, shown in Figure 13.3, is designed specifically for lique-faction evaluation using the SPT-based method by Youd *et al.* (2001), and thus, all the formulas of the limit state model (Equation (13.13) along with

Equations (13.1)–(13.10)) are implemented. For users of this spreadsheet, the input data consists of four categories:

1 the depth of interest ($d = 2.9$ m in this example),
2 the mean and standard deviation of each of the six random variables, $N_{1,60}$, FC, σ_v', σ_v, a_{max}, and M_w,
3 the matrix of the correlation coefficients (use of default values as listed in Table 13.3 is recommended; however, a user-specified matrix could be used if it is deemed more accurate and it satisfies the positive definiteness requirement, as described previously), and
4 the model factor statistics, μ_{c1} and COV (note: COV is set to 0 here as explained previously).

Figure 13.3 shows a spreadsheet solution that is adapted from the spreadsheet originally designed by Low and Tang (1997). The input data for this example, along with the intermediate calculations and the final outcome (reliability index and the corresponding nominal probability), are shown in this spreadsheet. It should be noted that the solution shown in

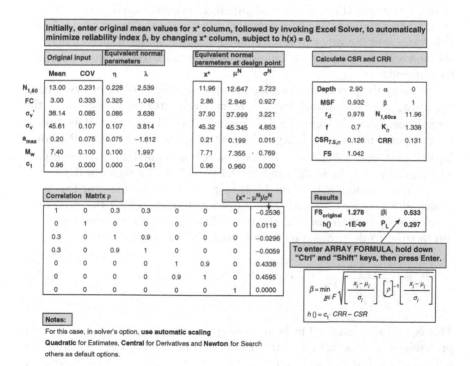

Figure 13.3 A spreadsheet that implements the FORM analysis of liquefaction potential (after Low and Tang, 1997).

Figure 13.3 was obtained using the mean model factor $\mu_{c1} = 0.96$. Because the spreadsheet uses the "Solver" feature in Excel™ to perform the optimization procedure as part of the FORM analysis, the user needs to make sure this feature is selected in "Tools." Within the "Solver" screen, various options may be selected for numerical solutions, and the user may want to experiment with different choices. In particular, the option of "Use Automatic Scaling" in Solver should be activated if the optimization is carried out in the original variable space (Figure 13.3). After selecting the solution options, the user returns to the Solver screen, and chooses "Solve" to perform the optimization and, when the convergence is achieved, the reliability index and the probability of liquefaction are obtained. Because the results depend on the initial guess of the "design point," which is generally assumed to be equal to the vector of the mean values of the input random variables, the user may want to repeat the solution process a few times using different initial trial values to make certain "stable" results have indeed been obtained. Using $\mu_{c1} = 0.96$, the spreadsheet solution (Figure 13.3) yields a probability of liquefaction of $P_L = 0.297 \approx 0.30$.

As a comparison, the spreadsheet solution that is adapted from the spreadsheet originally designed by Phoon (2004) is shown in Figure 13.4. Practically identical solution ($P_L = 0.30$) is obtained. However, experience with both spreadsheet solutions, one with optimization carried out in the original variable space and the other in the standard variable space, indicates that the solution with the latter approach is significantly more robust and is generally recommended.

Following the procedure described previously, the FORM analyses using $\mu_{c1} \pm 1\sigma_{\mu c1}$ as the model factors, respectively, are performed with the spreadsheet shown in Figure 13.3 (or Figure 13.4). The standard deviation of the computed mean probability is then estimated to be $\sigma_{P_L} = 0.039$ according to Equation (13.21). If the three sigma rule is applied, the probability P_L will approximately be in the range of 0.18–0.42, with a mean of 0.30.

As noted previously (Section 13.3), a preliminary estimate of the mean probability may be obtained from empirical models. Using the procedure developed by Juang *et al.* (2002), the following equation is developed for interpreting *FS* determined by the adopted SPT-based method:

$$P_L = \frac{1}{1 + \left(\frac{FS}{1.05}\right)^{3.8}} \tag{13.22}$$

This equation is intended to be used *only* for a preliminary estimate of the probability of liquefaction in the *absence* of the knowledge of parameter uncertainties. For this non-liquefied case, $FS = 1.28$, and thus $P_L = 0.32$ according to Equation (13.22). This P_L value falls in the range of 0.18–0.42 determined by the probabilistic approach using FORM. As noted previously

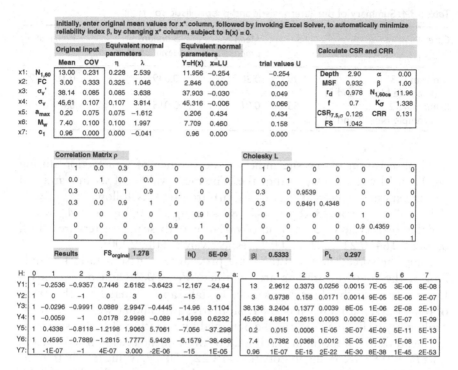

Figure 13.4 A spreadsheet that implements the FORM analysis of liquefaction potential (after Phoon, 2004).

in Section 13.3, the models such as Equation (13.22) require only the best estimates (i.e. the mean values) of the input variables, and thus, the calculated probabilities might be subject to error if the effect of parameter uncertainties is significant.

Table 13.4 summarizes the solutions obtained by using the deterministic and the probabilistic approaches for Example No. 3, and for Example No. 4 which is presented later. The results obtained for Example No. 3 from both approaches confirm field observation of no liquefaction. However, the calculated probability of liquefaction could still be as high as 0.42, even with a factor of safety of $FS = 1.28$, which reflects significant uncertainties in the parameters as well as in the limit state model.

13.3.7 Example No. 4: probabilistic evaluation of a liquefied case

This example concerns a liquefied case that was analyzed previously using the deterministic approach (See Example No. 2 in Section 13.2.3).

Table 13.4 Summary of the deterministic and probabilistic solutions.

Case	FS	Mean probability P_L	Standard deviation of P_L, σ_{P_L}	Range of P_L
San Juan B-5 (Example Nos. 1 and 3)	1.28	0.30	0.039	0.18–0.42
Ishinomaki-2 (Example Nos. 2 and 4)	0.55	0.91	0.015	0.86–0.95

Field observation of the site indicated occurrence of liquefaction during the 1978 Miyagiken-Oki earthquake. The mean values of seismic and soil parameters at the critical depth (3.7 m) are given as follows: $N_{1,60} = 5$, $FC = 10\%$, $\sigma'_v = 36.28$ kPa, $\sigma_v = 58.83$ kPa, $a_{max} = 0.2$ g, and $M_w = 7.4$, and the corresponding coefficients of variation of these parameters are 0.180, 0.200, 0.164, 0.217, 0.2, and 0.1, respectively (Cetin, 2000).

Using the spreadsheet and the same procedure as described in Example No. 3, the following results are obtained: $P_L = 0.91$ and $\sigma_{P_L} = 0.015$. Estimating with the three sigma rule, P_L falls approximately in the range of 0.86–0.95, with a mean of 0.91. Similarly, a preliminary estimate of the probability of liquefaction may be obtained by Equation (13.22) using only the mean values of the input variables. For this liquefied case, $FS = 0.55$, and thus $P_L = 0.92$ according to Equation (13.22). This P_L value falls within the range of 0.86–0.95 determined by the probabilistic approach using FORM. Although the general trend of the results obtained from Equation (13.22) appears to be correct based on the two examples presented, use of simplified models such as Equation (13.22) for estimating the probability of liquefaction should be limited to preliminary analysis of cases where there is a lack of the knowledge of parameter uncertainties.

Table 13.4 summaries the solutions obtained by using the deterministic and the probabilistic approaches for this liquefied case (Example No. 4). The results obtained from both approaches confirm field observation of liquefaction.

Finally, one observation of the standard deviation of the estimated probability obtained in Example Nos. 3 and 4 is perhaps worth mentioning. In general, the standard deviation of the estimated mean probability is much smaller in the cases with extremely high or low probability ($P_L > 0.9$ or $P_L < 0.1$) than those with medium probability ($0.3 < P_L < 0.7$). In the cases with extremely high or low probability, the potential for liquefaction or no-liquefaction is almost certain; it will either liquefy or not liquefy. This lower degree of uncertainty is reflected in the smaller standard deviation. In the cases with medium probability, there is a higher degree of uncertainty as to whether liquefaction will occur, and thus it tends to have a larger standard deviation in the estimated mean probability.

13.4 Probability of liquefaction in a given exposure time

For performance-based earthquake engineering (PBEE) design, it is often necessary to determine the probability of liquefaction of a soil at a site in a given exposure time. The probability of liquefaction obtained by reliability analysis, as presented in Section 13.3 (in particular, in Example Nos. 3 and 4), is a *conditional* probability for a given set of seismic parameters a_{max} and M_w at a specified hazard level. For a future case with uncertain seismic sources, the probability of liquefaction in a given exposure time ($P_{L,T}$) may be obtained by integrating the conditional probability over all possible ground motions at all hazard levels:

$$P_{LT} = \sum_{\text{All pairs of } (a_{max}, M_w)} \{p[L|(a_{max}, M_w)] \cdot p(a_{max}, M_w)\} \qquad (13.23)$$

where the term, $p[L|(a_{max}, M_w)]$, is the conditional probability of liquefaction given a pair of seismic parameters a_{max} and M_w; and the term, $p(a_{max}, M_w)$, is the joint probability of a_{max} and M_w. It is noted that the joint probability $p(a_{max}, M_w)$ may be thought of as the *likelihood* of an event (a_{max}, M_w), and the conditional probability of liquefaction $p[L|(a_{max}, M_w)]$ as the *consequence* of the event. Thus, the product of $p[L|(a_{max}, M_w)]$ and $p(a_{max}, M_w)$ can be thought of as the *weighted consequence* of a single event. Since all mutually exclusive and collectively exhaustive events [i.e., all possible pairs of (a_{max}, M_w)] are considered in Equation 13.23, the sum of all weighted consequences yields the total probability of liquefaction at the given site.

While Equation (13.23) is conceptually straightforward, the implementation of this equation is a significant undertaking. Care must be exercised to evaluate the joint probability of the pair (a_{max} and M_w) for a local site from different earthquake sources using appropriate attenuation and amplification models. This matter, however, is beyond the scope of this chapter, and the reader is referred to Kramer *et al.* (2007) and Juang *et al.* (2008) for detailed treatment of this subject.

13.5 Summary

Evaluation of liquefaction potential of soils is an important task in geotechnical earthquake engineering. In this chapter, the simplified procedure based on the standard penetration test (SPT), as recommended by Youd *et al.* (2001), is adopted as the deterministic model for liquefaction potential evaluation. This model was established originally by Seed and Idriss (1971) based on in-situ and laboratory tests and field observations of liquefaction/no-liquefaction in the seismic events. Field evidence of liquefaction generally consisted of surficial observations of sand boils, ground fissures, or lateral spreads.

Case histories of liquefaction/no-liquefaction were collected mostly from sites on level to gently sloping ground, underlain by Holocene alluvial or fluvial sediments at shallow depths (< 15 m). Thus, the models presented in this chapter are applicable only to sites with similar conditions.

The focus of this chapter is on the probabilistic evaluation of liquefaction potential using the SPT-based boundary curve recommended by Youd *et al.* (2001) as the limit state model. The probability of liquefaction is obtained through reliability analysis using the first-order reliability method (FORM). The FORM analysis is carried out considering both parameter and model uncertainties, as well as the correlations among the input variables. Procedures for estimating the parameter uncertainties, in terms of coefficient of variation, are outlined. In particular, the estimation of the seismic parameters a_{max} and M_w is discussed in detail. The limit state model based on Youd *et al.* (2001) is characterized with a mean model factor of $\mu_{c1} = 0.96$ and a standard deviation of the mean, $\sigma_{\mu c1} = 0.04$. Use of these mean model factor statistics for the estimation of the mean probability P_L and its standard deviation σ_{P_L} is illustrated in two examples. Spreadsheet solutions specifically developed for this probabilistic liquefaction evaluation using FORM are presented. The spreadsheets can facilitate the use of FORM for predicting the probability of liquefaction considering the mean and standard deviation of the input variables. It may also be used to investigate the effect of the degree of uncertainty of individual parameters on the calculated probability to aid in design considerations in cases where there is insufficient knowledge of parameter uncertainties.

The probability of liquefaction obtained in this chapter is a *conditional* probability at a given set of seismic parameters a_{max} and M_w corresponding to a specified hazard level. For a future case with uncertain seismic sources, the probability of liquefaction in a given exposure time ($P_{L,T}$) may be obtained by integrating the conditional probability over all possible ground motions at all hazard levels.

References

Abrahamson, N. A. and Silva, W. J. (1997). Empirical response spectral attenuation relations for shallow crustal earthquakes. *Seismological Research Letters*, 68, 94–127.

Ang, A. H.-S. and Tang, W. H. (1984). *Probability Concepts in Engineering Planning and Design, Vol. II: Design, Risk and Reliability*. John Wiley and Sons, New York.

Baecher, G. B. and Christian, J. T. (2003). *Reliability and Statistics in Geotechnical Engineering*. John Wiley and Sons, New York.

Bazzurro, P. and Cornell, C. A. (2004a). Ground-motion amplification in nonlinear soil sites with uncertain properties. *Bulletin of the Seismological Society of America*, 94(6), 2090–109.

Bazzurro, P. and Cornell, C. A. (2004b). Nonlinear soil-site effects in probabilistic seismic-hazard analysis. *Bulletin of the Seismological Society of America.* 94(6), 2110–23.

Berrill, J. B. and Davis, R. O. (1985). Energy dissipation and seismic liquefaction of sands. *Soils and Foundations*, 25(2), 106–18.

Building Seismic Safety Council (1997). *NEHRP Recommended Provisions for Seismic Regulations for New Buildings and Other Structures, Part 2: Commentary, Foundation Design Requirements.* Washington, D.C.

Campbell, K. W. and Bozorgnia, Y. (2003). Updated near-source ground-motion (attenuation) relations for the horizontal and vertical components of peak ground acceleration and acceleration response spectra. *Bulletin of the Seismological Socoety of America*, 93, 314–31.

Carraro, J. A. H., Bandini, P. and Salgado, R. (2003). Liquefaction resistance of clean and nonplastic silty sands based on cone penetration resistance. *Journal of Geotechnical and Geoenvironmental Engineering, ASCE,* 129(11), 965–76.

Cetin, K. O. (2000). *Reliability-based assessment of seismic soil liquefaction initiation hazard.* PhD dissertation, University of California, Berkeley, CA.

Cetin, K. O., Seed, R. B., Kiureghian, A. D., Tokimatsu, K., Harder, L. F., Jr., Kayen, R. E. and Moss, R. E. S. (2004). Standard penetration test-based probabilistic and deterministic assessment of seismic soil liquefaction potential. *Journal of Geotechnical and Geoenvironmental Engineering, ASCE,* 130(12), 1314–40.

Choi, Y. and Stewart, J. P. (2005). Nonlinear site amplification as function of 30 m shear wave velocity. *Earthquake Spectra*, 21(1), 1–30.

Dai, S. H. and Wang, M. O. (1992). *Reliability Analysis in Engineering Applications.* Van Nostrand Reinhold, New York.

Davis, R. O. and Berrill, J. B. (1982). Energy dissipation and seismic liquefaction in sands. *Earthquake Engineering and Structural Dynamics*, 10, 59–68.

Dief, H. M. (2000). *Evaluating the liquefaction potential of soils by the energy method in the centrifuge.* PhD dissertation, Case Western Reserve University, Cleveland, OH.

Dobry, R., Ladd, R. S., Yokel, F. Y., Chung, R. M. and Powell, D. (1982). *Prediction of Pore Water Pressure Buildup and Liquefaction of Sands during Earthquakes by the Cyclic Strain Method.* National Bureau of Standards, Publication No. NBS-138, Gaithersburg, MD.

Duncan, J. M. (2000). Factors of safety and reliability in geotechnical engineering. *Journal of Geotechnical and Geoenvironmental Engineering, ASCE,* 126(4), 307–16.

Figueroa, J. L., Saada, A. S., Liang, L. and Dahisaria, M. N. (1994). Evaluation of soil liquefaction by energy principles. *Journal of Geotechnical Engineering, ASCE,* 120(9), 1554–69.

Frankel, A., Harmsen, S., Mueller, C. et al. (1997). USGS national seismic hazard maps: uniform hazard spectra, de-aggregation, and uncertainty. In *Proceedings of FHWA/NCEER Workshop on the National Representation of Seismic Ground Motion for New and Existing Highway Facilities,* NCEER Technical Report 97-0010, State University of New York at Buffalo, New York. pp. 39–73.

Green, R. (2001). *The application of energy concepts to the evaluation of remediation of liquefiable soils.* PhD dissertation, Virginia Polytechnic Institute and State University, Blacksburg, VA.

Gutierrez, M., Duncan, J. M., Woods, C. and Eddy E. (2003). *Development of a simplified reliability-based method for liquefaction evaluation.* Final Technical Report, USGS Grant No. 02HQGR0058, Virginia Polytechnic Institute and State University, Blacksburg, VA.

Haldar, A. and Tang, W. H. (1979). Probabilistic evaluation of liquefaction potential. *Journal of Geotechnical Engineering, ASCE,* 104(2): 145–62.

Hasofer, A. M. and Lind, N. C. (1974). Exact and invariant second moment code format. *Journal of the Engineering Mechanics Division, ASCE,* 100(EM1), 111–21.

Harr, M. E. (1987). *Reliability-based Design in Civil Engineering.* McGraw-Hill, New York.

Hynes, M. E. and Olsen, R. S. (1999). Influence of confining stress on liquefaction resistance. In *Proceedings International Workshop on Physics and Mechanics of Soil Liquefaction.* Balkema, Rotterdam, pp. 145–52.

Idriss, I. M. and Boulanger, R. W. (2006). "Semi-empirical procedures for evaluating liquefaction potential during earthquakes." *Soil Dynamics and Earthquake Engineering,* Vol. 26, 115–130.

Ishihara, K. (1993). Liquefaction and flow failure during earthquakes, The 33rd Rankine lecture. *Géotechnique,* 43(3), 351–415.

Juang, C. H., Rosowsky, D. V. and Tang, W. H. (1999). Reliability-based method for assessing liquefaction potential of soils. *Journal of Geotechnical and Geoenvironmental Engineering, ASCE,* 125(8), 684–9.

Juang, C. H., Chen, C. J., Rosowsky, D. V. and Tang, W. H. (2000). CPT-based liquefaction analysis, Part 2: Reliability for design. *Géotechnique,* 50(5), 593–9.

Juang, C. H., Jiang, T. and Andrus, R. D. (2002). Assessing probability-based methods for liquefaction evaluation. *Journal of Geotechnical and Geoenvironmental Engineering, ASCE,* 128(7), 580–9.

Juang, C. H., Yang, S. H., Yuan, H. and Khor, E. H. (2004). Characterization of the uncertainty of the Robertson and Wride model for liquefaction potential. *Soil Dynamics and Earthquake Engineering,* 24(9–10), 771–80.

Juang, C. H., Fang, S. Y. and Khor, E. H. (2006). First-order reliability method for probabilistic liquefaction triggering analysis using CPT. *Journal of Geotechnical and Geoenvironmental Engineering, ASCE,* 132(3), 337–50.

Juang, C. H., Li, D. K., Fang, S. Y. Liu, Z., and Khor, E. H. (2008). A simplified procedure for developing joint distribution of a_{max} and M_w for probabilistic liquefaction hazard analysis. *Journal of Geotechnical and Geoenvironmental Engineering, ASCE,* in press.

Kramer, S. L. (1996). *Geotechnical Earthquake Engineering.* Prentice-Hall, Englewood Cliffs, NJ.

Kramer, S. L. and Elgamal, A. (2001). *Modeling Soil Liquefaction Hazards for Performance-Based Earthquake Engineering.* Report No. 2001/13, Pacific Earthquake Engineering Research (PEER) Center, University of California, Berkeley, CA.

Kramer, S. L. and Mayfield R. T. (2007). "Return period of soil liquefaction" *Journal of Geotechnical and Geoenvironmental Engineering, ASCE*, 133(7), 802–13.

Liang, L., Figueroa, J. L. and Saada, A. S. (1995). Liquefaction under random loading: unit energy approach. *Journal of Geotechnical and Geoenvironmental Engineering*, 121(11), 776–81.

Liao, S. S. C. and Whitman, R. V. (1986). Overburden correction factors for SPT in sand. *Journal of Geotechnical Engineering, ASCE*, 112(3), 373–7.

Liao, S. S. C., Veneziano, D. and Whitman, R. V. (1988). Regression model for evaluating liquefaction probability. *Journal of Geotechnical Engineering, ASCE*, 114(4), 389–410.

Low, B. K. (1996). Practical probabilistic approach using spreadsheet. In *Uncertainty in the Geologic Environment: from Theory to Practice*, Geotechnical Special Publication No. 58, Eds C. D. Shackelford, P. P. Nelson and M. J. S. Roth. *ASCE*, Reston, VA, pp. 1284–302.

Low, B. K. and Tang, W. H. (1997). Efficient reliability evaluation using spreadsheet. *Journal of Engineering Mechanics, ASCE*, 123(7), 749–52.

Low, B. K. (2005). Reliability-based design applied to retaining walls. *Géotechnique*, 55(1), 63–75.

Marcuson, W. F., III (1978). Definition of terms related to liquefaction. *Journal Geotechnical Engineering Division, ASCE*, 104(9), 1197–200.

Melchers, R. E. (1999). *Structural Reliability, Analysis, and Prediction*, 2nd ed. Wiley, Chichester, UK.

NEHRP (1998). *NEHRP Recommended Provisions for Seismic Regulations for New Buildings and Other Structures, Part 1 – Provisions: FEMA 302, Part 2 – Commentary FEMA 303*. Federal Emergency Management Agency, Washington DC.

National Research Council (1985). *Liquefaction of Soil during Earthquake*. National Research Council, National Academy Press, Washington, DC.

Ostadan, F., Deng, N. and Arango, I. (1996). *Energy-based Method for Liquefaction Potential Evaluation. Phase 1: Feasibility Study*. Report No. NIST GCR 96-701, National Institute of Standards and Technology, Gaithersburg, MD.

Phoon, K. K. (2004). *General Non-Gaussian Probability Models for First Order Reliability Method (FORM): A State-of-the Art Report*. ICG Report 2004-2-4 (NGI Report 20031091-4), International Center for Geohazards, Oslo, Norway.

Phoon, K. K. and Kulhawy, F. H. (1999). Characterization of geotechnical variability. *Canada Geotechnical Journal*, 36, 612–24.

Phoon, K. K. and Kulhawy, F. H. (2005). Characterization of model uncertainties for laterally loaded rigid drilled shafts. *Géotechnique*, 55(1), 45–54.

Sadigh, K., Chang, C. -Y., Egan, J. A., Makdisi, F. and Youngs, R. R. (1997). Attenuation relations for shallow crustal earthquakes based on California strong motion data. *Seismological Research Letters*, 68, 180–9.

Seed, H. B. and Idriss, I. M. (1971). Simplified procedure for evaluating soil liquefaction potential. *Journal of the Soil Mechanics and Foundation Division, ASCE*, 97(9), 1249–73.

Seed, H. B. and Idriss, I. M. (1982). 'Ground Motions and Soil Liquefaction during Earthquakes. Earthquake Engineering Research Center Monograph, EERI, Berkeley, CA.

Seed, H. B. and Lee, K. L. (1966). Liquefaction of saturated sands during cyclic loading. Journal of the Soil Mechanics and Foundations Division, ASCE, 92(SM6), Proceedings Paper 4972, Nov., 105–34.

Seed, H. B., Tokimatsu, K., Harder, L. F. and Chung, R. (1985). Influence of SPT procedures in soil liquefaction resistance evaluations. Journal of Geotechnical Engineering, ASCE, 111(12), 1425–45.

Skempton, A. K. (1986). Standard penetration test procedures and the effects in sands of overburden pressure, relative density, particle size, aging, and overconsolidation. Géotechnique, 36(3), 425–47.

Stewart, J. P., Liu, A. H. and Choi, Y. (2003). Amplification factors for spectral acceleration in tectonically active regions. Bulletin of Seismological Society of America, 93(1), 332–52.

Yang, S. H. (2003). Reliability analysis of soil liquefaction using in situ tests. PhD dissertation, Clemson University, Clemson, SC.

Youd, T. L. and Idriss, I. M. Eds. (1997). Proceedings of the NCEER Workshop on Evaluation of Liquefaction Resistance of Soils, Technical report NCEER-97-0022, National Center for Earthquake Engineering Research, State University of New York at Buffalo, Buffalo, NY.

Youd, T. L., Idriss, I. M., Andrus, R. D., Arango, I., Castro, G., Christian, J. T., Dobry, R., Liam Finn, W. D., Harder, L. F., Jr., Hynes, M. E., Ishihara, K., Koester, J. P., Laio, S. S. C., Marcuson, W. F., III, Martin, G. R., Mitchell, J. K., Moriwaki, Y., Power, M. S., Robertson, P. K., Seed, R. B. and Stokoe, K. H., II. (2001). Liquefaction resistance of soils: summary report from the 1996 NCEER and 1998 NCEER/NSF workshops on evaluation of liquefaction resistance of soils. Journal of Geotechnical and Geoenvironmental Engineering, ASCE, 127(10), 817–33.

Zhang, L., Tang, W. H., Zhang, L. and Zheng, J. (2004). Reducing uncertainty of prediction from empirical correlations. Journal of Geotechnical and Geoenvironmental Engineering, ASCE, 130(5), 526–34.

Index

Printed in the United States
by Baker & Taylor Publisher Services

Printed in the United States
by Baker & Taylor Publisher Services